CHEMICAL EFFECTS
OF NUCLEAR TRANSFORMATIONS IN
INORGANIC SYSTEMS

CHEMICAL EFFECTS OF NUCLEAR TRANSFORMATIONS IN INORGANIC SYSTEMS

EDITORS

G. HARBOTTLE

Department of Chemistry, Brookhaven National Laboratory
Upton, N.Y. 11973, USA

and

A.G. MADDOCK

University Chemical Laboratory, Cambridge, UK

1979

NORTH-HOLLAND PUBLISHING COMPANY
AMSTERDAM · NEW YORK · OXFORD

ISBN: 0 444 85054 6

PUBLISHERS:

NORTH-HOLLAND PUBLISHING COMPANY
AMSTERDAM · NEW YORK · OXFORD

SOLE DISTRIBUTORS FOR THE USA AND CANADA

ELSEVIER NORTH-HOLLAND, INC.
52 VANDERBILT AVENUE
NEW YORK, N.Y. 10017

Library of Congress Cataloging in Publication Data

Main entry under title:

Chemical effects of nuclear transformations in
inorganic systems.

Bibliography: p.
Includes indexes.
1. Radiation chemistry. 2. Chemistry, Inorganic.
I. Harbottle, G. II. Maddock, A.G.
QD636.C43 541′.38 78-10479
ISBN 0-444-85054-6

PRINTED IN THE NETHERLANDS

PREFACE

There is, at present, no collected account of the considerable body of information that has accumulated over the years on the chemical effects of nuclear transformations in inorganic systems. In this collaborative work the editors and their colleagues try to provide such an account. Each chapter has been contributed by a scientist who has been directly concerned in research in the relevant area.

The editors would like to thank their collaborators for their patient and painstaking cooperation in the production of this book.

CONTENTS

CHAPTER 1

INTRODUCTION

Garman HARBOTTLE
Department of Chemistry, Brookhaven National Laboratory, Upton, New York 11973, USA

Chemical Effects of Nuclear Transformations in Inorganic Systems
Edited by G. Harbottle and A.G. Maddock

If there is one field of chemical research that is uniquely related to the nuclear age, and especially to the neutron, it is the field of hot-atom chemistry. Although recoil effects were already seen as early as 1904 and hot-atom experiments are still carried out today, making use only of naturally radioactive isotopes, it was the production of neutron-induced radioactivity that gave the initial impetus to the research that is summarized in this book. There were a number of reasons for the intense, and quite international interest in hot-atom chemistry in the 1930s and '40s, and we shall attempt to trace out some of these in the historical portion of this introduction, but it is safe to guess that an important one was the fascination which chemists felt at being able to witness for the first time chemical events at essentially infinite dilution – that is, to have in their hands a kind of chemical reaction analog of the cloud chamber. A second stimulus lay in the early realization that some of these 'one-by-one' chemical reactions were quite probably occurring at very elevated energies, corresponding to kinetic 'temperatures' of 10^4–10^5 degrees Kelvin. Thus, a hitherto inaccessible region of reaction energetics had also opened.

If the growth of the field of 'Chemical Effects of Nuclear Transformations' may be likened to the growth of a tree, then the tree grew alone for a long time, and only near maturity did its branches begin to touch and intertwine with those of its neighbors in the forest (or jungle?) of Science. The nature of this interaction will be more clear as the individual chapters of this book are unfolded, but even at this date the editors can still lament a general lack of effective contact between hot-atom chemists dealing for example with recoil reactions in crystals, and solid-state physicists. It is the hope of the editors that this book will assist in building the kinds of interdisciplinary bridges, so necessary to the further development of hot-atom chemistry on the one hand, and to the dissemination of its insights relating to other fields of research, in turn. It is their belief that the interaction could be far more fruitful on both sides than it has been. But before discussing the situation as it stands today, let us begin at the beginning.

The first observation of the physical recoil of an atom following nuclear transformation was made by Miss Harriet Brooks, working with Rutherford at McGill University, in 1904 (Brooks 1904). She found that ^{218}Po atoms which had been deposited on a copper plate formed a 'volatile' activity which was transferred to the walls of her ionization chamber; it was later realized that such transfers arose not from volatility but rather from conservation of momentum in the transformation

$$^{218}\text{Po} \rightarrow {}^{214}\text{Pb (with a recoil energy of ca. 100 keV)} + {}^{4}\text{He} \ (E_\alpha \text{ ca. 6 MeV})$$

and that recoil atoms could even be observed following β decay in ^{214}Pb where the momentum transferred corresponded to recoil energies of only 0.6 eV (Rutherford 1951).

The study of hot-atom *chemistry*, that is to say of the chemical reactions of atoms which had undergone nuclear transformation, properly began with the experiment of Szilard and Chalmers (1934a). In 1934 they showed that if the organic liquid, ethyl iodide, was bombarded with neutrons it was afterwards easy to separate chemically a large fraction of the radioactive isotope ^{128}I produced by neutron capture from the great bulk of the target liquid. It is interesting that Szilard and Chalmers' explanation of this separation was based upon a partial misconception: in their words "we might expect those atoms of the element which are struck by a neutron to be removed from the compound", implying that transfer of momentum from the neutron to the ^{127}I atom was responsible. This explanation is of course true in the case of fast-neutron induced reactions; however, in their case a more important nuclear mechanism was soon provided by the ingenuity of Fermi (Amaldi et al. 1935), namely, that upon capture of a thermal neutron, which in itself carried inconsequential momentum, gamma rays were emitted from the excited state [in the case of ^{127}I + n, at 6.8 MeV (Bartholomew and Groshev 1968a, b)] and the recoil from these rays was more than sufficient to break the carbon–iodine bond. A convenient expression for the recoil energy of an atom of mass number A which has emitted a gamma ray of energy E_γ (in MeV) is

$$E_{\text{recoil}} \text{ (in eV)} = \frac{537\, E_\gamma^2}{A} \quad \text{or in kilocal/mol} = \frac{12\,360\, E_\gamma^2}{A} \tag{1.1}$$

In the case of the ^{128}I atom and the ground-state gamma transition this amounts to 194 eV or 4465 kilocal/mol: the carbon–iodine bond energy is only 2.0 eV.

In 1939 another important mode of coupling of a chemical reaction to a nuclear event was discovered independently by Segre et al. (1939) and by DeVault and Libby (1939). They reported the chemical separation of nuclear isomers; i.e., when ^{80m}Br (4.5 h) was incorporated, for example, in an organic molecule (t-butyl bromide), it was possible to water-extract atoms of the ^{80}Br (18 min) daughter following the isomeric transition process. It was soon demonstrated by Seaborg et al. (1940c) that the chemical effect of an isomeric transition was ultimately linked not to recoil energy but to the internal conversion of the isomeric level: such a mechanism, it was later recognized, could also apply to the n,γ reaction, in cases where levels populated during the capture-gamma cascade were themselves internally converted (Wexler and Davies 1952).

The Szilard–Chalmers effect, as the chemical reaction attending neutron capture came to be called, had an important practical sequence: it permitted, at a time when there were only feeble neutron sources, the preparation of radioisotopes of high specific activity, since it was possible to devise separations which in many elements strongly favored the isolation of the transmuted atoms from the bulk of the target material. This practical result was put to immediate use by Szilard and Chalmers themselves (Szilard and Chalmers 1934b); they used ethyl iodide as a sensitive neutron detector to discover the photoneutron reaction in beryllium, $^{9}Be(\gamma,n)\,^{8}Be$. A later application of the Szilard–Chalmers effect by Kurchatov and coworkers led to the discovery of the isotope ^{82}Br, in extracts from irradiated ethyl bromide (Kurchatov et al. 1935).

This 'enrichment' phenomenon was also studied by one of Fermi's Roman coworkers, D'Agostino, who found that inorganic crystalline compounds such as sodium bromate, chlorate, perchlorate and iodate and potassium permanganate, showed the Szilard–Chalmers effect as well as organic liquids. Since our present book deals with inorganic systems, it is really in D'Agostino's work that we find our beginning (D'Agostino 1935). In a typical experiment, potassium permanganate was irradiated with neutrons and dissolved in water. A little manganous ion was then added and precipitated as manganous carbonate: 80% of the ^{56}Mn (2.6 h half life) was carried down and separated from the permanganate, which remained in solution. D'Agostino's conclusion from this experiment was, that the complex ion (MnO_4^-) having been 'broken up' ('spezzato') the atom of 'transmuted manganese' in the crystal, upon dissolution, was carried "in an ionic state different from the original" and was thus separable. Presumbly, then, many of the atoms of manganese, initially in the formal valence VII(MnO_4^-) had, in the process of recoil, with perhaps a few hundred electron volts energy, either reduced their valence (captured electrons from the lattice) or else had formed metastable species which, although still in the VII state in the crystal, were reduced upon contact with water. Similar arguments held for chlorine atoms in chlorate and perchlorate, iodine atoms in iodate crystals, etc. Considering that in the great majority of nuclear events (n,γ reactions) the original Mn–O bonds must have been broken, the interesting question arose: "By what mechanism was the original *type* of complex ion *re-formed*, in roughly 20% of the events?" This was the question asked by Libby, who in 1940 at the University of California published a long paper (Libby 1940) in which, significantly, no enrichments were measured. Rather, the entire attention was focussed upon the chemical properties of the recoil atom in the crystal. In the case of potassium permanganate, Libby found that the observed "retention" (which he defined as the percentage of ^{56}Mn radioactivity found in the parent (MnO_4^-) form after analysis of the crystals, i.e., a figure of 20%, in D'Agostino's experiment cited above) depended upon the pH of the water in which the irradiated crystals were dissolved. He concluded that, had the recoil manganese been present in the crystal in only two stable forms, for example as MnO_2 and MnO_4^- no such dependence would have existed. Libby also investigated ^{32}P recoil atoms in phosphate and phosphite crystals, ^{76}As in arsenic acid (H_3AsO_4) and arsenious oxide, ^{80m}Br in sodium bromate and in the organic crystal CBr_4, and numerous solutions and organic liquids. Finally, he suggested some plausible mechanisms for the formation of the various hypothetical recoil fragments and their reactions with water at the moment of dissolution: although some of Libby's conclusions have been modified, his ideas led to very extensive additional research by other investigators (see, for example, chapters 6–9). It is not too soon to point out one conceptual interaction of hot-atom chemistry with another discipline: the solid-state, radiation damage physicist will note that the process which led to the 20% "retention" seen in 1935 by D'Agostino in $KMnO_4$ bears a kind of speaking familiarity with the concept of the "replacement collision" of Kinchin and Pease (1955b). To quote the latter authors, "Some of the collisions between moving atoms and stationary atoms at lattice sites will displace the initially stationary atom, but leave the original moving atom with insufficient energy to escape from the vacancy thus created. Such collisions may be called replacement collisions, since they change the identity of the displaced atoms but not the total number. Replacement collisions are clearly of little consequence in monatomic solids, but in compounds they may be important." Similarly, the species chemically separable after the recoil reaction must necessarily be compared to "interstitial" atoms in the alkali halides of radiation damage research, or more nearly to the paramagnetic fragments visible in ESR spectra of irradiated solids.

Already in 1939 Lu and Sugden (1939) in order to explain the high probability (30–50%) of re-entry of a recoil halogen atom in liquid organic halides such as C_4H_9Br – i.e., high retentions, invoked the "liquid cage" model (for photochemical reactions) of Franck and Rabinowitch (1934). Libby's 1940 paper referred to this idea and introduced the "billiard-ball concept" in these words: "Br atoms or Br^- ions of high velocity . . . on collision with atoms or groups of nearly equal mass lose large fractions of their energy to be left in the same reaction 'cage' with the fragments of the struck molecule." In 1947 Libby published a second paper in which the billiard-ball concept was explicitly introduced into a theoretical calculation of the retention (Libby 1947). The liquid cage was also used, hard-sphere elastic collisions were assumed and equations analogous to those developed by Placzek (1946, Miller et al. 1950) for the slowing-down of neutrons were adapted to the billiard-ball slowing-down of heavy recoil atoms. We see that, although derived for reactions in the liquid phase (organic halides) *mutatis mutandis*, Libby's "billiard-ball" model bears a quite close similarity to the "replacement collision" of Kinchin and Pease for crystals. For example, Libby's cage energy ε (the energy which is just sufficient to ensure that the struck atom will escape from the "cage" of solvent molecules) is similar to Kinchin and Pease's threshold replacement-collision energy E_α, etc.

To return to the history of experimental hot-atom chemistry, the most significant event after 1940 was the development of the nuclear reactor, which made available neutron fluxes many orders of magnitude higher than the best of those of the 1930s. It was only natural that Szilard–Chalmers reactions were immediately tried in the reactor, in the hope of preparing radioisotopes of both high total and high specific activity. However, two distinct effects emerged from these experiments, both of which tended to limit the usefulness of the method (Williams 1948a, b, Williams et al. 1951). The first effect, which could have been anticipated, was simply the radiation-induced decomposition of the substance while under reactor bombardment, and the net result was just the dilution of the highly-enriched recoil atoms with inert, and chemically indistinguishable, atoms from the target. This led to the specific activity declining with total neutron exposure even though the retention (probability of appearance of radioactive atoms in the parent chemical form) remained constant. Since this decomposition of the bulk target material occurred under the influence of radiation, for example in the "spikes" created by fast neutron impact, it was clear that there was a link, though not a very surprising one, to the fields of radiation damage and radiation chemistry.

The second effect was new and unexpected, and, we now realize, directly linked in quite a different way to the study of the creation and the properties of lattice defects. It was found that a salt such as ammonium fluometaantimonate (NH_4SbF_6) was suitable for Szilard–Chalmers separation of antimony activity. After a short reactor exposure the salt was dissolved in water and an SbIII fraction separated from the bulk SbF_6^- target ions by precipitation: around 25% of the ^{124}Sb could be recovered. However when the reactor exposure was increased, it was found that the yield of $^{124}SbIII$ decreased (i.e. retention increased) as shown in table 1.1. Quite clearly there was a radiation-induced back-reaction which returned the once-separable ^{124}Sb atoms to a form indistinguishable from the parent. A surprising feature of this back-reaction was that it operated even in the presence of a simultaneous forward reaction of ordinary radiation decomposition or radiolysis. Effects of this sort involving the increase of retention during exposure to the mixed radiations of the reactor have come to be known as "pile annealing" in hot-atom chemistry: examples will be discussed later in this volume.

In 1949 Green and Maddock (1949) found that the same type of back- (or annealing) reaction could be produced in neutron-irradiated solid potassium chromate crystals simply by heating, i.e., without any radiation exposure. They

Table 1.1
Recovery of ^{124}Sb by the Szilard–Chalmers process in NH_4SbF_6 as a function of reactor exposure (Williams 1948b, Williams et al. 1951)

Relative reactor exposure	Activity yield (%)	Retention (%)	SbIII[a] (μg/g of salt)	Decomposition (%)
1	25	75	30	
~25	3	97	1000	0.2
120	1	99	5500	1.0

[a]Formed by radiolysis.

bombarded K_2CrO_4 crystals with neutrons to produce ^{51}Cr recoil atoms via the n,γ reaction. At room temperature retentions amounted to 55%, but if, after neutron exposure, the crystals were heated in an oven to 480° C for a few hours the retention rose to 93%. Similar annealing effects were soon found in potassium permanganate (Aten and Van Berkum 1950) and in 1952 Cobble and Boyd (1952) published a detailed study of the isothermal annealing of neutron bombarded potassium bromate containing ^{82}Br recoil atoms.

The observations made on these systems, and the various theoretical explanations, will be described in the chapters to follow: let us note here, however, that in the annealing of recoil species both by gamma radiation and by simple heating we find a close analogy to the saturation and fading of color centers in irradiated crystals, and the conversion of an ESR-active species into an inactive one accompanying a rise in temperature in solids irradiated at low temperatures.

Despite all that has been said about the difficulties of making the Szilard–Chalmers reaction 'go' in the nuclear reactor, the fact remains that many useful radioisotope production processes have been developed that overcome these problems, and allow the preparation of medical, biological and agriculturally interesting species having high specific activity, in the research reactor (Harbottle and Hillman 1971).

The decade of the 1950s brought a flood of research in all the facets of what might be termed "classical" hot-atom chemistry – ingenious separations based on both wet chemistry and non-aqueous solvents, chromatography, electrophoresis, and ion exchange columns, the results generally tending to present an ever more complex picture of the distribution of recoil species, the more sophisticated the technique of separation. It also gradually became evident that the perturbing effect of the dissolution step constituted a serious limitation on the validity of the 'wet-chemistry' approach to analysis of recoil species. For this reason, beginning roughly at the time of the first IAEA Symposium on Chemical Effects of Nuclear Transformations (IAEA 1961) there was an increased interest in in situ techniques, the first to be investigated being the Mössbauer effect. Since that time there has been interest in perturbed angular correlations, time differential angular correlations, nuclear resonance fluorescence and half-life alteration: several of these will be explored in detail in the following chapters. It is, however, safe to guess that because of the generally limited resolving power of these techniques in distinguishing one molecular species from another in situ, their use will be more important in demonstrating the time scale of devolution of the chemical event following the nuclear event, than in establishing exactly what species were formed. By implication, the purely wet-chemistry analysis will continue to be important as an adjunct to the purely physical techniques.

It is our hope, then, to show that the study of the chemical effects of nuclear transformation has developed to the point where its results ought to be of interest not only to its own specialists but to the scientific community at large, by reason both of its intrinsic content and of its connections with many other fields of physics and chemistry.

CHAPTER 2

PRIMARY PROCESSES IN HOT-ATOM CHEMISTRY

Thomas A. CARLSON

Oak Ridge National Laboratory, Oak Ridge,*
Tennessee 37830, USA

*Operated by Union Carbide Corp. for the US Atomic Energy Commission.

Chemical Effects of Nuclear Transformations in Inorganic Systems
Edited by G. Harbottle and A.G. Maddock

Contents

1. Introduction

In this chapter we shall discuss the start of the long journey that a hot atom undergoes from the initial nuclear event to its final resting place where the labeled atom becomes part of the chemical environment. As will be obvious from the remainder of the book, knowledge of the details of the initial excitation process is insufficient for a complete description of the hot atom process. However, this knowledge is in general a necessary condition, and the information is important in itself for a better understanding of different types of atomic and molecular excitation processes.

An extensive and excellent review on the primary physical and chemical effects associated with nuclear events was published in 1965 by Wexler (1965b). In this chapter we shall be far more concise in our treatment, and will make no attempt at a complete literature survey. There has not been a substantial change in overall viewpoint of the nature of the primary processes since Wexler's review, but we have tried to incorporate some of the recent findings which have helped expand and clarify the knowledge of the basic nature of the primary processes.

There are three sources of excitation that are important to the primary processes in hot atom chemistry. They are (1) the recoil energy that is imparted to the hot atom; (2) the readjustment to a vacancy in an inner shell of an atom by a series of Auger processes (called a vacancy cascade) and (3) electronic excitation and ionization that arise from the process known as electron shake-off. (A fourth self-evident occurrence that plays an important role in the final chemistry of the hot atom is the creation of a new element as the result of a change in nuclear charge.) These processes may be found concomitant in a given nuclear event. In principle they can interact giving rise to cooperative effects, but in practice these effects can be studied separately; and when they occur together the results are generally a simple additive function. In any case, each of the nuclear events is usually dominated by one of the forms of excitation. For example, nuclear events involving heavy particles are concerned primarily with recoil, internal conversion and electron capture with vacancy cascades, and low-energy beta decay with electron shake-off.

The presentation will first begin with a discussion of the basic nature of the three excitation processes. Then, each type of nuclear process will be examined for its specific characteristics. Our main concern will be with whether the primary excitation will lead to bond breakage between the hot atom and the rest of the parent molecule; and if such rupture occurs, what will be the state of excitation of the hot atom in terms of its kinetic energy and ionization state. (We define here the *hot atom* as the atom undergoing nuclear decay, the *parent molecule* as the molecule containing this atom before decay, the bond between the hot atom and the rest of the parent molecule as the *primary bond*, and the remainder of the parent molecule after severance of this bond as the *parent ion*.) Some attention will be paid, however, to the state of the parent ion, both because this information is important in terms of types of excited species in close vicinity to the hot atom at the time of rupture, and because it gives us some insight into the basic nature of the excitation process.

The general viewpoint taken of the primary process will be one of dealing with an isolated molecule. This viewpoint is taken so one may treat the problem theoretically in simple terms, and because studies with low-pressure gaseous systems are usually required for deriving unambiguous experimental results on the primary processes. However, most of the general conclusions reached are applicable to condensed as well as gaseous systems. In a few instances, wherever possible, specific inferences to condensed systems will be made.

2. Excitation due to Recoil Energy

The most important form of excitation imparted in nuclear decay or nuclear reactions that involve heavy particles is recoil energy. The magnitude of the recoil energy is usually far in excess of bond energies; and rupture takes place in nearly every decay. In contrast, the recoil received in γ decay, electron capture, internal conversion or in beta decay is often insufficient to cause bond rupture. Table 2.1 gives some idea of the order of magnitude of recoil energies

Table 2.1
Approximate recoil energies expected with various nuclear events

Nuclear process	Range of recoil energy (eV) [a]
β^- decay	10^{-1} to 10^2
β^+ decay	10^{-1} to 10^2
α decay	$\sim 10^5$
Isomeric transition	10^{-1} to 1
Electron capture	10^{-1} to 10^1
n,γ (thermal)	$\sim 10^2$
n,p	$\sim 10^5$
Fission	$\sim 10^8$

[a] Based on the mass of the hot atom $\sim$ 100 and considering the most probable kinetic energy for a given nuclear process and a range of nuclear energies that are most frequently encountered.

encountered in a variety of nuclear processes. The formulae for the recoil energy received by the hot atom for the various types of nuclear processes have been worked out from kinematics in essentially every instance, and the results are given in the section covering the individual decay processes.

Only a portion of the recoil received by the hot atom goes into the breaking of the bond which attaches the hot atom to the rest of the parent molecule. Conservation of momentum requires that a portion of the energy be taken up by translational motion of the molecule as a whole. In the case of a diatomic molecule Suess (1940) has calculated that the energy available for internal excitation, E_i, is

$$E_i = \frac{M_2}{M_1 + M_2} E_R \quad , \tag{2.1}$$

where E_R is the recoil energy of the hot atom whose mass is M_1, while the mass of the attached atom is M_2. For $M_2 \ll M_1$ (for example, HI where the iodine receives the initial recoil) the energy available for bond rupture is only a small fraction of the recoil energy. In the case of polyatomic molecules E_i may be approximated by eq. (2.1) by assuming the molecule to which the hot atom is attached acts as a whole unit with mass M_2. Hsiung and Gordus (1962) have developed a more sophisticated model for polyatomic molecules in which each of the atoms is considered to be a mass point constrained by the motion for a perfect rotor and harmonic oscillator, where the rotational and vibrational motions of the molecule are independent. In this treatment it is possible to estimate the portion of energy that goes into internal excitation of the parent radical. For a series of methyl and ethyl halides this fraction ranged from 2 to 16%. The values of the minimum energy needed to effect bond rupture, as calculated for these same compounds by Hsiung and Gordus (1962), are on the average about 25% higher than estimates based on (2.1).

If the recoil energy is sufficiently high, ionization can occur as sort of a whiplash phenomenon. An estimate of the recoil energy that must be reached before ionization occurs can be obtained from Bohr's (1940, 1941) postulate that ionization will occur when the velocity of the recoil atom, v_R, is greater than that of the orbital electron, v_e.

For a Bohr orbit the kinetic energy of the outermost orbital electron is

$$\tfrac{1}{2} m_e v_e^2 = I_p , \tag{2.2}$$

where I_p is the ionization potential or the negative value of the total energy. From the condition $v_e = v_R$,

$$I_p = \frac{m_e}{M} E_R \quad , \tag{2.3}$$

where m_e and M are the masses respectively for an electron and the recoiling atom. From calculated ionization potentials (Carlson et al. 1970b) of highly charged ions one estimates for example for iodine ($Z = 53$, $A = 127$) that at a recoil of 100 MeV the average charge would be between +16 and +17. The estimate of the charge derived from recoil may be better understood if one examines the recent studies on the charge state of ions passing through matter in which Bohr's approximation may also be expected to hold. From the data of Datz et al. (1971) we find that the average charge of iodine with a recoil of 100 MeV is +17 in gases in good agreement with the estimate given above; but the charge is +26 for solids. The possible reason for the higher average charge of an ion passing through solids than in gases has been discussed by Betz and Grodzins (1970). An overall view of the average charge as a function of kinetic energy for elements from $Z = 20$ to 90 is given in fig. 2.1.

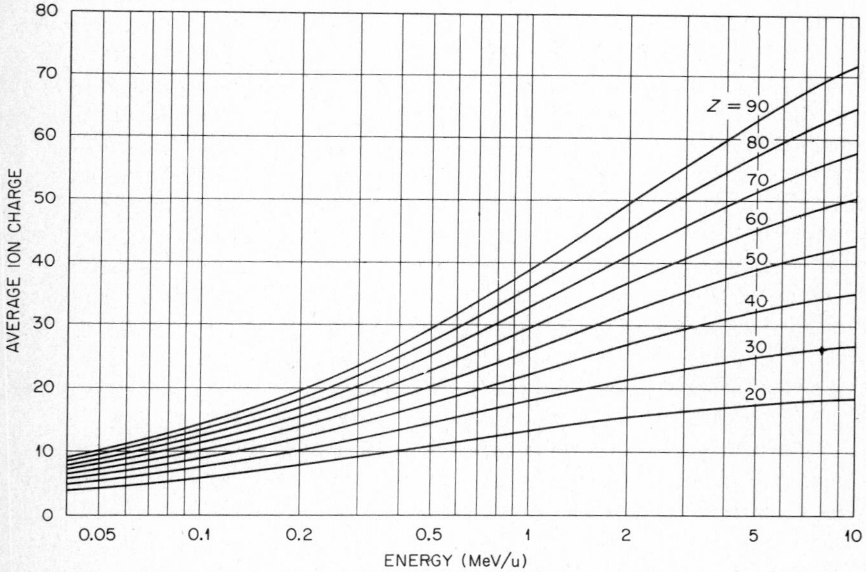

Fig. 2.1. The average charge of ions passed through a solid stripper according to the empirical formula of Nekolaev and Dmitriev, plotted by Livingston (1970). Kinetic energy equals (MeV/u) · atomic number.

Bohr's postulate must not be taken as a sharp step function but is a more gradual one where the probability for ionization is given as

$$P_e \approx (v_R/v_e)^2. \tag{2.4}$$

Equation (2.4) has been experimentally tested (Winther 1952) in the study of the β decay of ^{6}He. At a recoil energy of 1.0 keV, ionization due to recoil was noted to be 0.4%, which compares with 0.1% obtained from eq. (2.4). More detailed analysis (Carlson et al. 1963) of the problem shows that the exact solution of the problem is difficult, but eq. (2.4) at least represents a rough estimate.

3. Excitation due to Vacancy Cascades

When a vacancy in the inner shell of an atom is formed as in the cases of electron capture or internal conversion, this vacancy is generally filled by a series of Auger processes. In each Auger process a hole is filled by a nonradiative process in which one electron from an outer shell drops into the hole while a second electron goes into the continuum carrying away the excess energy represented by the difference in the initial and final states. Each Auger process creates one extra vacancy. The new vacancies are filled by still other Auger processes and the accumulation of these ionization processes is known as a vacancy cascade (Pleasonton and Snell 1957). Radiative processes may also occur, but the probability for Auger processes to occur is substantially greater than X-ray emission except in the case of the K and L shells of heavy elements. Figure 2.2 shows a typical example of a vacancy cascade in Xe. It has been possible to treat each Auger process independently, and evaluate the net charge spectrum by means of a Monte Carlo calculation (Carlson and Krause 1965a). Figure 2.3 shows a comparison (Krause and Carlson 1967) between experiment and theory for Kr. (The calculation also includes the effects of electron shake-off, which will be discussed in the next section.) The time for the whole cascade to take place is about 10^{-14} sec. The average charge resulting from a single initial vacancy in a free atom is given in fig. 2.4, determined empirically (Carlson et al. 1966) as a function of Z and the shell in which the initial vacancy was formed. Recent data on the decay of heavy elements (Gunter et al. 1966, de Wieclawik and Perrin 1969) give reasonable agreement with the extrapolated values.

What occurs when an atom undergoing a vacancy cascade is attached to a molecule? Early work (Wexler and Anderson 1960, Wexler 1962, Carlson and White 1963c) on gases containing ^{80m}Br and ^{125}I helped supply some qualitative answers. However, it was found (Carlson and White 1966, 1968) experimentally desirable to produce the initial inner shell vacancy by X-ray photoionization rather than by radioactive decay, since a greater variety of compounds could be studied; and, most important, the initial ionization could be confined to a smaller volume, which in a specially designed mass spectrometer allowed for a collection efficiency that was independent of recoil energy. This enabled measurements to be made on the kinetic energy of the various fragment ions as well as the identification and determination of their

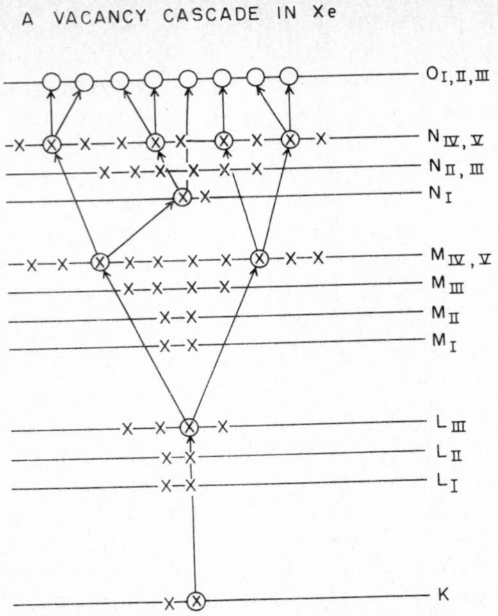

Fig. 2.2. Typical vacancy cascade in Xe as the result of an initial L vacancy. The ×'s represent electrons; ○'s vacancies and ⊗'s the cases where a vacancy has been formed and then filled by a subsequent Auger process.

relative intensities. From these studies the following picture arose: the atom in which an inner shell vacancy is formed acquires a high positive charge via a vacancy cascade; this charge is partially neutralized by extraction of electrons from other atoms in the molecule, as a consequence of which further autoionization occurs increasing the total number of electrons lost over that expected for an isolated atom. This whole charging process takes place in a sufficiently short time that it is complete before the molecule has a chance to fragment, and what remains is a collection of positive ions spaced at positions that are not substantially different from the initial internuclear distances. This collection of ions then literally explodes due to Coulombic repulsion, as illustrated in fig. 2.5. By use of this model and a calculation based on a computerized Coulombic explosion it was possible to calculate recoil energies that were in reasonable agreement with experimental values for some simple gaseous molecules (Carlson and White 1966, 1968). It was also found that as the molecular size increases, the degree of total electron ejection and fragment ion formation increases, but that the atom in which the initial vacancy is formed receives a lower charge and lower recoil energy. This suggests that in a condensed media the hot atom which has undergone a vacancy cascade may not recoil from its initial site but rather becomes the center for a large number

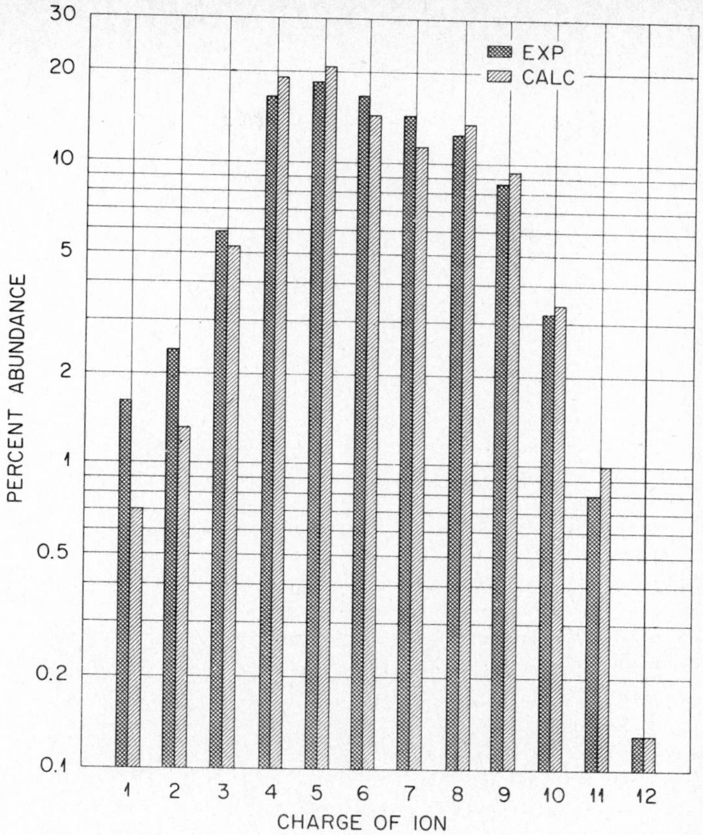

Fig. 2.3. Experimental charge distribution of Kr ions resulting from irradiation of krypton by Mo K X-rays compared with calculated spectrum. The initial vacancy distribution is taken as 1s = 86.5%; 2s = 8.3%; 2p = 3.7%; 3s = 0.7%; 3p = 0.4%; 3d = 0.2% and remaining shells 0.2% (cf. Krause and Carlson 1967).

of highly excited species. The possible effects of recoil from a vacancy cascade in solids have been discussed by Kazanjian and Libby (1965).

The Mössbauer effect has been employed to evaluate the results of vacancy cascades in relatively short times after the cascade (Wickman and Wertheim 1968). For example, electron capture in ^{57}Co precedes the emission of the Mössbauer γ by 10^{-7} sec. The chemical state of ^{57}Fe can thus be recorded by the Mössbauer effect 10^{-7} sec after K capture and its subsequent vacancy cascade. This time is still of course much larger than the time for the vacancy cascade itself.

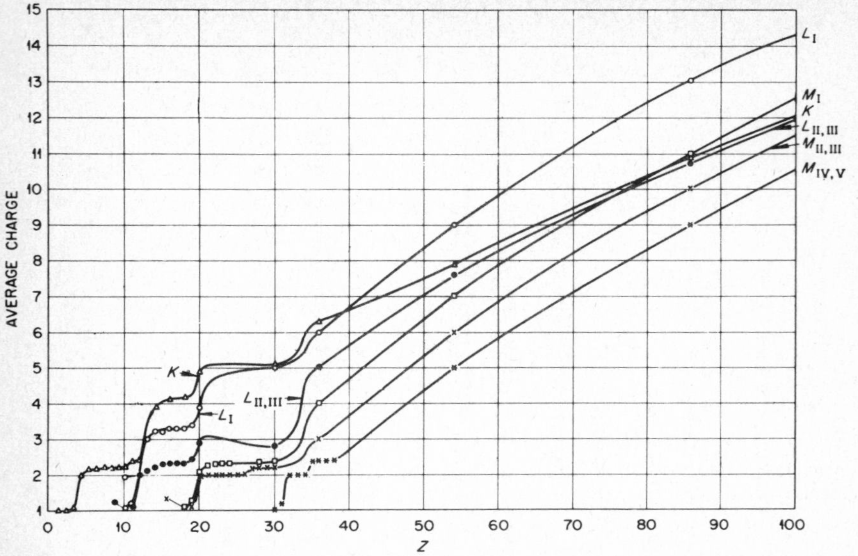

Fig. 2.4. Estimated average charge as the result of a sudden vacancy in the K, L and M shells as a function of atomic number (cf. Carlson et al. 1966).

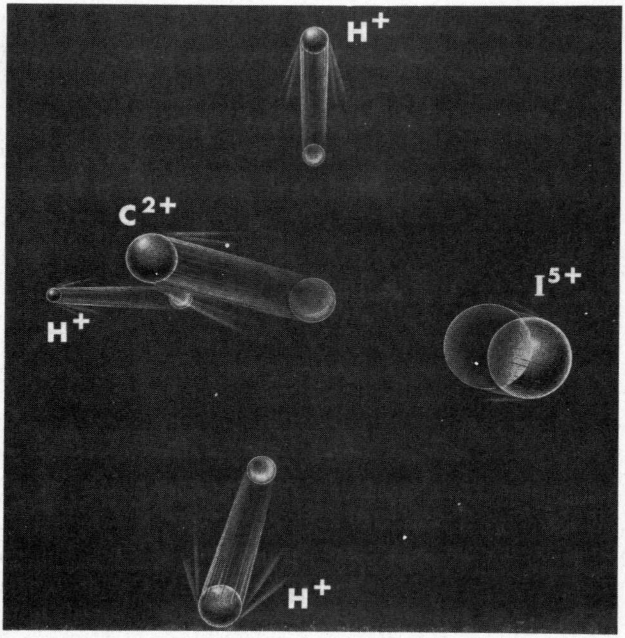

Fig. 2.5. A pictorial description of the 'Coulombic explosion' of ions resulting from a vacancy in the L shell of iodine belonging to CH_3I (cf. Carlson and White 1968).

4. Excitation due to Electron Shake-off

When there is a sudden change in the central potential of an atom, there is a finite probability that an electron in a given orbital will find itself in either an excited state or the continuum. This process has been discussed (Feinberg 1941, Migdal 1941, Levinger 1953, Schwartz 1953) in terms of the sudden approximation; and the shake-off probability P_{nlj} for the removal of an electron from orbital nlj is given by

$$P_{nlj} = 1 - [|\int \psi'_{nlj} \, \psi_{nlj} \, d\tau|^2]^N, \tag{2.5}$$

where ψ_{nlj} and ψ_{nlj} are the single electron wave functions for the initial state and final states of the atomic orbitals nlj. N is the number of electrons in the orbital of the initial state. The overlap $\psi^*\psi$ in eq. (2.5) is not unity since the final state has a different central potential. For example, in the case of beta decay the nuclear charge has changed by one (plus one for β^- decay and minus one for positron decay) and in the case of α decay it has decreased by two charges. Electron shake-off can also occur when a sudden change is created in the electron configuration since the potential seen by each electron is given as

$$Z_{eff} = Z - \sigma, \tag{2.6}$$

where Z_{eff} is the effective charge, Z is the nuclear charge and σ is the screening of the other electrons. If there is a sudden change in the screening as a consequence, for example, of the promotion of an inner vacancy by internal conversion, photoionization or an Auger process, the effect will be similar to a sudden change in the nuclear charge.

Systematic calculations using nonrelativistic Hartree–Fock and relativistic Hartree–Fock–Slater wave functions have been made (Carlson et al. 1968) for elements from $Z = 2$ to 92. Agreement with experimental values both for the inner and outer shells has been in general quite satisfactory (Freedman 1974). An example of one of the calculations is given in table 2.2. Note that this calculation includes transitions to both the continuum and discrete states. The term shake-off has sometimes been limited to transitions to the continuum (i.e., ionization) while shake-up is used for transitions to excited but discrete states. The terms monopole ionization and monopole excitation have also been employed, since as can be seen from eq. (2.5) the operator is unity and the transitions obey monopole selection rules. In general, electron shake-off is much more probable than electron shake-up as long as inner shell orbitals are involved, but in the case of the valence shell, transitions to discrete states are of comparable magnitude with those to the continuum. It is interesting to evaluate the average energy using the shake-off probability from table 2.2 for $Xe \xrightarrow{\beta^-} Cs^+$ and the estimated energy for the shake-off process as determined

Table 2.2
Probability for electron shake-off (%)[a] for Xe $\xrightarrow{\beta^-}$ Cs^+

Shell	% probability
1s	0.0348
2s	0.0791
$2p_{1/2}$	0.0494
$2p_{3/2}$	0.0819
3s	0.160
$3p_{1/2}$	0.144
$3p_{3/2}$	0.258
$3d_{3/2}$	0.205
$3d_{5/2}$	0.300
4s	0.376
$4p_{1/2}$	0.417
$4p_{3/2}$	0.923
$4d_{3/2}$	2.03
$4d_{5/2}$	3.17
5s	2.09
$5p_{1/2}$	3.88
$5p_{3/2}$	8.51
Total	21.3

[a] Carlson et al. (1968).

from the binding energies of the different orbitals. This estimate (70 eV) is in reasonable agreement with that obtained by Serber and Snyder (1952) (93 eV) using the overall change in the Thomas–Fermi potential. [See Carlson et al. (1968) for a more comprehensive comparison.] The point I wish to stress here is the large difference between the average and the most probable energy. Although the average energy is large compared to bond energies, most of the time *no* excitation arises as the result of a sudden change in nuclear charge. The high average comes from the few times that large amounts of energy are expended in the ejection of tightly bound electrons.

The energy distribution of the shake-off electrons has been estimated from hydrogenic wave functions (Levinger 1953). The most probable energy is zero, and this probability decreases monotonically with energy so that about two thirds of the shake-off electrons have kinetic energies below the value of the binding energy of the shell from which the electron was removed. This description has been verified by data derived from photoionization (Krause et al. 1968). Suzor and co-workers (Spighel and Suzor 1962, Suzor 1960) measured low-energy electron spectra following β decay. Some of their results agreed with theory but others were considerably different. However, their experimental procedures have been criticized (Erman et al. 1968).

At the beginning of this section electron shake-off was described as occurring as if there were a *sudden* change in the central potential. How sudden is sudden? Studies on the extent of electron shake-off as a function of the energy (or velocity) of the ejected inner-shell electron have been made (Erman et al. 1968, Carlson and Krause 1965b). The probability was found to be independent of the energy of the ejected electron (as the sudden approximation implies) until that energy declined below a value that was approximately three times the threshold energy for ejecting an extra electron via the shake-off process from a given orbital. It can be said that the sudden approximation failed at this point.

Another important consequence of the experiments mentioned in the previous paragraph is the light they shed on direct collision. That is, what is the relative probability of a β particle or photoelectron directly colliding with an orbital electron as it emerges from the given atom, as compared to ionization arising from electron shake-off? This probability has been considered theoretically (Feinberg 1941, 1965, Weiner 1966) and it was thought to be generally minor compared to electron shake-off, but this conclusion was reached with some reservation. If the direct collision process is present, it should have an energy dependence that is approximately equal to v_e^2 where v_e is the velocity of the emerging β particle or photoelectron. The absence of such a behavior in experimental data (Erman et al. 1968, Carlson and Krause 1965b) has allowed lower limits to be set on the relative importance of electron collisions, which can be said to be negligible at higher electron velocities. At lower electron velocities, just above threshold energy the description of ionization collapses in terms of the simple model of electron shake-off and direct collision, and a more complex description involving electron correlation is needed.

What is the molecular consequence of electron shake-off? This question has been studied using mass spectroscopy to collect fragment ions formed from molecular decomposition as the result of β^- decay (Snell et al. 1957, Snell and Pleasonton 1958, Wexler and Hess 1958, Wexler et al. 1960, Carlson 1960, Carlson and White 1962, 1963a, b). Some information has also been gleaned (Carlson et al. 1970a) from photoelectron studies where monopole excitation is detected by characteristic losses in the photoelectron spectra. From these studies it is apparent that the extent of excitation is roughly the amount expected from atomic calculations. The probability for going to an excited state depends on the overlap of bound electrons with the unoccupied molecular orbital for which transitions are allowed. The closer in energy these levels are, the greater the probability for excitation. As with atoms, in most cases no excitation is observed at all. But when it does occur, it can lead to a high degree of molecular fragmentation.

5. Primary Process for Specific Cases in Nuclear Decay

Before beginning our discussion of the specific problems involved for each type of nuclear event, the reader is reminded that several events often occur in the same decay so that the final consequence is a sum total of these events. In nuclear decay and particularly in nuclear reaction and fission a nucleus may emit more than one particle. In addition, excited states of a given nucleus (or isomers) are often formed that may decay by γ emission or internal conversion. In consideration of the time involved, the three types of excitation processes discussed in section 2 usually take place instantaneously in comparison with the normal time between nuclear decay processes. The time it takes a recoiling atom to break a bond is less than the order of a vibrational frequency, 10^{-14} sec. Vacancy cascades are complete in about 10^{-14} sec. The time for electron shake-off is that needed to remove the β particle from the atom, $\sim 10^{-18}$ sec. Thus, the excitation received from each nuclear event can often be considered separately without consideration of cooperative effects. (Important exceptions to this generalization are in the γ cascade produced in n, γ reactions and in particle emission in nuclear reactions.) The time it takes before an atom reaches its final resting place as a radioactively labeled but cold atom can sometimes be quite long compared to subsequent nuclear events and, in this sense the primary excitations do add their effects, such as in the case of $^{125}\text{I} \xrightarrow{\text{EC}} {}^{125}\text{Te}$ where a vacancy cascade arising from internal conversion is added to one coming from electron capture; or in the situation when a vacancy cascade from internal conversion is added to a recoiling ion as often occurs in fission and nuclear reactions.

5.1. BETA DECAY

Beta decay involves the simultaneous ejection of a neutrino and an electron ($n \rightarrow p^+ + e^- + \bar{\nu}$). The nucleus thus does not receive a discrete recoil energy but a spectrum depending on how the kinetic energy is shared between the electron and the neutrino and the angle between the two particles. The expression for the distribution of recoil energy, $N(E)$ is given (Johnson et al. 1963, Wu and Moszkowski 1966) as

$$N(E) = AF(E)\,[N_1\,(E) + \alpha N_2(E)] \tag{2.7}$$

where

$$N_1(E)\,\mathrm{d}E = Q^2\left[\frac{(\omega_0^2 - Q^2 - 1)^2}{(\omega_0^2 - Q^2)^3}[(\omega_0^2 - Q^2)^2 + 2\omega_0^4 + 3\omega_0^2 - 2\omega_0^2 Q^2 + Q^2]\right]\mathrm{d}Q$$

and

$$N_2(E)\,\mathrm{d}E = \left[N_1(Q) - 6Q^2 \, \frac{(\omega_0^2 - Q^2 - 1)^2}{(\omega_0^2 - Q^2)} \right] \mathrm{d}Q$$

where A is a proportionality constant, Q is the recoil momentum, ω_0 is the maximum beta energy, $F(E)$ is the Fermi–Coulomb function and α is the correlation coefficient. For allowed decay $\alpha = -\frac{1}{3}$ for Gamow–Teller transition and +1 for Fermi transitions. For forbidden and super allowed decays, see Wu and Moszkowski (1966). Examples of recoil spectra from beta decay are shown in fig. 2.6.

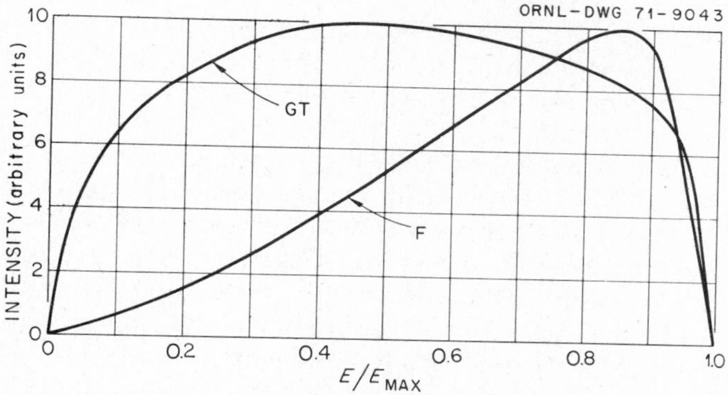

Fig. 2.6. Recoil energy spectra for allowed β decay. E is the energy of the beta particle whose maximum value is E_{max}. GT shows the shape for pure Gamow–Teller interactions and F for pure Fermi interactions. Beta decay having a mixture of GT and F will yield a spectral shape that is a mixture of the two cases. For further discussion, cf. Wu and Moszkowski (1966).

The maximum recoil is given (Edwards and Davies 1948) as

$$E_{max}\,(\mathrm{eV}) = 537 \times 10^{-6}\, E \left(\frac{E_\beta + 1.02}{M} \right), \tag{2.8}$$

where E_β is the maximum beta energy in MeV. The average recoil energy in beta decay can generally be taken as approximately $\frac{1}{2}\,E_{max}$.

In addition to the excitation received from recoil energy, electron shake-off may occur. For low-energy beta emitters, e.g., ^3H, ^{35}S and ^{14}C the recoil is often insufficient to rupture a normal chemical bond of about 1 to 5 eV, and electron shake-off may be the most important source of excitation. In most of the decays the electron shake-off process will actually provide no excitation, the molecule finding itself in the ground electronic state. However, in 20 to 40% of the decays some excitation will occur.

As discussed earlier electron shake-off arises from a sudden change in the nuclear charge. Under most circumstances the β particle leaves with a velocity that is much greater than the velocity of the orbital electrons, and the sudden approximation is valid. The agreement between experiment and theoretical calculations based on the sudden approximation is generally quite good (Freedman 1974). The binding energy of the K shell may sometimes, however, be comparable or larger than the low-energy portions of the beta decay spectra, in which case electron shake-off is diminished or energetically impossible. Stephas and Crasemann (1967, Crasemann and Stephas 1969) have outlined a complete treatment of the shake-off process that includes taking the neutrino into account, although Erman et al. (1968) have suggested that the separation of the beta–neutrino interaction from the electron shake-off process is a valid approximation and is supported by experiment. A still more recent appraisal has been given by Fischbeck et al. (1971). In any case, shake-off in the K shell is a rare event and as far as the overall consequences of beta decay are concerned the sudden approximation can be employed without concern.

The distribution of excited states has been studied in the case of beta decay (Micklitz and Luchner 1969) and in photoionization (Krause et al. 1968, Carlson et al. 1970a, Siegbahn et al. 1969). The formation of a K vacancy due to photoionization gives rise to electron shake-off in the valence shell that should be very similar to that arising from beta decay (Carlson et al. 1968). In the photoionization of neon about 78% of the time no excitation occurs, 8% of the time transitions occur to excited but bound states, requiring 37 to 46 eV, and 14% of the time transitions occur to the continuum requiring energy greater than 46 eV. A similar description was found to fit the data on the photoelectron spectra of molecules (Carlson et al. 1970a) except that the energies required for excitation and ionization were less, while the probability for going to an excited state was slightly greater. An illustration of the net effect of electronic excitation in beta decay is given in fig. 2.7 which shows the charge spectrum of Rb ions resulting from the beta decay of ^{85}Kr. Most of the ions (80%) are the singly charged Rb^+ which is the species formed following beta decay in which *no* shake-off occurs, but there is also a complex spectrum of more highly charged ions decreasing in intensity as the charge increases. The charge spectrum has been calculated (Carlson 1963) by considering multiple shake-off as a product of single shake-off probabilities and adding the effects of vacancy cascades when electron shake-off occurs in the inner shells. The agreement between theory and experiment is quite good.

As an example of the large amount of energy that can be fed into the parent ion from monopole excitation and ionization, the fragment ion spectrum from the decay of tritiated benzene (Carlson 1960) is shown in table 2.3. Note that the recoil in the beta decay of ^{3}H is too small to cause any molecular fragmentation of $C_6H_5^+$. The $C_6H_5He^+$ is unstable and the He will separate from $C_6H_5^+$ whether excitation is derived from the nuclear decay or not, the question being how much excitation energy can be delivered to $C_6H_5^+$. From

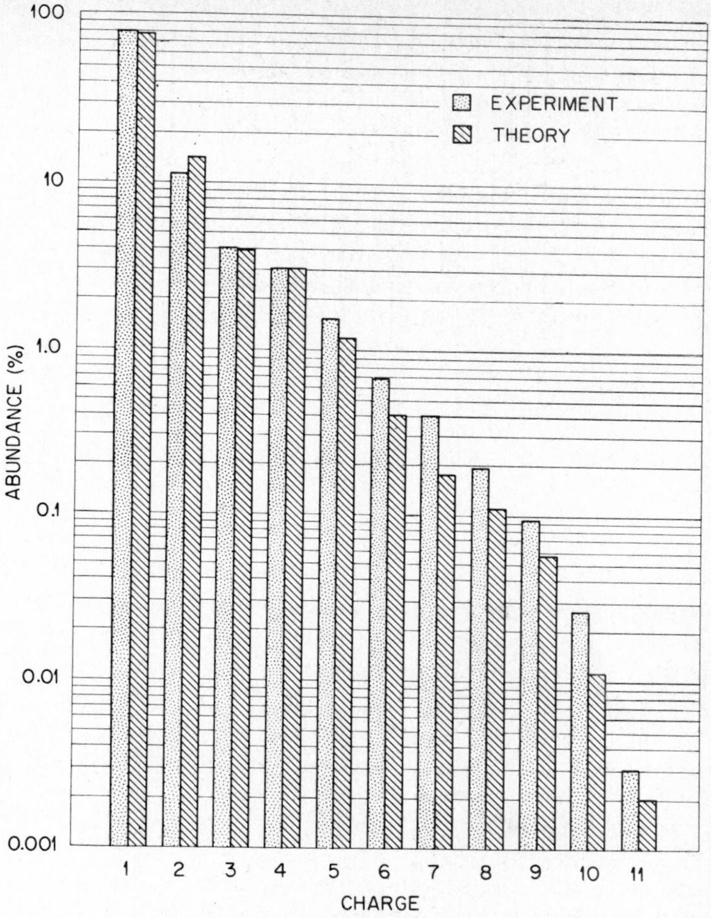

Fig. 2.7. Comparison of charge spectrum of Rb ions following $^{85}Kr \rightarrow {}^{85}Rb$ taken from experimental data of Snell and Pleasonton 1957, together with calculations based on shake-off probability, effects of vacancy cascade and consequences of multiple ionization (cf. Krause and Carlson 1967).

table 2.3 it is seen that when excitation can be delivered (about 70% of the time $C_6H_5^+$ does not receive enough energy to decompose) it is considerable, since the formation of such ions as C^+, for example, requires at least 40 eV. Similar conclusions have been reached in the study of the fragment ions of other molecules containing an atom undergoing beta decay (Snell et al. 1957, Carlson and White 1963a).

Recoil energy manifests itself primarily in the severance of the bond between the recoiling atom and the parent molecule. Electron excitation that arises from

Table 2.3
Fragment ion spectrum[a] following $C_6H_5T \xrightarrow{\beta^-} C_6H_5He^+$

Ion	% abundance	Ion	% abundance
$C_6H_5He^+$	0	C_3H_3	0.4
$C_6H_5^+$	71.6	$C_3H_2^+$	2.7
$C_6H_5^{2+}$	0.6	C_3H^+	3.7
$C_6H_3^+$	0.5	C_3^+	1.2
$C_6H_3^{2+}$	0.1	$C_2H_3^+$	0.4
$C_6H_2^+$	2.5	$C_2H_2^+$	0.5
C_6H^+	1.3	C_2H^+	0.5
C_6H^{2+}	0.4	C_2^+	0.4
C_6^+	0.3	CH_3^+	0.1
C_5H^+	0.2	CH_2^+	0.1
C_5H^+	0.7	CH^+	0.2
C_5^+	0.5	C^+	0.3
$C_4H_3^+$	2.6	H_2^+	0
$C_4H_2^+$	3.4	H^+	0.4
C_4H^+	3.4	He^+	0.1
C_4^+	0.6		

[a] Carlson (1960).

electron shake-off can feed back into the parent molecule causing extensive decomposition, as illustrated above, so that while in a majority of decays no excitation is derived from electron shake-off, when it does occur it can be quite destructive. Table 2.4 shows the relative abundance of fragment ions formed in

Table 2.4
Comparison of the fragment and charge spectra of ions resulting from the radioactive decay of CH_3I^{131} and CH_3I^{130}

	Abundance (%)	
Ion	Ch_3I^{131}	CH_3I^{130}
Ch_3Xe^+	69.4	34.0
CH_3^+	2.0 ± 0.2	38.8 ± 5.0
CH_2^+	2.4 ± 0.3	3.3 ± 0.7
CH^+	2.2 ± 0.4	3.4 ± 0.7
C^+	1.9 ± 0.3	3.9 ± 0.6
C^{2+}	0.07 ± 0.04	0.1 ± 0.07
Xe^+	14.6 ± 0.6	9.6 ± 0.6
Xe^{2+}	2.7 ± 0.3	2.0 ± 0.4
Xe^{3+}	0.9 ± 0.1	1.4 ± 0.3
Xe^{4+}	0.37 ± 0.07	0.29 ± 0.06

[a] Carlson and White (1963b).

the beta decay of $CH_3I \xrightarrow{\beta^-} CH_3Xe^+$. The main difference in the two beta decays is that ^{130}I gives a substantially greater recoil than ^{131}I. CH_3^+ is formed from simple C–Xe rupture, while the remaining hydrocarbon ions result from subsequent unimolecular decomposition. From table 2.4 it appears that the recoil energy influences primarily the rupture of the C–Xe bond, transforming CH_3Xe^+ to $CH_3^+ + Xe$, while the electronic excitation derived from monopole excitation and ionization gives rise to the remaining fragment ions. One might expect some interplay between the two modes of excitation, but to a fair approximation their effects seem to be separable.

5.2. POSITRON DECAY

The nature of positron decay runs a close parallel to beta (or negatron) decay except for a reversal in charge ($p^+ \rightarrow n + e^+ + \nu$). Electron shake-off occurs as the consequence of a sudden change in charge, but whereas in β^- decay the final charge is $Z + 1$, in β^+ the final charge is $Z - 1$. The final nuclear charge plays a profound effect in the final chemical form of the hot atom simply as the consequence of determining what element is created; but the extent of electron shake-off is similar, regardless of the direction of the change in charge. The more important consideration in electron shake-off is $|\Delta Z_{eff}|$, the amount of change in effective charge, not the sign. The probability for electron shake-off is in fact approximately proportional to ΔZ^2_{eff}. A comparison of electron shake-off probability for positron and negatron decay computed for Kr is shown in table 2.5. For the inner shells shake-off probabilities are essentially identical, although differences do occur for the outer shells.

The expressions for recoil energy for positron decay are similar to those given for β^- decay (Wu and Moszkowski 1966). One important consideration is that the threshold for positron decay is 1.02 MeV higher than that for

Table 2.5
Comparison of electron shake-off in β^- and β^+ decay

Shell	Probability for electron shake-off[a] (%) $Kr \xrightarrow{\beta^-} Rb^+$	$Kr \xrightarrow{\beta^-} Br^-$
1s	0.070	0.074
2s	0.175	0.200
2p	0.324	0.415
3s	0.460	0.525
3p	1.41	1.76
3d	3.39	3.73
4s	2.89	10.0
4p	14.1	19.8

[a] Carlson et al. (1968).

electron capture. Even when positron decay is energetically possible, electron capture predominates as an alternative mode of decay for a nucleus to decrease its charge by one, until the energy for positron emission is above approximately 1 MeV. In addition, the Fermi correction for positron decay due to nuclear charge will be of opposite sign to that of beta decay. Though small at higher β energies, the Fermi correction will strongly affect the kinetic energies near the threshold, the β^+ particle receiving more, the β^- particles less recoil.

Since a positron will annihilate upon interaction with an electron, there is the finite possibility that such annihilation will occur with one of the atomic electrons of the hot atom causing a vacancy in an atomic shell and a recoil from the annihilation radiation. Such an event is, however, highly improbable.

5.3. α DECAY

The recoil energy derived from the decay of a heavy particle is considerably larger than in beta and γ decay. For α decay the recoil energy is simply

$$E_r = \frac{m}{M} E_\alpha \quad , \tag{2.9}$$

where E_α is the kinetic energy of the α particle, and m and M are respectively the masses of the α particle and hot atom. Since the energy for α decay is usually in excess of 1 MeV, the recoil energy is sufficient to not only sever the primary bond in every case but to cause extensive damage to the parent molecule and its surroundings.

If the α particle has kinetic energy in excess of 400 keV, its velocity will be in excess of any orbital electron that might be attached to helium and thus can be considered as a free particle whose ejection may give rise to electron shake-off as the result of a sudden change in nuclear charge of −2. Since the shake-off probability is roughly proportional to ΔZ^2_{eff}, this probability would be roughly four times as great as that observed in beta decay. As will be discussed below this does not hold for the innermost atomic shells.

Additional electron loss may occur from the recoil itself. The velocity of electrons in the K orbital of the hot atom is generally considerably greater than the α particle velocity, and thus, as far as the change in the central potential is concerned for these K shell electrons, this is an adiabatic process and no electron shake-off occurs. On consideration of only excitation from the recoil effect, or the sudden acceleration of the center of charge from rest to a finite velocity, Levinger (1953) calculated the probability for electron ejection from the K orbital to be 2 to 3×10^{-7} for the decay of ^{210}Po. Experimentally, (Rubinson 1963) the value was found to be approximately 1.5×10^{-6}, higher than predicted by Levinger but still considerably below what one would expect from a simple electron shake-off due to a nonadiabatic change in nuclear charge.

5.4. GAMMA TRANSITION

Gamma decay can occur any time there is an excited isomer and the transition is not strongly forbidden. Gamma decay thus usually accompanies nuclear processes that do not go directly to the ground state of a given isotope. The chemical effects of gamma decay are small. There is no chance for electron shake-off or a vacancy cascade and the recoil energy is (Wahl and Bonner 1957)

$$E_r = 537\ E_\gamma^2/M, \tag{2.10}$$

where E_r is the recoil energy in eV, E_γ the γ-ray energy in MeV and M is the mass of the nucleus in atomic mass units. Bond rupture may not occur if the transition energy is much below 1 MeV.

5.5. INTERNAL CONVERSION

As an alternate route for deexcitation of an excited state of an isotope a nucleus may decay by internal conversion rather than gamma emission. The recoil energy derived solely from the ejection of the internally converted electron is (Wahl and Bonner 1957)

$$E_r = \frac{537\,(E_e)^2}{M} + \frac{537 \times 1.02\,(E_e)}{M} \tag{2.11}$$

where E_r is expressed in eV, and E_e expressed in MeV is the kinetic energy of the internally converted electron, which in turn equals $E_\gamma - E_b$, where E_γ is the transition energy and E_b the binding energy of the atomic shell in which internal conversion occurs. M is the mass of the recoil atom in atomic units. Internal conversion rather than γ decay occurs when the transition is highly forbidden and when the transition energy is low. Conversion usually takes place in the atomic shell having the lowest principal quantum number that is permissible on consideration of the energy of the transition. Comprehensive calculations on the probability for internal conversion as a function of the atomic shell and transition energy are available (Hager and Seltzer 1968).

The main chemical consequence of internal conversion is the vacancy cascade which is the result of the hole formed by the ejection of the internally converted electron. A typical charge spectrum for an atom undergoing internal conversion is shown in fig. 2.8.

One of the concomitant consequences of internal conversion is electron shake-off due to the sudden change in shielding caused by ejection of an inner-shell electron. In terms of the sudden approximation the shake-off probabilities will be identical with the effect of photoionization in the same shell. Recent ex-

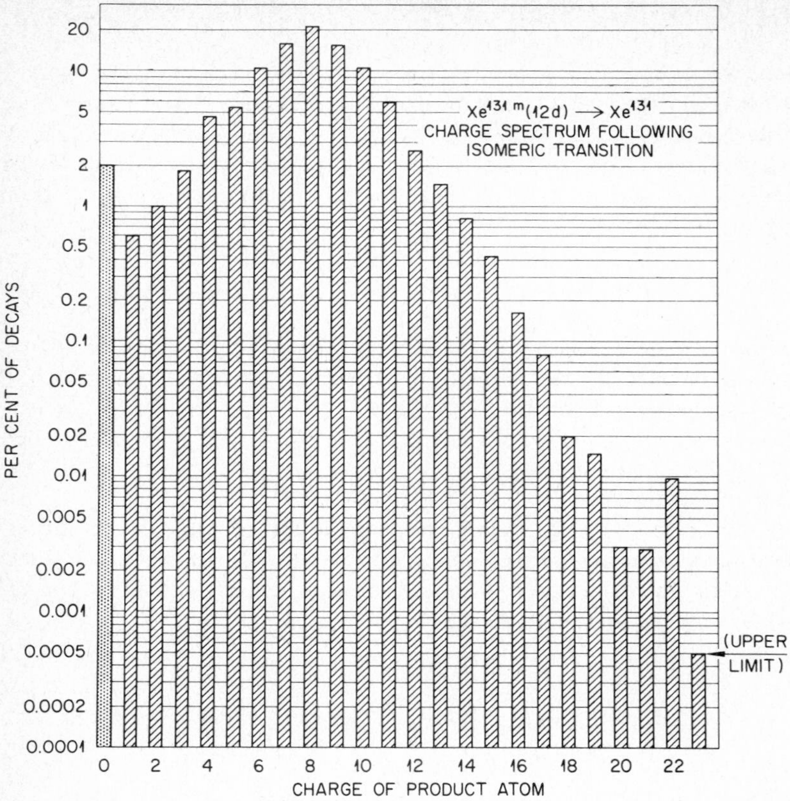

Fig. 2.8. Charge spectrum of ions following $^{131m}Xe \xrightarrow{EC} {}^{131}Xe$ (Pleasonton and Snell 1957).

perimental results (Porter et al. 1971) agree very well with electron shake-off calculations.

5.6. ELECTRON CAPTURE

As with internal conversion the principal chemical consequence of electron capture is the formation of an inner-shell vacancy which can then undergo a vacancy cascade.

The recoil energy in eV is given as (Allen 1958)

$$E_r = 537\ E_\nu^2/M, \tag{2.12}$$

where E_ν is the neutrino energy in MeV (assuming the rest mass of the neutrino is zero) and in which $E_\nu = Q - E_b$ where Q is the transition energy and E_b is the atomic binding energy. In the case of the decay of ^{37}Ar Snell and Pleasonton (1955) were able to measure the recoil, 9.63 ± 0.06 eV, in excellent agreement with expectation.

Electron shake-off as the result of an inner-shell vacancy is negligible since the effective charge for the outer shell electrons remains essentially unchanged, as the decrease in nuclear charge is compensated by a change in shielding. Only for electrons in the same atomic shell, or lower than that in which electron capture occurs will there be a significant change in effective charge. The probability for electron shake-off from the K shell as the result of K capture has been studied theoretically and experimentally (Stephas 1969), but it was found to be very small. For example, the probability is only 2.5×10^{-5} for K capture in ^{131}Cs (Lark and Perlman 1960).

For all practical purposes the net chemical excitation for electron capture is very similar to internal conversion except that in the former case one is dealing with an element of final nuclear charge $Z-1$.

5.7. NUCLEAR REACTIONS

The chemical consequences of nuclear reactions are primarily related to the recoil energy imparted to the hot atom. The recoil energy is generally in great excess of energies involved in chemical bonding, and the important question is not whether the hot atom will separate from the parent molecule, but how high will be the kinetic energy, since the recoil will determine the range of the hot atom and its equilibrium charge during the slowing down process.

On the assumption of isotropic emission of the ejected nuclear particle Libby (1947) derived an expression for the average recoil energy $\langle E_r \rangle$ for the nuclear reaction $A + x \rightarrow A' + y + Q$ where A is the initial nucleus; x is the bombarding nuclear particle with mass μ and kinetic energy E_μ; A′ is the recoiling atom with mass M and recoil energy E_r; y is the emitted nuclear particle with mass m, and Q is the overall mass difference in terms of energy (+ for a decrease in mass, exoergic reaction; − for an increase in mass, endothermic reaction).

$$\langle E_r \rangle = E \left\{ \frac{M}{(M+m)^2} + \frac{m(m+M-\mu)}{(m+M)^2} \left[1 + \frac{Q}{E} \left(\frac{m+M}{m+M-\mu} \right) \right] \right\}. \qquad (2.13)$$

For any angle θ between the incident particle and the emitted nuclear particle, the recoil energy is

$$E_r = \langle E_r \rangle - A \cos\theta \quad ,$$

where

$$A = \frac{2E}{(m+M)^2}\left\{Mm\mu\,(m+M-\mu)\;1 + \left(\frac{m+M}{m+M-\mu}\right)\right]\right\} \tag{2.14}$$

and because of the isotropic distribution, E_r is distributed uniformly over a range $\pm A$ about $\langle E_r \rangle$.

The above expressions are for single particle emission such as (n,p), (n,α), (p,n), (d,p) reactions. Also important are reactions where more than one particle is emitted. Under this category are the (n,2n) reactions and spallation reactions in which a very energetic massive projectile is absorbed by a nucleus and several nuclear particles are ejected. Heavier and heavier bombarding particles are being examined in nuclear reactions, and eventually, one anticipates that uranium–uranium collisions will be initiated for experiments leading to the production of superheavy elements.

The recoil energies involved in multiple particle emission are a complex function of their timing and direction. Some simplifications are possible. For example, in the case of (n,2n) reactions the two ejected neutrons can be considered as being evaporated in succession from a compound nucleus so that cancellation of momenta between the neutrons does not occur and the total kinetic energy is the sum of the energies associated with the successive events. If one is near the threshold energy for the formation of a compound nucleus, the kinetic energy of the ejected nucleon will be negligible compared to the recoil received from the incident particle, and the recoil is given (Winsberg and Alexander 1971) as

$$E_r \simeq E_\mu \mu\; m/(\mu + M)^2 \quad , \tag{2.15}$$

where the symbols are the same as in eq. (2.14). When the energy of the bombarding particle is considerably above the threshold for spallation, E_r will depend on the direction and energy of the emitted particles. The final distribution in energy can be computed by Monte Carlo calculations (Donovan et al. 1960), based on the formation of a compound nucleus followed by isotropic evaporation of the particle.

Of special interest is the n,γ reaction, particularly those reactions that take place with thermal neutrons, because of all the nuclear reactions this has been the one most studied by hot atom chemists. In general, 6–8 MeV of energy is released in neutron capture usually by a cascade of γ rays. See, for example Bartholomev and Groshev (1967) for a listing of the complex decay schemes to be expected in n,γ reactions. When thermal or low energy neutrons are involved in the capture process, most of the recoil is derived from γ decay [cf. eq. (2.10)]. Since in n,γ reactions there is usually a cascade of γ rays, the recoil energies can in some instances cancel each other so that insufficient energy is present for bond rupture. The probability for this has been investigated by several authors (Cobble and Boyd 1952, Hsiung et al. 1961, Schweinler 1961).

In general it amounts to about a few parts in 1000. Hsiung et al. (1961) for example have calculated this probability for $^{35}Cl(n,\gamma)^{36}Cl$, and, $^{79}Br(n,\gamma)^{80}Br$, and $^{127}I(n,\gamma)^{128}I$, and have had some success in correlating their calculations with experimental results.

In addition to readjustment of the excited nucleus by γ emission there is the possibility for internal conversion with the accompanying effects of a vacancy cascade. Chemical evidence and measurement of positive ions suggest that internal conversion may occur in a substantial number of decays following neutron capture. For example, Wexler and Davies (1952) estimate that vacancy cascades occur in 44% of $^{127}I(n,\gamma)^{128}I$ reaction.

Another category of nuclear reactions is Coulomb excitation in which a charged particle (from protons to heavy elements) undergoes inelastic collisions with another nucleus creating excited isomeric states. Coulomb excitation requires only a grazing collision; therefore the hot atom does not receive the full impact of a head-on collision. Knowledge of the recoil can be obtained experimentally from line broadening in γ emission due to a Doppler shift. Typical recoil energy spectra from Coulomb excitation are shown in fig. 2.9. The chemical consequences of Coulomb excitations have been studied in

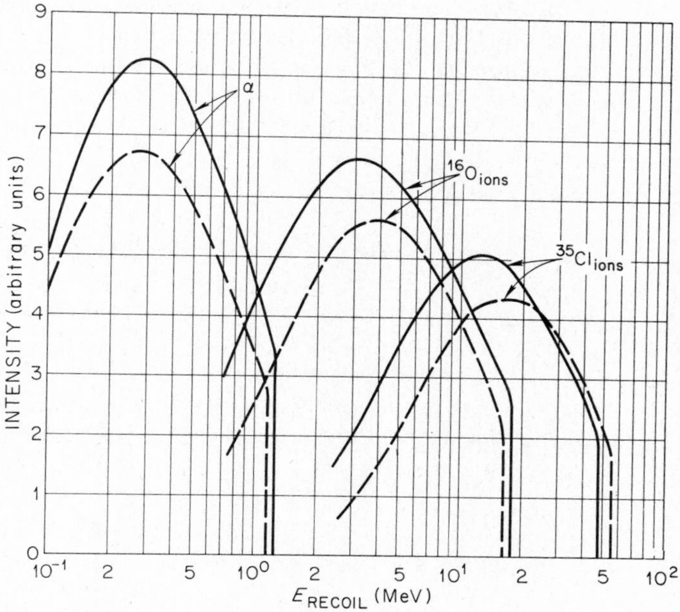

Fig. 2.9. Recoil energy spectrum of ^{50}Cr and ^{120}Sn ions following Coulomb excitation of first 2+ state. Dotted lines are for ^{120}Sn which has been excited by (1) 10 MeV α, (2) 40 MeV ^{16}O and (3) 80 MeV ^{35}Cl. Solid lines are for ^{50}Cr which has been excited by (1) 5 MeV α, (2) 25 MeV ^{16}O and (3) 50 MeV ^{35}Cl (Stelson and Milner, private communication).

special cases by the observation of the subsequent Mössbauer radiation (Obenshain 1968).

Finally, γ,n reactions should be mentioned. The average recoil energy is (Libby 1947)

$$\langle E_r \rangle = \frac{E_\gamma^2}{2(M+m)C^2}\left(\frac{M-m}{M+m}\right) + \frac{m(E_\gamma + Q)}{M+m}\,, \tag{2.16}$$

where $2(M + m)C^2$ is the rest mass in terms of MeV, and the other symbols have the same definition used in previous equations. Since the threshold for γ,n reactions is greater than the neutron binding energies, the recoil energy is very large.

5.8. FISSION

Perhaps the most drastic chemical effect as the result of nuclear decay comes from fission, for not only is the nucleus torn apart, but the atom is likewise. Still the two resulting elements separate with the orbital electrons of the inner shell intact, since even the high kinetic energies encountered in fission are insufficient to prevent an adiabatic readjustment to these shells. Prompt K X-rays are found in abundance, however, in fission, but these are due to subsequent internal conversion (John et al. 1967).

The total recoil energy encountered in fission $E_r(T)$ can be estimated (Jungerman and Wright 1949) from a simple Coulomb model

$$E_r(T) = Z_l Z_h e^2/(r_l + r_h), \tag{2.17}$$

where Z_l, Z_h and r_l, r_h are the nuclear charges and radii of the light and heavy fragment ions. For example, for ^{235}U using $Z_l = 38$ and $Z_h = 54$ and estimating r from Elton (1968), one finds a value of $E_r(T) = 279$ MeV, somewhat higher than experiment. Extensive studies have been made on the recoil energy spectra for the various fragment ions found in fission (Schmitt et al. 1966). An example of one such contour plot is shown in fig. 2.10. In the search for superheavy elements the presence of abnormally high recoil fission fragments could be used as proof for their existence.

Addendum (November 1976)

The central problems concerned with the primary processes in hot atom chemistry have changed very little with the passage of time. However, special areas of interest, connected in a peripheral manner, have received considerable attention in the literature since the chapter was written in 1971.

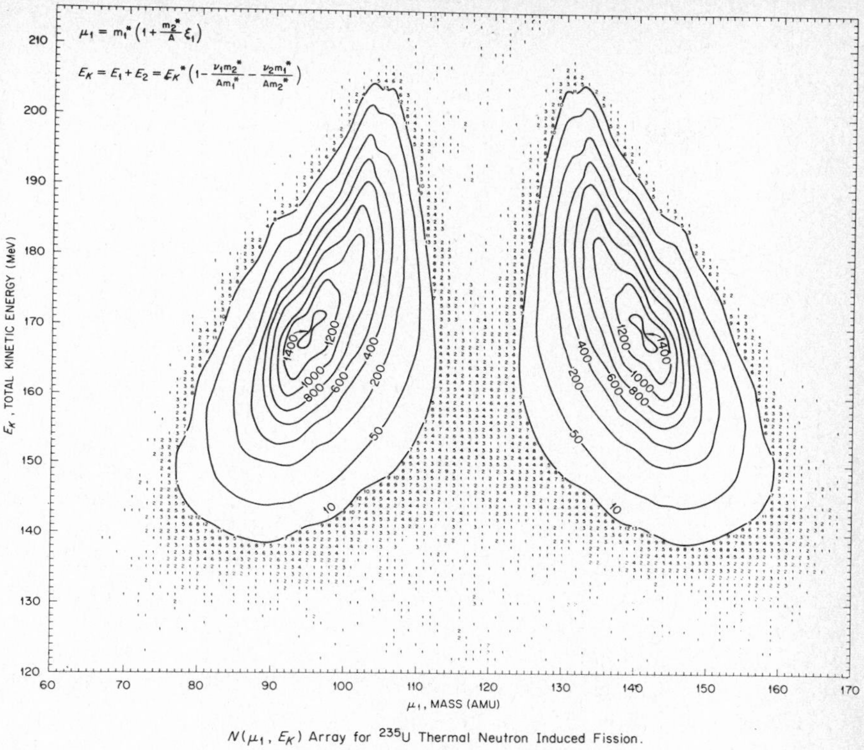

Fig. 2.10. Provisional mass, μ_f, vs. total kinetic energy for ^{235}U thermal-neutron induced fission. The contour lines give the relative intensities. μ is approximately equivalent to the atomic mass, A. For example, mass 100 will be most probably associated with E_T of 175 MeV. Its kinetic energy will be approximately $E_A = (236\text{-}A)/236\ E_K = 101$ MeV (cf. Schmitt et al. 1966).

(1) More sophisticated treatments of electron shake-off, particularly for the production of K shell vacancies, have been carried out by a number of authors, supported by additional experimental studies. See, for example, reviews by Freedman (1974, 1976).

(2) Special problems concerned with electron shake-up revealed by photoionization studies are reviewed in Carlson (1975).

(3) Extensive studies have been carried out on the behavior of high energy ions passing through matter. For a recent overview of this subject, see Saris and Van den Weg (1976).

CHAPTER 3

EFFECTS OF NUCLEOGENESIS PRECEDING CHEMICAL REACTION: DISSIPATION OF EXCITATION BEFORE CHEMICAL REACTION

Garman HARBOTTLE

Chemistry Department, Brookhaven National Laboratory, Upton, New York 11973, USA

Chemical Effects of Nuclear Transformations in Inorganic Systems
Edited by G. Harbottle and A.G. Maddock

Contents

1. Energy-loss Processes

1.1. RANGE OF RECOIL ATOMS

The present chapter should be thought of as a continuation of the previous one, in that, having by one or more nuclear and/or nuclear electron-shell coupled processes energized the 'hot atom', one may ask by what mechanisms it loses its excess energy, what happens to the atom-plus-environment during this energy loss, and immediately afterward, and what implications these events may have for determining the ultimate chemistry of the affected atom. Whereas a previous study of this sort (Harbottle and Sutin 1958, 1959) was forced to rely heavily on theoretical arguments, which seem in some cases to have been misinterpreted, the present chapter will contain a much higher proportion of empirical evidence drawn from the large volume of experimentation in the intervening time.

The number of inorganic, gas-phase hot-atom studies is relatively small (see ch. 4) and the theoretical analysis of these has followed in general the lines laid down by Wolfgang (1965a, b) who described the energy-loss processes of carbon and tritium recoil atoms in the gas-phase. In this chapter we shall first consider crystals and then give an indication of the modifications necessary to describe events in liquids.

Hot-atom chemists have not generally appreciated, until recently, how short the ranges of (n,γ)-generated recoil atoms really are. Libby (1947) attempted to calculate approximate ranges on the basis of a simple hard-sphere collision model – i.e. a model in which the collision cross section and hence the mean free path does not vary with energy. With such a model the energy is dissipated too rapidly and the ranges are underestimated. Thereafter, the extensive development of theory for the behavior of energetic atoms in solids fell to the Danish school of solid-state physics. To calculate ranges, we employ the comprehensive method of Lindhard, Scharff and Schiøtt (1963) (hereafter LSS). Their work is erected upon the firm foundation laid by Bohr (1948): LSS however employ the screened Coulomb potential $U(r) = (Z_1 Z_2 e^2/r)\varphi_0(r/a)$ where Z_1 and Z_2 are the atomic numbers of the moving and stationary (medium) atoms respectively, r the distance between atoms and $a = a_0 \times 0.8853\,(Z_1^{2/3} + Z_2^{2/3})^{-1/2}$. The function φ_0 is the Fermi function, and

a_0 is the Bohr radius (0.529×10^{-8} cm). The use of the screened Coulomb potential, having only one screening parameter a, led LSS to the 'natural measure of range and energy', the dimensionless quantities ρ and ε. These are

$$\rho = RN4\pi a^2 \frac{M_1 M_2}{(M_1 + M_2)^2}, \qquad \varepsilon = E \frac{a}{Z_1 Z_2 e^2} \frac{M_2}{M_1 + M_2}, \qquad (3.1, 3.2)$$

where M_1 and M_2 are the masses of the atoms of charge Z_1 and Z_2 and N is the number of atoms per unit volume.

These quantities ε and ρ are then the universal range and energy for fast atoms (subscript 1) interacting with a set of stationary atoms (subscript 2). The actual ranges and energies are R and E, connected to the universal quantities by the scaling factors constituting the remainders of eqs. (3.1) and (3.2). It should be noted that the equations for ε and ρ refer only to *nuclear* stopping, which is by far the most important contributor to the stopping of low-energy recoil atoms. At higher energies *electronic* stopping (the inelastic interaction of the interpenetrating electron shells of the moving and stationary atoms) must also be taken into account: at very high energies its contribution dominates the rate of energy loss dε/dρ. LSS show that the Thomas–Fermi picture of the atom leads to a simple velocity-dependent electronic stopping

$$(\mathrm{d}\varepsilon/\mathrm{d}\rho)_\mathrm{e} = k\varepsilon^{1/2} \qquad (3.3)$$

expressed in universal units, where

$$k = \xi_\mathrm{e} \frac{0.0793\, Z_1^{1/2} Z_2^{1/2} (A_1 + A_2)^{3/2}}{(Z_1^{2/3} + Z_2^{2/3})^{3/4} A_1^{3/2} A_2^{1/2}} \qquad (3.4)$$

and $\xi_\mathrm{e} \approx Z_1^{1/6}$. Except for very light fast atoms impinging on heavy stationary atoms k usually falls in the range 0.1 to 0.2.

The energy dependences of the two modes of energy loss, electronic and nuclear stopping, are shown in fig. 3.1, taken from LSS, who treat the two contributions as independent, add them and integrate to obtain the universal range–energy curves shown in fig. 3.2. The curve 'Th.F.' is the pure nuclear stopping case, $k = 0$, while the second curve is calculated for a k having the relatively high value of 0.4. There exists, of course, a family of ρ–ε curves for various k values, but for most (n,γ) hot-atom problems the curve 'Th.F.' is a satisfactory approximation. We shall see, below, how these curves may be used to calculate ranges in actual cases of hot-atom interest.

LSS show that the ρ–ε curve (fig. 3.2) is in good accord with experimentally measured ranges: however, several cautionary points must be noted. First, the 'range' calculated is the *path length* of the fast atom, including its various

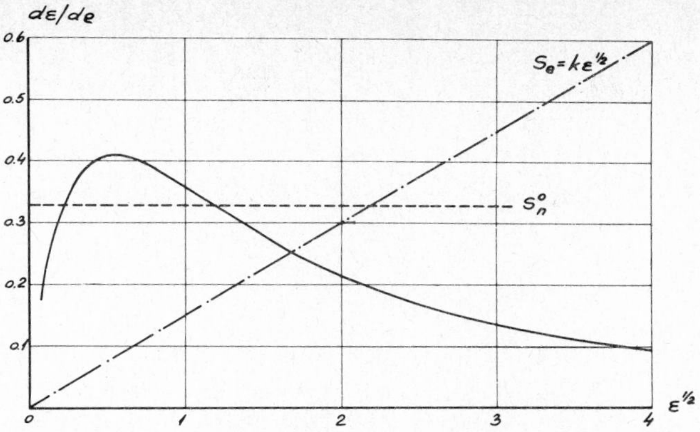

Fig. 3.1. Plot of energy loss $d\varepsilon/d\rho$, vs. $\varepsilon^{1/2}$, showing contributions from nuclear (full curve) and electronic (dot-and-dash) stopping. Taken from Lindhard et al. (1963).

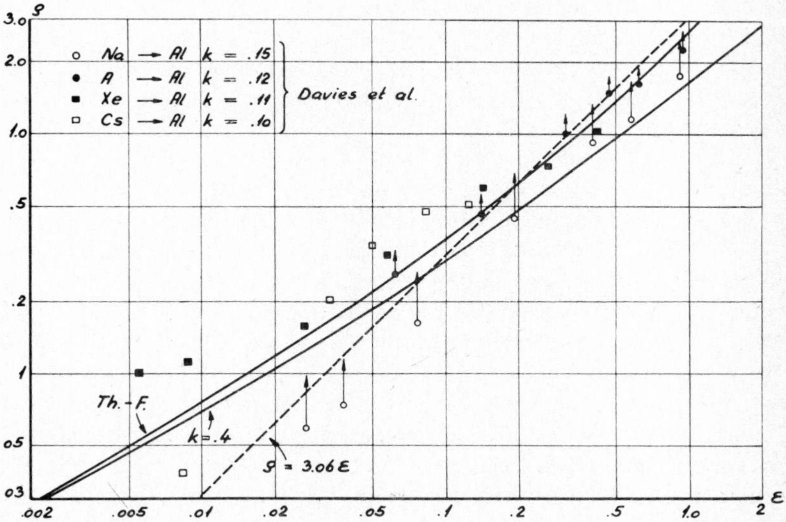

Fig. 3.2. Universal range–energy curve, ρ vs. ε, with some experimental points. Taken from Lindhard et al. (1963).

scattering twists and turns, whereas the 'projected range', i.e. the projection of the actual path length onto the line of initial direction of motion of the fast atom, may be somewhat smaller. A table of range corrections, based on the work of Schiøtt, will be found in Mayer et al. (1970); the projected range of course gives a better measure of straight-line displacement of the hot-atom from its starting to final resting point. Secondly, it is not clear how far down in energy the curve 'Th.F.' in fig. 3.2 may be used. LSS state that "It should be emphasized that for extremely low energies, $\varepsilon \lesssim 10^{-2}$, the theoretical curve is not too well-defined." Obviously in the limit, where the hot atom may either move or not move from its lattice point to a nearby interstitial position, depending on small increments of energy around the critical 'displacement energy' (ca. 25–35 eV), and on the direction of motion, the continuity of ρ and ε must break down.

1.2. DIRECTIONAL EFFECTS

Even at higher energies directional effects are very important in crystal lattices. An atom moving in certain 'open' directions through stationary lattice atoms may find itself in a tunnel or 'channel' where it loses energy very much more slowly than in passing through a random, uncorrelated assemblage of the same atoms at the same density. Such 'channelled' atoms may therefore have ranges greatly exceeding the estimates given by the ρ–ε calculations above. However, the conditions for channelling to take place are rather stringent: the initial direction of motion must be such that the atom to be channelled impinges on the open space in the lattice travelling with an alignment to the open direction good to better than ca. 5°. Since hot atoms generated by nuclear transformation in a crystal usually start out *on lattice sites* it will be difficult for them to achieve a channelling direction without scattering at least once and one would expect the fraction channelling to be rather small (Chadderton and Torrens 1966, Torrens and Chadderton 1967). These arguments do not, of course, apply to the penetration of externally accelerated atoms into crystals, where channelling effects are readily observed. Research on the channelling of atoms implanted in crystals is actively proceeding, and the reader may pursue the subject in its varied aspects in several recent reviews (Nelson 1968, Palmer et al. 1970).

Even with all these cautions, it is found that the measured 'ranges' of low-energy ions in amorphous crystal are in reasonable agreement with LSS theory and surprisingly this holds true even in the very low-energy region (30 eV–1 keV) where the LSS parameter ε has values less than 0.001. 'Ranges' seen in computer simulation studies are also in rough accord with LSS, bearing in mind the displacement threshold and straggling effects (Kornelson 1964, Domeij et al. 1964, Beeler and Besco 1963).

1.3. COMPUTER SIMULATIONS

Of the greatest interest to hot-atom chemists is the work of Torrens and Chadderton (1967) who applied the Brookhaven computer calculation technique (Gibson et al. 1960, Erginsoy et al. 1964, Erginsoy 1964) to simulate the dynamic situation in solids. However, instead of metals like copper and iron, they studied the ionic crystals PbI_2, KCl and NaCl. They found, as in the metal systems, that the focussing collision chain is an effective way of carrying off large amounts of kinetic energy originally concentrated on one recoil atom, and depositing it at a distance from the original site. This process, however, worked better in KCl, where the atoms are of nearly equal mass, than in NaCl or PbI_2 where unequal masses are involved, and their results leave one with the impression that in complex crystals like K_2CrO_4, K_2ReCl_6, etc. the removal of energy via 'focusons' would be relatively unlikely.

With regard to the appearance of ions having channelled trajectories ('channelons'), Torrens and Chadderton report that these were seen in the case of 80 eV Na^+ ions, for example, moving initially in a [112] direction, deflected by a collision with a neighboring Cl^- ion "into a field of low-interaction potential between adjacent [001] rows of ions, becoming effectively axially channelled."

This work provides additional confirmation of the extreme dependence of the threshold for displacement E_d upon the initial direction of motion. For example, in KCl, E_d for a K^+ in the $\langle 110 \rangle$ direction (i.e., across the diagonal of a cube face in the direction of the nearest K^+) was 25–30 eV, but in the $\langle 100 \rangle$ direction (i.e., along a $K^+Cl^-K^+Cl^-$ row) was greater than 150 eV, 'probably nearer to a value in the neighborhood of 200 eV.' This astonishingly high value is of great significance to hot-atom chemistry since recoil atoms in solids also invariably start off from lattice points. The work of Torrens and Chadderton on thresholds also makes clear the significance of 'ranges' of low-energy ions in solids, namely that they should be treated as averages over various initial directions of motion, with a very high degree of range straggling. We will return to this paper in our later discussion of thermal spikes and molten zones.

At this stage however, two points should be noted; first, that the high peak values of E_d are due to the efficient removal of energy in these cases via focusons, and that this also contributes to high apparent or effective values of the thermal diffusivity, and second that while channelons readily occur with aligned beams in open crystal 'tunnels', they are less favored in hot-atom experiments, in which recoil atoms start out on lattice points.

Similarly, the computer study of Bunker and Van Volkenburgh (1970), the first directed at a system specifically of interest to hot-atom chemists, further demonstrated the essentially short range of low-energy atoms in solids. Their work followed the trajectories of ^{32}P atoms initially in solid sodium phosphates, given initial impulses ranging from 74 to ca. 1000 eV. They found that up until the first collision the fast ^{32}P acted very much like a free atom, and that P–O bonds were "almost never carried away from the site." As would have been anticipated from LSS theory (difficult to apply in this case), or indeed from a simple billiard ball calculation of energy lost in successive collisions in a condensed medium, "most of the penetrating power of the phosphorus was exhausted by escape from the original cage."

In computer studies, and in the measurement of ranges of accelerated ions implanted in solids, the initial energy is specified and has a unique value. But in cases of (n,γ) recoil of interest to hot-atom chemists a very different situation prevails. The spectra of gamma rays emitted following neutron capture are often extremely complex (Bartholomew and Groshev 1967). In general a number of parallel cascades deexcite the initial (6–9 MeV) state to ground, and recoil momenta must be summed vectorically for each path, leading to partial cancellation, and averaged over the different paths weighted by their probabilities of occurrence. The net effect of this is, typically, a very broad spectrum of atomic recoil energy with a reasonable proportion of low-energy events (Hsiung et al. 1961, Cifka 1963). Calculations based on recoil from the gamma rays of maximum energy alone are bound to give serious over-estimates.

1.4. CONCLUSIONS

The inescapable conclusion to be drawn from the presently available data on neutron capture-gamma spectra and on the trajectories and ranges of atoms in the energy range 0 to a few hundred eV encountered in typical (n,γ) processes is that the atom affected does not and cannot travel far before coming to rest in the condensed phase, liquid or solid, in which it originated. At most its 'range' (disregarding, of course, subsequent diffusion processes) can amount to a few lattice layers and can be estimated, at least in order-of-magnitude, by an extrapolation of LSS theory. There appears to be every reason to expect that in a fair number of (n,γ) events the atom affected either does not displace from its initial position, or displaces and immediately returns. In liquids, many (n,γ) recoil atoms will be found immediately adjacent to their original partners because of the short ranges of the lowest-energy recoils: experimental evidence of this is seen in the phenomenon of 'geminate' recombination (Iyer and Martin 1961, Gennaro and Collins 1970).

The phenomenon of 'channelling', which so greatly extends the range of the fraction of beam-implanted atoms which enter a crystal in alignment with a favorable 'open' lattice axis can, of course, like 'focussing collisions', not occur at all in liquids. In crystals, where the (n,γ)-activated atoms start out on lattice points, it will also in general not be very probable because of the need of initial alignment.

Recoil following β^-, β^+, electron capture and isomeric transition decay processes (with or without internal conversion) is also a low-energy phenomenon, if we consider the primary events. However, as discussed in ch. 2, there are attendant secondary sources of excitation (charge-change leading to electron shakeoff, Auger electron emission, etc.) which can influence the chemistry of the hot atom. With the possible exception of an inner-shell vacancy, it is difficult to see how the energy released in any of these secondary processes can be translated into kinetic energy. The possibility of producing enough kinetic energy through Auger charging to enable an atom to exceed the threshold for displacement was long ago discussed by Varley (1962) and others (Durup and Platzman 1961, Smoluchowski and Wiegand 1961, Smoluchowski 1964, Vosko and Smoluchowski 1961, Howard and Smoluchowski 1959): this mechanism will be mentioned again later.

Hot-atom chemists also deal with much more energetic atoms produced by (n,2n), (γ,n) and α-decay nuclear reactions: their initial energies are easily calculated and their ranges in solids may be estimated from LSS theory, keeping in mind the phenomenon of channelling which, in single crystals, greatly extends the maximum ranges observed.

2. Slowing-down Experiments

We now examine the experimental work relevant to the slowing-down of a recoil atom in slightly more detail, first turning to the arguments of solid-state physics to improve our understanding. The velocity of an atom of energy E and mass M in cm/sec is given by

$$V(\text{cm/sec}) = 1.39 \times 10^6\ [E\ (\text{in eV})/M\ (\text{in amu})]^{1/2}. \tag{3.5}$$

Thus most interesting (n,γ)-generated hot atoms have velocities in the range of a few $\times\ 10^{14}$ Å/sec. Since at low energies collision cross sections of atoms in liquids or solids are so large that the mean free path is not much more than the separation of neighboring atoms, it follows that the time between collisions is in the 10^{-14} second range and the total time for slowing down to thermal energies is of the order of 10^{-13} sec. Even if we consider rather energetic atoms there is simply no way that this time could be increased even by one order of magnitude. If we wish to check this estimate of slowing-down time experimentally, studies of Mössbauer effect in the recoil atom do not really help much because the shortest Mössbauer time scales are ca. 10^{-9} sec, three or four orders of magnitude too slow. What the Mössbauer studies do give us is information on the appearance of the environment of the recoil atom immediately after the slowing-down atom has reached thermal energies: that will be dealt with in more detail in ch. 24. In the present chapter we will return to a discussion of the Mössbauer results for the light they can cast on the nature of the thermal spike or hot zone.

What is more germane is study of perturbed angular correlation (PAC), nuclear resonance fluorescence (NRF) and Doppler-shifted gamma spectroscopy (DSGS) in recoiling nuclei. We may dispose of the first type of measurement, PAC, or really time-differential PAC, by noting that the time scale in one case studied was in the 10^{-12} second range and, as discussed in a recent review by Jones (1970a), quoting a discussion by Grodzins (1969), what was seen was evidence of an anomalous nuclear precession in recoiling ^{70}Ge nuclei which flew into an iron backing, produced by a time-dependent perturbation lasting less than 2×10^{-12} sec and apparently having its origin in the slowing-down of the recoil atom. Here again, we will revert to PAC in our discussion of the time scale of the post-recoil, thermal spike phase.

Experiments in nuclear resonance fluorescence yield far more direct information on the 10^{-14} to 10^{-13} sec time region (Metzger 1959, Adloff 1971). The experiments of Ilakovac (1954) and Ofer and Schwarzschild (1959a, b) showed that slowing-down times of atoms in condensed phases, initially activated by β^-, β^+ or EC decay and having energies of the order of 50–100 eV, did not exceed a few $\times$ 10^{-14} seconds and that the distances traversed were only a few Angstroms. In particular the work of Cumming et al. (1960) is called to the reader's attention, because of the excellent analysis and discussion. They measured the fluorescence of two γ-rays of 669 and 963 keV fed by the EC and β^+ decay of ^{63}Zn (38 min) to ^{63}Cu. Both solid copper and dilute aqueous solution sources were studied. Since the mean lifetimes of the two gamma-emitting levels in ^{63}Zn are 3 and 7×10^{-13} sec respectively, it is clear that an appreciable number of recoiling atoms will emit while 'in flight' or, in condensed phases, while they are slowing down. The observation of the resonance in fact depends on this in-flight emission, which Doppler shifts the emitted gamma to an energy high enough to allow it to fluoresce a ground state nucleus in the ^{63}Cu scattering target. Their data allowed a direct comparison with predictions of the yield of fluorescent scattering based on 'empirical' computer-generated slowing-down results for copper atoms in copper (Gibson et al. 1960). Good agreement was obtained.

Of great interest to the hot-atom chemist was their measurement of the fluorescence produced by a source consisting of ^{63}Zn in aqueous solution. From this and with the assumption of a simple collision model they would conclude that the recoil copper atom having 30–60 eV initial energy required ca. 7 Å to slow to 1/e of this value. Good agreement was also obtained for ^{52}Cr coming from ^{52}Mn(β^+)^{52}Cr with an assumed 1/e path length of 6 Å (Ofer and Schwarzschild 1959a, b).

A different type of experiment in NRF, but of potential importance in hot-atom work, is that of Langhoff (1971), Abel and Kalus (1969), and Langhoff et al. (1969). Here the recoil atoms (^{187}Re in tungsten metal, carbide, and oxide, ^{152}Sm in Eu metal and oxide, and ^{131}I in Te) have so little energy from the nuclear decay ($E < 15$ eV) that they do not leave the lattice site at all, but lose their energy merely by exciting lattice vibrations. Fluorescence is achieved by spinning the sources in a centrifuge rapidly enough to Doppler-shift the emission line into resonance – it needs to shift a few electron volts to achieve this. The line profiles could be well explained by the Debye model of the lattice using Debye temperatures found by other techniques. According to the authors "this might be surprising for the case of ^{152}Sm, where the electron capture... (feeding the resonant 963 keV line)... with the subsequent emission of Auger electrons should influence the binding of the radioactive atom in the lattice." It is indeed surprising, since the lifetime of the 963 keV level is only 4×10^{-14} sec. and although the back-flow of electrons to neutralize Auger-charged atoms could occur that rapidly in metals, it seems unlikely that it could in insulating solids like europium oxide.

The technique of Langhoff et al. seems interesting in another way. Their sources, $^{187}W \rightarrow {}^{187}Re$, $^{152m}Eu \rightarrow {}^{152}Sm$ and $^{131}I \rightarrow {}^{131}Xe$ can all be prepared by neutron irradiation of the corresponding compounds. Since the 'centrifuge' technique can probe the interaction of the low-energy recoil atom with the lattice in cases where the recoil energy is below the threshold for displacement, it might be employed to yield information on the lattice environment following the (n,γ) event – for example – by a comparison of the resonance profiles in annealed and non-annealed neutron-irradiated source crystals. No work seems to have been done on this line, to date.

Adloff (1971) has recently written a review of these NRF experiments from the point of view of hot-atom chemistry: the field would appear to contain much untapped potential for the study of the slowing-down processes.

We close the discussion of slowing-down experiments by mentioning a relatively new and quite interesting technique: that of Doppler-shifted gamma-ray measurement. There are actually two procedures, the recoil distance method and the Doppler-shift attenuation method (DSAM), and both are discussed by Schwarzschild and Warburton (1968). In the first technique the atom containing a nucleus in an excited state recoils across a variable path length, and the ratio of Doppler-shifted gamma rays (i.e. those emitted by decay in flight) to the number of unshifted gammas is observed as a function of the flight distance. In the second technique, which yields information on solid-state energy-loss processes, the fast atom is generated in a solid and thus immediately begins to slow down, with no free flight path. The spectrum of Doppler-shifted radiation thus reflects both the lifetime of the emitting state and the rates of energy loss of the emitting atom by the electronic and nuclear stopping already mentioned above in connection with the LSS procedure. Although the technique has been principally employed to measure short lifetimes, it seems probable that it can also yield fundamental information on slowing-down reactions. The connections are traced out in Appendix B of the paper of Warburton et al. (1967).

3. The Thermal Spike

3.1. HISTORY AND GENERAL CONSIDERATIONS

We now turn to a consideration of that thorny problem, the 'thermal spike', and what it may lead to, if anything, in the hot-atom chemistry of the recoil atom which initiates it. Our procedure will be first, to review briefly the history of the spike, then to discuss the present solid-state experimental evidence and theory, and finally the hot-atom results that bear on the question.

It will surprise some to learn that the concept of a thermal spike makes its first clear appearance in the 1947 paper of Libby who saw that the rapid loss of translational energy by a fast atom in a solid would necessarily lead to a heated volume: "...we shall think of the situation immediately after trapping of the radioactive atom as though it were contained in a small liquid droplet of about a dozen ions or atoms at most, resting in a cavity in the crystal about the size of one coordination sphere. This hot zone then transmits heat vibrationally to its crystalline wall until it freezes and the immediate fate of the radioactive atom is determined." Libby's work dealt with atom recoil in organic crystals: the interests of solid-state physicists at the end of the war lay more with metals and graphite, because of the obvious practical necessity of understanding structural changes produced by the mixed radiation fields inside nuclear reactors. In the work of Siegel (1949) on the disordering of ordered Cu_3Au alloys by reactor irradiation the concept of a 'temperature spike' is introduced: it is a small volume of a crystal, along the track of a fast atom, the atoms of which are momentarily heated to a high temperature. The historical discussion of Seitz (1952) mentions the thermal spike as a widely accepted concept and gives the approximate figures 10^4 °C and 10^{-11} sec for the temperature and duration.

The 1954 paper of Brinkman introduces a new idea, the 'displacement spike'. Brinkman suggested that in the region of heavy damage, including secondary and tertiary knock-on atoms, the volume affected is actually melted for a short time and resolidified, epitactically following the crystalline orientation of the matrix. This idea is of course very close to the fast-atom engendered 'molten droplet' of Libby and his collaborators in molecular crystals (Libby 1947, Fox and Libby 1952, Rowland and Libby 1953). Koehler and Seitz (1954, Seitz and Koehler 1956) examined the 'thermal spike' in more detail, calculating the temperature and duration of the heated zone. The long review of Kinchin and Pease (1955b) questioned the thermal spike concept in the sense that over very short distances and times it would be unsafe to apply macroscopic concepts of 'heat', but concluded that a genuine thermal spike is probably generated in the stopping of a fission fragment in uranium. Shortly thereafter Yankwich (1956) and the present author and Sutin (1958) proposed that the thermal spike (resulting from the slowing down of the recoil atom) could have chemical consequences: following the model of Koehler and Seitz attempts were made to estimate the probability of rate processes occurring in the very short times at high temperature in the 'hot zone'.

Before discussing the hot-atom chemistry of the thermal spike it would be well worthwhile to complete our survey of the evidence from solid-state physics. As mentioned above the earliest instance adduced to support the thermal spike was the observation of radiation-induced disordering of initially ordered Cu_3Au alloy (Siegel 1949). In the 1961 Faraday Society paper Koehler and Seitz (1961) review much additional work on this alloy and mention that the disordering "cannot be due to focussing since the differences in the masses of copper and gold make focussing ineffective." This is a very significant point for the hot-atom chemist, who almost invariably deals with compound crystals containing two or more atoms of unequal masses. Introducing the same series of papers Vineyard (1961) illustrated the formation of thermal spikes in copper resulting from knocked-on copper atoms of initial energy of the order of 100 eV. 'Isothermal' contours were drawn for various times, based upon the computer-generated description of the motion of copper atoms in the crystallite. 'Hot spots' were seen, merging in times of the order of 10^{-12} sec: at the beginning (3×10^{-13} sec) temperatures (defined by kT = average kinetic energy of the atoms) in excess of 4500° C are evident but these vary erratically both in distance and time. Vineyard comments that "by 1.65×10^{-12} sec most of the atoms are below 100 K (above the ambient temperature) and only isolated hot spots. . .remain. There is a certain overall resemblance to cooling according to classical laws of heat conduction for a continuum but there are also striking differences. . .very roughly, a thermal diffusivity of about 0.01 cm^2/sec seems operative. This is about an order of magnitude higher than values that have commonly been used in thermal-spike calculations. . .". The result of this is that cooling is about an order magnitude faster than usually calculated, in this monatomic metal.

It is interesting that these computer-generated kinetic pictures of a copper lattice which has been agitated by an initial fast atom do not contain anything that looks like a 'molten droplet' of copper metal. Indeed, the structure of the copper crystal is preserved to a surprising degree, even in cases where a large number of the contiguous copper atoms have kinetic temperatures far in excess of the melting point. Perhaps this is due to the rapid energy loss by focussing collisions leading to an unexpectedly large effective thermal diffusivity. The same is true of ionic crystals: the work of Torrens and Chadderton (1967, Chadderton and Torrens 1966) on computer-generated trajectories leads to the same conclusion. They state that "the severity of any localized damage is therefore reduced by efficient mechanisms of energy removal away from the thermal/displacement volume." However it should be noted, as mentioned earlier, that in complex crystals, or those having atoms of widely differing masses, correlated collisions will be less effective in removing energy.

Vineyard (1961) concluded, on the basis of the computer studies of copper, that "the very ragged distributions of effective temperature, spreading extremely rapidly and according to complex laws, that are revealed in these figures do not inspire confidence in the significance of conventional discussions of thermal spikes", while Torrens and Chadderton (1967) said that "Agitations following damage events bear some resemblance to thermal spikes as they are conventionally understood. . ." It appears from all this that the thermal spike model becomes more realistic, the more complex the crystal.

A third, and very important line of evidence comes from sputtering studies. These have been recently reviewed by Nelson (1968): each atom in the thermal spike has an average energy

$$E = E_T + \tfrac{3}{2}kT_0, \tag{3.6}$$

where

$$E_T = E_p/(\tfrac{4}{3}\pi r_2^3 N_0) = \tfrac{3}{2}kT_s$$

is the excess energy over that due to the ambient temperature T_0 and where E_p is the energy of the primary particle, input to the spike. N_0 is the atomic density and $\tfrac{4}{3}\pi r_2^3$ the spike (spherical) volume. Then T_s is the spike 'temperature'. Using Maxwell–Boltzmann statistics to describe the energy spectrum within the spike Thompson and Nelson (1962) derive an expression for the energy spectrum of evaporated atoms and show that this fits the results of sputtering studies in quite satisfactory detail. The effective spike temperature comes out of this fit, and from the above equation the radius: for gold bombarded by 45 keV Xe^+ ions, the spike temperature T_s was 900 K above ambient, had a radius of 100 Å and lasted 3×10^{-12} sec.

Although most of the research in solid-state physics has dealt with the effects produced by thermal spikes in metals, there is a growing interest in the corresponding study of insulators and semiconductors (Brack and Schwuttke 1971), arising of course from the strongly felt need to understand the basic physics of the ion implantation technique. An interesting paper on the effects of heavy-ion bombardment of zirconium oxide is that of Naguib and Kelly (1970). A crystalline solid, they point out, may under bombardment either remain crystalline or become amorphized. In their work, the opposite effect is studied: an amorphous ZrO_2 film bombarded with 20 keV Kr^+ ions is observed first to form small crystalline regions, then develop complete crystallinity. They analyzed their results in terms of a thermal-spike model similar to that of Seitz and Koehler and were able to predict successfully which oxides would become amorphous and which would remain crystalline under heavy-ion bombardment. This kind of result is probably explicable only by the thermal-spike promoted ordering of the ZrO_2 lattice: a displacement spike would be a disordering influence.

3.2. NUCLEAR PHYSICS EXPERIMENTS

We now turn to results in the field of nuclear physics to see what bearing they may have on the thermal spike question. The study of nuclear resonance fluorescence mentioned above as an indicator of slowing down times will in general be ineffective here, precisely because the gamma energies must be shifted by at least a few electron volts to achieve fluorescence, and in the thermal spike we are dealing with energy spectra running over a few tenths of an electron volt. It seems then that NRF samples too short a time range after the recoil even to give much information on thermal spikes.

The studies of chemical effects of nuclear transformation via Mössbauer effect measurements have been recently reviewed by Jones (1970a) and by Maddock (1972) and will be examined in a later chapter: here again one might hope to find some trace of the thermal spike. This is particularly true inasmuch as the Mössbauer level can be fed, and Mössbauer atoms generated, in several different ways including some in which the initial energy of the atom involved is very high (Mullen 1965).

One might mention first the experiment which has constituted the most direct approach to the short time-span after nuclear decay and recoil, that of Triftshauser and Craig (1967). They doped cobalt oxide with ^{57}Co, triggered their Mössbauer apparatus with the arrival of the 122 keV gamma ray which precedes the 10^{-7} sec 14.4 keV Mössbauer line, and recorded the spectrum as a function of the 'age' of this level, the time periods sampled being from a (4–43) up to a $(146–200) \times 10^{-9}$ sec bracket. Experiments were carried out at room temperature.

They also tried nickel oxide, ferrous ammonium sulfate, ferrous sulfate and ferrous chloride, similarly doped with ^{57}Co. Although the spectra revealed the presence of both Fe(III) and (II) as decay products of the ^{57}Co, in no case did the relative proportion Fe(III)/Fe(II) change with time. Since the recoil energy of the ^{57}Co (electron capture) was small, it would have been a little surprising to have seen a shift of Fe(III)/Fe(II) occurring in the $4–200 \times 10^{-9}$ sec region, which could be attributed to the thermal spike. One could not exclude, however, that the observed Fe(III)/Fe(II) ratios had already been influenced by a thermal spike whose effective period had long since passed: we will take this point up again later in the discussion of chemical effects attributable to the hot zone. A final possible difficulty in the interpretation of the Triftshauser–Craig results is that, in the compounds other than CoO, one cannot be sure of the initial chemical form and location of the dopant cobalt ions in the host lattices. In low-recoil events the nature of the dopant species and the place it lodges in the host crystal structure is certain to have a profound effect on the decay-induced chemistry, however one can probably assume that the cobaltous ion was substituted for ferrous in the lattices investigated. Subsequent experiments by Hoy and Winterstein (1972), using ^{57}Co labelled $CoSO_4 \cdot 7H_2O$ have, however, revealed positive evidence for time dependence of the ^{57m}Fe yield.

Although instances are seen in which the Mössbauer recoil-free fraction f is lower following a very energetic recoil than in the corresponding static or low-energy case, these differences also can be shown not to be due to the presence of a thermal spike, but rather to the alteration of the immediate lattice environment. One such case which has received considerable attention (Mullen 1965, Stone and Pillinger 1964, 1966, Kaplan 1966) is that of the Mössbauer effect in ^{237}Np fed by ^{241}Am (α) giving a high-energy recoil atom, or by ^{237}U (β^-) giving one of low-energy. The recoil-free fraction in the 59.6 keV gamma decay ($t_{1/2} = 63 \times 10^{-9}$ sec) was found to be about four times larger for the ^{237}Np fed by ^{237}U (β^-) than for the same isotope fed by the α decay of ^{241}Am. But even in this case it can be demonstrated by a simple calculation that the lifetime of the Mössbauer level is much too long to allow any reflection of the thermal spike in the recoil-free fraction. That such a large difference in f requires a thermal spike is conclusively disproved by the interesting results of Czjzek et al. (1968). They coulomb-excited ^{73}Ge nuclei by bombarding their target with energetic oxygen ions, the germanium nuclei being simultaneously driven by recoil into catcher targets (implantation) where the Mössbauer decay occurred. Two targets of GeO_2 differed only in that one had a hexagonal, the other a tetragonal crystal structure and yet the recoil-free fractions f were 6% and 32% respectively. One would assume that the thermal spikes would be fairly similar in the two cases, and that the large difference in f reflected a fundamentally different average environment (differing numbers of vacancies or perhaps simply differing probabilities of substitution) leading to an effect on the coupling of the Mössbauer emitter to the lattice. The work of Czjzek and Berger (1970) explores this idea in greater detail, in metals, and arrives at the same conclusion.

One gets the impression that the Mössbauer effect will give valid information (subject to questions of interpretation) about the lattice environment and the chemical state of the recoil atom at a time very short compared to chemical procedures, in fact, at the shortest possible times following the cooling of the hot zone in well-bound solids, but will yield relatively little information about the spike-cooling period itself. These remarks are made subject to one reservation, that if one measured the Mössbauer spectrum of a recoil atom projected with high energy (e.g. by alpha decay or heavy-particle excitation) and having a short lifetime, of the order of a nanosecond or less, into a 'soft' crystal – one having a low melting point and heat of fusion, for example a suitable organic crystal, one might just possibly see an effect due to the 'molten zone.' It would be interesting to study such a system, and if there was no effect, to set a better upper limit on the duration of the hot zone than now exists.

The time-differential angular correlation (TDAC) technique has also been applied to nuclei having metastable gamma-emitting levels, excited by heavy-particle reactions and implanted by the recoil from those same reactions. The 'chemical' information is not as easy to extract as that contained in the Mössbauer spectra but the results, recently reviewed by Jones (1970a) and by Vargas (1972) indicate that, as with the Mössbauer effect, the TDAC technique will be of greatest value in giving information on chemical forms in situ following radioactive decay or neutron capture, and not on the cooling-down period per se.

3.3. SEITZ–KOEHLER THEORY

In order to set the stage for the discussion of the thermal spike or hot zone in hot-atom chemistry we first consider the basic Seitz–Koehler (1956) treatment of the spherical spike. The classic equation for the spread of heat outward from a point source is

$$T(r,t) = \frac{Q}{\pi^{3/2} cd(4Dt)^{3/2}} \exp\left(-\frac{r^2}{4Dt}\right), \tag{3.7}$$

where $T(r,t)$ is the increment in temperature at time t, radius r from the sphere center due to a heat input Q to the point source, the heat being released instantaneously at time $t = 0$, c is the specific heat and d the density such that cd is the specific heat per unit volume. It seems logical that c is actually c_v: the difference $c_p - c_v$ is however small in solids (Lewis 1907) so either could be used. In any case, the Dulong–Petit (high temperature) value of c is appropriate since that is our region of interest. The quantity (cd) then has the value $3K$ per atom (K = Boltzmann's constant). D is the thermal diffusivity: its units are cm^2/sec and it may be related to tabulated values of the heat conductivity k by the relation

$$D = k/cd. \tag{3.8}$$

Values of k are ordinarily tabulated (Forsythe 1954) in units of $(cal/cm^2 \; sec)/(K/cm)$. It should be noted however that the ordinary thermal diffusivity calculated from macroscopic thermal conductivity in this way is not the thermal diffusivity which is used in thermal spike calculations. The difference will be discussed in more detail below. The temperature distribution given by eq. (3.7) is that of a Gaussian function whose effective radius grows with the square root of the time. Thus the hot zone spreads rapidly at first, at high temperature, then ever more slowly. This simultaneous spread and drop in T allows one to speak at least in approximate terms of the 'duration' of a temperature spike.

If the heat Q were considered to be released at time $t = 0$ this would lead to a singularity in eq. (3.7): to avoid this a fictitious starting time of

$$t_0 = r_s^2/4D \tag{3.9}$$

is taken by Seitz and Koehler. Here r_s, the mean atomic radius, is defined by

$$\tfrac{4}{3}\pi r_s^3 = 1/n_0, \tag{3.10}$$

where n_0 is the number of atoms per unit volume. The temperature cannot, of course, be infinite since Q is finite and at worst if Q is focussed on one atom initially, one could obtain a fictitious starting temperature T_0:

$$3K\,T_0 = Q.$$

This value of T_0 could be set equal to $T(r,t_0)$ and eq. (3.7) solved for t_0 as an alternative procedure.

The quantity $(4Dt)^{1/2}$ is of interest, in that it represents the effective radius of the hot zone over which the temperature is relatively uniform: actually at $r = (4Dt)^{1/2}$ the temperature has fallen by 1/e of the central value. Seitz and Koehler employ $(4Dt)^{1/2}$ as a criterion for the applicability of eq. (3.7): if $(4Dt)^{1/2}$ is not large compared to atomic dimensions the medium could not be regarded as continuous and the equations of heat flow might not be applicable. In fig. 3.3 is reproduced a set of curves from Dienes and Vineyard (1957) showing the behavior of the idealized temperature spike with $Q = 1000$ eV in copper, as a function of the time. The diffusivity D was taken as 10^{-3} cm^2/sec: if $t = 10^{-11}$ sec then $(4Dt)^{1/2} = 2 \times 10^{-7}$ cm $= 20$ Å. However for $t = 10^{-13}$ sec $(4Dt)^{1/2}$ falls to 2 Å, clearly close to the limit for the continuity approximation though perhaps still meaningful.

The temperature rise at the center of the spherical spike is given by

$$T(t) = Q/cd(4\pi Dt)^{3/2}. \tag{3.11}$$

It is interesting that in this as in the original equation [eq. (3.7)] the quantities D and t enter multiplicatively everywhere. This allows a considerable simplification in the discussion, since in every way a change in the thermal diffusivity by a given factor is exactly equivalent to an opposite change in the time t by the same factor.

Seitz and Koehler also give the corresponding equation for the cylindrical (i.e., heat deposited along a track at a rate Q' per unit length) case:

$$T(\rho,t) = \frac{Q'}{4\pi cdDt}\exp(-\rho^2/4Dt), \tag{3.12}$$

where $T(\rho,t)$ is the temperature at time t and distance ρ normal to the cylinder axis and the other constants have their previous significance. It will be seen that

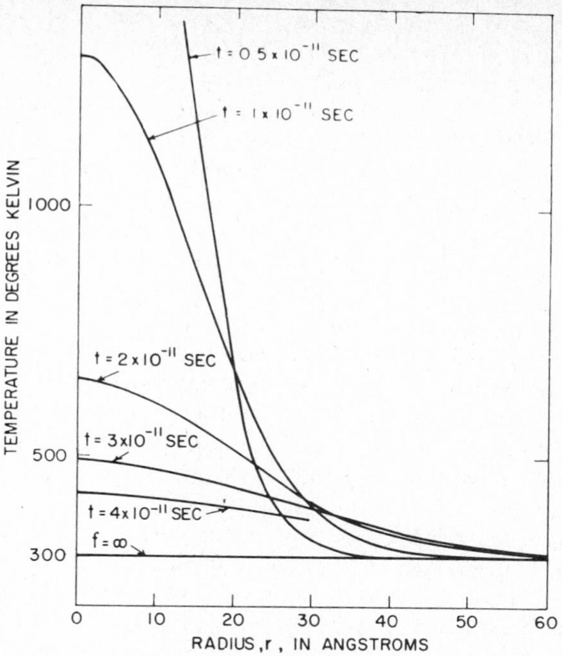

Fig. 3.3. Temperature history of an idealized thermal spike in copper. Input 1000 eV. Taken from Dienes and Vineyard (1957).

the exponential radial dependence is exactly as in the spherical case, the main difference being that the center of the cylindrical spike cools off more slowly, as $1/t$, than the sphere, which goes as $1/t^{3/2}$. At the end of a cylindrical spike, where an initially high-energy hot atom would be found, one would expect to have a kind of blending of the two cases and it seems from a study of the heat diffusion equations that the great bulk of heat Q' released along the track of an initially fast atom would be ineffective in heating the zone at the end of the track, the final trapping point. The relatively long mean free path of atoms moving with energy more than a few hundred eV in solids ensures that the energy deposited will be 'strung out' along the track: only when the atom slows to energies a few times larger than the displacement energy, ca. 25 eV, does its mean free path fall to the order of mean atomic radii r_s, and the remaining energy is deposited in essentially the few last collisions. Reference to computer studies of fast atoms in metals and salts reminds us that the hot zones are ragged in outline, and far from looking like spheres or cylinders. Nevertheless, the results suggest that the final deposition of energy produces a fairly compact temperature spike.

Thus the hot-zone theory strongly suggests that the thermal spike surrounding the final trapping place of an initially fast recoil atom does not in any serious way reflect the magnitude of the initial energy. We will later discuss the experimental evidence bearing on this point, in our sampling of hot-atom studies which have been employed to question or confirm the hot-zone theory.

The quantity D, the thermal diffusivity, is seen to be of crucial importance in determining the temperature history of the idealized thermal spike, and for this reason it is worth examining somewhat more closely. If eq. (3.8) is used to calculate D from tabulated values of the heat conductivity k, the specific heat c and density d, one finds that for typical metals $D = 0.05–1$ whereas many inorganic crystals such as CaF_2, MgO, NaCl and SiO_2 have $D \approx 0.026$ and typical organic liquids and crystals have $D \approx 4–6 \times 10^{-4}$, all in cm^2/sec. The high value of D in metals is, however, illusory for spike calculations as Seitz and Koehler point out that much of its magnitude is due to the effect of conduction electrons, and even when these are excited by ionizing radiation, their energy does not couple effectively with the lattice motion of heavy atoms, and consequently does not go to increase the kinetic 'temperature' of atoms in the thermal spike zone. Rather, they state "In a typical lattice we should expect the time required for thermal energy to be transmitted from one atom to its neighbor to be of the order of 10^{-13} sec, which is a reasonable magnitude to employ for t_0. This corresponds to values of D of the order of 0.001 cm^2/sec." If the equation (Seitz and Koehler 1956)

$$D = \tfrac{1}{3}\lambda v, \tag{3.13}$$

where λ is the mean free path of atoms in crystals and v the velocity, is used to calculate the diffusivity due to atomic motion, we find that for atoms having an energy equivalent to a melting temperature of 1000 K ($3KT_m = 0.086$ eV), in KCl taking $\lambda = 1.4$ Å, the mean atomic value of r_s [eq. (3.10)] is $D \approx 4 \times 10^{-4}$ cm^2/sec. At 1 eV energy, which seems reasonable for the earlier, more effective stage of the hot zone $D = 1.1 \times 10^{-3}$, about the value used by Seitz and Koehler. It will be recalled that the behavior of the computer-generated thermal spikes in copper required a diffusivity of very roughly 10^{-2} cm^2/sec: this corresponds to an effective atom energy in copper of ca. 150 eV, which seems too high. However, in a monatomic solid like Cu, the ease of diffusion of energy outward via focussing and other correlated collisions probably increases the effective diffusivity: these considerations do not apply to the complicated structures usually studied by hot-atom chemists.

We see, then, that the somewhat higher values of D obtained for *macroscopic* heat diffusion in metals compared to ionic crystals is really without much effect on the thermal history of temperature spikes in the two media. What then is the situation in organic crystals (and liquids)? Here the macroscopic heat diffusivity is roughly a factor of two lower than the 10^{-3} figure arrived at above. But because the quantities D and t are always multiplied this has only the effect of stretching out the duration of the hot zone by the same factor.

3.4. THE QUESTION OF MELTING

We now return to the question, already touched upon during the historical introduction to the thermal spike (above), as to whether actual melting takes place in the spike zone. The question is of importance since, if melting occurs, we are dealing with a high-temperature liquid medium in which diffusion or random Brownian motion must readily take place and will influence both the rate and nature of reactions involving the recoil atom.

In our earlier article, the present author and Sutin (1958) following Seitz and Koehler (1956) noted that in an ionic lattice having a melting point $T_m = 1000$ K, and a recoil-atom energy input $Q = 300$ eV, the quotient $Q/3KT_m \approx 1000$, which simply expresses the truism that at such a time as 300 eV is partitioned equally among roughly 1000 atoms, the 'kinetic temperature' of each is 1000 K. We then noted that such a hot zone of 1000 atoms "remains at a temperature above the melting point for $\sim 10^{-11}$ sec." In retrospect it is clear that this phrase was poorly worded: this has led to its meaning being misinterpreted by a number of authors. In turn the misinterpretation has been used to call into question the entire hot-zone model. It is time to set the record straight: we never intended to assert that there was actual melting to produce a zone remaining molten for 10^{-11} seconds, then recrystallizing. In fact, we stated that "Seitz and Koehler point out that the hot spot in metals (and ionic crystals) does not attain the true equilibrium disorder of a liquid throughout, and should probably be likened to a superheated solid which finally cools below the melting point without ever having absorbed, or given up, the latent heat of fusion." Despite this explicit disclaimer of melting and its accompanying flow and disorder, in ionic crystals, the hot-zone model for hot-atom reactions has been attacked by some workers on the grounds that it predicts randomization of ligand species (which it does not) which is not observed experimentally. In the case of organic or other molecular crystals, we felt that because of the smaller thermal conductivity, lower melting point and, one would now add, lower latent heat of fusion, something more akin to real melting might occur in the hot zone. This latter question, now as then, remains open.

It must be admitted, however, that Seitz and Koehler were themselves not always consistent with this concept, since they speak in some places of a "sphere of melting" of radius r_m. In particular, toward the end of their Section 23, they speak of the re-establishment of crystalline order "once 'freezing' begins" etc. A careful reading, however, of their Sections 21 and 22 reveals statements such as that already cited, and this: "In fact, the following discussion will show that only the central portions of the heated regions appear to attain thermal equilibrium" introducing the argument we will mention below for contributions to n_j. The paper of Tucker and Senio (1956) also treats the question of the nature of thermal spikes and points out that although the atoms in question have plenty of kinetic energy, the words "heated" and "melted" are not strictly appropriate.

It is worth inserting a word here about the latent heat of fusion (Forsythe 1954). For metals this appears to run about 0.1–0.2 eV/atom while for ionic crystals the figure is comparable, 0.2–0.4 eV/molecule. For organic substances it is about 0.02–0.1 eV/molecule. However, since we are discussing the possible formation of a molten *zone* we should also consider the latent heat of fusion per unit volume. Here the values are somewhat more widely separated, metals such as Al, Cu and Pt being 0.3–0.6 kcal/cm^3, ionic crystals 0.08–0.17 and organic crystals 0.013–0.046 in the same units. The substantially lower volume latent heat for organic substances reinforces the idea that here actual melting might indeed take place, especially with energetic recoil atoms. A second point is that energy removal from the hot zone via focussing, assisted focussing and correlated collision chains, which operate very well in metals and a few ionic crystals like KCl where all atoms have nearly the same mass would not operate at all in organic crystals.

We now consider the available evidence for melting. The computer studies of Torrens and Chadderton, Vineyard et al., Bunker et al. (cited above) and Harrison et al. (1963) give no indication of actual melting, turbulent flow or randomization, at energies typical of (n,γ) recoil. On the contrary, all the studies demonstrate how efficiently excess translational energy is carried away rather than remaining to heat the lattice. Studies of sputtering (Thompson and Nelson 1962, Brack and Schwuttke 1971, Nelson 1968) tend to confirm this picture: here, the incident fast atom either impinges upon a crystal surface or is transmitted through the crystal, in either case depositing its energy in the lattice, and causing the ejection of atoms of the host lattice. The sputtering event is, then, a thermal spike or collision cascade that intersects the surface of the crystal, and from measurement of the number, energies, and directions of the ejected atoms we should be able to learn something about the nature of the spike zone. Let us see where this leads. Consider, for example, the behavior of the self-sputtering ratio (the number of atoms of element M ejected per incident ion of M^+ at a standard energy) with atomic number (Nelson 1968). A dependence upon $1/E_b$, where E_b is the binding energy of the host atom to its surface, is predicted theoretically, and appropriate experiments are found to fit this dependence closely if E_b is taken equal to the energy of sublimation. These results strongly suggest that atoms leave the hot zone at the surface in a fashion akin to sublimation rather than by boiling or vaporization from a liquid. Again, a large proportion of the sputtering yield from the surface of a single crystal goes in preferred directions which lie along crystallographic axes such as ⟨110⟩ in face-centered cubic crystals like gold and copper. This preferential ejection is the result of correlated or focussing collisions in the layers near the surface, but in any case the sputtering data strongly suggest that in most encounters the structure of the crystal must be preserved.

But objections may immediately be raised to taking either of these experimental observations from sputtering as evidence against melting. For the first part, the energy of binding (sublimation from the solid) and of vaporization (from the liquid) differ only by the energy of fusion, which is usually small by comparison (Forsythe 1954). For the second part, the retention of crystal structure must refer only to the events in which very energetic atoms are expelled from the surface, 'knocked-on' or 'out' and hence to a time scale short compared to the formation of a molten zone. The clearest evidence for a thermal spike comes from sputtering studies of *slower* ejected atoms, i.e. those in the 0.05–0.4 eV region, as mentioned above. Here, the atoms seem to emerge in a fashion exactly characteristic of evaporation; in fact, the phenomenon is termed 'thermal sputtering.' When monocrystalline copper was bombarded with 45 keV Xe^+ ions, as the temperature of the copper was increased from 20° C, at first the copper atoms were ejected predominantly along crystallographic axis directions such as ⟨110⟩, but finally a point was reached (ca. 700–800° C) where the uniform, unstructured background due to 'thermal sputtering' began to increase and overwhelm the directional ejection. This increase was as predicted by the thermal spike model. A complete discussion of this and other cases will be found in the review of Nelson (1968).

Although Thompson and Nelson's (1962) work does not give a clear indication of surface molten zones under impact of sputtering ions at high energy on targets at high temperature, it does demonstrate that the overall processes alter drastically as one nears the melting point of the host. This point may again be of considerable importance in the hot-atom chemistry of organic crystals: if one imagines an organic crystal held just below the melting point, then virtually the only thing an energetic recoil atom can do in the crystal is melt it, and the energy required is only the rather low heat of fusion. In the case of some typical organic halides that have been studied, for example, the temperature of bombardment may have been close enough to the melting point that one could expect that this type of local melting might occur.

Finally, it is worth recalling the experiment of Naguib and Kelly (1970) cited earlier in which impact of 20 keV Kr^+ produced recrystallization of amorphous ZrO_2 films. This begins to look very much like local melting and crystallization upon solidification, although one could not be sure that the enhanced diffusion in thermal spikes in unmelted ZrO_2 might not produce the same reordering effect.

In summary, then, it appears that sputtering experiments at high energies (20–50 keV impact energy, heavy ions) in metal targets confirm the reality of thermal spikes but do not settle the question of molten or superheated solid zones. The ZrO_2 experiment suggests a molten zone (but does not prove it) while the computer studies do not agree, and suggest more strongly retention of crystal structure even at high vibrational or kinetic temperatures.

3.5. THE EFFECT OF INNER-SHELL VACANCIES

We have thus far largely avoided, for the sake of clarity in discussion, considering the role of inner-shell vacancies produced by electron capture or by internal conversion in hot-atom chemistry. There is plenty of evidence that low-lying levels, fed by decay of upper states produced by neutron capture, or other nuclear reactions, are often internally converted, and that such conversion events, leading to inner-shell vacancies, produce charged ions (de Wieclawik and Perrin 1969, Yosim and Davies 1952, Thompson and Miller 1963, Cardito and Diethorn 1970, Tumosa and Ache 1970). In ch. 2 these mechanisms and their results have been discussed, and ch. 15 will take up the chemistry of atoms which have undergone isomeric transition on a 'laboratory' time scale. Chapter 24 will review Mössbauer evidence on the effect of inner-shell vacancies. Such studies tell us that the internal conversion event by itself can produce highly reactive atoms which often behave surprisingly similarly to those activated by other nuclear reactions. It is therefore important to consider the participation of internal conversion in the sequence of events immediately following, for example, neutron capture.

A recent paper by Jones (1970a) has treated the question in some detail: he points out that the internally converted decay of low-lying levels not only produces a locally high positive charge concentration (in insulators) but also that this event may be appreciably delayed in time, tending to occur after the hot atom slowing-down is completed. He cites some examples of experiments in angular correlation which show a 'hole-recovery effect' traceable to this short-term disruption of charge distribution following an inner-vacancy.

Recovery of the ejected Auger charge in metals is very fast but in insulating crystals may be quite slow: Stratton (1941) gives the equation

$$\tau = \varepsilon/\sigma, \qquad (3.14)$$

where τ is the relaxation time for the decay of an initially unstable charge distribution in a medium of dielectric constant ε and conductivity σ. He notes that in "sea water the relaxation time is $\sim 2 \times 10^{-10}$ sec," in distilled water 10^{-6} sec but in insulators such as quartz, perhaps more than 10^{6} sec. Vineyard (1961) calculates for metals τ is ca. 10^{-19} sec. On the other hand, aliovalent ions generated in some crystals may be stable almost indefinitely, provided the Lidiard (1957) criterion is met.

In the gas phase, the consequences of inner-shell vacancy production are well-known (see ch. 2) but in solids much speculation has been given to the question of the time scale for charge relaxation. This point bears heavily on the Varley (1962) mechanism, a postulated process by which ionization of an anion, for example X^- in an alkali halide, to a positive charge state $+n$, whether by inner-shell vacancy plus Auger cascade, or multiple outer-shell ionization, renders the ion X^{+n} unstable with respect to coulombic displacement into an interstitial position. The question is whether the positive charge on the halogen atom can persist long enough to allow the ion X^{+n} to move off its original lattice site, leaving a vacancy behind. To date much evidence and discussion concerning Varley's proposal has been presented in the literature, but conclusions hardly seem firm (Cruz-Vidal and Gomberg 1970, Cruz-Vidal et al. 1970). A paper by Chadderton et al. (1966) on computer-simulation of 'Varley' events, however, suggests that through processes of charge relaxation and ionic motion more complex than envisioned by Varley, interstitial–vacancy pairs might still be formed.

There is, of course, no reason to believe that the decay of the original charge distribution should follow a simple one-term exponential law with the Stratton τ: in the defect zones generated by a hot atom both the dielectric constant ε and the conductivity σ will vary according to the degree of local damage, and in any case this equation refers to classic conduction in a continuum, not to a sequential capture of electrons by a single ion of initial charge $+n$ embedded in a lattice. One would guess that as long as the ion's electron affinity remains high compared to the band gap of the crystal, electrons will be drawn in via the conduction band, rapidly reducing the value of $+n$. At the point where these energies become roughly equal, quantum-mechanical tunnelling processes, and thermal untrapping of electrons from defect sites not too far from the positive ion will be rate-determining and finally, when the electron affinity of the ion places its electron capture level somewhere in the band gap, the ion becomes quasi-stable.

The initial rapid return of electrons to neutralize (in part) the Auger ion of charge $+n$ will inevitably produce a thermal spike located at the origin, and the heat Q released [eq. (3.7)] may well be of the order of the inner-shell electron binding energy, reduced by the energy radiated as X-rays, where the original hole was produced by internal conversion or electron capture. Thus an internal conversion in an atom like ^{80m}Br may produce a thermal spike of considerably greater total energy than the (n,γ) recoil process itself.

There is however one great difference between the Auger-ion neutralization and (n,γ) recoil thermal spikes, considered in their pure forms. Whereas in the latter the energy content is entirely kinetic energy, leading to rather low heat diffusivity coefficients (see above) but with ample opportunity for atom displacement, in the former the initial deposit of energy, though larger, is entirely into the electronic system, which is said (Seitz and Koehler 1956) to couple only weakly with the translational motions of atomic nuclei. Also, this electronic excitation would be expected to 'cool off' with an effective heat diffusivity much larger than the translational value of ca. 0.001 cm^2/sec. In fact, the high electronic diffusivity may account for most of the observed macroscopic heat conduction, leading to observed bulk values of D of 0.026 in NaCl, for example, and 0.05–1 in metals, in units of cm^2/sec.

Some time ago Walton (1964) mentioned the possibility that this electronic excitation effect could produce chemical effects following neutron capture or other nuclear reactions: he coined the expression 'hot-electron chemistry' to describe the phenomenon and pointed out that the radiation chemistry (radiolysis) of solids is dependent upon it. Earlier, Geissler and Willard (1963) discussed the possibility of 'auto-radiolysis' of local zones of liquid organic halides to account for observed yields of hot-atom reactions. Both the Walton and Willard concepts are obviously closely related to the electronic excitation spike described here. What is added here is the idea that following the rather rapid charge relaxation the affected zone will be still at a rather high local 'temperature' owing to the heat released in the neutralization of the local positive charge excess. It is interesting that the same phenomena may be involved in the storage of charged-particle or heavy-ion tracks in solids (Van Vliet 1970).

3.6. HOT ATOM CHEMISTRY IN THERMAL SPIKES

We now turn at last to a consideration of hot-atom results and the various attempts that have been made at interpretation in terms of a thermal spike model. The basic problem is, of course, that in hot-atom research the techniques employed do not permit the observation of the 'primitive' distribution of chemical forms: this is true even of the Mössbauer and other in situ procedures. In fact, considering the opportunities for short-term in-spike or low-temperature annealing reactions, it may even be meaningless to imagine a primitive distribution. For example, in $KMnO_4$ (n,γ), the yield of $^{56}MnO_4^-$ is found to drop steadily toward zero as the irradiated crystals are stored and dissolved under increasingly cold conditions. Similarly drastic changes with temperature are seen in other crystals (Veljkovic and Harbottle 1961a, 1962). How do we know, then, that most of the crystals investigated by conventional hot-atom procedures might not likewise reveal utterly different chemical distributions, if only one could work at sufficiently low temperatures? How can we discuss mechanisms to produce observed distributions, if we cannot be sure that the distributions are anything more than the end-products of complex sequences of annealing reactions?

Despite these uncomfortable reservations, the hot-zone model has been invoked to explain observations in both solid and liquid-phase hot-atom chemistry, and we can review this evidence for sake of completeness. The expression 'a hot-zone reaction' may be nothing more than the way of rationalizing our observation of at least one of the pathways in which a hot-atom's fate is influenced by the environment in which it de-excites. For example, Rieder et al. (1950) found that in $KMnO_4$–$KClO_4$ mixed crystals, permanganate retentions increased with decreasing mol fraction. The hot-zone interpretation is obviously that if Mn^{++} is the primary recoil product (Apers and Harbottle 1963, Cogneau et al. 1967, 1968) it is easier for it to be oxidized to MnO_4^- in a hot zone of perchlorate ions than in one of permanganate, where it is hardly oxidized at all.

The hot-zone idea actually had its inception in an attempt to explain the very low retentions of ammonium chromate (Harbottle 1954), dichromate and iodate (Cleary et al. 1952) compared to the alkali–metal analogs. For bombardment at reduced temperatures retentions of Li, Na or K chromate or dichromate fall in the range 48–75% while those for ammonium chromate and dichromate are ca. 11 and 26% respectively (Veljkovic and Harbottle 1962, Ackerhalt and Harbottle 1972). Virtually identical results are obtained whether thermal or fast neutrons are used to produce the ^{51}Cr. It is interesting that Ikeda et al. (1970), irradiating $(Co(NH_3)_6)_2(CrO_4)_3 \cdot 5H_2O$ with thermal neutrons, also found the characteristically 'ammonium' value of 10.3% for the chromate retention. Somewhat the same effect – a lowering of retention with ammonium cations – was seen by Facetti et al. (1969) with ammonium perrhenate, but DeKimpe et al. (1969) showed by low-temperature techniques that NH_4ReO_4 is again, unusually susceptible to low-temperature thermal annealing. Aten and Kapteyn (1968b) did not find such a large effect of ammonium anions with perrhenate but did note that the substitution of ammonium for alkali metal reduced K_2ReCl_6 and (slightly) K_2OsCl_6 retentions. Müller and Cramer (1970) found a similar small effect in K_2ReBr_6, where the nuclear reaction was isomeric transition rather than (n,γ). Since here the Br^- ions are in their lowest oxidation state both before and after the nuclear transformation, it looks as though ammonium ions in the hot zone interfere in some way with the processes normally leading to high retentions.

The possibility remains that most of these ammonium effects are due to nothing more than the bulk-radiolytic generation of reducing species such as hydrazine etc. which preferentially react with the recoil atoms at the moment of dissolution of the crystals. Or, the nitrogenous reducing species would be formed locally relative to the recoil atoms by Auger self-radiolysis of the Geissler–Willard pattern. The first possibility can be examined by studies at lower and lower total gamma doses, i.e. studies of the 'intrinsic' retention* and the second checked by reference to isotopic effects in recoil with crystals containing ammonia or ammonium cations, hopefully with isotopes where the probability of inner-shell vacancy production following nuclear transformation is known.

Studies of the effect of initial recoil energy of the hot atom on its subsequent

*'Intrinsic retention' is used here in the sense defined by Collins and Harbottle (1964).

chemistry have been undertaken, and the results sometimes related to the hot-zone model on the theory that more energetic atoms, resulting from for example (n,2n), (γ,n) and alpha-decay reactions, ought to produce larger, hotter, and longer-lived thermal spikes, in which rate processes would have a better chance to take place. It is important to note that such comparisons, for example between (n,γ) and (n,2n) generated hot-atom reactions, could only hope to have validity if the same isotope was generated, for example ^{80m}Br from ^{79}Br(n,γ) and ^{81}Br(n,2n), and even here under the assumption that the decay scheme of the excited nucleus, especially as regards probability of internal conversion, is fairly independent of mode of formation. However, as we have already noted above in the description of the thermal spike, within the context of this model it appears that variation of initial energy will not much alter the crucial parameters of the hot zone at the terminal site. These remarks do not, of course, preclude finding substantial energy effects in the vicinity of the threshold for displacement. We will discuss later the experimental work on recoil energy effects.

In the case of ^{32}P recoil reactions in crystalline phosphates (see ch. 9 below) the presence of condensed polyphosphates such as pyrophosphate and tripolyphosphate might be taken as evidence for a high temperature reaction in the hot zone, as these species can be prepared by heating orthophosphates. But it has been found that the phosphate system is extraordinarily complex, with some reaction, dependent on carrier concentration, occurring at the moment of dissolution (Ujimoto et al. 1970).

In the understanding of hot-zone reactions, the extensive and detailed work of Müller (1964a, b, 1965, 1966, 1967, 1968, 1970, Müller and Martin 1969) has played a leading role. Müller has neutron-irradiated isomorphous mixed crystals of the type K_2ReBr_6–K_2SnCl_6 of varying compositions and separated the resulting hot-atom products $(ReBr_nCl_{6-n})^{-2}$ by electrophoresis. His results show that the hot atom's original ligands tend to be found in the recoil product and that the production of mixed-ligand entities does not follow a random-recombination pattern, as one would have predicted with a large molten zone and complete mixing of all ligands present on a statistical basis. From his data he calculates a radius of reaction zone of about 5 Å and a content of about 20 atoms (i.e. around 3 complex ions). Unfortunately, owing to a misunderstanding over the concept of 'melting' contained in our original paper (Harbottle and Sutin 1958) he concludes that his results disprove the hot-zone model, and hence he invokes, unnecessarily, a new model, that of 'little disorder' in the hot zone. His results, taken at face value without reservations concerning low-temperature annealing etc. certainly support the concept of little disorder. But the argument and example given in our original paper (taken from Seitz and Koehler) which he cites for comparison are not only misconstrued, but also not even very similar to his case. Consider first the recoil energy Q deposited in the hot zone. The original example took $Q \approx 300$

eV, but in $^{185}Re(n,\gamma)^{186}Re$ the maximum capture gamma-ray energy is only 6.1 MeV (Bartholomew and Groshev 1967) leading to a maximum recoil energy of only 102 eV for ^{186}Re. Examination of the very complex decay scheme of capture gamma rays from rhenium indicates that many parallel deexcitation cascades occur with values of 4–5 MeV for the highest energy gamma ray not uncommon. A value of 5 MeV, even without any momentum cancellation, leads to a value of Q of only 72 eV, and the average value is certainly considerably smaller than this. Taking $Q = 72$ eV and a melting point $T_m = 600°$ C (estimated) the partitioning of $3KT_m$ to each atom in the hot zone leads to a zone of ca. 300 atoms, not 1000. If the ratio of the radius of the hot zone to the average atomic radius, r_m/r_s, be evaluated, it is only 6.8.

However, there is an even more serious misconstruction of the Seitz–Koehler theory implicit in Müller's conclusions. To undertand this we consider again here, as in the original paper of Sutin and the author, the Seitz–Koehler calculation of rate processes in the hot zone.

Seitz and Koehler considered the occurrence of reactions such as the formation of Frenkel defects and the exchange of atoms in order–disorder processes in the hot zone in alloys: in our paper we extended this to include chemical reactions, ligand exchange, bond breaking, redox, etc. They showed how the quantity n_j, the number of jumps for a particular atom involved in a rate process of activation energy F in the hot zone, could be calculated by evaluating the integral

$$n_j = \int_{t_0}^{\infty} \gamma_0 e^{-F/kT(t)} \, dt, \tag{3.15}$$

where $T(t)$ is the temperature history of the hot zone at the particular atomic site, given by eq. (3.7) or (3.12) above. (It should be noted that $T(t)$ contains the additive term T_0, the ambient temperature, but this is usually neglected with respect to the temperature rise produced by the heat source Q.)

We also discussed the steric and statistical factors which could influence the value of n_j. The conclusion was that n_j might in typical cases range from 1–5 for $Q = 50$–300 eV and $F = 1$ eV, in other words, that typical rate processes of activation energy 1 eV (or even more) had an excellent chance of taking place during the very short but very hot spike duration.

However, one may advance a step farther. Equation (3.15), relating the temperature history $T(t)$ and the number of jumps n_j, in fact allows us to take a rather more fine-grained look at reactions occurring in the thermal spike, and leads to a result of fundamental importance to hot-atom chemistry of the solid state. *It is not from later stages involving long times, $\geqslant 10^{-11}$ sec, and large hot-zone volumes, 300–1000 atoms at temperatures near T_m, that the major contributions to the integral for n_j arise, but from the much hotter, earlier 'core' of the hot zone.* Seitz and Koehler define a quantity α

$$\alpha = \frac{Q}{F}\,\frac{k}{cd(4\pi D)^{3/2}} = \frac{4}{9\pi^{1/2}}\left(\frac{Q}{F}\right)t_0^{\,3/2}, \tag{3.16}$$

where the other quantities are as defined earlier (k = thermal conductivity) and show that most of the integral for n_j, and consequently most of the hot-zone reactions, arise in fact from the time interval between t_0 and a time of the order of $\alpha^{2/3}$, or about 10^{-12} sec, and of course in a correspondingly smaller, hotter spike.

Let us see now what happens in Müller's case at these important shorter times. Seitz and Koehler, again discussing the integral for n_j, give the equation

$$r = \frac{4}{(324\pi)^{1/3}}\left(\frac{Q}{F}\right)^{1/3} r_s \tag{3.17}$$

where r_s is the mean atomic radius, ca. 1.8 Å in K_2ReCl_6 for example, and the other quantities have their usual meanings, to define the radius r not of a zone heated to any particular temperature, but what is much more important, of a zone outside of which the probability that an atom engages in a rate process of energy of activation F is very small. If we take $Q = 100$ eV and $F = 1$ eV then $r = 3.3$ Å. For the case of $Q = 70$ eV (more realistic for ^{186}Re recoil) and $F = 2$ eV (since the two ligands must be exchanged, two bonds broken) $r = 2.3$ Å. These numbers are certainly not wildly discordant from Müller's estimate of 5 Å for the reaction radius in his zone of 'little disorder.' It follows from the extended LSS range calculations (mentioned earlier) that 5 Å is also quite reasonable for the range of a heavy atom of energy 50–100 eV. It certainly follows from all this that the concept of 'little disorder' is already firmly embedded in the Seitz–Koehler theory, and that there is no reason to create a new hypothesis or model to explain Müller's excellent experimental results. The chemical hot-atom results on which Müller's conclusions were based are at present being extended via an interesting computer simulation technique (Robinson et al. 1974) which will be discussed in ch. 25.

Another type of experiment along the same lines is that of Saito et al. (1962e) who studied hot-atom reactions of cobalt in mixed anion-cation complex crystals such as $Co(NH_3)_6Fe(CN)_6$ and of Ikeda et al. (1970) already cited. Results in these studies agree with Müller's findings, that there is no large, molten, or randomly mixed hot zone. One should also note that all these studies exclude the 'billiard-ball' substitution mechanism.

In conclusion it should be reiterated that each hot-atom system ought to be studied, calculations made and conclusions drawn on an individual basis, and the theory applied as exactly as possible to that particular case.

At this point it would be well to mention that, although we have focussed our attention on the passage of the fast atom through the condensed phase, and the thermal agitation produced by its deceleration, we have neglected to describe an additional result of that process that appears in solids, and that is the production of interstitial atoms and vacancies. It was shown early (Seitz and Koehler 1956, Snyder and Neufeld 1955, Neufeld and Snyder 1955, Kinchin and Pease 1955a, b) that roughly half the energy of fast atoms is used up in the production of displaced atoms, at a cost of E_d per displacement. We keep in mind that E_d, as mentioned above, depends sharply upon the direction of motion taken in the lattice. One should not, however, assume that this half of Q the input energy is lost to the thermal spike: the energy stored as a result of displacement production is not more than a few eV per interstitial–vacancy pair, and all the remainder eventually goes to heat. In the case of complex crystals, some energy is probably also used up as a result of bond-breaking processes.

It would appear then that the final resting place of a recoil atom in a crystal may often have interstitial atoms and vacancies in its vicinity. There is some physical evidence of this (Czjzek et al. 1968, Holmes 1970, Pronko and Kelly 1970, Carter et al. 1971, Nelson 1969, Hinman et al. 1962, Seyboth 1969). In complex crystals we may add additional defects such as free radicals, complex ions with one ligand displaced etc. In general it is not clear what role these may play in determining the immediate fate of the hot atom. In cases however where the recoil atom itself is interstitial, it does seem likely that one form of the annealing process may be a complex analog of interstitial–vacancy recombination. It is important not to confuse these defects and their possible interaction with the hot atom with the defects introduced by 'doping' a crystal before bombardment with aliovalent ionic species, whose presence requires associated defects such as vacancies in order to preserve charge balance: the density of the recoil self-created defects will be far higher, in the vicinity of the recoil atom.

In the thermal spike model, it seems clear that defects will provide a possible source of reactants for the hot atom in the spike cooling period. Also, if internal conversion occurs, the high local concentration of defects will constitute a rich variety of potential electron and hole-trapping sites, which can later thermally release their charges and so influence the course of subsequent thermal annealing.

3.7. EFFECT OF INITIAL RECOIL ENERGY

We now return to the discussion of the effect of changes in initial recoil energy on the hot-atom chemistry. It was shown above that the Seitz–Koehler spike theory, combined with what is known of changes in mean free path of atoms slowing down in condensed phases leads to the expectation that the temperature history $T(t)$ of the terminal hot zone ought not to reflect in any significant way the initial energy of the recoiling atom. Although this should be true of 'hard' crystals, an exception might have to be made in the case of organic crystals, having low melting points, low heat diffusivity and low volume heat of fusion.

There is, by now, an abundance of research, much of it carried out with reduced-temperature precautions, in which initial hot-atom yields have been compared for differing nuclear processes such as (n,γ), (γ,n) and (n,2n), which produce atoms having initial recoil energies differing by a factor as large as 10^3 (Ackerhalt and Harbottle 1972, Saito et al. 1965a, Jones 1970b, Van Dulmen and Aten 1971, Dupetit 1967, Ambe et al. 1968, Hillman et al. 1968). In some of these studies the same isotope was produced by low- and high-energy processes, for example ^{63}Cu(n,γ)^{64}Cu and ^{65}Cu(n,2n) or (γ,n)^{64}Cu, thereby reducing uncertainty related to isotopic vis-a-vis energy effects. The overwhelming conclusion is that, although isotopic effects produce some yield differences (presumably through their differing probabilities of internal conversion), the effect of initial recoil energy is relatively minor, even in cases where one examines the distribution of species found in some detail (Ackerhalt and Harbottle 1972). This is very much in accord with the picture of the terminal thermal spike outlined above.

Additional evidence is afforded by studies of the recoil chemistry of *cis* and *trans* crystalline compounds. Early studies had shown that the *cis* or *trans* configuration is preserved under (n,γ) recoil (Zuber 1954a, b, Rauscher et al. 1961, Jagannathan and Mathur 1968), but recently Wolf (1966) showed that this is also true for the reaction ^{59}Co(n,2n)^{58}Co in *cis* and *trans* dichloro-bis-ethylenediamine cobaltic nitrate. All this is in agreement with the limited-range recoil energy-insensitive terminal spike concept of the present model, and, like Müller's results, incompatible with a large molten zone leading to randomization.

Examination of isotopic differences in hot-atom yields following (n,γ) reactions which produce two or more isotopes in the same crystal, for example Br(n,γ)^{80m,80,82}Br, Ru(n,γ)^{97,103,105}Ru, Mo(n,γ)^{91,93m,99,101}Mo etc. is in general a poor way to study the effect of recoil energy on the hot-atom chemistry: calculation of the recoil energy from the emission of the ground-state gamma ray following capture is misleading inasmuch as this transition generally has a trivial probability. Also, because of the complexity of capture decay schemes, all isotopes will in general show broad recoil energy spectra, effectively concealing differences based on recoil energy effects. Where differences are seen, they are much more likely to be due to differing probabilities of inner-shell vacancy production during capture-state decay, or passage through a longer-lived isomeric state after capture but before analysis, as in ^{82m}Br (6.2 min), than to differences in average recoil energy.

There are, however, two rather more direct ways of exploring recoil energy effects with better control over the energy: these are ion implantation, which will be discussed in ch. 23 and the Yoshihara 'appearance energy' technique (Yoshihara et al. 1970, Yoshihara and Kudo 1970, Yoshihara and Mizusawa

1972). The difficulty with ion implantation is that the energy range easiest to study is rather high, 10–80 keV, and for these high energies the effect of initial energy ought to be negligible. If ions are implanted at low energies (20–1000 eV), they are necessarily found close to the surface of the host crystal and their chemistry is probably influenced by surface effects, radiation damage, adsorbed layers of ambient gases etc. Nevertheless the ion implantation technique is interesting in that it appears to reproduce many hot-atom effects (Andersen et al. 1968, Andersen and Ebbesen 1971) without the uncertainty of neutron capture decay schemes and internal conversion.

The 'appearance energy' technique measures the yield of hot atoms activated by the (γ,γ') reaction, recoiling out of the parent compound, as a function of the energy of the bombarding gamma rays. Various assumptions must be made concerning the decay of the γ-excited levels, since the momenta of the incoming and outgoing gamma rays must be added vectorially, but within these limitations, which do not seem too serious, the interesting result is obtained that in crystals of indium and lutetium EDTA complexes, the recoil yield is zero up to a certain 'appearance energy' (60 eV with ^{115m}In, 40 eV with ^{176m}Lu) then rises and flattens out on a plateau, which extends all the way up to the (γ,n) recoil energies. The yield of ^{115}In$(n,\gamma)^{116m}$In is on the same plateau, for an estimated recoil energy of several hundred eV.

The 'appearance energy', as pointed out by Yoshihara and Kudo (1970), is analogous to the displacement energy E_d. One may ask, as the initial recoil energy moves up, first to values high enough to break the atom away from its original ligands, ca. 60 eV then higher, "Why does the percent yield saturate at only ca. 20%? When 500 eV, or 5×10^4 eV are available to the recoil indium atom, why does the yield not go to 100%?" The answer given by hot-zone theory is clear, that the re-entry is due to generation and recombination of In and EDTA radicals, and In EDTA exchange reactions occurring in the hot zone, and these zones look the same and have roughly similar thermal histories $T(t)$ regardless of the initial recoil energy once it has reached a sufficiently high value, in this case ca. 100–200 eV. Note that the 'range' of such an atom would not exceed a few Angstroms. Thus the results of the 'appearance energy' experiments not only yield information on the primary displacement energy, but also confirm the hot-zone model.

It is quite interesting that many cases of β^- decay show a very high probability of the daughter atom being found in the parent atom–ligand (Perlow and Perlow 1965, 1968, Shiokawa et al. 1969a) configuration: such decays in the rare earth region, where the chemistry of neighboring elements is so similar, would seem to correspond in energy and lack of chemical effect to the region below the 'appearance energy' in the Yoshihara experiments. As mentioned earlier, the physical picture of the interaction of these 'sub-displacement' atoms with their lattices is given by the Langhoff (1971) NRF technique.

4. Summary

To summarize the author's present view of some of the important factors in the overall process of energy loss of recoil atoms and its relation to hot-atom chemistry:

(1) Because of the existence of numerous parallel pathways of cascade deexcitation of the excited state formed by neutron capture, the spectra of recoil energies of atoms which have undergone the (n,γ) process are typically very broad and peak at energies somewhat lower than generally supposed. It will in general be very difficult to discern chemical effects due solely to recoil energy differences in studies of a series of isotopes of the same element, all of which are activated by the (n,γ) process.

(2) Ranges of recoil atoms generated by (n,γ), β^- or β^+ decay processes are very short, typically not more than one or two molecular layers in solids and the equivalent in liquids. Both these, and the longer ranges resulting from (n,2n), (γ,n) and α-decay processes, can be approximately calculated using the LSS theory. Because of the very high dependence of range on initial direction of motion, there will be excessively high straggling in the case of (n,γ) recoils.

(3) There is little argument concerning the *existence* of a hot zone produced by a recoil (or implanted) atom: the short mean free path of 50–500 eV atoms in crystals ensures that the final slowing-down and energy transfer to the lattice will be rapid compared to the 'cooling off' period. But there is ground for argument concerning the chemical effects, if any, traceable to this thermal spike.

(4) Although higher-energy atoms produced by implantation or by (n,2n), (γ,n), etc. nuclear reactions or by α-decay have longer ranges, the decrease in mean free path with decreasing energy, in condensed phases, leads to a terminal hot zone not greatly differing in size or thermal history from the (n,γ) case and consequently having similar chemical effects. This, of course, neglects any effect of subsequent inner-shell vacancy production.

(5) The consequences of an internal conversion event in an atom located in a crystal are apt to be a local radiolysis of the surroundings on a distance scale appropriate to the ranges and rates of energy loss of the Auger electrons and, on a somewhat smaller distance scale, a spike of electronic excitation at the site of the neutralized Auger-daughter atom. This electronic spike will be characterized by a much higher effective heat diffusivity and by lack of coupling to atomic motion. It could be expected to lead to chemical reactions including short-term annealing of the decayed hot atom, and the high local concentration of defects such as F centers, trapped holes, etc. could well influence subsequent thermal annealing reactions, this being particularly so in cases where an (n,γ), (n,2n), etc. recoil preceded the Auger event.

(6) Up to a certain small recoil energy, which may be related to the displacement energy E_d of solid-state theory, the lattice can take up the energy

and chemical effects which are seen may be explained by charge-change, 'shake off' etc. processes. Many β^-, β^+ and some unconverted isomeric decays fall in this category. This region of energy is explorable by the 'appearance energy' as well as conventional hot-atom chemical techniques.

(7) If we follow the Seitz–Koehler calculations for n_j, they indicate that the heat spike following (n,γ), (n,2n), α-decay, isomeric transition etc. processes should be able to bring about chemical reactions such as ligand bond breaking, ligand exchange, generalized displacement, redox, and bond reformation. The last is one instance of a general class of short-term annealing reactions, which can also be promoted by the thermal spike.

(8) The essential hot zone of recoil chemistry is shown to be small by both Seitz–Koehler theory and numerous hot-atom chemical studies. But theory also shows that it is in the earlier stages, when the core of the zone is smaller and hotter (from t_0 to ca. 10^{-12} sec) that most of the contribution to the integral for n_j arises, and not when the thermal agitation has spread to a volume of the order of hundreds of atoms at 'temperatures' nearer the melting point. There is no theoretical evidence for an actually molten zone, with random mixing of atoms or ligands in 'hard' crystals, and computer and chemical studies also negate the concept of a definite fusion volume.

(9) There is good evidence that 'billiard ball' mechanisms do not function effectively in solids.

(10) Although discussion has centered on solids, there is no reason to think that the slowing-down of fast atoms in liquids will not also lead to a hot zone, with attendant fragmentation and recombination reactions. The major difference, of course, is that in a liquid diffusion can lead to recombination or 'liberation' of reactive entities at any temperature, including recombination of oppositely charged ions. Thus the essential spike phenomena of solids are repeated in liquids but with an added stage involving relative movement of the affected entities. This second stage can, however, be studied by means of scavenger reactions in liquids, which is not possible in solids.

Addendum (December 1976)

For further developments see Adloff (1975), Anselmo (1973), Fantola-Lazzarini and Lazzarini (1973a, b), Glentworth and Nath (1975), Newton (1975), Grimm et al. (1975), Harbottle (1975), Hauser et al. (1974), Maddock (1975a, b, c), Robinson et al. (1974), Rössler and Pross (1974), Rössler and Robinson (1974), Saeki and Tachikawa (1973), Siekierska et al. (1973), Stocklin (1975) and Wiles (1974).

CHAPTER 4

INORGANIC HOT-ATOM CHEMISTRY IN GASEOUS AND ONE-COMPONENT LIQUID SYSTEMS

Peter P. GASPAR and Michael J. WELCH

Department of Chemistry, Washington University, Saint Louis, Missouri 63130, USA

Chemical Effects of Nuclear Transformations in Inorganic Systems
Edited by G. Harbottle and A.G. Maddock

Contents

1. Introduction

While 'hot-atom chemistry' has for many years been synonymous with 'the chemical effects of nuclear transformations' it is clear that nuclear recoil is but one process which can give rise to a highly energetic atom (Stöcklin 1969, Wolf 1964, Wolfgang 1965a, 1969). The convenient production of atoms with high kinetic energies by nuclear recoil provides opportunities for the chemist to attack problems unapproachable by other means. In this chapter we shall limit ourselves to a discussion of nuclear recoil experiments in gaseous, and one-component liquid, inorganic systems. The chapter is divided into three sections. Section 2 presents a brief review of the problems to which the techniques of gas and liquid phase hot-atom chemistry have been addressed. Some of the more important contributions to the elucidation of these problems are discussed in section 3. Section 4 consists of a brief element-by-element review of recoil atom studies in inorganic gaseous, and single-component liquid, systems.

Interest in hot-atom chemistry has focussed upon measurement of yields and the interpretation of these data in terms of mechanisms. The field has also, however, been exploited for its purely synthetic possibilities. Nuclear recoil offers an experimentally convenient technique for the formation of free atoms under conditions sufficiently mild that their reactions with normal molecules may be studied. There are many refractory elements whose atomisation by normal chemical methods requires conditions so vigorous that most covalently bonded substrate molecules would be destroyed. Carbon, silicon, germanium, and most metals are among the elements whose atomic reactions can be studied most conveniently by recoil methods although flash photolysis (Brown et al. 1969), discharge flow systems (Michael and Weston 1966, Lam et al. 1971, Gaspar et al. 1971) and monatomic vapor beams provide alternative techniques. Since recoil experiments provide only microscopic quantities of free atoms – usually 10^5 to 10^{15} atoms per experiment – their use in synthesis is limited to labelling with radioactive atoms. However, recoil experiments can serve as prototypes for synthetic sequences utilizing macroscopic amounts of free atoms produced by methods such as thermal evaporation (compare Mackay and Wolfgang 1962, Skell et al. 1967).

Most recoil studies have concerned themselves with mechanistic questions involving both physical and chemical phenomena. Nuclear recoil often produces atoms with such high kinetic energies (10^3 to 10^6 eV) that considerable energy must be lost before bond-making processes can occur. So-called 'hot-atom' reactions occur in an energy region considerably higher than the 1–2 eV region which is typical of thermal reaction threshold energies. In the energy region of hot-atom chemistry there is competition between non-reactive collisions in which energy is transferred from the recoiling atom, and reactive collision in which the recoiling atom is captured with the formation of a chemical bond or is altered in charge state by electron transfer.

It should be emphasized that recoil atom reactions are generally followed by separation of the stable products in which the recoil atoms are incorporated. The changes in the yields of the radioactive products with changes in reaction conditions are thus the primary data of recoil atom experiments. Direct kinetic measurements are precluded by the fact that instantaneous concentrations of the hot-atoms are often of the order of *one atom per liter*.

Since reaction probabilities for typical hot-atom reactions are vanishingly small both at very high and at thermal energies, an important long-term goal has been the determination of recoil reaction probabilities as a function of energy in the intermediate region. The mechanisms by which recoil reactions occur are also of interest. Ideally, a reaction mechanism would include knowledge of the electronic structure, geometry and energy changes which occur during a bondmaking or electron exchange collision.

In the case of a polyvalent recoil atom, a stable molecule normally cannot result from a single reactive collision of the free atom. Thus the chemically stable products observed are end-results of reaction sequences, each step of which involves unstable, unusual and interesting species as reactants and products. Thus for recoil atoms of high valency such as carbon, determination of the chemical identities of the intermediates is a challenge which must be met before more detailed elucidation of reaction mechanism can be achieved.

At present it is not possible in general to compute or measure reaction probabilities for recoil atoms directly as functions of energy. Nevertheless kinetic theories of hot-atom reactions have been developed which correlate product yields with reaction probabilities integrated over all energies, taking into account competing collisional energy loss. The refinement and testing of such kinetic theories is one means of increasing our insight into the nature of hot-atom reactions, and inasmuch as these kinetic theories have been most completely formulated for gaseous systems, we shall first consider these.

2. Scope of Gas-Phase Hot-Atom Chemistry

One may ask: "To what questions does gas-phase hot-atom chemistry address itself?"

Since hot-atom reaction systems, by virtue of their excess kinetic energy, are not constrained to the lowest lying valleys and passes of their potential energy surfaces, they can follow reaction pathways whose high threshold energies would render them unobservable in normal experiments. Therefore our study of hot-atom reactions must begin at a primitive starting point: namely, the simple identification of the reactions taking place. One can distinguish the mechanism, the energetics and the dynamics of the hot reaction.

Reaction mechanism or *systematics* refers to the identification of the reactants and products in each step of a reaction sequence which converts a recoil atom into a chemically stable species. The individual steps are the elementary or partial reactions which together comprise a classical reaction mechanism.

If one appeals to a classical potential surface, then the term *reaction energetics* would imply simply the mapping of the terrain, the determination of the depths of valleys and the heights of passes. The energetics of hot-atom reactions encompasses such questions as the variation of reaction probabilities as functions of kinetic and electronic energy of the recoiling atom, and the extent to which energy is interconverted among internal and external modes during the reactive collision. Excited products can undergo either collisional de-excitation or unimolecular decomposition.

Dynamics denotes the path of a specific reaction system over a potential surface. In the general case dynamics refers to a detailed description of the relative motion of reactants and products before, during and after reactive collisions.

We next ask, "How do gas-phase hot-atom experiments attack or answer the questions posed?"

Reaction mechanisms (systematics) are elucidated by the identification of reaction products and reactive intermediates formed from the recoiling atoms. Microscopic quantities of the chemically stable end products in which radioactive recoil atoms are incorporated are generally identified by comparison of their physical properties with those of macroscopic quantities of authentic samples. For low boiling liquid and gaseous products, the most convenient and widely used technique for quantitative estimation is radio-gas-liquid-partition chromatography (Wolfgang and Rowland 1958, Lee et al. 1962, Welch et al. 1967). Here a flow radiation counter is used as an auxiliary detector for a conventional vapor chromatograph, and the chromatographic retention times of the radioactive products detected by the flow counter are compared with the retention times of authentic samples detected by some form of mass detector. If authentic samples are unavailable, or if the identity of a particular product cannot be guessed, then variation of the product retention time with the character of the chromatographic stationary phase can be used to deduce the unknown structure.

The identification of reactive intermediates in multi-step reaction sequences is a difficult problem: often the structures of the reaction products provide some clues. For instance, the formation of next-higher homologs of various silanes in the reactions of recoiling silicon atoms has provided evidence for the intervention of the intermediate silylene: $^{31}SiH_2$ (Gaspar et al. 1966). Often intermediates are identified (or precluded) by the use of scavengers, species known to undergo efficient reactions with specific types of reactive intermediates. An example is the *exclusion* of $^{31}SiH_3$ as a reactive intermediate in the reactions of recoiling silicon atoms by the failure of nitric oxide NO, a molecule known to combine rapidly with free radicals of all kinds, to eliminate the formation of reaction products (Gaspar et al. 1968a).

Unfortunately, the reactivities of many suspected intermediates toward scavengers are unknown. Therefore identifications based on scavenger experiments may be rather insecure. A useful complement to recoil atom experiments is the preparation of suspected intermediates by chemical means. If similar products are formed, positive evidence is provided for the nature of the intermediate in the recoil experiment, while if different products appear, a given intermediate may usually be regarded as definitively excluded.

The energetics of recoil atom reactions have been investigated by experiments designed to determine the kinetic energy ranges in which recoiling atoms undergo reaction and the distribution of internal energies of product molecules. The presence of excess kinetic energy carried by the recoil atom at reaction can be demonstrated by moderator experiments. Addition of inert gas to a reaction mixture increases the ratio of non-reactive collisions (in which kinetic energy transfer de-excites the recoiling atom) to reactive collisions. A decrease in product yields with increased moderator concentration is taken as an indication of the occurrence of a 'hot' reaction, i.e. a reaction of a recoil atom at higher than thermal kinetic energies.

Kinetically excited atoms have been regarded by some workers as being indiscriminate in their reactions, and the occurrence of hot reactions has been diagnosed by the demonstration of 'indiscriminate reactivity', e.g. by the failure of low concentrations of certain substrates to act as scavengers. The present authors believe that it is not reasonable to regard kinetically excited atoms as being *necessarily* indiscriminate in their reactions: lack of selectivity appears to be a poor criterion for occurrence of hot reactions.

The question of the internal excitation of reaction products has been attacked by measurement of the effect of total pressure and therefore collision frequency on the suppression of unimolecular decomposition (Rabinovitch and Setser 1964, Maccoll 1969, Frey and Walsh 1969). If the standard enthalpy changes for the primary hot reaction and the secondary decomposition reactions are known, and if the entire decomposition mechanism can be experimentally characterized, then at least crude information on energy distributions of products can be gleaned from recoil experiments (Krohn et al. 1971). Information about reaction cross sections and reaction dynamics can, however, be extracted indirectly from gas-phase recoil atom experiments. The usual method is to employ a kinetic or dynamic model to calculate such average quantities as total reaction yields as a function of moderator pressure, the calculated quantities being such that they can be compared with experiment.

When parameters in the theoretical model such as the shape of the cross section versus energy curve are determined by reference to experiment, the value of the information on the parameters gained depends upon the quality of the model. Several models are presented in section 3. It is important to realize, however, that these models are tested experimentally only in forms which are integrated or averaged over a broad spectrum of reaction energies. Thus while it is possible in principle to invert such a quantity as the total yield given by the integrated cross section to obtain the cross section function itself, this has yet to be carried out. The use of atomic beams provides a much more direct technique for obtaining differential reaction cross sections (see e.g. Datz and Taylor 1970).

3. Theory and Experiments with Gas-Phase Systems Involving Hydrogen

We can now proceed to discuss the major contributions that have been made to the field of gas-phase hot-atom chemistry. The overwhelming majority of these experiments have not penetrated beyond the study of reaction mechanisms or systematics. Only for the simplest and most intensively investigated systems have we progressed to a knowledge of reaction energetics and dynamics. In hot-atom chemistry as in other branches of chemical kinetics the reaction of hydrogen atoms with molecular hydrogen has been singularly important as a meeting place for experiment and theory.

The first hot-atom experiments on the reactions of hydrogen atoms with hydrogen molecules were only reported in 1960 by Lee et al. These workers found an intermolecular isotope effect for the reactions of tritium atoms recoiling from the nuclear transformation $^3He(n,p)T$ in H_2–D_2 mixtures. The ratio of products HT to DT from an equimolar mixture of H_2 and D_2 was $1.55 \pm 0.06 : 1$.

Table 4.1
Exchange reactions of tritium atoms with isotopic variants of hydrogen molecules (see Jones 1949)

$T + H_2 \rightarrow TH + H$	$\Delta H_0 = -0.048$ eV
$T + D_2 \rightarrow TD + D$	$\Delta H_0 = -0.006$ eV
$T + HD \rightarrow TH + D$	$\Delta H_0 = -0.013$ eV
$T + HD \rightarrow TD + H$	$\Delta H_0 = -0.048$ eV

The isotope effect was later found by Seewald et al. (1966) to be 1.14 ± 0.06 in iodine scavenged mixtures of hydrogen and deuterium molecules. These workers also reported the changes in the hot reaction product yields caused by the presence of inert argon moderator in quaternary mixtures $^3He + Ar + H_2 + I_2$ and $^3He + Ar + D_2 + I_2$. While HT is favored over DT in reactions of recoiling tritium atoms with H_2–D_2 mixtures, DT is favored over HT in reactions of tritium atoms with HD. This isotope effect is $I_{HT/DT} = 0.62 \pm 0.06$ (Seewald and Wolfgang 1967). The effect of argon moderator on this reaction was also studied. We shall see that insight has been gained into the source of these isotope effects from trajectory calculations of the reaction cross sections (vide infra).

Root and Rowland, in experiments designed to test the validity of the elastic nonreactive collision model for recoil atoms, found that methane is a more effective moderator for recoiling tritium atoms than is deuterium (Root and Rowland 1963, Root et al. 1965). The greater efficiency of methane as a moderator for hot tritium atoms over hydrogen was attributed to the inadequacy of a hard sphere elastic collision model for nonreactive 'moderating' collisions in which kinetic energy is transferred from recoil atom to moderator.

The kinetic theory of hot-atom reactions developed by Estrup and Wolfgang (1960) has provided a basis for the interpretation of these experimental results, which may also be compared with predictions of the theory. The parameters extracted in averaged form from the experimental data by means of the Wolfgang kinetic theory and the isotopic effects directly measured by experiment have been compared with the results of complete calculations of reaction rates carried out by semi-classical trajectory methods (Wolfgang 1963).

The Estrup–Wolfgang theory of hot-atom reactions is a purely *kinetic*, as opposed to a mechanistic theory which assumes no model of the actual reaction mechanism. The Estrup–Wolfgang theory expresses experimental data in terms of parameters describing intrinsic reactivity and collisional energy loss. It may be regarded as an analog of the classical collision theory of reaction rates which summarizes experimental data on *thermal* reactions in terms of the phenomenological parameters 'activation energy' and 'steric factor' in the Arrhenius equation. Implicit in the Estrup–Wolfgang theory is the assumption of central forces. A hard-sphere model for nonreactive collisions was first assumed, and later Estrup published an extension of the theory to soft-sphere scattering.

The starting point for development of the kinetic theory is an expression for the total reaction probability as an integral, over the energy range in which a recoil reaction can take place, of the collisional probability for the reactants, times the reaction cross section at a given energy, times the collision density at a given energy. This expression was originally due to Miller and Dodson (1950, Miller et al. 1950)

$$P = \sum_j \int_{E_2}^{E_1} f_j P_j(E) n(E) \mathrm{d}E,$$

where P is the total probability for hot reaction, f_j is the probability for collision of the recoiling atom with component j of the reaction system, $P_j(E)$ is the probability for reaction of the recoiling atom with component j at energy E, and $n(E)\mathrm{d}E$ is the number of collisions made by a hot-atom between energies E and $E + \mathrm{d}E$, the collision density. The range of integration is the energy range in which the recoiling atom can react with component j and the summation is over the components of the reaction system, the chemical species present.

For a mixture of a single reactant and moderator one can write

$$-\frac{1}{\ln(1-P)} = \frac{\alpha_{\text{reactant}}}{I} + \frac{\alpha_{\text{moderator}}}{I}\left(\frac{1-f}{f}\right) \tag{1}$$

since

$$\alpha = f\alpha_{\text{reactant}} + (1-f)\alpha_{\text{moderator}}, \tag{2}$$

I is the reactivity integral, the cross section integrated over the logarithm of energy, and α is as defined below.

It is clear that in a series of moderator experiments, change in the moderator concentration changes the collision distribution function (collision density). But the reaction cross section and collision density are also related to each other since loss of recoil atoms by reaction affects the collision density. By assuming a convenient form for the collision distribution function a useful expression is derived for the total reaction probability in terms of an integrated reaction cross section and a parameter α which describes the average collisional energy loss and which can be calculated for an elastic collision model from the kinetic theory of gases.

In practice $-1/\ln(1-P)$ is plotted against $(1-f)/f$. The ratio of slope to intercept gives the quantity $\alpha_{\text{moderator}}/\alpha_{\text{reactant}}$. Since $\alpha_{\text{moderator}}$ can be calculated using a hard-sphere, elastic collision model, the ratio yields a value for α_{reactant} and the α for the reaction mixture is computed from eq. (2) above, the collision probability f being known. Alternatively α_{reactant} and I can be expressed in terms of $\alpha_{\text{moderator}}$ without resort to a calculation of $\alpha_{\text{moderator}}$.

Application of kinetic theory analysis to data from the systems $T + D_2$ and $T + H_2$ gave values $I_{D_2} = (7.1 \pm 0.7)\alpha_{Ar}$ and $I_{H_2} = (6.9 \pm 0.7)\alpha_{Ar}$ whose ratio is the isotope effect 1.03 ± 0.04 (Seewald et al. 1966). Wolfgang has also obtained values for I_{D_2} in terms of α_{D_2} and α_{CH_4} (Wolfgang 1963). For the system T + HD a kinetic analysis plot $-1/\ln(1-P)$ versus $(1-f)/f$ gives a good straight line yielding $I_{HD} = (7.8 \pm 0.8)\alpha_{Ar}$ and $\alpha_{HD} = (3.8 \pm 0.8)\alpha_{Ar}$. Plots of $P_j(\alpha/f)$ versus f/α were also linear, yielding as intercepts individual reactivity integrals for the products TH and TD, whose ratio is the observed isotope effect $I_{HT}/I_{DT} = 0.62 \pm 0.06$ (Seewald and Wolfgang 1967).

Root and Rowland applied the Estrup–Wolfgang kinetic theory to oxygen-scavenged mixtures of methane and D_2, both molecules acting as moderators and reactants (1967). The data were accommodated successfully by the theory, with the proviso that the greater moderating ability of CH_4 compared to D_2, $\alpha_{CH_4}/\alpha_{D_2} = 2.1 \pm 0.3$, indicates the inadequacy of the elastic hard sphere model for nonreactive collisions at high energies.

It should be noted that in all these reactions of tritium atoms with hydrogen the more exoergic process predominates (see table 4.1). Wolfgang points out however that the energy differences are too small to explain by themselves the observed isotope effects.

The assumption of the Wolfgang kinetic theory that nonreactive collisions of recoiling atoms may be treated by an elastic scattering model has been questioned (Rowland and Coulter 1964, Baer 1969, Thomsen 1968), and tested by means of Monte Carlo calculations involving computer simulation of recoil atom reactions. For systems of high reactivity deviations are expected from its predictions due to variation of the energy loss parameter α. Other assumptions of the Estrup–Wolfgang theory are that the initial energy of the recoil atom is sufficiently high that enough collisions have occurred above the energy range of chemical reactions to establish the energy distribution $n(E)\mathrm{d}E = -(\mathrm{d}E/\alpha E)$. It is also assumed that the hot reaction region is well above the thermal region and that the collision probability for a recoil atom with a given component of the reaction mixture does not vary with energy. Various numerical calculations have vindicated the internal consistency of the Estrup–Wolfgang theory (Baer and Amiel 1970, Malcolme-Lawes 1969).

The parameters extracted from experimental data on the reactions of tritium atoms with hydrogen molecules with the aid of the Estrup–Wolfgang theory have been compared with theoretical predictions of the reaction rates.

The theoretical prediction of reaction rates involves the calculation of potential surfaces for reaction systems and the attempt to relate actual reaction rates to these potential surfaces. The theoretical calculation of absolute reaction rates received serious attention as early as the 1930's (see Glasstone et al. 1941). Already in 1936 trajectories for the reaction system $H + H_2$ upon a potential surface were calculated (Hirschfelder et al.). These calculations were extremely tedious before the advent of the digital computer, since analytical solutions cannot be obtained for the three-body equations of motion. Recent calculations of absolute reaction rates using computers allow one to obtain numerical solutions for the classical equations of motion applied to a model of atoms and molecules colliding in space, subject to a quantum-mechanically calculated potential energy. Wall and coworkers studied the collinear collisions of $H + H_2 \rightarrow H_2 + H$ in 1958, using a London–Eyring–Polanyi potential function and assuming a constant ratio of coulombic to total energy for the reaction system (Wall et al. 1958, 1961). The apparent activation energy for the exchange reaction was found to vary with the internal vibrational energy of the hydrogen molecule. Later, noncollinear collisions were included and the effect of rotational energy on the reaction was examined. The reaction probability for the exchange reaction $H + H_2 \rightarrow H_2 + H$ was found to be inversely proportional to rotational energy and the activation energy was at a minimum for collinear collisions.

The effect of the form of the potential surface on the $H + H_2$ exchange reaction probability was later investigated (Wall and Porter 1963). Use of a potential surface with a single saddle point between reactant and product valleys led to the prediction of unit reaction probability for all energies from the reaction threshold, the effective activation energy, to a factor two to four times greater, the range depending on the position of the saddle point. The predicted reaction probability was zero for energies above and below this range. No long-lived collision complexes were predicted. Thus the direct reaction model was endorsed for the dynamics of the $H + H_2$ exchange reaction.

Weston has used a Sato potential surface (Sato 1955a, b) and the equilibrium formulation of transition state theory to calculate the $H + H_2$ exchange reaction rate (Weston 1959) for hydrogen atoms with a Boltzmann distribution of kinetic energies.

The first complete quasiclassical calculation of an atom–molecule exchange reaction rate for a realistic potential without restrictive approximations was carried out for the system $H + H_2$ by Karplus et al. (1964). The potential surface was obtained from a semi-empirical valence-bond calculation in which coulomb and exchange integrals were expressed in terms of empirical molecular electronic state energies in order to increase accuracy over that achievable with calculated matrix elements (Porter and Karplus 1964). Cross sections for the exchange reaction were evaluated by integration of the appropriate equations of motion for the reaction system. From the cross sections, rate constants were determined by averaging over the distribution of initial conditions for the hydrogen atom and hydrogen molecule. Analysis of the calculated trajectories yielded information about the details of reactive collisions including collision time, configuration of the collision complex in the neighborhood of the H_3 saddle point, and the dependence of the reaction probability on the impact parameter (the shortest distance between the hydrogen molecule and the velocity vector of the hydrogen atom). The total reaction cross section was calculated for various rotation–vibration states of the hydrogen molecule and for various relative velocities of the atom with respect to the molecule. Thus the dependence of the cross section on these parameters was determined. Integration of the cross section as a function of the rotational and vibrational quantum numbers of the hydrogen molecule and the relative velocity over the equilibrium distributions of these parameters gave the thermal reaction rate constant as 11×10^{11} cm^3 $mole^{-1}$ sec^{-1} at 1000 K compared to the experimental estimates of 11 or 22×10^{11} (Farkas and Farkas 1935, Boato et al. 1956). A plot of the logarithm of the calculated rate constant versus the reciprocal of temperature was linear and yielded an activation energy of 7.49 kcal $mole^{-1}$ in good agreement with experiment (Weston 1959). The familiar Arrhenius temperature dependence was *not* built into the calculation, and thus its occurrence lends confidence to the results.

Calculations on the reaction $D + H_2 \rightarrow DH + H$ indicated that the exchange takes place via a direct interaction (stripping) mechanism rather than via a compound state or long-lived collision complex. The reaction cross section was found to be a smoothly rising function of relative velocity, corresponding neither to a step function nor to a dependence on the kinetic energy along the line of centers connecting the incoming atom and the molecule (Karplus et al. 1965). The minimum calculated relative translational energy required to cross the potential barrier was found to be almost twice the difference between the barrier height and the zero-point vibrational energy of the hydrogen molecule. Thus it seems that not all of the zero point vibrational energy and relative translational energy is available for crossing of the barrier between the reactant and product region of the potential surface. At high relative kinetic energies the reaction cross section decreases as expected, finally approaching zero.

Comparison was made of the rate constant obtained from the trajectory calculations with that calculated using the classical equilibrium equations of absolute rate theory. The same potential surface was used for both calculations, with unit transmission coefficient assumed for the absolute rate theory calculation and experimental moments of inertia employed in the transition state calculation. The discrepancies in the rate constants – the collisional treatment gives higher values than the transition state calculation – are explained by pointing out the apparently greater amount of zero-point vibrational energy available to surmount the barrier in the collision treatment. Therefore the absolute rate theory treatment yields a higher activation energy than the collision treatment. The average H–H–H angle for reactive collisions was calculated to be much greater (ca. 160°) than for nonreactive collisions (ca. 95 to 125°). Thus the nearer a collision is to collinearity, the more effective it is likely to be in promoting the exchange reaction. The interaction (collision) time was calculated to be no more than that required for the atom to pass unimpeded past the molecule, indicating the absence of long-lived collision complexes in the $H + H_2$ system.

The theoretical calculations of the reaction rates so far described treat thermal reactions in which the initial conditions and kinetic energies are defined by the Boltzmann distribution for a given temperature. Theoretical studies of hot-atom reactions employ the reaction cross section functions calculated in these studies, but are complex since competition between reactive collisions and nonreactive collisions involving energy loss makes necessary the calculation of cross sections for nonreactive scattering as well as reaction cross sections.

Porter has developed a general theoretical treatment of hot-atom reactions (1967). From the classical laws of conservation of energy and momentum, Porter derived an expression for the final laboratory translational energy of a particle after a collision with a second particle as a function of the initial kinetic energy of the first particle, the velocity of the second particle, the motion of the center of mass of the two particles, and the internal energy difference between the initial and final states. Then the general form of the distribution function was derived for the final translational energy of a particle after a single collision with a second particle whose velocity distribution is Maxwellian. This is obtained in terms of reactive and nonreactive collision cross sections. The distribution function for a collision was then derived and used to obtain the collision density (the average number of collisions per unit energy for final and initial energies). The expression for the collision density as a function of the reaction probability per collision obtained was identical to that in the Estrup–Wolfgang kinetic theory. The collision density takes into account loss of the incident particle by reactive collisions. From the collision density the integral reaction probability (total reaction yield) is calculated as a function of the reaction probability (cross section). The equation for the integral reaction probability can be solved for reaction probability as a function of energy. Thus the reaction cross section can be evaluated from the measurement of hot reaction yield as a function of *the initial energy of the hot-atoms.*

In the Porter formulations the integral reaction probability, i.e. the *yield* of reaction product, is calculated in terms of the reaction probability as a function of energy and the collisional energy distribution function. The integral reaction probability equation has the following form:

$$A(E) \quad = \quad P(E) \quad + \quad [1 - P(E)] \int_0^{\infty} P(E,E')A(E')\mathrm{d}E'$$

integral reaction probability as a function of *initial* energy E	probability for reaction in initial collisions at energy E	probability for reaction in subsequent collisions at energies E' less than E

This equation can be rearranged to give an expression for the reaction cross section $P(E)$ in terms of the integral reaction probability $A(E)$ and the collisional energy distribution function $P(E,E')$. Thus if measurements of the yields of recoil atom reactions could be made as a function of the initial energy of the recoil atom, the reaction cross section could be calculated from the equation above, *if* the collisional energy distribution function could be obtained either experimentally or theoretically. This remains a hope for the future.

Although no direct determination of reaction cross section as a function of energy has yet been carried out from hot-atom experimental data using the analysis outlined above, comparisons of theory and experiment using integrated cross sections have recently been made.

Karplus et al. (1966) carried out a trajectory calculation on the cross sections for the reactions $T + H_2 \rightarrow TH + H$ and $T + D_2 \rightarrow TD + D$ as a function of the tritium atom energy over the energy range of interest to hot-atom chemistry, 0 to 80 eV. Not only were threshold energies and upper energy limits for hot reactions obtained but also the integrals of the reaction cross section over the logarithmic energy range were directly calculated. This latter quantity is related to the reactivity integrals of the Estrup–Wolfgang theory. The ratio of the theoretical reaction integrals is an intermolecular isotope effect which can be compared with that obtained experimentally. The isotope effect directly measured by the ratio of HT to DT yields from 1:1 H_2–D_2 mixtures can be shown to be equal to the ratio of the reactivity integrals of the Estrup–Wolfgang theory in the limit of high moderator concentrations. Wolfgang determined the experimental ratio to be $I_{H_2}/I_{D_2} = 1.15 \pm 0.04$ (Seewald et al. 1966). The theoretical calculation yielded an isotope effect of 1.37 (Karplus et al. 1966).

Trajectory calculations of reaction cross sections have been carried out for other reaction systems. Strictly inorganic examples are given below (Anlauf et al. 1967, Bunker 1970). Several organic systems have been treated, but these lie outside the province of this chapter.

$$Cl + H_2 \rightarrow HCl + H \qquad Br + H_2 \rightarrow HBr + H$$
$$I + H_2 \rightarrow HI + H \qquad I + I + H_2 \rightarrow I_2 + H_2$$

Recently Porter and Kunt have applied their integral equation methods to calculate the integral reaction probabilities for the exchange reactions of hot tritium atoms with hydrogen and deuterium molecules (1970). The reaction cross sections previously calculated by the trajectory method were employed (Karplus et al. 1964, 1965, 1966). Several models were tested for the nonreactive collision cross sections. Anisotropic nonreactive scattering models gave theoretical results in better agreement with experiment than did a hard-sphere isotropic model. The integral reaction probabilities were calculated for each model as a function of the tritium initial kinetic energy in reaction systems: pure hydrogen, hydrogen plus helium moderator, and hydrogen plus deuterium plus helium. The soft-sphere nonreactive scattering models led to higher integral reaction probabilities than did the hard-sphere model. The exchange reaction yield increases monotonically as the nonreactive scattering becomes more anisotropic. The integral reaction probability calculated by the integral equation method was compared with the yield calculated from the equations of the Estrup–Wolfgang kinetic theory using calculated reaction cross section functions. Agreement was quite good. The small difference between the two theories increases with decreasing mole fraction of moderator. The largest differences accompany the use of the most realistic soft-sphere model for nonreactive scattering. The intermolecular isotope effect calculated by Porter using the integral equation method is in excellent agreement with the hot-atom experiments of Lee et al. on H_2–D_2–^3He mixtures (1960), and with the photochemical experiments of Chou and Rowland (1967). Good agreement with experiment is *not* obtained for argon moderator at high moderator mole-fraction.

For all the theoretical effort, much insight remains to be gained into the causes of the intermolecular isotope effect in the exchange reaction of tritium atoms with H_2–D_2 mixtures. The cross-section calculations of Karplus et al. (1966) have indicated that the threshold energy is lower for the $T + D_2 \rightarrow DT + D$ reaction (0.54 eV) than for the $T + H_2 \rightarrow HT + H$ reaction (0.63 eV), but the upper boundary for the hot reaction is much higher for H_2 (76.0 eV) than for D_2 (54 eV), and the cross-section maximum is also higher for H_2 than D_2 (ca. 8.6 versus ca. 6.9 Å^2). No simple reason can be given, however, for the higher upper boundary for the hot-atom reaction with hydrogen. Lee et al. (1960) have pointed out that the fraction of the collisional kinetic energy available as internal energy, as determined by the conservation of energy and momentum, is greater for D_2 than for H_2. This might permit a larger isotope effect in the decomposition of the less excited H_2T collision complex than in the more highly excited D_2T collisions complex. That this is the cause of the intermolecular isotope effect has been rendered unlikely by the agreement of *all* calculations on the lack of a long-lived collision complex along the reaction coordinate. Theoretical calculations of hot-atom reaction rates have been most valuable in providing explicit functions for the variation of reaction cross section with energy, and in validating the fundamental ideas about recoil atom reactivity implicitly and explicitly included in the Estrup–Wolfgang kinetic theory.

Inversion of the integral equation for integral reaction probability to yield reaction cross sections has been promised and will be extremely useful when it has been achieved. Kuppermann has discussed this problem (1967). He pointed out the extraordinary accuracy needed in the numerical inversion of a Laplace transform, the process required to obtain reaction cross sections as a function of energy from knowledge of the *rate constant* as a function of temperature. To determine the cross section to $\pm 10\%$ accuracy, the rate constant and the temperature must be known to *seven significant figures*! Such accuracy problems should not be so severe in obtaining a reaction cross section from an integral reaction probability equation. Nevertheless the promise made in 1967 that reaction cross sections would soon be obtained from photochemical experiments on hydrogen atom reactions with hydrogen molecules, experiments in which a specified initial energy for the hydrogen atom can be directly controlled, remains to be fulfilled five years later.

Progress is being made on quantum mechanical as opposed to classical calculation of the hydrogen atom plus hydrogen molecules exchange reaction rate (Tang and Karplus 1968, Truhlar and Kuppermann 1970, Saxon and

Light 1971, Rankin and Light 1969). It is clear that greater accuracy can be obtained from a quantum mechanical treatment (Miller and Light 1971), but at the cost of even more elaborate computations.

4. Review by Element of Other Inorganic Gas-Phase Hot-Atom Chemistry

4.1. HYDROGEN (TRITIUM) WITH OTHER ELEMENTS

The discovery by Rowland and coworkers that the yield of HT from the reaction of energetic tritium atoms with carbon–hydrogen bonds correlates well with the bond dissociation energies of the bonds involved in hydrogen abstraction (Breckenridge et al. 1963, Root et al. 1965) has been exploited for the estimation of bond dissociation energies for nitrogen–hydrogen bonds.

The correlation of HT yields with nitrogen–hydrogen bond strengths in the reactions of tritium atoms with ammonia, methylamine and dimethylamine was found to be similar to that for hydrocarbons (Tominaga and Rowland 1968). It was concluded that the mechanism for hydrogen abstraction by hot tritium atoms is similar for C–H and N–H bonds, the mechanisms in both cases involving factors which are important in controlling the bond dissociation energy. Unfortunately, very few pertinent bond dissociation energies are known with a probable error less than the differences between the bonds being compared, and therefore such abstraction yield-bond strength correlations involve very large uncertainties.

Cetini and coworkers have examined the reaction of tritium atoms recoiling from the $^3He(n,p)T$ transformation with silane (1963, 1967). Hydrogen abstraction predominates over displacement in the ratio 68:28. A small amount of labelled disilane is also formed, but is eliminated by the presence of scavengers indicating that its source was a multiple displacement reaction:

$$T + SiH_4 \rightarrow \cdot SiTH_2 + H_2 \qquad \text{or}$$
$$T + SiH_4 \rightarrow :SiTH + H + H_2.$$

Moderator experiments with helium were analyzed with the aid of the Estrup–Wolfgang kinetic theory. The total reactivity integral (vide supra) for silane was found to be twice as large as that for methane (0.53 vs 0.27), indicating the higher reactivity of silane. The predominance of abstraction in the reactions of tritium atoms with silane was attributed to the lower bond strengths in silane compared with methane. Experiments with disilane have shown that silyl group displacement as well as hydrogen abstraction predominate over hydrogen replacement (Cetini et al. 1965). It should be pointed out that TH and SiH_3T

$$T + Si_2H_6 \rightarrow TH + SiH_3T + Si_2H_5T$$

can also arise from disilane by secondary decomposition of vibrationally excited primary displacement product:

$$T + Si_2H_6 \rightarrow Si_2H_5T^*$$

$$Si_2H_5T^* \xrightarrow{\text{stabilization}} Si_2H_5T$$
$$Si_2H_5T^* \xrightarrow{\text{dissociation}} HT,\ SiH_3T$$

Wilkin and Wolfgang examined the reactions of recoiling tritium atoms with methylsilane, trimethylsilane and tetramethylsilane (1968). The principal reactions were hydrogen abstraction, accounting for ca. 60% of the observed activity, and hydrogen replacement, accounting for 25 to 31% of the volatile radioactivity. Replacement of a methyl group occurred in only ca. 2% yield, while replacement of a silyl radical to give labelled methane occurs in ca. 5% yield. All products are obtained in higher yields from the methylsilanes than from ethane (Urch and Welch 1966), but most of the increase is due to hydrogen abstraction. The inefficiency of oxygen scavenging of thermal tritium atoms in the silane systems made it difficult to decide whether hydrogen abstraction was a hot or thermal process. It was noted that capture of the tritium atom by a methyl group is favored in all alkyl vs silyl substitutions in the methylsilanes. However, the ratio $CH_3T : Si(CH_3)_2HT$ from trimethylsilane is *smaller* than the ratio $CH_3T : C(CH_3)_2HT$ from isobutane. This is rationalized by pointing out the larger inertial moment of $(CH_3)_2SiH$- vs $(CH_3)_2CH$-, in accord with the Wolfgang 'golden rule' of hydrogen hot-atom chemistry. This hypothesis states that reactions which can only provide a strong bonding orbital for the incoming tritium atom by an intrinsically slow rotational–vibrational relaxation tend to be disfavored (Urch and Wolfgang 1961a, b, Wolfgang 1965a, b, Odum and Wolfgang 1963).

Rowland and coworkers have also investigated the reactions of recoiling tritium atoms with trimethylsilane and tetramethylsilane (Tominaga et al. 1969). They found lower yields of HT than were observed by Wilkin and Wolfgang and a marked scavenger effect due to the presence of oxygen. A lower yield of HT was obtained from tetramethylsilane than from neopentane (ca. 0.8:1), and only 15% more hydrogen abstraction was noted for trimethylsilane than for isobutane. This latter finding could nevertheless indicate that silicon–hydrogen bonds are twice as reactive toward abstraction as are carbon–hydrogen bonds.

Daniel and Tang have also studied the reactions of tritium atoms with dimethylsilane, trimethylsilane, and tetramethylsilane (1969). Product yields indicated that hydrogen abstraction is dominant over substitution reactions, in agreement with the findings of Wilkin and Wolfgang. In analogous silane–alkane competition experiments, tetramethylsilane was found to be 1.17 times as reactive as neopentane in tritium-for-hydrogen substitution, trimethylsilane was 1.26 times as reactive as isobutane, and dimethylsilane 1.55 times as reactive as propane. The silanes are also roughly three times as reactive in tritium-for-methyl substitution as are the alkanes. These trends were rationalized by bond strength differences. The lower tritium-for-silyl to tritium-for-methyl substitution ratios were explained by steric arguments, the relative sizes of carbon and silicon atomic radii, and the availability of 3d-electronic orbitals, as well as by the relative electropositivity of silicon vs. carbon.

4.2. CARBON

The reactions of carbon atoms have been studied with many gaseous and liquid substrates. Yang and Wolf studied the reactions of carbon-14 produced by the $^{14}N(n,p)^{14}C$ nuclear reaction with anhydrous ammonia in the gas phase (Yang and Wolf 1960). They observed that ^{14}C-methane was produced in high yield as the sole significant product, its yield being unaffected by moderators or scavengers. The authors pointed out that the concomitant radiation dose to the system was high, and that the question of the effect of this dose required investigation. To this end Cacace and Wolf (1965) studied the same system, but with carbon-11 (20 min half-life) produced by the $^{14}N(p,\alpha)^{11}C$ nuclear reaction. At high radiation dose methane was the only product, but at low radiation dose a mixture of methane (40%), methylamine (35%), and methyleneimine (20%) was observed. This study pointed out the need in hot-atom work to separate the primary hot-atom steps from the secondary radiolytic effects.

Wolfgang et al. (MacKay et al. 1961, 1962, Dubrin et al. 1964) have studied the reactions of carbon-11 with a series of inorganic gases (O_2, CO, CO_2, SO_2, N_2, N_2O, NO and NO_2). With oxygen, carbon monoxide, carbon dioxide, sulphur dioxide, and nitrogen dioxide, carbon-11 labelled carbon monoxide was the only observable product, even in liquid oxygen. The authors postulate that the carbon attack is an *end-on attack on the molecule rather than a π-complex formation.* Extended Hückel three-dimensional LCAO calculations have been performed on the system carbon atom plus oxygen molecule (Kaufman et al. 1969) and lead to a reasonable picture of the end-on attack. These calculations show that insertion into the oxygen–oxygen bond would not be expected.

Carbon atoms were observed to attack either end of molecules containing a nitrogen atom at one end. Thus from nitric oxide and nitrous oxide a mixture of $^{11}CN^-$ and ^{11}CO was formed while only cyanide was formed from nitrogen. Moderator studies in nitrous oxide showed that ^{11}CN formation was a hot-atom reaction, whereas ^{11}CO formation occurred with both hot and thermal carbon atoms.

Ache and Wolf (1965, 1966) have investigated more thoroughly the reactions of carbon-11 with nitrogen, and obtained the relative reactivities of carbon atoms in nitrogen–oxygen mixtures. They found that the carbon atom is nine times more reactive with oxygen than with nitrogen. These authors also studied nitrogen–hydrogen mixtures and observed mixtures of labelled methane and hydrogen cyanide formed from all mixtures studied.

In the liquid phase Stenström (1970, 1971) has studied the fate of carbon atoms in water as a function of radiation dose and scavenger concentration. At low radiation dose he observed $^{11}CO_2$ (23%), ^{11}CO (32%), $H^{11}COOH$ (17%), $H^{11}CHO$ (21%), $^{11}CH_3OH$ (6%) and $^{11}CH_4$(6%). At medium dose ranges $^{11}CO_2$ and $H^{11}COOH$ predominated, while at high dose rate only $^{11}CO_2$ (> 97%) was formed.

Using scavengers he determined the following yields of primary carbon-11 products:

$$^{11}CO_2 + {}^{11}COOH = 9\%; \qquad {}^{11}CO = 44.5\%;$$
$$H^{11}COOH + H^{11}C(OH)_2 = 16\%; \qquad H^{11}CHO + {}^{11}CH_2OH = 26\%;$$
$$^{11}CH_3OH + {}^{11}CH_3 = 4\%; \qquad {}^{11}CH_4 = 0.06\%.$$

Stenström suggests that the ^{11}CO results from a true hot reaction. This conclusion was based upon the non-nuclear work of Skell et al. (1967) who reacted carbon atoms, probably in excited singlet states, with water and observed no deoxygenation of water. Recently, however, Husain and Kirsch (1971) have measured the rate of deoxygenation of water in the gas-phase by ^3P carbon atoms and found it to have an upper limit $< 3.6 \times 10^{13}$ cm^3 $molecule^{-1}$ sec^{-1}, at 300 K.

4.3. SILICON

Studies of silicon recoil atom chemistry are emerging from the 'systematics' stage of the elucidation of mechanism (see section 2) and are beginning to yield information about the deposition of energy in reaction products.

When silicon atoms with ca. 10^5 eV recoil energy are formed by the fast-neutron induced $^{31}P(n,p)^{31}Si$ transformation in phosphine–silane gaseous mixtures the radioactive products are silane $^{31}SiH_4$, disilane $H_3{}^{31}SiSiH_3$, silylphosphine $H_3{}^{31}SiPH_2$ and trisilane $^{31}SiSi_2H_8$, with disilane formed in greatest yield (Gaspar et al. 1966, 1968a, 1972, Gaspar and Markusch 1970). The major products from mixtures of phosphine with higher silanes are the next higher silane homologs – $^{31}SiSi_2H_8$ from disilane, and *iso-* and *normal* $^{31}SiSi_3H_{10}$ from trisilane (Gaspar and Markusch 1970, Gaspar et al. 1972). Scavenger experiments have precluded $\cdot{}^{31}SiH_3$ as a dominant reaction intermediate (Gaspar et al. 1968, 1972). The product structures together with scavenger experiments indicate that $:{}^{31}SiH_2$ is the principal product-forming intermediate. The formation of trisilane from phosphine–silane mixtures suggests that a primary reaction of recoiling atoms may be direct insertion into silicon–hydrogen bonds (Gaspar and Markusch 1970, Gaspar et al. 1972). The decrease in the yields of higher silanes and the increase in the silane yield with decrease in total pressure of the reaction mixture indicate that higher silanes are formed in vibrationally excited states capable of unimolecular dissociation to monosilane and other products. The proposed mechanism for the reaction of silicon atoms from $^{31}P(n,p)^{31}Si$ in phosphine–silane mixtures is:

$^{31}Si + SiH_4 \rightarrow (H^{31}SiSiH_3)^*$ (an asterisk indicates a vibrationally excited molecule)

$$^{31}Si + PH_3 \rightarrow (H^{31}SiPH_2)^*$$
$$(H^{31}SiSiH_3)^* + M \rightarrow H^{31}SiSiH_3$$
$$(H^{31}SiSiH_3)^* \rightarrow H^{31}Si + SiH_3$$
$$(H^{31}SiPH_2)^* \rightarrow H^{31}Si + PH_2$$
$$H^{31}Si \rightarrow \rightarrow :^{31}SiH_2$$
$$H^{31}SiSiH_3 + SiH_4 \rightarrow (SiH_3{}^{31}SiH_2SiH_3)^*$$
$$(SiH_3{}^{31}SiH_2SiH_3)^* + M \rightarrow SiH_3{}^{31}SiH_2SiH_3$$
$$(SiH_3{}^{31}SiH_2SiH_3)^* \rightarrow SiH_2 + {}^{31}SiH_3SiH_3$$
$$(SiH_3{}^{31}SiH_2SiH_3)^* \rightarrow {}^{31}SiH_2 + SiH_3SiH_3$$
$$^{31}SiH_2 + SiH_4 \rightarrow (^{31}SiH_3SiH_3)^*$$
$$^{31}SiH_2 + PH_3 \rightarrow (^{31}SiH_3PH_2)^*$$
$$(^{31}SiH_3SiH_3)^* + M \rightarrow {}^{31}SiH_3SiH_3$$
$$(^{31}SiH_3PH_3)^* + M \rightarrow {}^{31}SiH_3PH_2$$
$$(^{31}SiH_3SiH_3)^* \rightarrow {}^{31}SiH_4 + SiH_2$$
$$(^{31}SiH_3PH_2)^* \rightarrow {}^{31}SiH_4 + PH$$

This mechanism may require revision in view of the possible importance of ionic species, e.g. Si^+ in the recoil chemistry of silicon.

Moderator experiments have thus far been equivocal as to the energy range in which the primary reactions of recoiling silicon atoms occur (Gaspar et al. 1972).

The $^{30}Si(n,\gamma)^{31}Si$ transformation has also been applied to the study of silicon atom recoil chemistry. Radioactive silane, disilane and trisilane are obtained from the thermal neutron irradiation of silane, the highest yield being of disilane (Gaspar et al. 1968b). Cetini and coworkers have examined the thermal neutron irradiation of silane, disilane, and trisilane (1969). Trisilane was the major radioactive product from disilane and disilane from silane, leading Cetini to propose $^{31}SiH_2$ as an important intermediate. Moderator experiments were interpreted as supporting the occurrence of hot as well as thermal reactions.

A single investigation of the reactions of recoiling silicon atoms in an organosilicon compound (tetramethylsilane) has been reported (Snediker and Miller 1968).

4.4. GERMANIUM

Recoiling germanium atoms of convenient half-life (82 min) are available from the $^{76}Ge(n,2n)^{75}Ge$ and $^{74}Ge(n,\gamma)^{75}Ge$ nuclear transformations. The former reaction has a threshold energy of 9.35 MeV and delivers ca. 4×10^4 eV recoil energy.

Fast-neutron irradiation of germane gives radioactive germane $^{75}GeH_4$, digermane $^{75}GeH_3GeH_3$ (major product) and trigermane $^{75}GeGe_2H_8$ (Gaspar et al. 1969, Gaspar and Frost 1971). Germylene $^{75}GeH_2$ is believed to be an important product-forming intermediate.

Insertion by germanium atoms into germanium–hydrogen *and* silicon–hydrogen bonds is indicated by the products formed from fast-neutron irradiation of germane–silane mixtures. Trigermane, disilylgermane $^{75}GeSi_2H_8$ and digermanylsilane $^{75}GeGeSiH_8$ are believed to result from two consecutive insertions by germanium atoms into, respectively, two germane, two silane, or one germane and one silane molecules. Radioactive germane, digermane and silylgermane $^{75}GeH_3SiH_3$ are also formed. The proposed mechanism is analogous to that presented for silicon atoms in phosphine–silane mixtures:

$$^{75}Ge + GeH_4 \rightarrow (H^{75}GeGeH_3)^*$$
$$^{75}Ge + SiH_4 \rightarrow (H^{75}GeSiH_3)^*$$
$$(H^{75}GeGeH_3)^* + M \rightarrow H^{75}GeGeH_3$$
$$(H^{75}GeSiH_3)^* + M \rightarrow H^{75}GeSiH_3$$
$$(H^{75}GeGeH_3)^* \rightarrow \rightarrow {}^{75}GeH_2$$
$$(H^{75}GeSiH_3)^* \rightarrow \rightarrow {}^{75}GeH_2$$
$$H^{75}GeGeH_3 + GeH_4 \rightarrow GeH_3{}^{75}GeH_2GeH_3$$
$$H^{75}GeGeH_3 + SiH_4 \rightarrow SiH_3{}^{75}GeH_2GeH_3$$
$$H^{75}GeSiH_3 + GeH_4 \rightarrow SiH_3{}^{75}GeH_2GeH_3$$
$$H^{75}GeSiH_3 + SiH_4 \rightarrow SiH_3{}^{75}GeH_2SiH_3$$
$$^{75}GeH_2 + GeH_4 \leftrightarrows ({}^{75}GeH_3GeH_3)^*$$
$$^{75}GeH_2 + SiH_4 \leftrightarrows ({}^{75}GeH_3SiH_3)^*$$
$$({}^{75}GeH_3GeH_3)^* + M \rightarrow {}^{75}GeH_3GeH_3$$
$$({}^{75}GeH_3SiH_3)^* + M \rightarrow {}^{75}GeH_3SiH_3$$
$$({}^{75}GeH_3GeH_3)^* \rightarrow {}^{75}GeH_4 + GeH_2$$
$$({}^{75}GeH_3SiH_3)^* \rightarrow {}^{75}GeH_4 + SiH_2$$

Moderator studies are equivocal about the importance of kinetic energy in the primary reactions of recoiling germanium atoms (Gaspar and Frost 1971). Thermal neutron irradiation of gaseous germane also yields radioactive germane, digermane, and trigermane (Gaspar and Frost, unpublished).

4.5. NITROGEN

Amiel and Yellin (1964) studied nitrogen atom reactions in nitrogen–oxygen mixtures using 7.35 sec half-lived nitrogen-16 and concluded that

$$^{16}N + O_2 \rightarrow {}^{16}NO + O$$

was the major reaction, with competition from the reaction

$$^{16}N + N_2 \rightarrow N^{16}N + N.$$

The production of ^{16}NO is somewhat surprising as this reaction has not been observed with nitrogen atoms produced by non-nuclear means (Brown and Winkler 1970). In studies in moderated oxygen–carbon monoxide mixtures (Welch and Ter-Pogossian 1968) producing nitrogen-13 by the $^{12}C(d,n)^{13}N$ reaction, all the nitrogen was observed in N_2, leading Welch and Ter-Pogossian to assume a very high rate constant for the $^{13}N + N_2 \rightarrow N^{13}N + N$ reaction.

Welch (1968), studying the reaction of nitrogen-13 in carbon dioxide containing a trace of nitrogen, observed all the volatile activity as either $N^{13}N$ or $N^{13}NO$ with $< 1\%$ appearing as ^{13}NO. This result has been confirmed by Stewart (1971), and both these authors postulate an excited species of molecular nitrogen as the precursor of $N^{13}NO$.

Statnick et al. (1969) produced ^{13}N from carbon deposited on the walls of the irradiation chamber. Only $N^{13}N$ and $C^{13}N^-$ were observed as products of reaction with molecular oxygen. They deduced a low reactivity to form ^{13}NO and suggested that the $C^{13}N^-$ was formed in the solid carbon and then diffused out.

Dubrin et al. (1966) studied the fate of nitrogen atoms induced by $^{14}N(\gamma,n)^{13}N$ reaction in nitric oxide and found only $N^{13}N$ (79%) and ^{13}NO (21%). They used NO as a scavenger in hydrocarbon systems and concluded that the reacting species were $N(^2D)$ and $N(^4S)$ with the $N(^4S)$ state scavenged by nitric oxide in all systems investigated.

The reactions of nitrogen in water have been studied employing the $^{16}O(n,p)^{16}N$ reaction. Schleiffer and Adloff (1964) deduced that the 7.35 sec isotope formed NH_4^+ (30%), NH_2OH (16%), NO_3^- (25%), NO_2^- (10%) and NO (10%). This result is very different from the products observed when nitrogen-13 atoms are introduced into solid lattices and dissolved. Upon dissolution no oxygen-containing product has been observed (Hudis 1960, Dostrovsky et al. 1961, Welch and Lifton 1971). The reaction of nitrogen-13 in water has been studied by suspending diamond dust in water (Straatmann and Welch 1973) and inducing the nitrogen-13 by the $^{12}C(d,n)^{13}N$ reaction. In this work only ammonia and $^{13}CN^-$ were observed, the $C^{13}N^-$ presumably being formed in the solid and diffusing out.

4.6. PHOSPHORUS

The reactions of recoiling phosphorus-32 atoms have been studied with PH_3, (Halmann 1964) PF_3 and PCl_3, (Stewart 1971, Halmann 1964, Halmann and Kugel 1965, Cann and Hein 1957, Setser et al. 1959, Henglein et al. 1963) in the gas phase; in all of these systems the labelled parent was the only volatile product observed with yields as high as 78% for PH_3 (Stewart 1971). In PH_3 the remaining activity was deposited on the walls as phosphorus oxyacids. Halmann and Kugel concluded that phosphorus was scavenging thermal phosphorus atoms. In PCl_3 however, Henglein et al. (1963) concluded that the labelled PCl_3 was formed by both hot and thermal atoms.

In the liquid systems of PCl_3 and PBr_3 high yields of labelled parent were obtained (> 90%), while in $POCl_3$ mixtures of labelled PCl_3 (57%) and $POCl_3$ (30%) were formed (Clark and Moser 1963). When radioactive phosphorus was formed by the $^{32}S(n,p)^{32}P$ reaction in S_2Cl_2, $PSCl_3$ was found to be the major product (Korshunov and Shafiev 1958).

The work with recoil phosphorus has to date been mainly qualitative, due to the difficulty in finding a suitable scavenger for thermal phosphorus atoms.

4.7. OXYGEN

Statnick et al. (1969) have shown that in nitrogen the reaction $^{15}O(^3P) + N_2 \rightarrow N^{15}O + N$ does not occur. They observed mainly $O^{15}O$ together with traces of $CO^{15}O$ and $N_2{}^{15}O$, the latter compounds being formed from the reaction of ^{15}O with impurities in the nitrogen. This work has been substantiated by the studies of Welch and Ter-Pogossian (1968) who showed that it was possible to form pure $O^{15}O$ in very high yield from the reactions of oxygen-15 with 'pure' nitrogen.

4.8. HALOGENS

Although much work has been done on the behaviour of halogen atoms in organic media, very little work has been carried out in inorganic gaseous and simple liquid systems. Welch et al. (1971) have investigated the reactivity of fluorine-18 formed by the $^{20}Ne(d,\alpha)^{18}F$ reaction with nitric oxide and observed very high yields of labelled nitrosyl fluoride.

The behavior of iodine-125 formed by the decay by electron capture of xenon-125 in xenon, hydrogen, oxygen and water vapor have been investigated (Kuzin et al. 1970b), and it was shown that in hydrogen and water vapor all the iodine was I^-, in xenon 80% was I_2 and 20% I^- while in oxygen the distribution obtained was I^- 80%, I_2 10%, and IO_3^- 10%. Other workers (Yajima et al. 1963a, b, 1965) studying the chemical form of fission iodine diffusing from uranium found a mixture of I^-, IO_3^-, and IO_4^- with the percentages varying. Lambrecht et al. (1971) studied the decay of xenon-123 (E.C. β^+) in chemically reactive inorganic gases. They found that in Cl_2 the yield of ^{123}ICl was $85 \pm 5\%$ and was independent of pressure over the range 10–760 torr, while in NOCl the yield was $60 \pm 5\%$ (90 torr pressure).

Addendum (December, 1976)

Several important reviews of gas-phase hot-atom chemistry have appeared recently (Urch 1972, 1975, Rowland 1972, IAEA Report 1975). Much theoretical effort has been directed toward a comparison of exact quantum mechanical (Light 1971) with quasiclassical trajectory (Bunker 1971, 1974, Porter 1974) calculations of reaction probabilities for the simple gaseous systems $H + H_2$ (Tang and Karplus 1971, Truhlar and Kuppermann 1971, 1972, Bowman and Kuppermann 1971, 1973a, b, Diestler et al. 1972, Saxon and Light 1972a, b, Careless and Hyatt 1972, Tyson et al. 1973; Truhlar et al. 1973, Schatz and Kuppermann 1973, Wolken and Karplus 1974, Kuppermann et al. 1974, Altenburger-Siczek and Light 1974, Malcolme-Lawes 1975, Kuppermann and Schatz 1975) and $F + H_2$ (Muckerman 1971, 1972a, b, Jaffe and Anderson 1971, 1972, Wilkins 1972, Blais and Truhlar 1973, Schatz et al. 1973, Bowman et al. 1974) and their isotopic variants. For $H + H_2$ at energies substantially above threshold, the quantum reaction probabilities oscillate around the corresponding classical ones (Bowman and Kuppermann 1971). Complex formation, for which little evidence has been found in trajectory calculations, has been found to contribute to the quantum oscillations of the reaction probability as a function of energy (Schatz and Kuppermann 1973). Recently the quantum mechanical calculations have been extended to three dimensions (Kuppermann and Schatz 1975). For the $F + H_2$ system the shape of the reaction probability versus relative kinetic energy curve is very different for the quantum mechanical and trajectory calculations (Schatz et al. 1973). This difference is more significant for the $F + H_2$ than for the $H + H_2$ systems. The $Cl + H_2(D_2,T_2)$ reaction systems have also been treated by exact quantum mechanical calculations (Persky and Baer 1974), which were found to be in good agreement with the predictions of transition state theory and with experimental results. Vibrational excitation in molecular hydrogen was found to greatly enhance reactivity toward Cl atoms.

Since quantum mechanical reaction probability calculations are both difficult and expensive for the complex systems of interest to experimental hot-atom chemists, trajectory calculations (Bunker 1971, Porter 1974), which can be adapted to the study of atomic collisions with polyatomic molecules, continue to be of great importance. In addition to the frequently studied $H + H_2$ and $F + H_2$ systems, trajectory studies of the systems $H + Br_2$, $H + I_2$ (Porter et al. 1975), $Br + H_2$, $I + H_2$ (Porter et al. 1973), $Br + I_2$, and $Cl + Br_2$ (Borne and Bunker 1971), have also been carried out, and the calculated scattering functions for the halogen atom–molecule exchange reactions have been compared with beams results. In the halogen atom–hydrogen molecule reactions the importance of the vibrational energy of the hydrogen molecule in facilitating reaction may be judged from the theoretical conclusion that the reaction is dynamically forbidden for the ground vibrational state (Porter et al. 1973).

Bond energy effects in hot-atom reactions have been investigated by trajectory calculations (Chapman and Suplinskas 1974). Bond energy effects are believed to be due to a correlation of bond energies with reaction barriers (Chapman et al. 1974). Porter has included quantum-mechanical oscillations in his model for nonreactive scattering (Adams and Porter 1973).

Among the most significant recent applications of trajectory calculations have been the study of an organic reaction $H + CH_4$ (Bunker and Pattengill 1970, Raff 1974, Valencich and Bunker 1973a, b). Walden inversion has been predicted to occur as a mechanism for displacement of a hydrogen atom, and is predicted to be quite significant at energies near threshold.

The simple hard-sphere collision model has been used to study isotope effects in hydrogen abstraction reactions of hot tritium and fluorine atoms (Malcolme-Lawes 1972a, 1974a, b), and a high degree of internal excitation has been predicted in the products of F atom substitution reactions. An even simpler stochastic computer model has been applied to the study of competition by scavengers in hot-atom reactions (Malcolme-Lawes 1972b).

Our understanding of the variation with electronic state of the reactivity of carbon atoms with hydrogen molecules has been expanded by an ab initio calculation of the potential surfaces (Blint and Newton 1975). Facile insertion forming CH_2 is predicted for thermal ^1D carbon atoms, with no reaction whatever for ^3P and ^1S carbon atoms at thermal energies. Energetic ^1D and ^3P carbon atoms can produce CH radicals on reaction with H_2, while translationally excited ^1S carbon atoms are predicted to cause collisional dissociation of H_2 or suffer collisional deactivation to the ^1D state.

A new Steady State Theory of hot-atom reactions has been formulated by Keizer (1972, 1973). Kinetic equations governing the time behavior of the hot-atom probability distribution as well as the concentration of hot-atoms are derived, based on Porter's stochastic theory of hot-atom reactions (Keizer 1972). The probability distribution can relax to a steady-state form before an appreciable number of hot-atoms have reacted, and if this is the case the steady-state distribution can be used to define hot-atom reaction rate constants that can be deduced from product yields of competing reactions. The time evolution of the hot-atom energy probability distribution has also been obtained directly from the Boltzmann equation (Keizer 1973). The steady-state distribution is defined in terms of a hot-atom temperature.

The Estrup–Wolfgang Kinetic Theory has been extended to the analysis of product yields from reactions of recoiling tritium atoms with mixtures of two components (Johnstone and Urch 1974).

Root and coworkers have successfully analyzed yield data from reactions of recoiling ^{18}F atoms in $H_2(D_2)$-perfluoropropylene mixtures in terms of energy-distribution-dependent rate constants which were found to be temperature dependent and to fit the Arrhenius equation (Grant and Root 1976). Since rate constants for the reactions of thermalized recoiling ^{18}F atoms in this system had been determined (Grant and Root 1974), and since the analysis does not lead to temperature-independent relative rates for any assumed fraction of thermal reaction, it was concluded that for the first time a true ambient temperature dependence had been found for a hot-atom reaction (Grant and Root 1976). From the temperature dependence of the relative rate constants the fraction of thermal reaction (ca. 0.2) was estimated. A relationship was also developed between the intermolecular kinetic isotope effects in the reaction $^{18}F + H_2$ vs D_2, and the average energy of the hot-atom at reaction (Feng et al. 1976). The same value 50 ± 10 eV was obtained regardless of whether it was postulated that the hot-atoms were monoenergetic, or instead had a steady-state distribution à la Keizer. Root has employed a simple model for hot-atom translational energy relaxation to calculate product yields in the $^{18}F + H_2(D_2) + C_3F_6$ reaction system as a function of both time and hot-atom temperature (Root 1975). The calculation indicated, in agreement with the conclusion from experiment, that 90% of the ^{18}F atoms reacted while hot. The calculation also supported Keizer's theory by predicting that substantial fractions of the yield of $H^{18}F(D^{18}F)$ in unmoderated hot-atom experiments may well arise from reactions within a high energy quasi-steady-state distribution.

Experimental work in inorganic gas-phase hot-atom chemistry is a very active field. In the reactions of photochemically produced hot D atoms (2.9 eV) with HBr, the ratio of DH to DBr formation, 0.7 to 1.5 (Su and White 1975), was compared with the 0.24 ratio predicted by a trajectory calculation (Su et al. 1975). The collisional dissociation of hot substitution products from T + HD has been estimated to be $70 \pm 12\%$ (Hawk and Moir 1974), employing the Amiel–Baer model (Alfassi et al. 1971). The possible role of hot hydrogen atom reactions in interstellar space and planetary atmospheres has evoked interesting speculation (Hong et al. 1974). The reactions of recoiling tritium atoms with ammonia have been carefully investigated (Castiglioni and Volpe 1975). The high substitution yield (> 40%) was rationalized by postulating attack at the unshared electrons of nitrogen in a quasi-inversion process.

The reactions of ^{11}C atoms with mixtures of nitrogen and hydrogen, ultimately producing $H^{11}CN$, $^{11}CO_2$, ^{11}CO and $^{11}CH_4$ with high specific activity have been optimized for large-scale production (Christman et al. 1975).

Reactions of silicon ions Si^+ with SiH_4 have been found which produce trisilane, also obtained in recoil systems, and the possible role of silicon ions in the chemistry of recoiling silicon atoms has been discussed (Stewart et al. 1973), but no definitive conclusions have been reached. It has been proposed that recoiling silicon atoms can abstract fluorine atoms from PF_3, leading to the formation of $^{31}SiF_2$ which, in contrast to thermally generated SiF_2 (Gaspar and Herold 1971), can be trapped in the gas phase (Tang et al. 1972, Zeck et al. 1975a). Tang and coworkers have claimed the detection of both singlet and triplet $^{31}SiH_2$ from the reactions of recoiling silicon atoms with PH_3 (Gennaro et al. 1973), and have employed a recoil experiment to deduce that the ground state of $^{31}SiH_2$ is a singlet (Zeck et al. 1974a, b). The role of silylene in the reactions of recoiling silicon atoms has however recently been questioned (Hwang 1976).

The reactions of recoiling germanium in germane, digermane and germane–silane mixtures have received further scrutiny (Gaspar and Frost 1973), and direct insertion into Si–H bonds has been observed for thermally evaporated (Conlin et al. 1975) as well as nucleogenic germanium atoms (Cohen and Gaspar, unpublished).

The reactions of recoiling ^{13}N atoms with the following substrates have been examined: N_2, O_2, O_3, N_2O, and NO (Stewart et al. 1974, Parks et al. 1975). The end-products detected were ^{13}NN, ^{13}NNO, and $^{13}NO_2$. The reaction of recoiling nitrogen atoms with water has been studied (Welch and Straatmann 1973, Gersberg et al. 1976) and at low radiation dose the products are $^{13}NH_4^+$ and $^{13}NO_3^-$ and at high radiation dose $^{13}NO_3^-$. Labelled nitrate produced in this manner has been used as a tracer for direct quantitative measurements of denitrification rates in soils (Gersberg et al. 1976).

The hydrogen abstraction reactions of recoiling ^{32}P atoms with PH_3 (Stewart and Hower 1972, Zeck et al. 1974, Gennaro and Tang 1974), and SiH_4 (Zeck et al. 1974, 1975b), have been contrasted with hydrogen abstraction from hydrocarbons (Zeck et al. 1974, 1975b). Fluorine atom abstraction by recoiling ^{32}P atoms from PF_3 and PF_5 has also been studied (Stewart and Hower 1972, Gennaro and Tang 1973, 1974).

Study of the reactions of recoiling sulfur atoms, while technically difficult, shows promise (Kremer and Spicer 1975) and even recoiling tellurium atoms have been captured in gas-phase reactions (Strickert et al. 1974).

The reactions of ^{18}F atoms (Grant and Root 1975) and F^+ ions (Lin et al. 1974), Wendell et al. 1975) with molecular hydrogen have been studied. At low energies, complex formation occurs in the reaction $F^+ + H_2(D_2) \rightarrow FH^+(FD^+) + H(D)$, while at higher energies a stripping mechanism operates. The bulk of the exothermicity is converted into vibrational energy of FD^+. No $H^+(D^+) + FH(FD)$ is formed. The rates of reaction of thermalized recoiling ^{18}F atoms with O_2, NO, N_2 and CO are all less than 0.01 the rate of addition of ^{18}F to acetylene (Milstein et al. 1974).

The reaction of fluorine-18 in inorganic systems has been studied to investigate methods of producing useful fluorinating agents (Lambrecht and Wolf 1973, Rowland et al. 1973) and $SF_3{}^{18}F$, $OSF^{18}F$, $SiF_3{}^{18}F$, $F^{18}F$ and $H^{18}F$ have been prepared.

Acknowledgement

This publication was prepared in part with financial support from the United States Atomic Energy Commission. This is technical report COO-1713-34/70.

CHAPTER 5

CHEMICAL EFFECTS OF NUCLEAR TRANSFORMATIONS. INORGANIC COMPOUNDS IN SOLUTION

Manny HILLMAN

Department of Energy and Environment, Brookhaven National Laboratory, Upton, NY, USA

Chemical Effects of Nuclear Transformations in Inorganic Systems
Edited by G. Harbottle and A.G. Maddock

Contents

1. Introduction

This chapter treats the chemical effects of nuclear reactions involving an atom contained in an inorganic compound in solution. 'Inorganic compounds' in this context include all in which C–H bonds are not of major importance in the recoil chemistry, but exclude organometallic compounds in which the metal (or metalloid) is bound to a carbon atom, which are discussed in ch. 13. For completeness, though, reactions of recoil carbon atoms to produce simple carbon compounds are included. Solutions are defined as requiring some undefined dilution. Pure liquids and mixtures of liquids are not included; however, pure water is included because of the interesting reactions of recoil carbon (Stenström 1970), nitrogen (Schleiffer and Adloff 1964) and oxygen (Blaser 1970a, b) atoms that occur on irradiation of water with protons or neutrons. Also included are *quasi* solutions, that is, systems where recoil reactions occur partly in solutions; e.g., colloids, finely divided suspensions, and compounds adsorbed on resins but still in equilibrium contact with the solvent.

Unfortunately, some interesting systems may have been omitted through oversight. Indices to 'Recoil processes' often do not reveal whether or not they occur in solution, and an exhaustive search of every article in the literature was not possible. The literature was reviewed through Nuclear Science Abstracts *30*, No. 7, and in Radiochimica Acta, through Vol. 19, No. 4. The compilation of J.-P. Adloff and the files of G. Harbottle were of special help in accumulating the bibliography. All of the references found for all of the systems are listed in table 5.1, which summarizes the extant literature on effects of nucleogenesis in solution.

2. Reactions in Solution

Most studies have been made on aqueous solutions. In a few, for various reasons, solvents such as benzene, acetone, chloroform, pyridine, acetonitrile, nitromethane, phenol, carbon tetrachloride, carbon disulfide, quinoline, ether, and dimethylsulfoxide were used.

A nuclear reaction occurring in an element which is part of a molecule has

Table 5.1
Effects of nucleogenesis in solutions

Product element	Target compound	Nuclear reaction	Solvent	References
Antimony	Te on anion exchanger	β-decay	HCl	Moskvin (1962)
Arsenic	AsO_4^{-3}	n,γ	water	Broda and Müller (1950), Libby (1940), Müller and Broda (1951), Süe (1948)
	AsO_2^-	n,γ	water	ibid.
	cacodylate	n,γ	water	Süe (1948)
	$AsCl_3$	n,γ	benzene	Halpern et al. (1964)
	$GeCl_4$	β-decay	benzene	ibid.
	H_2SeO_3	β-decay	water	Halpern (1959)
Bromine	BrO_3^-	n,γ	water	Bakker (1937), Campbell (1959b, 1960), Libby (1940), Saito et al. (1959)
	BrO_3^-	IT	water	Campbell (1959a, 1960), DeVault and Libby (1939, 1941)
	Br_2	IT	CCl_4, CS_2, ethanol	DeVault and Libby (1941)
	Br^-	n,γ	water	Süe and Melander (1947)
	Br(<V)	IT	water	Campbell (1959a)
	$Co(NH_3)_5Br^{+2}$	n,γ	water	Saito et al. (1960a)
	$M(NH_3)_5Br^{+2}$ [a]	IT	water	Adamson and Grunland (1951), Herr and Schmidt (1962)
	MBr_6^{-2} [b]	IT	water	ibid.
	SeO_4^{-2}	β-decay	water	Burgus et al. (1948)
	SeO_3^{-2}	β-decay	water	ibid.
Carbon	C° colloid	γ,n	water	Morinaga and Zaffarano (1954)
	NH_4NO_3	n,p	water	Norris and Snell (1947, 1948, 1949), Yankwich et al. (1946)
	$(NH_2)_2CO$	n,p	water	Yankwich et al. (1946)

	C_5H_5N	n,p	water	ibid.
	water	p,3p3n	water	Stenström (1970)
Cerium	La^{+3}	β-decay	water	Burgus et al. (1948)
Chlorine	ClO_4^-	n,γ	water	Libby (1940)
	ClO_3^-	n,γ	water	Bakker (1937), Libby (1940)
Chromium	Cr(VI) on resin	n,γ	water	Matsuura (1968)
	Cr(VI)	n,γ	water	Fishman and Harbottle (1954), Ivanoff and Haissinsky (1956)
	$Cr(NCS)_6^{-3}$	n,γ	water	Kaufman (1960)
	$Cr(C_2O_4)_3^{-3}$ on resin	n,γ	water	Matsuura et al. (1965), Matsuura and Hashimoto (1966), Sensui and Matsuura (1965)
	$Cr(H_2O)_6^{+3}$ on resin	n,γ	water	Matsuura and Sasaki (1966, 1967a, b)
	CrL_3 [c]	n,γ	benzene	Tominaga and Nishi (1972a, b), Tominaga (1973a, b)
	MnO_4^-	β-decay	water	Burgus and Kennedy (1950)
	Mn^{+2}	β-decay	water	Burgus (1948), Burgus and Kennedy (1950)
Cobalt	$Co(C_2O_4)_3^{-3}$ on resin	n,γ	water	Matsuura and Hashimoto (1966)
	$Co(C_2O_4)_3^{-3}$	IT	water	Hoffman and Martin (1952)
	$Co(NO_2)_6^{-3}$ on resin	n,γ	water	Matsuura and Hashimoto (1966)
	$Co(CN)_6^{-3}$	n,γ	water	ibid.
	Co anion complex on resin	EC	water	Ablesimov and Bondarevskii (1973a, b)
	$Co(NH_3)_m(NO_2)_n^{+p}$ [d]	n,γ	solvent [e]	Ambe et al. (1972a, b)
	$Co(L)_n^{+3}$ [f]	n,γ	water	Süe and Kayas (1948)
	$Co(L)_n^{+3}$ [g] on resin	n,γ	water	Matsuura (1965, 1966, 1967)
	$Co(L)_3$ [c]	n,γ	benzene	Tominaga (1973a), Tominaga and Nishi (1972b), Tominaga et al. (1974)

Table 5.1 (contd.)
Effects of nucleogenesis in solutions

Product element	Target compound	Nuclear reaction	Solvent	References
	$Co(L)_3$ [c]	n,γ	solvent [h]	Tominaga et al. (1971a, b)
	$Co(L)_3$ [k]	n,γ	benzene	Tominaga and Sakai (1972)
	$Co(L)_3$ [c]	n,γ	solvent [l]	Tominaga (1973b)
	Ni(II)L [m]	EC,β-decay	water	Omori et al. (1970a, b)
Copper	Cu L [m]	n,γ	pyridine	Duffield and Calvin (1946)
	Cu L [m]	γ,n	pyridine	Holmes and McCallum (1950)
Gold	Au° colloid	n,γ	water	Majer (1937)
Hafnium	Hf L [p]	n,γ	dimethylsulfoxide	Hillman et al. (1968)
Holmium	Ho L [p]	IT	quinoline-ether	Stenström and Jung (1965)
	Er L [p]	β-decay	quinoline-ether	ibid.
Indium	In oxinate	IT	$CHCl_3$	Goldsmith and Bleuler (1950)
Iodine	IO_4^-	n,γ	water	Cleary et al. (1952)
	IO_3^-	n,γ	water	Bakker (1937), Cleary et al. (1952), Libby (1940)
	TeO_4^{-2}	β-decay	water	Bertet et al. (1964), Bertet and Muxart (1965), Burgus and Davies (1951), Davies (1948), Gordon (1967), Hashimoto et al. (1970)
	TeO_3^{-2}	β-decay	water	ibid.
	UO_2^{+2}	n,f	water	Burgus et al. (1948), Burgus and Davies (1951)
Iridium	$IrCl_6^{-2}$	n,γ	water	Bell and Herr (1966)

Lanthanum	BaF_2	β-decay	water	Vasudev and Jones (1972)
Lutetium	Yb EDTA	β-decay	water	Glentworth and Betts (1961)
Manganese	MnO_4^-	n,γ	water	Bakker (1937), Broda (1948), Cogneau et al. (1968), Davidenko and Kucher (1957), Dodson et al. (1946), Edge (1956), Erber et al. (1950), Libby (1940), Rieder (1951)
	MnO_2 colloid	n,γ	water	Nesmeyanov et al. (1959)
Neptunium	UO_2 L [m]	n,γ β-decay	pyridine	Melander (1947)
	UO_2^{+2}	β-decay	water, ether, TBP in synthine	Peretrukhin et al. (1967a, b)
Nickel	Ni(II)LL′ [q]	n,γ	solvent [r]	Ndiokwere and Elias (1973)
	Ni(II) L [s]	n,γ	solvent [r]	ibid.
	C_5H_5NiNO	n,γ	solvent [r]	ibid.
Nitrogen	water	n,p	water	Schleiffer and Adloff (1964)
Osmium	ReO_4^-	β-decay	water	Sato et al. (1966)
Oxygen	water	n,γ	water	Blaser (1970a, b)
Phosphorus	Graham salt	n,γ	water	Matsuura et al. (1969)
	$P_3O_{10}^{-5}$	n,γ	water	Matsuura et al. (1969), Matsuura and Lin (1971)
	$P_2O_7^{-4}$	n,γ	water	Matsuura et al. (1969)
	H_3PO_4,$H_2PO_4^-$, HPO_4^{-2},PO_4^{-3}	n,γ	water	Libby (1940)
	PO_4^{-3}	n,γ	water	Fenger and Pagsberg (1973)
	P	n,γ	tetralin	Pauly and Süe (1955, 1957)
	$S_2O_8^{-2}$	n,p	water	LoMoro and Frediani (1966)

Table 5.1 (contd.)
Effects of nucleogenesis in solutions

Product element	Target compound	Nuclear reaction	Solvent	References
	$S_2O_6^{-2}$	n,p	water	LoMoro and Frediani (1967)
	$S_2O_5^{-2}$	n,p	water	ibid.
	$S_2O_3^{-2}$	n,p	water	ibid.
	SO_4^{-2}	n,p	water	LoMoro and Frediani (1966)
	SO_3^{-2}	n,p	water	LoMoro and Frediani (1967)
	CNS^-	n,p	water	ibid.
	$(C_2H_5)_2NCSS^-$	n,p	water	LoMoro and Frediani (1969)
	$(NH_2)_2CS$	n,p	water	LoMoro and Frediani (1967, 1969)
	$CH_3NH(CS)NH_2$	n,p	water	LoMoro and Frediani (1969)
	$CH_3(CS)NH_2$	n,p	water	ibid.
	$HSCH_2COOH$	n,p	water	ibid.
	L-cysteine	n,p	water	ibid.
	S (finely divided)	n,p	water	Pauly (1955), Pauly and Süe (1955, 1957)
Platinum	$(Pt(en)_2)^{+2}$	n,γ	water	Haldar (1954)
Praseodymium	Ce AcAc	β-decay	water	Edwards and Coryell (1948)
	Ce DCTA	β-decay	water	Glentworth and Wiseall (1965), Glentworth and Wright (1969), Shiokawa et al. (1970)
	Ce DTPA	β-decay	water	Glentworth and Wiseall (1965), Glentworth and Wright (1969), Shiokawa and Omori (1969)
	Ce EDTA	β-decay	water	Cendales et al. (1965), Glentworth and Wiseall (1965), Shiokawa et al. (1965), Shiokawa and Omori (1965)
	Ce HEDTA	β-decay	water	Glentworth and Wright (1969)
Radium	$Th(OH)_4$ colloid	α-decay	water	Beydon and Gratot (1968)
	Ra^{+2}	α-decay	water	Flohr and Appleman (1968), Haseltine and Moser (1967)

Rhenium	ReO_4^-	n,γ	water	Schweitzer and Wilhelm (1956)
	nitron perrhenate	n,γ	pyridine	ibid.
	$ReCl_6^{-2}$	n,γ	water	Schweitzer and Wilhelm (1956), Herr (1952a, b, d)
	$ReCl_3$	n,γ	water	Herr (1952a, b, d)
Selenium	SeO_4^{-2}	IT	water	Langsdorf and Segre (1940)
	SeO_3^{-2}	n,γ	water	Apers et al. (1957)
	$UO_4 \cdot 4H_2O$	n,f	water	Stanley and Davies (1951)
Silver	Ag° colloid	n,γ	water	Parker (1962)
Sulfur	$S_2O_3^{-2}$	n,γ	water	Meyer (1971)
	SO_4^{-2}	n,γ	water	ibid.
	SO_3^{-2}	n,γ	water	ibid.
	S^{-2}	n,γ	water	ibid.
	S° colloid	n,γ	water	Pauly (1955), Pauly and Süe (1957)
	ClO_4^-	n,p	water	Meyer and Adloff (1967)
	ClO_3^-	n,p	water	ibid.
	Cl^-	n,p	water	ibid.
	Cl^-	n,p	methanol-water	Meyer (1970a)
Technetium	Mo blue hemicolloid	β-decay	water	Gratot and Beydon (1970)
Tellurium	TeO_4^{-2}	n,γ	water	Bertet and Muxart (1965)
	TeO_4^{-2}	IT	water	Adloff and Bacher (1962), Bertet and Muxart (1965), Burgus et al. (1948), Hahn (1963, 1964), Hillman and Weiss (1966a, 1971), Kirin et al. (1968), Murin et al. (1961d), Seaborg and Kennedy (1939), Seaborg et al. (1939, 1940a, b), Williams (1948a)
	TeO_3^{-2}	n,γ	water	Chanut and Muxart (1962)
	TeO_3^{-2}	IT	water	Williams (1948a), Burgus et al. (1948), Hahn (1964), Hillman and Weiss (1971)

Table 5.1 (contd.)
Effects of nucleogenesis in solutions

Product element	Target compound	Nuclear reaction	Solvent	References
	TeO_3^{-2}	IT	butanol	Dobici and Salvetti (1964)
	$(CH_3)_2Te(NO_3)_2$	IT	water	Murin et al. (1961d)
	UO_4	n,f	water	Stanley and Davies (1951)
	IO_4^-	β-decay,EC	water	Hillman et al. (1973)
	IO_3^-	β-decay,EC	water	ibid.
	I^-	β-decay,EC	water	ibid.
Thallium	Tl^+	n,γ	water	Frediani and LoMoro (1969)
	Bi^{+3}	α-decay	water	Aten et al. (1960)
Thorium	UO_2 L [n]	α-decay	pyridine	Govaerts and Jordan (1950)
	UO_2 L [u]	α-decay	acetone	Haissinsky and Cottin (1948)
	UO_2 L [v]	α-decay	acetone	ibid.
Thulium	ErL [t]	β-decay	water	Asano et al. (1974)
Tin	Sn^{+2}	n,γ	water	Dehmer and Wahl (1969)
	Sn^{+4}	n,γ	water	ibid.
	U(IV)	n,f	water	Brown and Wahl (1967)
	UO_2^{+2}	n,f	water	Brown and Wahl (1967), Erdal et al. (1969), Lin and Wahl (1973)
Uranium	UO_2L [n]	n,γ	pyridine	Govaerts and Jordan (1950)
	$UO_2(CH_3COO)_3^-$	n,γ	water	Irvine (1939)
	UO_2 L [u]	n,γ	benzene	Starke (1942b)
	U(IV)	n,γ	water	Saito and Sekine (1958)
	U(VI)	n,γ	water	ibid.

Xenon	XeO_3	n,γ	water	Heitz and Cassou (1969)
	XeO_3	IT	water	Heitz and Cassou (1970), Nefedov et al. (1966c)
	XeF_2	IT	CH_3CN	Zaitsev et al. (1968)
	IO_6^{-5}	β-decay	water	Gusev et al. (1967a, b, c)
	IO_4^-	β-decay	water	Gusev et al. (1967a, b, c), Murin et al. (1966)
	IO_3^-	β-decay	water	Kirin et al. (1966), Murin et al. (1966)
	I^-	β-decay	water	Murin et al. (1966)
	$C_6H_5IO_2$	β-decay	water	Nefedov et al. (1966d)
	$(C_6H_5)_2I^+$	β-decay	water	Nefedov et al. (1967)
Ytterbium	Yb L [t]	β-decay	water	Asano et al. (1974)
Yttrium	Nb° Zr° powder	660 MeV protons	solvent [w]	Zaitseva et al. (1974)
	Y L, Zr L [c]	660 MeV protons	solvent [e]	Zaitseva et al. (1973)
Zirconium	Zr L [p]	n,γ	dimethyl-sulfoxide	Hillman et al. (1968)
	Zr°, Nb° powder	660 MeV protons	solvent [w]	Zaitseva et al. (1974)
	Y L, Zr L [c]	660 MeV protons	solvent [e]	Zaitseva et al. (1973)

[a]M=Co, Ir, Pt. [b]M = Ir, Re, Os, Pt. [c]acetylacetone. [d]m=6, n=0, p=0; m=5, n=1, p=1; m=4, n=2, p=2 (cis and trans). [e]not specified in abstract. [f]L=NH_3,n=6; L=ethylenediamine,n=3; L=diethylenetriamine,n=2. [g]L=NH_3,n=6; L=ethylenediamine,n=3. [h]benzene, ethanol, acetone, acetic acid. [k]L=NH_3, ethylenediamine, ethylenediaminetetraacetic acid. [l]saturated solution in benzene or acetic acid in equilibrium with solid. [m]nitrosonaphtholato. [n]L=salicylaldehyde-o-phenylenediimine. [p]L=phthalocyanine. [q]L=acetylacetone, L′=ethylenediamine. [r]Solvent = methylcellosolve, dimethylcellosolve, dimethylformamide, benzene, pyridine, n-hexane, or alcohol. [s]L=haematoporphyrin dimethyl ester. [t]L=1,2-cyclohexanediamine-N,N,N′,N′-tetraacetic acid. [u]L=benzoylacetone, [v]L=dibenzoylmethane. [w]acetylacetone, benzoylacetone, trifluoroacetylacetone, or hexafluoroacetylacetone.

consequences which are often quite disruptive to the molecule, whether it occurs by virtue of the recoil energy imparted to the element as a consequence of momentum conservation (chs. 2 and 3) or whether it is because of Coulomb repulsion caused by charge buildup on the molecule as a consequence of Auger electron ejection (chs. 2, 3 and 15).

When nuclear reactions are carried out on molecules already in solution, the products observed may be: (a) the result of non-breaking of bonds as a result of too low a recoil energy; (b) the rapid reformation of bonds of species held together for a sufficient time by the solvent cage; (c) the interaction of the ions, radicals and molecules formed in the local hot zone; and (d) the interaction of activated species with the bulk of the solvent after diffusion out of the local hot zone. The separation of (a) and (b) is experimentally extremely difficult and has not been achieved satisfactorily to date. Experiments in radiation chemistry have successfully separated (c) and (d), but very few attempts have been made to do this in parallel fashion in investigations of chemical effects of nuclear transformations. One example of the latter is the investigation by Stenström (1970) of the interaction with water of the carbon atom species formed by the reaction of high-energy protons with water.

Almost all investigations have involved two themes: (a) to what extent is the parent compound retained following the nuclear reaction, and (b) qualitatively and quantitatively what are the products formed from the reaction of the active intermediates with the solvent? This second theme leads to the important question: what are the intermediates? An overlap exists between both themes in those cases where the parent compound or ion can be reformed by interaction of the active intermediates with the solvent.

It has long been considered that intrinsic retention, (i.e. non-breaking of bonds), is negligible except in low-energy isomeric transitions which are not accompanied by emission of Auger electrons as illustrated by ^{69m}Zn as diethylzinc (Seaborg and Kennedy 1939). In the case of (n,γ) reactions in relatively heavy elements, and especially where the spin change of the nuclear reaction is high, momentum cancelling may well give a recoil energy distribution with a significant probability of energies low enough to account for a non-negligible intrinsic retention (Hillman et al. 1968).

The beta decay of rare earth complexes has been extensively studied: these compounds seem to be excellent candidates for detecting intrinsic retention and fast recombination reactions because of the low energy of the transitions in some cases and the low abundance of Auger electron emission. The examples investigated are, however, complicated by exchange and hydrolysis reactions of the starting compound and the products. The principal studies were by Shiokawa et al. (1965, 1969a, b, 1970) and Glentworth et al. (1961, 1965, 1969) who found that decomposition of the complexes of $^{143,144}Ce$ or of ^{177}Yb was not due exclusively to momentum recoil but rather mainly to a pH-dependent decomposition of a charged complex formed as a result of the beta decay. Bond rupture following beta decay was much lower in the DTPA than in the EDTA and DCTA complexes: this was attributed to the octadentate structure of the first compared to hexadentate in the latter two. Differences in the results between ^{143}Ce and ^{144}Ce labelled complexes were explainable by differing probability of internal conversion and consequently different probability of charging of the activated complex.

The beta decay of ^{131}I (maximum recoil energy of 2 eV) in various ions and compounds has been studied extensively by a Russian group (Gusev et al. 1967a, b, c, Nefedov et al. 1966d, 1967, Kirin et al. 1968, and Murin et al. 1966). This decay leads to the formation of ^{131}Xe which under the different experimental conditions of the reaction may be found either as the oxidized form, XeO_3, or as Xe gas. It was found that XeO_3 was formed in higher yields from iodine compounds in which the iodine atom was already bound to oxygen, e.g. IO_3^-, IO_4^-, iodoxybenzene, $M_3H_2IO_6$, where M = H,Li,Na, and NH_4IO_4. Under neutral conditions, I^- and diphenyliodonium cation gave negligible yields of XeO_3, while in the presence of perchloric acid, the latter compound gave a 40% yield of XeO_3. (Xenon itself is not oxidized under those conditions.) Unfortunately, XeO_3 is reduced by I^-. However, the reducing properties of diphenyliodonium and the oxidizing properties of IO_3^- and of IO_4^- are apparently unknown. Further work is necessary to ascertain whether unless oxygen is bound to the iodine atom in the parent compound, the excited xenon species formed cannot interact in water to form XeO_3.

In complex molecules, and in condensed media, even reactions accompanied by Auger emission may result in non-breaking of the bonds of the element in question. In complex molecules, the net effect may be loss of electrons from elsewhere in the molecule (Gordon 1967), breaking other bonds, and in solutions, especially in polar solvents, the emitted Auger electrons may be rapidly replaced by others from the solvent. Hillman and Weiss (1971) have found evidence for this in an isomeric transition that is 100% converted. Thus the yield of $^{127,129}Te$ as tellurium(VI) is higher when starting with $^{127m,129m}Te$ as tellurium(VI) than as tellurium(IV). This indicates at least a partial 'memory' of the product for the number of oxygen atoms in the parent. Similar results were observed (Hillman et al. 1973) in the 91% electron capture decay of ^{121}I to ^{121}Te when starting with iodide, iodate and periodate. If complete disruption of the molecule always resulted from the emission of Auger electrons as in the gas phase, then no 'memory' effect would be possible. On the other hand, Heitz and Cassou (1970) summarized results suggesting an exact correlation between percent of bond breakage and percent of isomeric transitions accompanied by Auger electrons. This correlation was independent of the physical state of the compound. Finally, Adamson and Grunland (1951) suggested that the degree of disruption induced by an isomeric transition is dependent on the charge of the parent molecule, with disruptions inhibited by a negative charge.

In almost all of the other types of nuclear transformations studied, [e.g. (n,γ), (n,f) and α-decay] retention was not as important a consideration as the nature of the chemical species produced by the nuclear reaction and their interaction with the medium. In general, the results may be summarized by noting that in the presence of oxidizing agents oxidized products are produced in greater yield, while in the presence of reducing agents (methanol was frequently used) reduced products are produced in greater yields: the parent compound may in some cases act as the reducing or oxidizing agent, complicating the interpretation. Unfortunately, another complication that has been overlooked by most investigators is the oxidizing influence of oxygen from the air. In only a few reports is there an attempt made to eliminate air from the experimental medium. Hillman and Weiss (1971, Hillman et al. 1973) specifically investigated the influence of air on the distribution of products obtained in the Te system.

Despite the more than thirty years of research that have gone into solution studies, it is still not possible to interpret the observations with any greater generality. The reaction mechanisms are very complicated and the details are very elusive. Most of the research appears to have been performed with the narrower interest of specific systems in mind rather than in an attempt to elucidate general behavior. Experimental conditions have therefore been extremely varied and difficult if not impossible to correlate. To gain deeper insight it will probably be necessary, unfortunately, to repeat much of the work already done with much greater attention paid to details which may relate one type of system to another. To help, new techniques are now being tried. For example, Sato et al. (1966) suggested that in systems involving β-decay followed by coincident gamma transitions, information about the initially formed species in solution may be extracted by the measurement of perturbed angular correlation.

Libby (1940) pioneered the investigations of the effects of (n,γ) reactions of oxyanions in solution with studies of permanganate, orthophosphate, arsenate, arsenite, perchlorate, chlorate, bromate, and iodate. He interpreted his results in terms of successive oxygen atom or oxide ion ejection from the parent ion following the nuclear reaction.

Only a little additional work has been done on permanganate reactions (see, for example, ch. 6). Cogneau et al. (1968) found no pH dependence of retention in conflict with Libby (1940), and Erber et al. (1950) and Rieder (1951) found that retention increases with temperature and concentration of permanganate: the latter results may be attributed to oxidation of intermediates by permanganate. The (n,γ) induced decomposition of permanganate to manganese dioxide has found application as a neutron monitor (cf. Davidenko and Kucher 1957, Dodson et al. 1946, and Edge 1956). The neutron flux is determined by measuring the ^{56}Mn activity in the manganese dioxide.

Matsuura et al. (1969) have extensively studied the effects of (n,γ) reactions on phosphates and have found that the primary radioactive product is phosphate, independent of whether the starting anion is pyrophosphate, tripolyphosphate, or a higher polyphosphate. In those instances where radioactivity was found in the polymers, the activity was uniformly distributed among the phosphorus atoms. Lo Moro and Frediani (1966, 1967, 1969) obtained phosphate and reduced phosphorus anions from the (n,p) reaction with a variety of sulfur-containing anions and organic compounds. The formation of the reduced species was correlated with the lability of the parent compound toward radiolysis, that is the ability of the original molecule to scavenge the oxidizing agents formed in solution as a result of the radiation.

Müller and Broda (1951) found that arsenite and arsenate give the same retention following an (n,γ) reaction, but Süe (1948) had determined that the yield of arsenate from arsenate parent increased with increasing time involved in the separation from arsenite. At low separation times it is quite likely that the yield of arsenite was the same from either parent. Similar results were found starting with the cacodylate ion. Halpern (1959) reported that the decay of ^{72}Se as selenite gave 45% arsenite.

Chlorine anions have not been further studied since Libby's work, but bromine and iodine anions have been extensively investigated. Campbell (1959a, b, 1960) studied the effects on bromate of isomeric transition and of the (n,γ) reaction. He proposed the formation of a specific intermediate but was unable to confirm its existence with specific chemical tests. DeVault and Libby (1941) looked for evidence for the formation of bromine atoms in the reaction of bromate with neutrons, but their results did not give any clear indication of these. Süe and Melander (1947) found a small percentage of reactive bromine species both from the reaction of bromide with neutrons and from isomeric decay occurring in a bromide. The reactivity with phenol could be attributed to the species being bromine atoms; however, other active, though unknown, species were also conceivable.

The reactions of iodine species arising from various nuclear transformations have been investigated by a number of groups. Special attention has been devoted to the beta decay of tellurates and tellurites, particularly the work of Cummisky et al. (1961) and of Gordon (1967). Gordon interestingly found that the yield of periodate was dependent on the method of preparation of tellurate, the method resulting in the higher yield of periodate being one that is believed to form a polymeric form of tellurate. The neighbouring system involving isomeric decay of tellurium and the decay of iodine to tellurium has already been discussed above in another context. Earlier studies of the tellurium isomeric decay are found in the work of Williams (1948) and Hahn (1963, 1964), among others.

Meyer and Adloff (1967) and Meyer (1970, 1971) extensively studied the reaction of sulfur species arising both from the (n,p) reaction on ^{37}Cl and from the (n,γ) reaction on ^{36}S.

In a very interesting application, Brown and Wahl (1967) made use of the differences between β-decay and fission induced reactions to determine the independent fission yields of several tin isotopes. This was possible since tin formed directly from fission occurs mainly as tin(II), while tin(IV) results from β-decay.

3. Colloids and Suspensions

Colloids are similar to solutions in that many of the recoils enter the solution. It is possible that the charge state distribution of the recoils is changed on passing through the molecular aggregate of the colloid, and that the final oxidation states of the recoil may thus be affected, but this has not been studied. In suspensions this effect may even be greater depending on the particulate size. Nesmeyanov et al. (1959) investigated the effect of pH and carrier on the (n,γ) reactions on colloidal MnO_2, and Pauly and Süe (1955, 1957) the relative effects of thermal and fast neutrons on the (n,p) reaction in sulfur suspensions. Other work included (n,γ) studies on colloidal silver (Parker 1962) and gold (Majer 1937), the (γ,n) reaction on carbon as Aquadag (Morinaga and Zaffarano 1954), alpha decay in thorium hydroxide (Beydon and Gratot 1968), and beta decay in 'molybdenum blue' (Gratot and Beydon 1970). The (n,γ) reactions and beta decay give recoil yields between 0.3 and 15%, while nuclear reactions giving higher recoil energies give recoil yields of 25–95%. Parker (1962) suggested that recombination with the charged colloid can reduce the recoil yield.

4. Adsorbed Ions

Because of their intimate contact with the solvent, recoil reactions of ions adsorbed on a substrate are more closely related to those of ions in solution than to those of ions in a crystal lattice. There have been very few investigations of nuclear transformations in adsorbed ions. In an extensive series of articles, Matsuura (1965, Matsuura and Sasaki, 1966a, b, Matsuura and Hashimoto, 1966, Matsuura and Matsuura, 1968) and Sensui and Matsuura (1965) have reported the results of (n,γ) reactions on various cobalt and chromium ions adsorbed on different resins. The irradiations were conducted both under static conditions and during continuous elution of the recoil cobalt and chromium from the resin. The continuous extraction method inhibits the radiation-induced slow recombination of the recoil ions with the resin and has obvious practical applications. Hillman and Weiss (1966a) studied the isomeric decay of tellurium-129m adsorbed as tellurium(VI) on alumina or on a cation exchange resin, and Dobici and Salvetti (1964) studied the same decay but in tellurium(IV) on paper. In the former, the yield of

^{129}Te(IV) resembled that obtained in solid telluric acid, and in the latter, the recoils reacted to some extent with butyl alcohol in the solvent to give dibutyl tellurium. In one other investigation involving ions on resins, Moskvin (1962) found that ^{118}Te decays to pentavalent ^{118m}Sb.

CHAPTER 6

PERMANGANATES AND OTHER OXYANIONS

D. APERS

Université de Louvain, Laboratoire de Chimie Inorganique et Nucléaire,
B-1348 Louvain-la-Neuve, Belgium

Chemical Effects of Nuclear Transformations in Inorganic Systems
Edited by G. Harbottle and A.G. Maddock

Contents

1. Permanganate Salts

Because for a long time only neutron sources of low intensity were available, the study of the Szilard–Chalmers process was restricted to nuclides possessing comparatively high neutron capture cross sections: iodine, bromine, manganese, and a few others. The choice of potassium permanganate as a target allowed for the first time a thorough investigation of the chemical consequences of an (n,γ) event in a crystalline material.

This pioneering work was undertaken by Libby (1940). His findings and the interpretation he then proposed have for many years constituted the principal theoretical theme of most hot-atom studies on oxyanionic solids.

In a parallel set of experiments, Libby irradiated potassium permanganate either in crystalline form or dissolved in water and analysed the samples to obtain the distribution of the ^{56}Mn created by neutron capture. The irradiated crystals were dissolved in aqueous solutions containing manganese dioxide as an adsorbing carrier. The MnO_4^- fraction was separated by a simple filtration method. Libby called this fraction the total "retained" material. The amount of radioactive material which is carried on the manganese dioxide was attributed to more energetic bond breaking recoil events. As no accurate chemical identification was made it was proposed that the adsorbed fraction was present in the form of manganese dioxide, this compound being classically the most stable in the presence of excess permanganate ions over the whole pH range. The highly reproducible coprecipitation on added manganese dioxide was advanced as an additional proof of the identification.

A most striking fact was observed: the retention values varied with the pH of the solution: from a value of 30% in neutral media the retention increased to 45% in acids and to 70% in basic solutions. This pH dependence was even more evident when permanganate solutions were irradiated: 5% retention in neutral, 13% in acid and 100% in alkaline solution (fig. 6.1).

An elegant explanation was suggested, based on the assumption that the recoiling atom preserves its oxidation state unchanged even when it has become thermalised. Some bond breaking occurs, leaving the recoil atom linked to a reduced number of oxygen ions and trapped as such in the crystal lattice, from which it will be released only on solution. The possible fragments

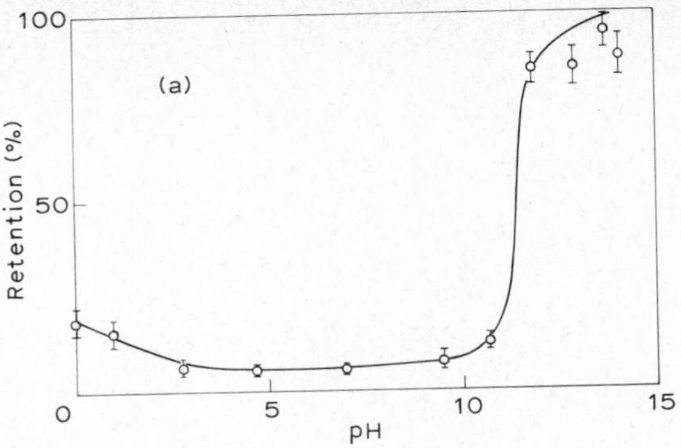

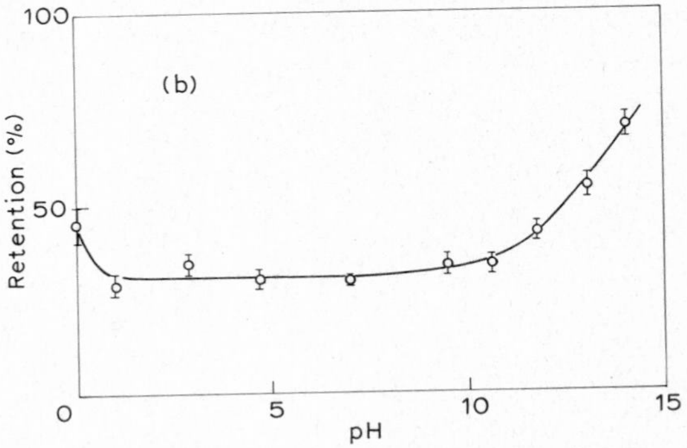

Fig. 6.1. (a) Retention vs. pH for MnO_4^- solutions. (b) Retention vs. pH for solid potassium permanganate exposed and dissolved later in solutions of various pH values.

were tentatively listed as follows:

MnO_3^+ $\quad MnO_2^{3+}$ $\quad MnO^{5+}$ $\quad Mn^{7+}$

As oxygen ions are readily available, the MnO_3^+ species was considered to be present in the largest proportion. The retention would thus be predicated in the lattice, but determined only on dissolution, according to the reactivity of the MnO_3^+ fragment as a function of pH conditions. Thus:

(i) hydrolysis in an alkaline medium leads to an increase of retention:

$$^{56}MnO_3^+ + OH^- \rightarrow {}^{56}MnO_4^- + H^+,$$

(ii) solution in a neutral aqueous medium leads to reduction:

$$2\ {}^{56}MnO_3^+ + H_2O \rightarrow 2\ {}^{56}MnO_2 + 2\ H^+ + 3/2\ O_2,$$

(iii) hypothetical exchange is catalysed in acids:

$$^{56}MnO_3^+ + MnO_4^- \rightarrow MnO_3^+ + {}^{56}MnO_4^-.$$

The overall balance of these reactions yields the experimental data if one supposes the following distribution in the solid:
– 30% of the recoil atoms recombine to give radioactive permanganate in the crystal. This fraction accounts for the minimum retention.
– 30% are already reduced in the solid and stabilised in situ as $^{56}MnO_2$ which is found over all the pH range.
– 40% are trapped as MnO_3^+ and react on solution according to the above reaction scheme.

When permanganate solutions are irradiated the overall distribution is altered: the pH effect is enhanced as a result of similar but immediate and straightforward reactions of the recoil atom with the water molecules without intermediate metastable trapping.

Broda and Erber (1950) suggested that the recoil manganese in the solution was present in a radiocolloidal form that appeared to be particularly stable in the presence of a phosphate buffer. Manganese dioxide thus resembled the heavy metal hydroxides, although it showed an abnormal behaviour during electrophoresis and at extreme pH values.

Rieder et al. (1950) accepted Libby's views and estimated that the MnO_4^- group undergoes scission during the gamma emission yielding O^{--} and MnO_3^+, the latter fragment carrying 13% of the total available recoil energy. They discard other explanations since reoxidation of the other species to MnO_4^- is improbable in alkaline solution, and MnO_4^-, once reformed, cannot give rise to MnO_2 in neutral non-reducing media. The retention they observed when the irradiated crystals were dissolved in neutral water only reached a value of about 20%. This value has been reproduced in all further work, provided irradiations are carried out at ambient temperature.

In solution the retention seemed to be dependent on the concentration of the salt above 0.3 g MnO_4^-/100 g solvent (retention 5%) up to 100 g MnO_4^-/100 g (retention 22%) (fig. 6.2).

These authors observed that the retention increased up to 62% when mixed crystals of $KMnO_4/KClO_4$ were used (fig. 6.3). The retention was however independent of the amount of foreign salt added at the moment of dissolution. Additives such as sulfate and nitrate have an identical effect. It seems however important to specify that all these experiments are carried out at ambient temperature. The decrease in retention was tentatively explained by transfer reactions that could occur within the mixed solid, such as

$$^{56}MnO_3^+ + ClO_4^- \rightarrow {}^{56}MnO_4^- + ClO_3^+.$$

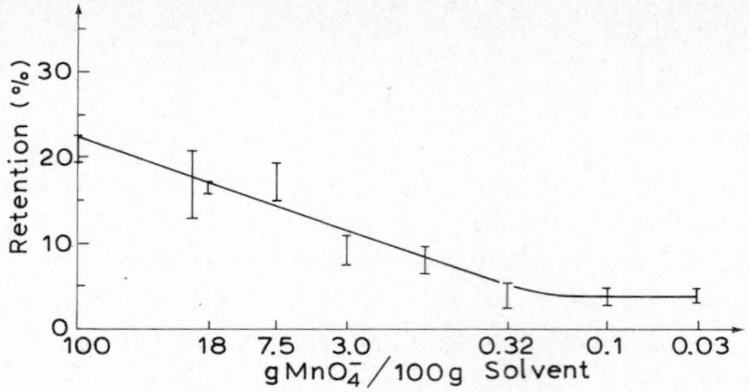

Fig. 6.2. Retention as function of concentration of $KMnO_4$

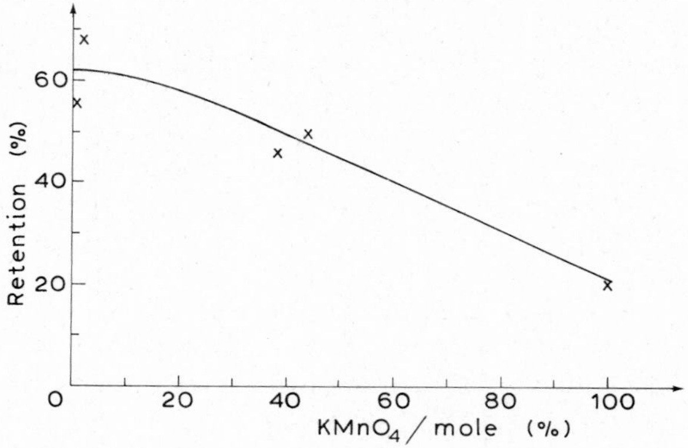

Fig. 6.3. Retention in mixed crystals, $KMnO_4/KClO_4$, as function of mole percentage of $KMnO_4$.

The experimental results for potassium permanganate were confirmed by McCallum and Maddock (1953) and extended to lithium, sodium and barium salts although the pH effect seems less pronounced for these compounds.

In 1951 Jordan (1951a, b) used a new approach to determine the oxidation state of the recoil manganese before isolation as manganese dioxide, by filtration. On chromatographic analysis on alumina of the freshly dissolved material he observed that a fraction (less than 1%) of the activity was eluted which was clearly distinguishable from both permanganate ions and manganese dioxide. Jordan therefore suggested that at least part of the recoil manganese is present in a reduced form, Mn(II) or Mn(III).

The original proposal and long accepted picture involving a MnO_3^+ fragment is based on two fundamental assumptions:

(i) during the recoil event the manganese maintains its heptavalent state and oxygen ions, O^{--}, are split off by the recoil;

(ii) the recoil energy is imparted to the whole ion and not to the manganese atom alone as predicted by dynamic considerations for a heavy molecule or a crystal.

Nowadays these hypotheses are questionable. On modest recoil an atom does not lose or gain any electrons. When such an atom is bound in what is conventionally called a covalent molecule, it will leave its site without further ionisation (Oleari et al. 1966). In this case, the effective charge of the manganese in a permanganate ion is close to zero and certainly less than unity (Mn^0 or $Mn^{<1+}$). After neutron capture, molecular oxygen would be left, as is observed in radiolytic decomposition and the manganese atom recoils in its original charge state.

If there exists any possibility for the recoil atom to remain trapped as such in the crystal lattice, or even if it loses one or two electrons, the possibility of recovering it on solution in a highly reduced state should not be excluded. In the presence of an oxidising agent such as permanganate at high concentration, the reduced manganese is liable to be converted into manganese dioxide by the reaction

$$3Mn^{++} + 2MnO_4^- + 2H_2O \rightarrow 5MnO_2 + 4H^+. \qquad \text{(a)}$$

However, this reaction is slow at room temperature and in the absence of catalysts, so that survival of the reduced state is still possible. Some of the recoil manganese may, because of this reaction, be recovered as the dioxide, but some reduced species may survive and be physically adsorbed on solid manganese dioxide carrier or any other adsorbing surface.

Keeping this in mind new analytical procedures have been developed (Apers and Harbottle 1963). On solution of the irradiated crystals, aluminium oxide or manganese dioxide is added. If manganous or manganic ions are present, some of them might adsorb, and some might not, according to pH conditions. The adsorbed fraction is then liable to extraction by a specific reagent such as 8-hydroxy-quinoline in chloroform solution. The fraction which is not adsorbed must necessarily be present in the filtrate together with the bulk permanganate. It can there be converted to the dioxide if conditions and concentration are made favourable by the addition of inactive manganous ions as a carrier. A repetition of the latter procedure enables one to correct for the presence of labelled manganese dioxide from radioactive permanganate. The experimental treatment is schematized in fig. 6.4.

The results then show that only a few percent of the activity is present as manganese dioxide. Most of it is recovered as permanganate (retention) and in a reduced form, presumably manganous ions. The distribution of the activity is still pH dependent but this influence is greatly reduced.

Electrophoresis experiments on glass fiber paper confirm the presence of positively charged manganese ions which can only be identified as bivalent Mn^{++}.

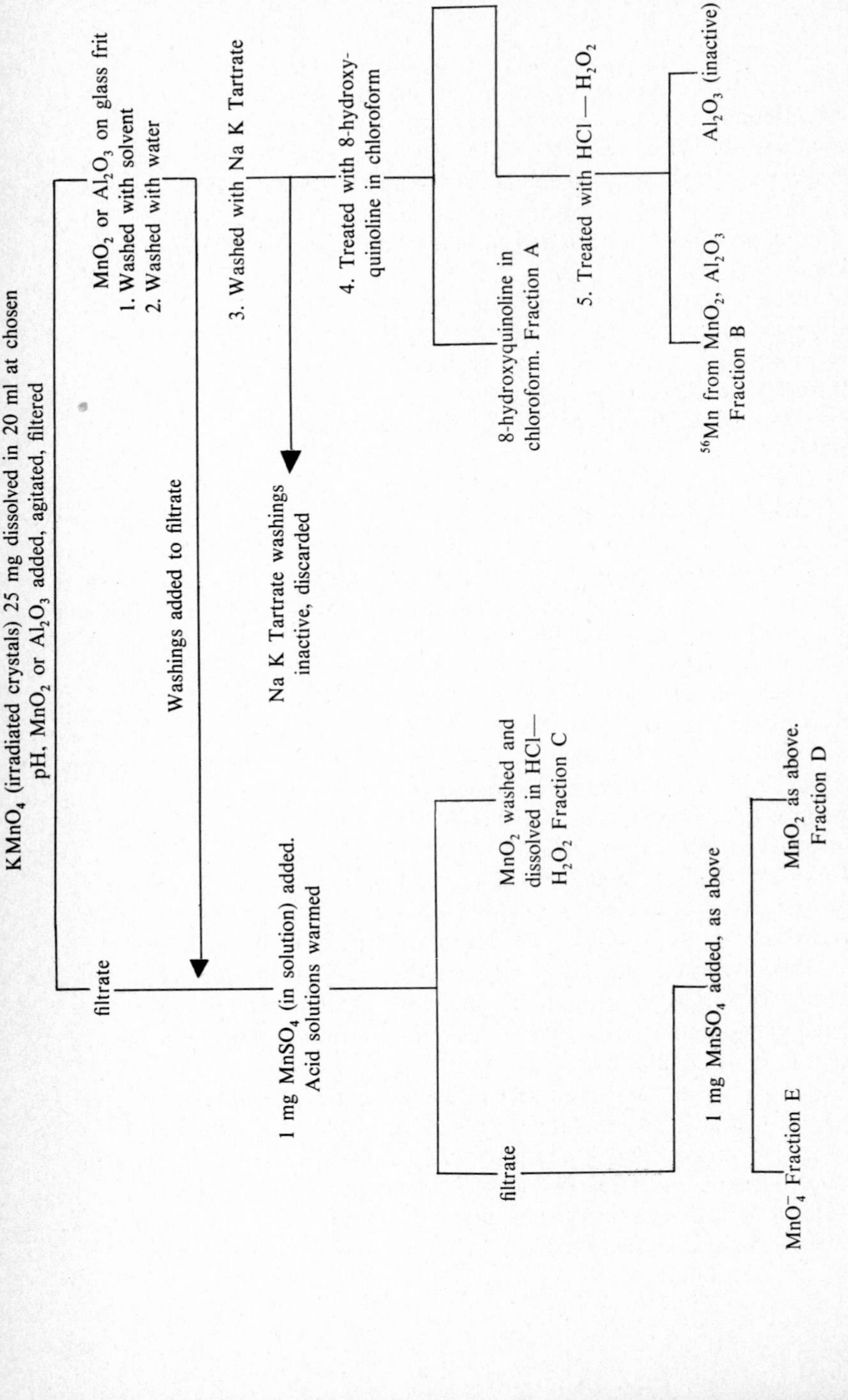

Fig. 6.4. Flow diagram of adsorption-elution procedure.

It was however suspected that the remaining labelled manganese dioxide might be formed by reaction (a) after dissolution and that the variations with pH still observed had nothing to do with recoil but were to be attributed to varying pH dependent adsorption coefficients. These doubts could be resolved by using a double tracer technique (Cogneau et al. 1967).

An identical adsorption and extraction procedure (Apers and Harbottle 1963) was followed, but at the moment of dissolving the crystals, a trace amount of a $^{54}Mn^{++}$ salt was added. If recoil manganese is present in the form of $^{56}Mn^{++}$ at the instant of solvation, it should throughout the experimental procedure undergo identical reactions to the tracer added. For instance: whatever fraction of $^{54}Mn^{++}$ is transformed into $^{54}MnO_2$, an equal fraction of $^{56}Mn^{++}$ will be transformed to $^{56}MnO_2$; whatever exchange the former undergoes will also involve the latter; failure of adsorption will be equally probable or improbable for both. Now corrections are possible, based on the following arguments:

After chemical separation ^{56}Mn may be distributed over three fractions with respective yields represented by:

A = $^{56}MnO_4^-$ or ^{56}Mn accompanying the MnO_4^- fraction,
B = $^{56}Mn^{++}$ = amount desorbed from the carrier,
C = $^{56}MnO_2$.

The ^{54}Mn activity, originally present as $^{54}Mn^{++}$ ions only, is also distributed over the three species with yields represented by:

a = $^{54}MnO_4^-$ or ^{54}Mn accompanying the MnO_4^- fraction,
b = $^{54}Mn^{++}$ = amount desorbed from the carrier,
c = $^{54}MnO_2$.

B only represents a fraction b of the total amount of $^{56}Mn^{++}$ initially present, so that the original amount of $^{56}Mn^{++}$ equals B/b. Fractions a and c of the total $^{54}Mn^{++}$ have been altered and go with permanganate and dioxide respectively. These modifications have also affected the $^{56}Mn^{++}$, so that the same fractions a and c have been converted into $^{56}MnO_4^-$ and $^{56}MnO_2$. Thus we have:

the original amount of $^{56}Mn^{++} = B/b$,
the original amount of $^{56}MnO_4^- = (A - aB/b)$,
the original amount of $^{56}MnO_2 = (C - cB/b)$.

Two striking facts are revealed when these corrections are applied:

(1) No $^{56}MnO_2$ was originally formed by recoil
(2) pH has no influence whatsoever on the recoil process and its consequences.

Figure 6.5 gives the corrected retention values as a function of pH.

Experiments have shown that these statements remain entirely valid for permanganate salts of other cations, such as lithium, sodium, silver and barium (table 6.1) (Cogneau et al. 1972) (fig. 6.6) and also when the compounds are neutron bombarded in solution over the whole pH range (Cogneau et al. 1968) (fig. 6.7).

The retention values lie close to the ones measured earlier in neutral

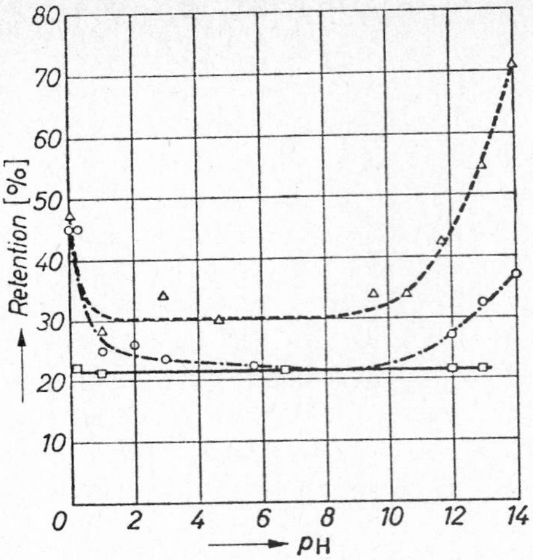

Fig. 6.5. Retention of ^{56}Mn as permanganate in potassium permanganate crystals, as a function of pH in aqueous solution: △ – – – Libby (1940), O·-·-·- Apers and Harbottle (1963), □— Cogneau et al. (1967).

Table 6.1

Compound	Retention	Mn^{++}	MnO_2
$LiMnO_4$	5.1 ± 0.1	94.9 ± 0.1	0
$KMnO_4$	22.3 ± 0.2	77.7 ± 0.2	0
$RbMnO_4$	46.9 ± 0.1	53.1 ± 0.1	0
$CsMnO_4$	46.4 ± 0.1	53.6 ± 0.1	0
$Ba(MnO_4)_2$	10.4 ± 0.1	89.6 ± 0.1	0
$AgMnO_4$	18.2 ± 0.2	81.8 ± 0.2	0

media, as far as irradiation and dissolution at ambient temperature are concerned. In these conditions the adsorption of manganous ions on alumina is quantitative, so that the untreated filtrate contains only the unaltered permanganate fraction.

All the published data relevant to the distribution of radioactive manganese following a nuclear reaction (Shiokawa et al. 1969c, Oblivantsev et al. 1971) will need to be corrected in this way.

The pH dependence of the retention in permanganates, which seemed to reflect an exceptional behaviour compared to all other oxyanionic species, now appears to be due only to peculiarities of the chemistry of manganese during the analytical procedure. The pseudo pH effect is at least partly to be ascribed to variations of the adsorption coefficient of manganous ions on solid surfaces, and this effect has been verified (Benes and Garber 1966, Roy 1972, Roy et al.

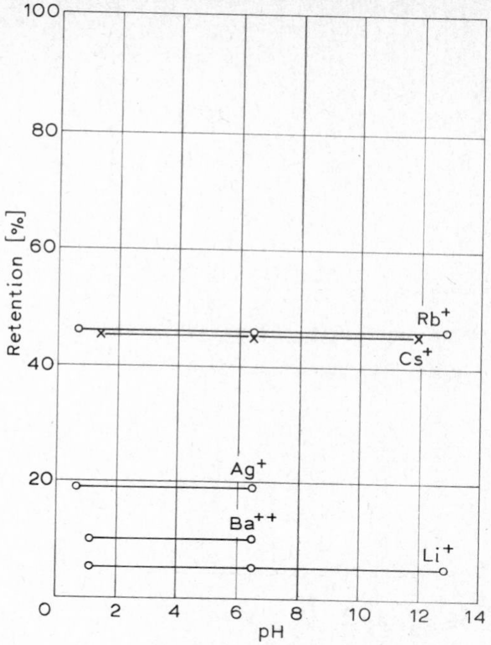

Fig. 6.6. pH effect on retention in permanganate salts.

1973). These adsorption experiments have not been performed under exactly the same conditions as the hot atom experiments and therefore accurate corrections are not feasible. Nevertheless, the pH dependence of the uncorrected retention values is qualitatively explained by the lowered adsorbability of manganous ions at low and high pH values. This might be attributed to a stabilization of the colloidal solution obtained after hydrolysis in approximately neutral solution, or more simply to a larger desorptive power in acids and a transformation into Mn^{+++} and MnO_2^- in strongly alkaline solutions. A better knowledge of manganese chemistry in the presence of adsorbants is necessary in order to clarify this point.

Although attractive, the picture of successive stripping of oxygen ions must now be considered as based on an inaccurate interpretation of the earlier experimental results. In chs. 25 and 26 a new model is proposed explaining the retention phenomenon and supported by the observation of transfer reactions in many solids. A verification of this exchange in permanganate crystals is hard to achieve, as the material cannot be doped by macroscopic amounts of reduced manganese species, because of oxidation during co-crystallisation. Some experiments showed promise: potassium permanganate crystals, contaminated $^{54}MnO_2$ traces, give evidence that a fraction of the

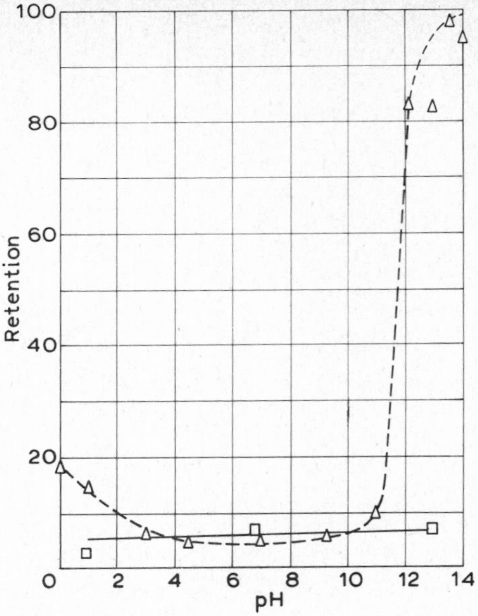

Fig. 6.7. Retention of ^{56}Mn as permanganate in potassium permanganate solutions as a function of pH. △ – – – Libby (1940), □ — Cogneau et al. (1968).

active manganese is recovered as active permanganate on heating (Van de Leest 1965). Doping by ion implantation with $^{54}Mn^{+}$ or $^{54}Mn^{++}$ or doping by vapours or lyophilisation might throw more light on the question.

The presence of reduced manganese species on recoil has received considerable confirmation in an experiment of Tumosa and Ache (1970). They use appropriate "catchers" for manganese recoil atoms ejected by (n,γ) recoil from a thin metallic manganese foil. These were placed close enough that the manganese atoms were caught before any internal conversion could alter their charge state as determined by the primary event. They were thus able to prove that in all cases at least 98% of the events lead to formation of Mn^{0} or Mn^{+}: the atom recoils without change of effective charge.

As was mentioned before more evidence has been obtained using paper electrophoresis (Apers and Harbottle 1963, Shiokawa et al. 1969c). The activity which still appears as manganese dioxide is clearly to be attributed to secondary parasitic reactions, as the results have not been corrected in the appropriate way. Also, one should be suspicious of the effect of organic ligands present in glass fiber paper, if no specific precautions are taken. A pertinent confirmation of the presence of bivalent manganese and possibly of other valency states has also been given by Nesmeyanov (Nesmeyanov et al. 1966) and Ache (Swordsma et al. 1971). The latter paper gives evidence for the creation of Mn^{++} within potassium permanganate which has been exposed to high doses of ionising radiation (over 500 Mrad) and submitted to magnetic susceptibility measurements.

All above-mentioned recoil experiments have been performed at normal reactor irradiation temperatures (~ 50° C) followed by a chemical treatment at room temperature. It is evident that some annealing of the samples has taken place leading to an enhanced recombination and to a higher retention. It could indeed be shown that, if the whole procedure is repeated at low temperature, the retention is markedly decreased. Use of non-aqueous solvents, such as acetone, is then required. The danger of bulk reduction by the latter is avoided by dissolution and treatment in a minimum time (Cogneau et al. 1967, Veljkovic and Harbottle 1962). Retention values as low as 3% are then observed, showing that 19% of the recoil atoms have already recombined to permanganate at 20° C.

1.1. THERMAL ANNEALING

Isothermal heating of neutron-irradiated permanganate salts follows the classical annealing pattern: a rather steep increase of the retention as a function of heating time, levelling off into a pseudo-plateau (Rieder et al. 1950, McCallum and Maddock 1953, Apers and Harbottle 1963, Aten and Van Berkum 1950, Shiokawa et al. 1970). The highest retention reached is about 85% (fig. 6.8). These experiments are limited by temperature because of interference by gross thermal decomposition with an induction time which is shortened because of the radiation damage. As the thermal decomposition is triggered by the presence of lattice defects, it is furthermore evident that the recoil site itself will act as a nucleation centre and this may lead to a dramatic change in the fate of the recoil atom itself.

The effect of the pH of the solution on the distribution of the ^{56}Mn after annealing is almost nil (Apers and Harbottle 1963). This might also be attributed to a local stabilisation induced by the above-mentioned nucleation mechanism.

On isochronal annealing (Shiokawa and Sasaki 1970, Roy 1972, Cogneau 1967) (fig. 6.9) it appears that the thermally induced recombination involves no clearly defined separate reaction steps, contrary to what has been observed in other compounds. It has also been proved that there is no intermediate formation of manganese dioxide when manganous ions are annealed into permanganate.

Yet it seems that two types of thermal annealing take place. One part is quite sensitive to radiation dose, the other fraction is not (Van Herk and Aten 1972). Although the nucleogenesis of ^{54}Mn and ^{56}Mn is very different, retention values and behaviour on annealing are identical within experimental error for both isotopes, provided the samples have received identical radiation doses.

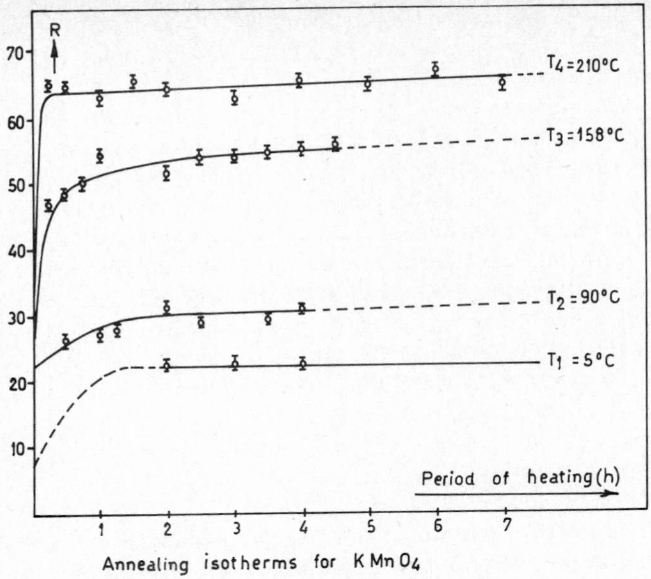

Fig. 6.8. Annealing isotherms for $KMnO_4$.

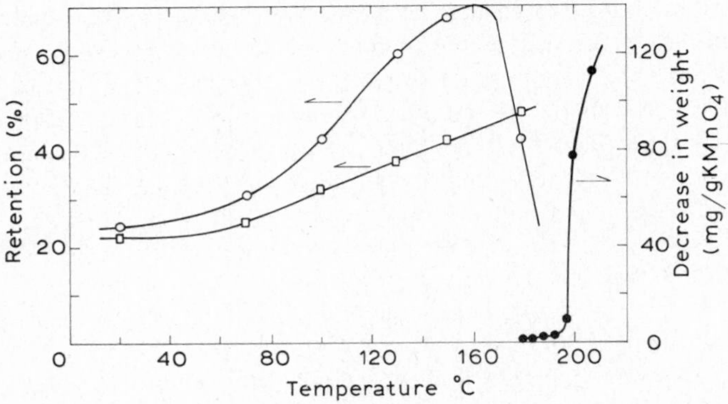

Fig. 6.9. Isochronal annealing curves (heating time: 5 h) and the results of thermal decomposition of potassium permanganate irradiated with fast neutrons. O: ^{54}Mn recoils, □: ^{56}Mn recoils, •: results for thermal decomposition (from Shiokawa and Sasaki 1970)

Preheating also has a marked influence. The retention of (n,2n) produced ^{54}Mn seems however to increase only very slowly at low annealing temperatures, but fast recombination is observed above room temperature (Owens and Lecington 1975).

A surprising observation which seems unique for the permanganate system is the difference in retention depending on whether the irradiated material is heated in a sealed or in an open tube. Annealing while exposed to air gives a greater increase in retention by at least 5 to 6%. The reason is not understood as yet. Possibly the presence of moisture on the surface of the crystal or on the silica tube, which evaporates on warming, plays a role. The phenomenon may be related to the effects of the annealing atmosphere reported for irradiated cobalt complexes (see ch. 11).

Recoil studies in the permanganate system are difficult and sometimes frustrating. A deeper understanding of the results requires a better knowledge of the chemistry of manganese itself in its different oxidation states and at tracer concentrations. More research in this area seems necessary.

2. Selenium Compounds

Selenites, selenates and selenium oxide present some interesting features in the recoil chemistry field: selenium has well-defined oxidation states of −2,0, +4 and +6; the element has several stable isotopes, all having reasonable cross sections for radiative neutron capture, thus permitting the observation of isotopic or isomeric effects; elemental selenium is sometimes found as a recoil product in quite substantial amounts.

Earlier work indicated bond breaking by an isomeric transition in ^{81}Se (Langsdorf and Segrè 1940). Daudel was the first to prove that on neutron irradiation of selenite elemental selenium was formed (Daudel 1941, 1942). A similar process was confirmed in bombarded selenium oxide, where a simple extraction by carbon disulfide could isolate a small but definite and reproducible part of the radioactivity (Apers et al. 1967).

It has also been shown that the distribution of the activity between the mother compound and the recoil products is different for the ^{81m}Se and ^{81}Se isomers which are simultaneously formed on neutron activation, this distribution varying as a function of the energy of the incident neutron (Apers et al. 1957).

The chemical consequences of recoil in selenium compounds have been the subject of renewed interest since the work of Constantinescu et al. (1966), Duplatre et al. (1972), Al-Siddique and Maddock (1972) and Al-Siddique et al. (1972).

Because of the appearance of the radio-selenium in several oxidation states, analysis methods had to be developed which guarantee quantitative separations while no exchange reactions disturb the results.

Elemental selenium isolated by adsorption on silica or on red selenium leads to somewhat irreproducible results; extraction by carbon disulfide or trichlorethylene is equally successful; an alternative way is to precipitate selenite and selenate as the respective barium salts, while elemental selenium is kept in solution as selenocyanide, with carrier added. The selenite–selenate separation can be performed by selective barium precipitation at controlled pH. Paper chromatography yields satisfactory results but is extremely slow. Thus it is not applicable for the shorter lived isomers. Complexation of selenium (IV) by sodium diethyl-dithiocarbamate and extraction into carbon tetrachloride at pH 5.5 also appears to be very successful. All these methods have been thoroughly tested for the absence of exchange reactions.

Potassium selenate gives a retention of 77% when irradiation and analysis are performed at room temperature. This value drops to 42.1% for 3 h irradiations at 77 K and to 32% for 1 min irradiations. It thus appears that radiation annealing during irradiation substantially favours recombination to the original compound.

Isothermal annealing at a temperature of 98.7° C increases the retention to nearly ninety percent, following the usual pattern, but involving some irregularities (fig. 6.10).

Isochronous annealing and heating of the samples with a linear temperature increase both show a rather complex structure of the annealing (figs. 6.11 and 6.12). Three to six reactions, according to experimental conditions, follow each other, they have activation energies ranging from 0.793 to 2.060 eV. The proportions involved in each of these processes differ and they respond differently towards ionizing radiation-induced defects.

Ion implantation experiments, with ^{75}Se implanted in a selenate matrix, including annealing of the doped samples yield results which are not in total agreement with the neutron irradiation experiments.

The recombination reactions may be summarized as follows:

$$Se^0 \rightarrow Se(IV) \rightarrow Se'(VI).$$

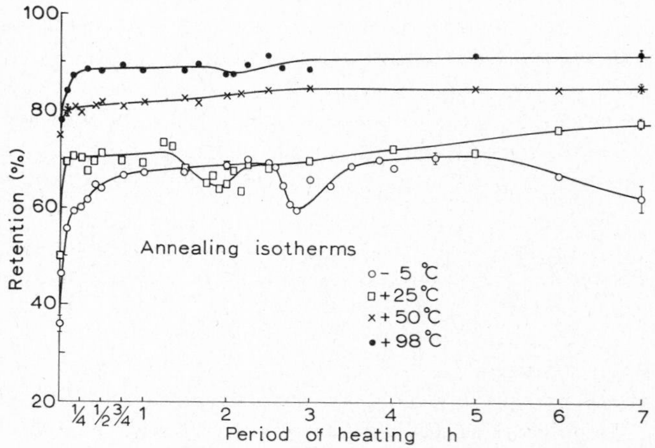

Fig. 6.10. Annealing isotherms (Duplatre 1969).

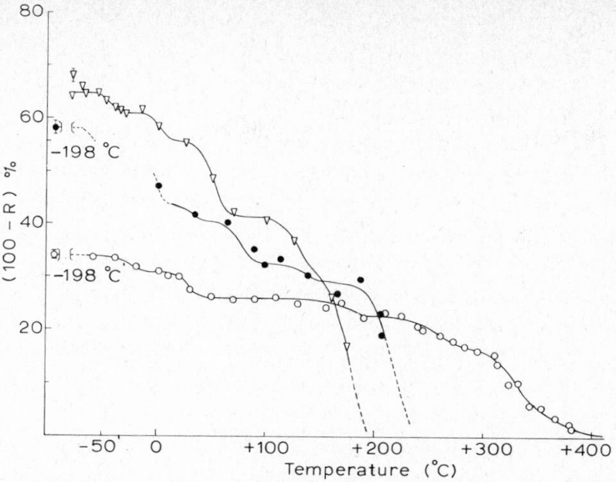

Fig. 6.11. Annealing isochronals. • Sample II (2 h anneals), O sample V (20 min anneals), ▽ sample VI (20 min anneal).

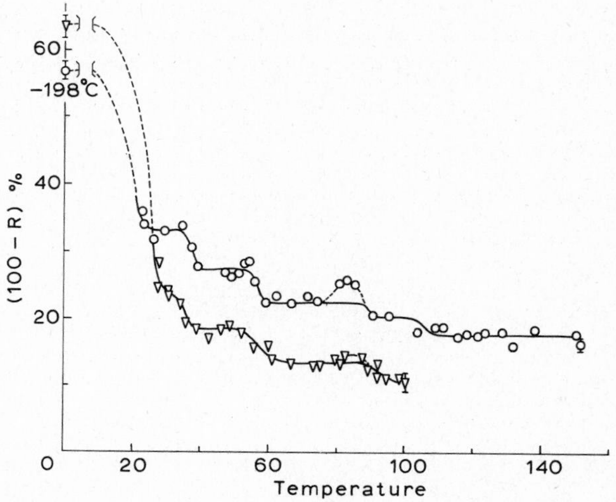

Fig. 6.12. Linear temperature annealing (Duplatre 1969). O – rate of heating = 1.08 ± 0.03° C min^{-1} (sample IV), ▽ – rate of heating = 0.711 ± 0.013° C min^{-1}.

Parallel electron spin resonance studies predict that these reactions may involve the species SeO_4^-, SeO_3^- and SeO_2^-.

Anhydrous sodium selenate behaves differently from the potassium salt. Only two kinds of centres are found and they do not seem to be the same as in the potassium salt. Differences are even more pronounced in the low-temperature region. The sodium salt anneals to an appreciable extent only above room temperature. Dehydration of the aquated salt does not greatly influence the retention (from 37.9 to 40.0%).

The results show that the process removes those centres with the lowest energies of activation for annealing, so that thermal annealing of the dehydrated material only proceeds at a higher temperature. The high-temperature annealing is enhanced by the presence of oxygen, but selenite also suffers some macroscopic oxidation under these conditions.

Sodium selenite shows some very surprising results. A sample of the hydrated salt irradiated at liquid nitrogen temperature and annealed at room temperature shows the following distribution of the activity: 24.8% Se(0), 21.7% Se(VI) and 53.5% Se(IV). Exposure to ultraviolet light transfers half of the Se(0) into Se(VI), while the amount of Se(IV) remains almost unchanged (fig. 6.13). These figures are also dependent on irradiation time or rather on the gamma dose during neutron bombardment.

Thermal annealing follows an identical scheme (fig. 6.14): elemental selenium is converted into selenate, while the amount of selenite, i.e. the retention, remains virtually unchanged. This behaviour is exceptional, and has not been observed in other compounds. It does mean that the reactions of the recoil atom, which are promoted by external agents as radiation or heat, do not necessarily convert the atom into its original compound. In this case a new

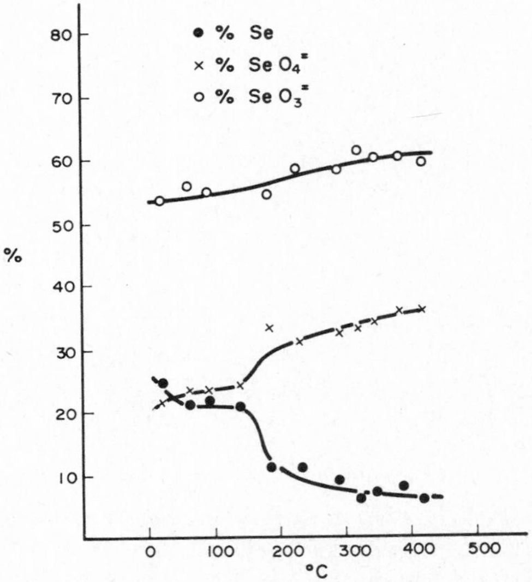

Fig. 6.13. (a) $Na_2SeO_35H_2O$ (2 h irradiation) 1 h isochronal.

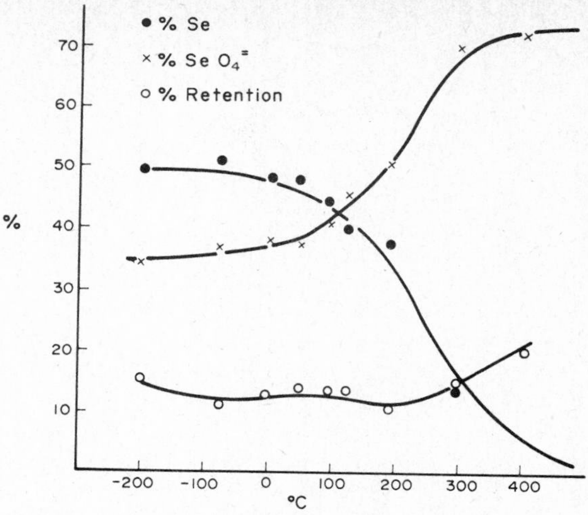

Fig. 6.13. (b) Na_2SeO_3 $5H_2O$ (10 min irradiation) 1 h isochronal.

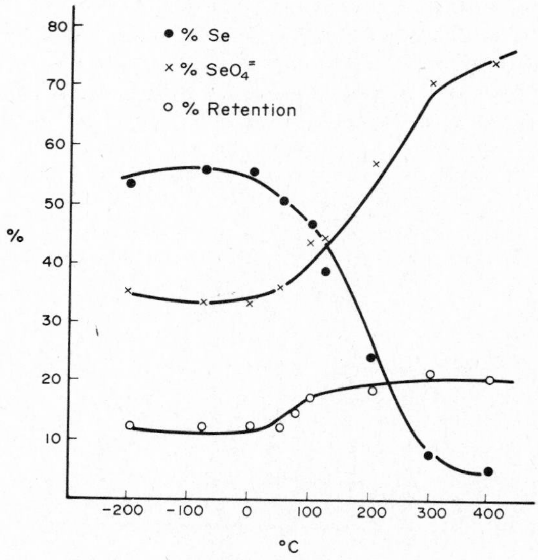

Fig. 6.14. (a) Na_2SeO_3 10 min irradiation.

type of molecular arrangement is preferred, which however can use as an intermediate a compound which on dissolution and analysis appears identical to the mother compound (selenite). It has been shown that the effect is not due to any thermal decomposition or radiolytic decomposition.

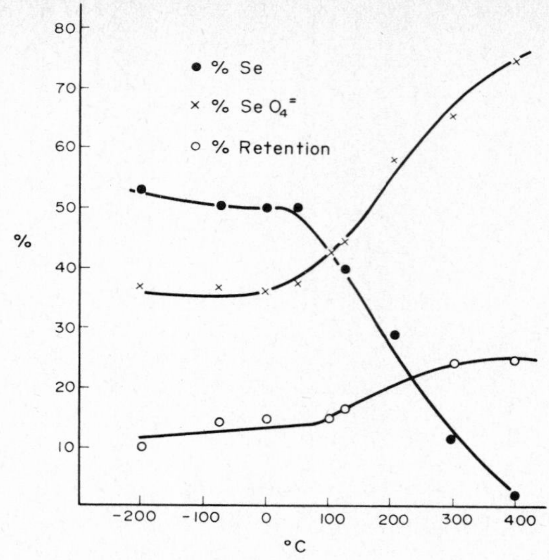

Fig. 6.14(b) Na_2SeO_3 30 min irradiation isochronal vacuum annealing.

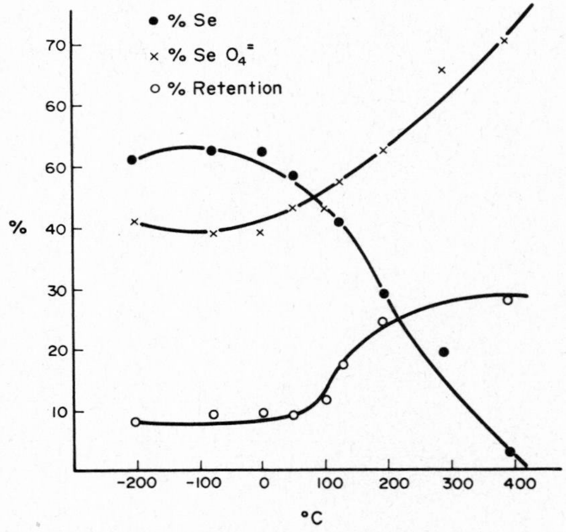

Fig. 6.14(c) Na_2SeO_3 1 h, irradiation.

The peculiar behaviour is explained by Al Siddique et al. (1972a) as due to a re-organisation of the immediate surroundings of the recoil atom which sits at zero-oxidation level in an interstitial position:

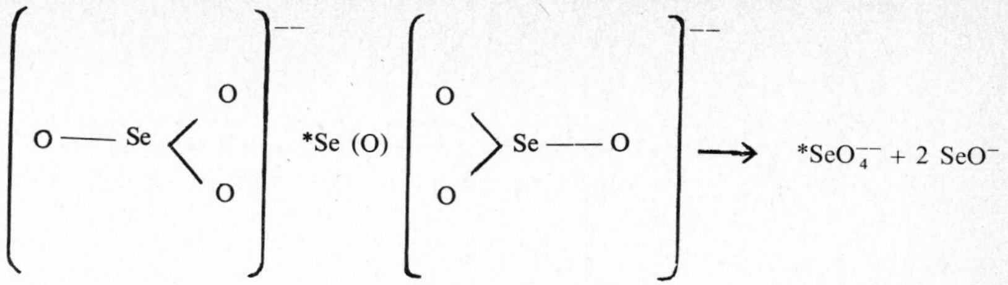

$$\left[O \!-\! Se \!\!\!\begin{array}{c} O \\ < \\ O \end{array} \right]^{--} \begin{array}{c} {}^{*}Se\,(O) \end{array} \left[\begin{array}{c} O \\ > \\ O \end{array}\!\! Se \!-\! O \right]^{--} \longrightarrow {}^{*}SeO_4^{-} + 2\ SeO^{-}$$

or by a straighforward oxidation of the elemental selenium by an O_3^- species which arises from radiolysis.

As only the fate of the radioactive atom can be followed, it is not possible to decide yet what the mechanism really is.

3. Perrhenate Salts

Earlier studies always concluded that the retention of ^{186}Re and ^{188}Re in neutron-irradiated potassium perrhenate reached 100% when irradiation and chemical treatment were performed at ambient temperature (Herr 1952a, b, Schweitzer and Wilhelm 1956). This could be explained in three ways: either the neutron capture is an almost recoilless event, so that no bond breaking occurs, or the recoil species readily reacts with water molecules reforming

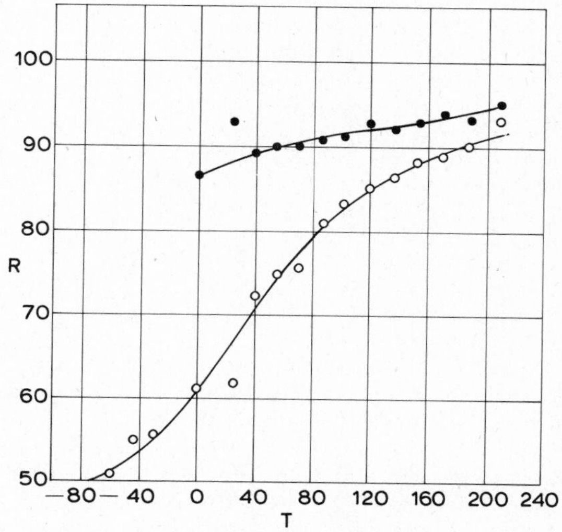

Fig. 6.15. Retention of ^{186}Re in NH$_4$ReO after 22 h annealing. • dissolution in water, o dissolution in liquid ammonia (De Kimpe et al. 1969).

perrhenates, because of the high stability of the heptavalent state rhenium. The latter suggestion seemed most likely, as even solutions of perrhenate exposed to neutrons show total retention.

In order to decide which hypothesis is correct, ammonium perrhenate crystals have been irradiated at liquid nitrogen temperature (De Kimpe et al. 1969). The material has been dissolved in liquid ammonia. A retention of about 50% is then obtained, which rapidly increases up to 100% on moderate annealing (fig. 6.15). This excludes the hypothesis of failure of bond rupture of the perrhenate ion. The presence of ammonium cations dramatically lowers the retention, as is also observed in numerous other systems.

The recoil rhenium is reduced and may appear as rhenium dioxide, which survives as such in a non-aqueous solvent. Its precursor is readily transformed into perrhenate in water and is particularly sensitive to annealing. A more elaborate analytical procedure, using Re(III) and Re(IV) carriers and specific extraction and precipitation (Facetti et al. 1969) shows that part of the reduced recoil species may be recovered as Re(III).

All rhenium compounds show a small isotope effect, as ^{188}Re invariably has a retention which is higher by a few percent.

A non 100% retention has been indicated by some angular correlation experiments which proved that the radioactive species is present in the crystal in a form that is different from ReO_4^- (Sato et al. 1966).

CHAPTER 7

EFFECTS OF NUCLEOGENESIS IN SOLIDS. NUCLEAR TRANSFORMATIONS INVOLVING NEUTRONS, BUT WITHOUT CHANGE IN ATOMIC NUMBER

Salts of Halogen Oxyanions

C.W. OWENS

Department of Chemistry, University of New Hampshire, Durham, New Hampshire 03824, USA

Chemical Effects of Nuclear Transformations in Inorganic Systems
Edited by G. Harbottle and A.G. Maddock

Contents

Introduction

The halogen oxyanions were among the earliest subjects of Szilard–Chalmers study, and they continue to be among the most-studied of the inorganic systems. As early as 1935 [within a year of the initial report by Szilard and Chalmers (1934a)], it was reported that the majority of the radiohalogens produced by neutron irradiation of $NaClO_3$, $NaBrO_3$, $NaIO_3$, and $NaClO_4$ could be found in the halide state when the salts were dissolved (Amaldi et al. 1935, D'Agostino 1935). A few years later Libby (1940), Daudel (1942), and Berne (1949, 1964) obtained similar results with these salts. Much of the early work was directed toward preparing high specific-activity radiohalogens, and it was during this work that the complications of thermal and radiation effects were first recognized. As Boyd later pointed out (1964), Szilard–Chalmers enrichment is lowered by two processes: (i) radioactive recoil atoms are 'lost' by recombination back into the parent state, and (ii) those recoils which do not recombine, and which would therefore otherwise be separable in high specific activity, are diluted by non-radioactive isotopes in chemically identical forms due to radiolysis of the target compound. Consequently, attention gradually shifted toward the study of the fundamental processes, especially toward identification of the recoil species and their reactions in the solids. The results of such studies show clearly that the ultimate chemical fate of atoms undergoing nuclear transformation in halogen oxyanions (and in most other solid systems) is determined by rather complex interactions of nuclear, thermal, radiation, and composition effects. An examination of the chemical effects of nucleogenesis in halogen oxyanions must be made within the context of all these considerations.

1. Nuclear Properties of the Halogens

No doubt part of the early popularity of halogen oxyanion systems can be ascribed to favorable nuclear properties. As table 7.1 shows, each of the halogens (except fluorine and astatine, which are excluded from consideration) has at least one isotope with a reasonable cross section for thermal neutron capture producing a radionuclide with a convenient half life. When the only available sources of neutrons were low-flux Ra/Be sources, this was a necessity. Moreover, the fact that each of the halogens can give rise to several radioisotopes contributes to their popularity in more recent work, since it is therefore possible to examine the importance of the physics of the nuclear process to the chemistry observed. As will be discussed later, one often observes an 'isotope effect' which many authors have attributed to differing details of the nuclear processes generating the different isotopes.

Table 7.1
Nuclear properties of the halogens

Reaction	Percent abundance target nuclide	Thermal neutron cross section (b)	Half-life recoil nuclide	Radiation emitted by recoil nuclide
$^{35}Cl(\gamma,n)^{34m}Cl$	75.53		32.4 min	β^+, γ
$^{35}Cl(n,\gamma)^{36}Cl$		30	3.0×10^5 y	β^-
$^{37}Cl(n,\gamma)^{38}Cl$	24.47	0.6	37.3 min	β^-, γ
$^{79}Br(n,2n)^{78}Br$	50.54		6.4 min	β^+, γ
$^{79}Br(n,\gamma)^{80}Br$		8.5	17.6 min	β^-, γ
$^{79}Br(n,\gamma)^{80m}Br$		2.9	4.5 h	IT(γ)
$^{81}Br(n,\gamma)^{82m}Br \rightarrow ^{82}Br$	49.46	3.3	35.3 h	β^-, γ
$^{127}I(n,2n)^{126}I$	100		13.1 d	β^-, γ
$^{127}I(\gamma,n)^{126}I$			13.1 d	β^-, γ
$^{127}I(n,\gamma)^{128}I$		7.0	25.0 min	β^-, γ
$^{129}I(n,\gamma)^{130}I$	*a)*	32	12.5 h	β^-, γ

a) Fission product, half life 1.6×10^7y

2. Systems Investigated

The vast majority of studies of halogen oxyanions have dealt with the (n,γ) reaction in alkali metal salts (see table 7.2).

3. Analytical Considerations

Attempts to understand the actual state-of-affairs in irradiated solids (including halogen oxyanions) suffer from a lack of sufficiently sensitive analytical techniques. In most of the early work (and some later work) the intention was only to separate the reduced halogen forms from the higher oxidation states, usually by solvent extraction or precipitation. But when more discriminating separation techniques, such as thin-layer chromátography (Seiler and Seiler 1967), anion-exchange (Berne 1964, Boyd and Larson 1968a, b, 1969), and paper electrophoresis (Apers et al. 1963, Harbottle 1960, Jach et al. 1958, Jach and Harbottle 1958) have been used, at least minor fractions of the recoil atoms have appeared in several oxidation states other than halide. It is thus likely that in many cases where only the retention or only the yield of halide has been measured, important information about the state of the recoil atom has been obscured by the separation technique.

Table 7.2
Studies of salts of halogen oxyanions

	1. Effects following (n,γ) reaction
$NaClO_2$	Vlatkovic and Aten 1962, Seiler and Seiler 1967
$KClO_2$	Boyd and Larson 1968a
$NaClO_3$	Amaldi et al. 1935, D'Agostino 1935, Libby 1940, Vlatkovic and Aten 1962, Seiler and Seiler 1967, McCallum and Holmes 1952, Sharman and McCallum 1955, Owens and Rowland 1962, Aras et al. 1965
$KClO_3$	Vlatkovic and Aten 1962, Boyd and Larson 1968a, Aras et al. 1965
$LiClO_4$	Aras et al. 1965
$NaClO_4$	D'Agostino 1935, Daudel 1942, Vlatkovic and Aten 1962, Seiler and Seiler 1967, Aras et al. 1965, Yaffe 1949
$KClO_4$	Vlatkovic and Aten 1962, Boyd and Larson 1968a, Owens and Rowland 1962, Aras et al. 1965
$LiBrO_3$	Apers et al. 1963, Campbell and Jones 1968a, Boyd and Larson 1968b
$NaBrO_3$	Amaldi et al. 1935, D'Agostino 1935, Boyd and Larson 1968b, Saito et al. 1959, Campbell 1959b, Campbell 1960, Saito et al. 1965a, b, 1967a, Hobbs and Owens 1966a, b
$KBrO_3$	Berne 1964, Apers et al. 1963, Boyd and Larson 1968b, Saito et al. 1959, 1965b, 1967a, Boyd et al. 1952, Cobble and Boyd 1952, Harbottle 1960, Veljkovic and Harbottle 1962
$CsBrO_3$	Boyd and Larson 1968b
$LiIO_3$	Cleary et al. 1952
$NaIO_3$	Amaldi et al. 1935, D'Agostino 1935, Libby 1940, Cleary et al. 1952, Bellido 1967, Ambe and Saito 1970, Shukla et al. 1970
KIO_3	Cleary et al. 1952, Ambe and Saito 1970, Aten et al 1965, Boyd and Larson 1969, Owens and Connor 1969, unpublished, Lin and Wiles 1970
$RbIO_3$	Ambe and Saito 1970
$CsIO_3$	Ambe and Saito 1970, Dupetit and Aten 1967
$NaIO_4$	Daudel 1942, Bellido 1967
KIO_4	Boyd and Larson 1969, Aten et al. 1956
$Mg(BrO_3)_2$	Saito et al. 1965b, 1967a, b, Ambe et al. 1968
$Ca(BrO_3)_2$	Maddock and Müller 1960, Müller 1961, Arizmendi and Maddock 1961
$Ba(BrO_3)_2$	Campbell and Jones 1968a, Saito et al. 1965b, 1967a
$Al(BrO_3)_3$	Saito et al. 1965b
$TlBrO_3$	Saito et al. 1965b, 1967a
Transition metal bromates	Saito et al. 1965b, 1967a, Ambe et al. 1968, Andersen and Maddock 1963a, Arnikar et al. 1970a, b, c
NH_4IO_3	D'Agostino 1935, Cleary et al. 1952, Ambe and Saito 1970
$Mg(IO_3)_2$	Ambe and Saito 1970
$Ca(IO_3)_2$	Ambe and Saito 1970, Kaučić and Vlatkovic 1963
$TlIO_3$	Ambe and Saito 1970
$Nd(IO_3)_3$	Ambe and Saito 1970
Transition metal iodates	Ambe and Saito 1970

contd.

Table 7.2 (contd.)

	2. Effects following (γ, n) and (n, 2n) reactions
Alkali salts	Saito et al. 1965a, Ambe and Saito 1970, Aten et al. 1965, Ambe et al. 1968, Dupetit 1967
	3. Related compounds mostly (n, γ) effect
HIO_3	Cleary et al. 1952, Ambe and Saito 1970, Muruyama and Idenawa 1967, Arnikar et al. 1969, 1970a, b, c
$KH(IO_3)_2$	Cleary et al. 1952, Ambe and Saito 1970, Saito and Ambe 1970
I_2O_5	Ambe and Saito 1970, Muruyama and Idenawa 1967, Saito and Ambe 1970, Vlatkovic and Aten 1961
HI_3O_8	Ambe and Saito 1970
I_4O_9, I_2O_4 and $I_2O_3 \cdot SO_3$	Vlatkovic and Aten 1961

But even the more discriminating methods do not necessarily reveal the actual situation in the irradiated crystal since for analysis the crystal must be dissolved. And one can often only guess at what recoil species might have generated the observed states upon reaction with the solvent. Ideally, it would be possible to examine the recoil species in situ without disturbing their environment. But in any practical experiment the concentration of recoil atoms is well below the detection limits of available instruments, and interpretation of experimental results must be based on indirect and often incomplete evidence. Despite this handicap, a good deal has been learned about the behavior of recoil species in halogen oxyanions, and at least a general picture of the important features is beginning to emerge.

4. Room Temperature Results

For all systems investigated the results vary widely depending on the exact experimental conditions employed. But clearly there is a tendency for the recoil atom to stabilize in an oxidation state lower than the parent state; and this tendency is most pronounced in the chlorine oxyanions, least in the iodine oxyanions. Considering sodium and potassium salts irradiated at room temperature and analyzed without specific post-irradiation treatment, the ^{38}Cl retention in chlorates is generally about 1.5–5.0% (Vlatkovic and Aten 1962, Seiler and Seiler 1967, Boyd and Larson 1968a, McCallum and Holmes 1952, Aras et al. 1965), the ^{80}Br and ^{80m}Br and ^{82}Br retentions in bromates are about 12–25% (Berne 1964, Apers et al. 1963, Boyd and Larson 1968b, Campbell 1959b, 1960, Saito et al. 1965a, b, 1967a, b, Boyd et al. 1952, Cobble and Boyd 1952, Harbottle 1960, Veljkovic and Harbottle 1962), while the ^{128}I

retention in iodates is about 67–75% (Bellido 1967, Cleary et al. 1952, Ambe and Saito 1970, Aten et al. 1965, Lin and Wiles 1970, Saito and Ambe 1970). Similarly, in perchlorates the $^{38}ClO_4^-$ yield is less than 1% and $^{38}ClO_3^-$ amounts to only 8–16% (Vlatkovic and Aten 1962, Seiler and Seiler 1967, Boyd and Larson 1968a, Aras et al. 1965). In periodates the $^{128}IO_4^-$ retention is only a few percent, but the major product (85–90%) is $^{128}IO_3^-$ (Bellido 1967, Boyd and Larson 1969, Aten and van Berkum 1950). In chlorate and perchlorate systems a few percent $^{38}ClO_2^-$ may be formed (Seiler and Seiler 1967, Boyd and Larson 1968a) and a few percent $^{82}BrO^-$ and $^{82}BrO_2^-$ in bromates (Boyd and Larson 1968b) but no $^{128}IO_2^-$ has been detected in iodine systems (Boyd and Larson 1969).

Moreover, very few of the recoils end up in a higher-than-parent oxidation state. Virtually no $^{38}ClO_4^-$ is found in irradiated chlorates (Vlatkovic and Aten 1962, Seiler and Seiler 1967, Boyd and Larson 1968a, McCallum and Holmes 1952, Aras et al. 1965) and no more than a few percent $^{128}IO_4^-$ occurs in irradiated iodates (Bellido 1967, Boyd and Larson 1969). Although Brown et al. (1969) observed the radiolytic formation of BrO_4^- in $CsBrO_3$, their attempts to find $^{82}BrO_4^-$ in neutron-irradiated $^7LiBrO_3$ were fruitless.

The question of the chemical state of the recoil species in the crystal (before dissolution) remains incompletely answered. As mentioned earlier, available analytical probes are not sensitive enough to observe directly the relatively small number of recoil atoms in the crystal. Several attempts have been made to obtain evidence on the nature of the crystal species by varying the solvent composition. Through such experiments Henry et al. (1957) estimated that about 50% of the ^{82}Br in $KBrO_3$ existed as $BrO^- + BrO_2^-$ at the end of the irradiation with another 20% existing as $Br^- + Br^0$. Apers et al. (1963) have found evidence for Br^0 in irradiated $KBrO_3$, but Boyd and Larson (1968b) argue that this is really BrO_2^-. The last authors propose that the recoil ^{82}Br in bromates stabilizes in the crystal in the solution-stable forms Br^-, BrO^-, BrO_2^-, and BrO_3^-.

In any case, the occurrence of a substantial fraction of the recoil atoms in lower oxidation states upon dissolution is consistent with the idea that nuclear transformations lead to rupture of at least one of the halogen–oxygen bonds and that retention in the parent state is a result of recombination reactions subsequent to the recoil. One of the most effective methods of promoting recombination is by heating the irradiated solid to an elevated temperature, i.e. by thermal annealing.

5. Thermal Annealing

The phenomenon of thermal annealing, first observed in K_2CrO_4 (Green and Maddock 1949), has been amply demonstrated in halogen oxyanions.

Numerous investigators have observed that heating irradiated halogen oxyanions to elevated temperatures produces changes in the relative chemical yields and, below the decomposition temperature, generally increases the yield of radioactivity in the parent state*.

Typical isothermal annealing curves show the retention rising rapidly initially and then more slowly to a pseudo-plateau whose level depends on the temperature, the nature of the sample, its previous treatment, and other experimental conditions. In a few cases 100% retentions have been reported: $^{7}LiBrO_3$ heated at the melting point for 1 h (Boyd and Larson 1968b), $Cd(BrO_3)_2 \cdot 2H_2O$ heated 40 min at 150° C (Arnikar et al. 1970a), and HIO_3 heated 80 min at 90° C (Arnikar et al. 1970b). But the pseudo-plateau usually occurs at a significantly lower level. Typical maximum retentions observed are about 10% in chlorates (Vlatkovic and Aten 1962), about 13% in $NaClO_4$ (Aras et al. 1965), 80–90% in alkali metal bromates (Apers et al. 1963, Campbell and Jones 1968a, Boyd and Larson 1968b, Campbell 1959b), 90–100% in iodates (Cleary et al. 1952, Boyd and Larson 1969, Lin and Wiles 1970, Simpson 1966, Kaučić and Vlatkovic 1963, Arnikar et al. 1970c) and 80–90% in periodates (Aten et al. 1956).

Annealing for identical times but at lower temperatures leads to lower plateau levels (Maddock and Müller 1960, Campbell and Jones 1968b). Maddock and Müller (1960) found the plateau level in $Ca(BrO_3)_2$ to be independent of the thermal history of the irradiated sample. A sample annealed to the pseudo-plateau at 161° C and then heated to 196° C gave the same plateau retention as a sample heated only at 196° C.

The increase in retention must be reflected in a decrease in the yield of some other chemical state, usually assumed to be the halide form. Indeed, Boyd and Larson (1968b) found the $^{82}BrO_3^-$ retention in $LiBrO_3$ increased upon thermal annealing almost solely at the expense of $^{82}Br^-$. They did, however, observe some decrease in the $^{82}BrO_2^-$ and $^{82}BrO^-$ yields: especially at high temperatures, the decomposition of oxyanions becomes important. Thus in $KClO_4$ and $NaClO_4$ the $^{38}ClO_4^-$ yield increases from essentially zero to as much as 13% after 20 min at 500° C, largely at the expense of $^{38}ClO_3^-$ (Apers et al. 1963). After 20 min at 500° C there is no $^{38}ClO_4^-$ in $LiClO_4$, which melts (with decomposition) at 236° C, and no $^{38}ClO_3^-$ in $KClO_3$, which melts (also with decomposition) at 368° C (Aras et al. 1965). In KIO_4, one hour annealing at various temperatures shows that $^{128}IO_3^-$ reaches a maximum of 95% at 75° C,

*See Vlatkovic and Aten 1962, Aras et al. 1965, Apers et al. 1963, Campbell and Jones 1968a, Boyd and Larson 1968b, Campbell 1959b, Cobble and Boyd 1952, Harbottle 1960, Veljkovic and Harbottle 1962, Cleary et al. 1952, Bellido 1967, Ambe and Saito 1970, Aten et al. 1965, Boyd and Larson 1969, Lin and Wiles 1970, Dupetit and Aten 1967, Aten et al. 1956, Simpson 1966, Maddock and Müller 1960, Müller 1961, Andersen and Maddock 1963, Arnikar 1970a, Kaučić and Vlatkovic 1963, Arnikar et al. 1969, 1970b, Jach et al. 1958, Jach and Harbottle 1958, Campbell and Jones 1968b, Jach and Waitz 1969, Arnikar et al. 1970c.

decreases to 22% at 250° C, and then increases to near 100% at 300° C. The last increase occurs at the expense of $^{128}IO_4^-$ which decreases to zero at 300° C. All $^{128}I^-$ is gone at 100° C (Boyd and Larson 1969).

Samples heated above the decomposition temperature actually exhibit 'inverse' annealing, the yield of radioactivity in the decomposing state decreasing rapidly at first, then more slowly. Such inverse annealing has been observed for the ClO_3^- fraction in irradiated $NaClO_4$ at 370° C and above, in $NaClO_2$ at 100° C and 154° C (Vlatkovic and Aten 1962), in $KClO_2$ at 200° C (Boyd and Larson 1968a), in $KBrO_3$ (Jach and Waitz 1969) and in $NaBrO_3$ (Jach 1968) at 319° C and above. By comparing the kinetics of recoil thermal annealing and of bulk crystal decomposition in bromates, Jach has concluded that the two processes may be intimately related (1968, Jach and Waitz 1969).

But it is clear that thermal annealing occurs well below decomposition temperatures. In particular Boyd and Larson (1968b) obtained an apparent threshold of 110° C for thermal annealing in $^7LiBrO_3$, $KBrO_3$, and $CsBrO_3$ irradiated at 35° C, while Maddock and Müller (1960) found a corresponding value of 87° C for $Ca(BrO_3)_2$ irradiated at 40° C.

Indeed, there is extensive evidence that thermal annealing occurs even well below room temperature. Retentions for samples irradiated, stored, and analyzed at low temperatures are always lower than for similar samples handled at room temperature*. For example, Apers et al. (1963) found a ^{80}Br retention of 16.2% in $LiBrO_3$ irradiated at −78° C compared to 27.2% in a sample irradiated at room temperature. A retention of 82.5% was reported for $^{128}IO_3^-$ in $LiIO_3$ irradiated at 23° C compared to 70% for a sample irradiated at −195° C (Arnikar et al. 1970c). And Saito et al. (1965a) found 9.1% $^{78}BrO_3^-$ in $NaBrO_3$ irradiated with fast neutrons at −196° C and dissolved at 0° C, compared to 12% for samples handled at room temperature. Andersen et al. (1966) found a ^{82}Br retention of 14% in $KBrO_3$ irradiated at −78° C, which increased to 17% after 24 h at −11° C and to 18–20% within 24 h at +21° C. In $Cd(BrO_3)_2 \cdot 2H_2O$ the room temperature retention of ^{80m}Br is 23% while the retention in samples irradiated at −196° C is 16% (Arnikar et al. 1970a).

6. Effect of Radiation

Recombination of the recoil atoms with oxygen is also promoted by exposure of the neutron-irradiated salt to electron, gamma, or X-radiation. Although perhaps not as thoroughly studied as thermal annealing, the effects of radiation annealing in bromates (Campbell 1960, Maddock and Müller 1960) and iodates (Cleary et al. 1952, Boyd and Larson 1969, Lin and Wiles 1970,

*See Apers et al. 1963, Saito et al. 1965a, Hobbs and Owens 1966a, b, Saito et al. 1967a, b, Cobble and Boyd 1952, Veljkovic and Harbottle 1962, Aten et al. 1965, Ambe et al. 1968, Arnikar et al. 1970a, Jach and Harbottle 1958, Arnikar et al. 1970c, Andersen et al. 1966a, b.

Arnikar et al. 1969, 1970b) have been well demonstrated: the retention increases upon exposure to radiation. For small doses, the extent of radiation-induced recombination depends linearly on radiation dose (Cobble and Boyd 1952, Boyd and Larson 1969, Maddock and Müller 1960), but recent work (Boyd and Larson 1968b, Lin and Wiles 1970) shows evidence for saturation at large doses.

In $K^{129}IO_4$ (where $^{130}IO_3^-$ is the major product found before radiation annealing) Boyd and Larson (1969) found that ^{60}Co γ-rays promoted a conversion of the $^{130}I^-$ yield to $^{130}IO_3^-$ and $^{130}IO_4^-$. The $^{130}IO_3^-$ fraction reached a plateau after about 5 h irradiation (at 1.03×10^{18}eV g^{-1} min^{-1}), while the $^{130}IO_4^-$ yield continued to increase after 10 h.

Radiation annealing is clearly related to the effect of radiation on the host crystal rather than directly on the recoil species, and attempts have been made to correlate radiolytic yields in halogen oxyanions with annealing results (Berne 1964, Boyd and Larson 1968a, Boyd et al. 1952, Brown et al. 1969, Boyd and Brown 1970, Andersen et al. 1966). An observation of particular interest is that of Cobble and Boyd (1952) in which they found a definite recombination of recoil products upon continued radiation exposure, but no recombination of radiolytic products. They concluded that some of the recoil atoms were in a more reactive state than the products of the radiation decomposition. Andersen et al. (1966) have proposed that the annealing proceeds only upon the thermal release of charge carriers from several radiolytic products including $(BrO{-}BrO_3)^-$, $(Br{-}BrO_3)^-$, and BrO_3^{2-}.

Indeed, it appears that radiation alone is not sufficient to produce recombination, i.e. radiation annealing is thermally activated. For example, Boyd and Larson (1968b) have reported that exposure of alkali metal bromates to gamma radiation at −78° C has no effect on retention, while exposure at 0° C and above induces a definite increase. Maddock and Müller (1960) found the ^{82}Br retention in $Ca(BrO_3)_2$, first thermally annealed 19 h at 149° C corresponding to a pseudo-plateau retention of 43.6%, increases only to 44.9% upon being subjected to 30 Mrad of electron irradiation. But when the sample is then reheated at 148° C for 15 h, the retention rises to 63.3%.

A few instances of post-neutron annealing by exposure of the sample to ultraviolet light have been reported (Arizmendi and Maddock 1961). Here, too, there is a thermal component: Lin and Wiles (1970) found appreciable annealing at −196° C but significantly greater retention at higher temperatures.

These results suggest that the radiation introduces a greater supply of defects (including radiolytic fragments) into the solid and that upon necessary thermal activation, these or their derivatives react with the recoil species.

If post-neutron irradiation, by whatever mechanism, promotes recombination of the recoil atoms with oxygen, then it would seem that radiation concomitant with the neutron bombardment should have a similar effect, particularly where the half-life of the recoil atoms is long relative to the duration of the neutron bombardment. Although some investigators, specifically looking for such effects in halogen oxyanions, have found none (Apers et al. 1963, Harbottle 1960, Bellido 1967), Boyd et al. (Boyd and Larson 1968b, Boyd et al. 1952, Cobble and Boyd 1952, Boyd and Larson 1969) have found significant increases in retention in both bromates and iodates with increasing duration of reactor bombardment. Muruyama and Idenawa (1967) reported that the $^{128}IO_3^-$ retention in NH_4IO_3, HIO_3, and I_2O_5 increases during the first 30 min of irradiation, then stabilizes. Bellido (1967) found a higher retention of both ^{128}I in $NaIO_4$ and ^{130}I in $Na^{129}IO_3$ at a higher neutron flux.

Further, the 12.5% retention observed for ^{36}Cl in chlorates (Sharman and McCallum 1955, Owens and Rowland 1962), compared to 1.5–5.0% for ^{38}Cl, is probably attributable to radiation annealing during neutron bombardment. Generation of convenient amounts of ^{36}Cl requires a much greater irradiation dose than ^{38}Cl, and a sizeable fraction of the ^{36}Cl recoils are subjected to radiation annealing for long periods after their generation.

Also probably the result of radiation annealing during neutron irradiation is the unusually high retention (23–29%) in $LiBrO_3$ (Apers et al. 1963, Campbell and Jones 1968a, Jach and Harbottle 1958). In this case the $^6Li(n,\alpha)T$ reaction, which has a large cross section, generates products which cause extensive radiation damage. The lack of a bromine isotope effect in $LiBrO_3$ (see below) has been explained by such self-radiation damage (Apers et al. 1963, Campbell and Jones 1968a). Retentions in $^7LiBrO_3$ are normal (Apers et al. 1963, Boyd and Larson 1968b).

But even in non-lithium salts, self-radiolysis, arising from the (n,γ) reactions on the halogens, must be recognized as a major source of radiation damage in the locality of the recoil atom (Boyd 1964, Campbell and Jones 1968b). Such damage is unavoidable in any experiment involving a nuclear transformation and, together with external radiation, generates solid-state defects which play a significant role in post-neutron annealing of all types.

7. Effects of Defects

Several attempts have been made to determine the importance of solid-state defects in halogen oxyanions by thermally annealing out the defects before neutron irradiation. Apers et al. (1963) observed no effect of preheating $KBrO_3$, Lin and Wiles (1970) observed none with KIO_3 heated at 103° C and 200° C for 20 min to several hours, and Campbell and Jones (1968a) reported little effect of preheating $LiBrO_3$ at 175° C. But the last authors found that preheating $LiBrO_3$ at 200° C reduces post-recoil annealing of both ^{80m}Br and ^{82}Br somewhat. They have also observed (1968b) that heating $NaBrO_3$ before neutron irradiation increases the initial retention of ^{80m}Br but lowers the level of the pseudo-plateau upon post-neutron thermal annealing. Assuming that pre-neutron heating removes intrinsic solid-state defects, they suggest that the initial retention increases because the recoil atom needs a reducing defect to stabilize in the reduced state. The oxidizing defects required for recombination upon thermal annealing are also presumably removed by pre-neutron heating, leading to the observed lower plateau values.

Campbell and Jones (1968b) were able to identify two different thermal annealing processes in $NaBrO_3$, one occurring between 100° C and 200° C, the other above 200° C. Their data indicate that the lower temperature process is suppressed by pre-neutron heating and may be attributed to intrinsic defects. The higher temperature process, being relatively insensitive to pre-heating, is thought to be mostly due to defects introduced during the neutron irradiation.

Apparently consistent with these results are those of Maddock and Müller (1960) with $Ca(BrO_3)_2$ subjected to pre-neutron irradiation with 1.8 MeV electrons. Pre-irradiation causes the initial retention of ^{82}Br to decrease, the extent depending on the dose. But pre-irradiated samples give higher pseudo-plateau retentions upon post-neutron thermal annealing. If the electron irradiation generates reducing and oxidizing defects, the reducing defects may act to stabilize the recoil in lower states as proposed by Campbell and Jones (1968b). And at elevated temperatures the increased concentration of oxidizing defects may increase the fraction of recoils ultimately recombined into the parent state.

In addition to the lack of effect of pre-heating KIO_3, Lin and Wiles (1970) also found no effect of pre-neutron exposure to UV light for times up to 84 h, nor of pre-neutron exposure to gamma doses of 2.7 Mrad h^{-1} for 40 min. This seems consistent with the hypothesis of Boyd and Larson (1969) that rapid annealing of disrupted iodate ions at least partially accounts for the high retentions observed in iodates. Defects introduced by pre-neutron irradiation may be essentially annealed out at room temperature before generation of the recoil atoms. The effects of low temperature pre-neutron irradiation might be revealing in this respect.

In contrast to KIO_3, Arnikar et al. (1969, 1970b) have found that pre-neutron heating of HIO_3 increases the retention of ^{128}I as iodate.

Vlatkovic and Aten (1962) found that although pre-neutron heating of $NaClO_4$ has little effect on the initial retention, it has a substantial effect on the thermal annealing behavior. The $^{38}ClO_3^-$ yield upon annealing at 425° C remains at about 12% after 80 min in samples pre-heated at 370° C. In samples merely dried at 160° C before neutron exposure, the $^{38}ClO_3^-$ yield drops to about 6% under the same thermal annealing conditions.

8. Composition Effects

In addition to solid-state defects, either intrinsic or induced, the final chemical state of the recoil atoms depends on the chemical composition of the solid. Specifically, it is affected by the nature of the cation, water of crystallization, and by the nature and abundance of other anions present.

8.1. CATION EFFECT

Several investigations concerned with the importance of the cation in halogen oxyanions have been reported*. Although meaningful comparisons can be made only among salts handled identically during the entire experiment, in general the retention among the alkali metal halogen oxyanions appears to be relatively insensitive to the specific cation, except lithium. However, Ambe and Saito (1970) have observed somewhat higher retention for rubidium and cesium iodates than for the sodium and potassium salts, a difference they attribute to differing crystal structures. Boyd and Larson (1968b) found that although the BrO_3^- retention depends little on the cation, the yields of $^{82}BrO_2^-$ and $^{82}BrO^-$ are strongly cation dependent in the series Li, ^{7}Li, Na, Cs and Ag bromates. They also have noted that the cation affects the ease and extent of both thermal and radiation annealing in this series. Some of their data suggest that recombination proceeds through the intermediate states.

Saito et al. (1965b, 1964, Ambe et al. 1968) have examined the retention of ^{80m}Br, ^{82}Br, and ^{78}Br in a variety of bromates, especially of the transition metals, and, with the exception of $TlBrO_3$, find a good correlation of the retention with the electron affinity of the cation. They argue that cations with large electron affinities may trap electrons, preventing the reduction of the recoil bromine species, and thus enhancing the retention. A similar explanation has been offered by Ambe and Saito (1970) for the observation of higher retention in hydrated transition metal iodates than in $NaIO_3 \cdot H_2O$. The exception with $TlBrO_3$ is thought to be related to oxidation-reduction reactions involving Tl(I) and Tl(III).

The room temperature retentions in $Ca(BrO_3)_2$ (Maddock and Müller 1960, Müller 1961, Arizmendi and Maddock 1961) and $Ca(IO_3)_2$ (Ambe and Saito 1970, Kaučić and Vlatković 1963) do not differ significantly from corresponding values of alkali metal salts. However, the retention of ^{128}I decreases successively in KIO_3, HIO_3, $KH(IO_3)_2$ and NH_4IO_3 (Cleary et al. 1952, Muruyama and Idenawa 1967), which suggests that cationic hydrogen in the crystal may act as a reducing agent.

8.2. WATER OF CRYSTALLIZATION

The presence of water in the crystal affects the retention in halogen oxyanions, but the nature of the effect is not clear. Ambe and Saito (1970) found that the

*See Vlatkovic and Aten 1962, Boyd and Larson 1968b, Saito et al. 1965b, 1967a, b, Cleary et al. 1952, Ambe and Saito 1970, Ambe et al. 1968, Andersen and Maddock 1963a.

retention for $(n,2n)^{126}I$ and $(n,\gamma)^{128}I$ in a variety of hydrated iodates is lower than that of the corresponding anhydrous salt for all pairs studied. But in $Cu(BrO_3)_2 \cdot 6H_2O$ the retention of both ^{80m}Br and ^{82}Br is significantly greater than in $Cu(BrO_3)_2 \cdot 2H_2O$ (Saito et al. 1965b, 1967a, b). Maddock and Müller (1960) reported the same ^{82}Br initial retention in $Ca(BrO_3)_2$ and $Ca(BrO_3)_2 \cdot H_2O$, but after thermal annealing the retention in the hydrate is much higher than in the anhydrous salt.

Deuteration of the water of crystallization causes no significant effect on the ^{126}I or ^{128}I retention in $NaIO_3 \cdot H_2O$ (Ambe and Saito 1970), while in $Zn(BrO_3)_2 \cdot 6H_2O$, the retention is 4–9% greater than in the deuterated salt (Andersen and Maddock 1963a).

An interesting result reported by Arnikar et al. (1970a) is that if irradiated $Cd(BrO_3)_2 \cdot 2H_2O$ is allowed to dehydrate during thermal annealing, the retention increases rapidly to 100%. But if the dehydration of the irradiated salt is done at room temperature before annealing, the retention increases more slowly upon thermal annealing to a limiting value of 53%. The authors' interpretation is that radiolytic oxygen is trapped by water of crystallization during irradiation and then released to react with recoil bromine species upon thermal annealing at 150° C. In the dehydrated salt, the radiolytic oxygen either escapes or is trapped at sites which do not release it upon thermal annealing.

Results reported by Müller (1961) show that part of the effect of water of crystallization may be due to differing crystal phases of the hydrated and anhydrous salts. He finds that the initial retention of ^{82}Br in $Ca(BrO_3)_2 \cdot H_2O$ is about 20%, while that in $Ca(BrO_3)_2$ is about 15%. Hydration of the latter at room temperature increases the retention to 20% and subsequent dehydration at 163° C increases it to 44%. On the other hand, when the irradiated hydrate is allowed to dehydrate at 163° C, the retention increases to 60.5% and even without dehydration at 113° C, the measured retention is 51%. Müller argues that such behavior is not consistent with the Harbottle and Sutin hot-zone model (1958) and instead suggests that there is no disruption of the bromate ion, only a high charge assumed by the central bromine atom. The idea of an initial high charge on the bromine recoil atom is consistent with the observation that almost all ^{82g}Br is formed via the highly internally converted isomeric transition from ^{82m}Br. It may be worth pointing out that part of Müller's results can be interpreted along the lines suggested by Arnikar et al. (1970a). The greater retention in $Ca(BrO_3)_2 \cdot H_2O$ at room temperature and upon thermal annealing at 163° C may be due to the trapping (and subsequent release) of radiolytic oxygen by the water of crystallization while the water added upon hydration of the anhydrous $Ca(BrO_3)_2$ may serve to transfer the otherwise inaccessibly trapped oxygen to sites available to the recoil bromine. The retention then increases from 15% to 20% at room temperature and to a plateau of 44% upon thermal annealing. The lower plateau of the originally anhydrous salt compared to that of the hydrated salt may indicate the fraction of radiolytic oxygen which escaped the lattice during irradiation.

8.3. OTHER ANIONS

Only a very limited study has been made of the effect of other anions in halogen oxyanion systems. But the importance of the environment is shown by the fact that ^{82}Br retention increases from about 20% in pure $NaBrO_3$ to 50–55% as the $NaClO_3$ content increases in a series of $NaBrO_3$–$NaClO_3$ mixed crystals (Boyd and Larson 1968a, Hobbs and Owens 1966a, b). This effect appears to be suppressed at low temperatures: the increase is from 10% to about 19% at −77° C (Hobbs and Owens 1966a, b).

8.4. ISOTOPE EFFECT

An important question is whether isotopes arising from different nuclear reactions in the same solid will show the same chemistry. As it turns out, differences in measured chemical yields of different isotopes are often observed. In a wide variety of bromates both the initial retention and the pseudo-plateau retention upon thermal annealing are found to be 2–6% greater for ^{82}Br than for ^{80m}Br (Apers et al. 1963, Campbell and Jones 1968a, Saito et al. 1965b, 1967a, b, Harbottle 1960, Veljkovic and Harbottle 1962, Andersen and Maddock 1963a, Jach and Harbottle 1958). Two exceptions are $LiBrO_3$ where no difference (Campbell and Jones 1968a, Jach and Harbottle 1958) or a slight reverse effect (Apers et al. 1963) is found and $TlBrO_3$ (Saito et al. 1965b, 1967a, b) where ^{80m}Br retention is greater than ^{82}Br. Essentially no difference has been found in bromate yields for ^{78}Br and ^{80}Br in $NaBrO_3$ (Saito et al. 1965a, Ambe et al. 1968).

Isotope effects are also observed in iodate systems, but the results are not as consistent. Retentions of ^{128}I in excess of ^{130}I retentions have been reported for $NaIO_3$ (and $Na^{129}IO_3$) (Bellido 1967) but in KIO_3 (and $K^{129}IO_3$) the ^{130}I retention is slightly greater (Boyd and Larson 1969). Retentions of ^{128}I exceed those of ^{126}I in KIO_3 (Ambe and Saito 1970, Aten et al. 1965, Saito and Ambe 1970), $CsIO_3$ (Dupetit and Aten 1967) and several other iodates (Ambe and Saito 1970, Saito and Ambe 1970). In $CsIO_4$ Dupetit (1967) reports IO_3^- yields of 86.7%, 89.4%, and 87.3% for ^{128}I, ^{126}I, and ^{130}I (from the ^{133}Cs(n,α) ^{130}I reaction). In KIO_4 (and $K^{129}IO_4$) Boyd and Larson (1969) found the $^{128}IO_3^-$ yield and the $^{128}IO_4^-$ yield greater than the respective ^{130}I yields.

Apparently the explanation for isotopic differences does not lie in differing recoil energies. The recoil energies of ^{80m}Br and ^{80}Br or of ^{128}I and ^{130}I should not be greatly different, and experiments comparing chemical yields of isotopes produced with recoil energies differing by orders of magnitude often have similar results (Saito et al. 1965a, Ambe and Saito 1970, Ambe et al. 1968). All current evidence points to different modes of nuclear de-excitation as the principal source of the isotope effect. Several authors have suggested that the electronic state of the recoil atom trapped in a region of more or less disorder is the principal factor determining its ultimate state*. Indeed, isomeric transition effects (Jones 1967, 1970b, Shiokawa et al. 1969b, 1971, Sasaki and Shiokawa 1970), are ascribed almost entirely to Auger charging, and Jones (1970b) finds isomeric transition results in $NaBrO_3$ not much different from (n,γ) or (n,2n) results. This strongly suggests that long-lived excited nuclear states whose decay is highly internally converted are of major importance in the chemistry of recoil halogens. The available evidence on such states has recently been reviewed by Jones (1970a). He points out that while there are no long-lived states in the (n,γ) cascade populating ^{80m}Br or ^{80}Br, nearly 100% of the ^{82g}Br is formed from the 6.2 min ^{82m}Br isomer which has an internal conversion coefficient of 382. Details are lacking for some reactions of interest, but it seems likely that isotope effects can be attributed to such details of de-excitation.

One different and somewhat puzzling isotope effect has been observed by Andersen and Maddock (1963a). They found the ^{80}Br and ^{82}Br retentions in $^{64}Zn(BrO_3)_2 \cdot 6H_2O$ greater than in $^{68}Zn(BrO_3)_2 \cdot 6H_2O$ by a factor of 1.06. The fact that this factor remains constant upon thermal annealing suggests that the difference is in the primary retention and not in the annealing process.

9. Summary

Taken together, the above facts suggest the following view of recoil reactions in halogen oxyanions. After the nuclear transformation the halogen recoil finds itself at rest in the crystal in a state of electronic excitation and ionization determined by the exact nature of the nuclear process. Initially it is associated with fewer ligand oxygen atoms than in the parent state, perhaps with none. The ultimate fate of the recoil atom is determined by thermally activated competing interactions with reducing and oxidizing defects, either intrinsic or induced during irradiation, including radiolytic fragments of the host lattice, or with the host ions themselves. The nature of the defects and their concentrations are functions of the chemical composition and physical structure of the solid.

All these effects are closely interrelated and untangling them is the great challenge in the study of recoil effects in halogen oxyanions. Success will require better control over each of the factors or at least detailed knowledge of the variations. More sensitive and more selective analytical techniques, detailed information on nuclear reaction mechanisms and on the mechanisms of

*See Apers et al. 1963, Campbell and Jones 1968a, Boyd and Larson 1968b, 1969, Saito et al. 1967a, b, Ambe et al. 1968, Müller 1961, Saito and Ambe 1970.

radiation and thermal decomposition of the host lattices, and information on the nature of the solid-state defects present and their chemistry are called for. Transfer annealing experiments, in which it has been found that radioactive non-recoil bromide or iodide introduced into halogen oxyanion lattices can be converted by heat or radiation to bromate or iodate (Boyd and Larson 1968b, 1969, Kaučić and Vlatkovic 1963) may be a step in the right direction. At least here, the initial state of the reacting atom is known, and it should be possible to study the solid-state reactions without the uncertainty introduced by the nuclear processes. When this is done, we may be better able to understand the reactions of recoil atoms in the solids.

Addendum (December 1976)

Ambe and Ambe (1973) have reported a study of the effects of the $^{37}Cl(n,\gamma)^{38}Cl$ and $^{35}Cl(n,2n)^{34m}Cl$ reactions in anhydrous and hydrated K, Cs, Ba, Ni, and Cu chlorates. For the (n,γ) reaction, the dependence of the results on the nature of the cation parallels that found previously for bromates (Saito et al. 1967a, b) and iodates (Ambe and Saito 1970). An isotope effect is observed: the oxidized yield of ^{34m}Cl is greater than that of ^{38}Cl in K and Ba salts. But again the effect appears not to be the result of differing recoil energies.

Retentions of ^{80m}Br following the (n,γ) transformation and subsequent thermal annealing have been measured for $NaBrO_3$ (Arnikar and Rao 1971, Arnikar and Patil 1971), $Ca(BrO_3)_2$ and $Ca(BrO_3)_2 \cdot H_2O$ (Arnikar et al. 1971a, b), and hydrated Sr and Ni bromates (Arnikar et al. 1973a, b). In the last systems, the effects of γ-radiation during neutron-bombardment and of dehydration of the crystals were observed.

The retention of ^{82g}Br in $NaBrO_3$–$NaClO_3$ solid solutions has been interpreted in terms of recombination reactions of the recoil atoms with trapped radiolytic fragments produced by self-radiolysis (Owens and Boyd 1976). A method for preparing carrier-free radiobromine from neutron-irradiated $KBrO_3$ has been described (Alfassi and Feldman 1976).

The chemical effects of neutron capture in Li, Na, K, Cs, and NH_4^+ perbromates have been studied (Hasan and Heitz 1975, Gavrilov et al. 1975a, b). For the alkali metal perbromates, the oxidized yields are intermediate between those for perchlorates and for periodates. Reducing fragments produced upon decomposition of NH_4^+ ions account for the significantly lower oxidized yields observed with NH_4BrO_4.

The effects of irradiation temperature, thermal annealing, pre-neutron gamma irradiation, and pre-neutron thermal treatment on the retention of ^{128}I in $NaIO_3$ have been studied (Arnikar et al. 1971a, b). Interestingly, pre-neutron gamma irradiation reduces the retention, while pre-neutron thermal treatment shows no significant effects.

Further work (Hull and Owens 1975) on transfer annealing in KIO_3 suggests that a volatile intermediate may play an important role in the conversion of radioiodide to iodate.

CHAPTER 8

THE HOT-ATOM CHEMISTRY OF CRYSTALLINE CHROMATES

Carol H. COLLINS and Kenneth E. COLLINS

Universidade Estadual De Campinas, Department of Chemistry,
S.P., Brazil

Chemical Effects of Nuclear Transformations in Inorganic Systems
Edited by G. Harbottle and A.G. Maddock

Contents

Introduction

Chromates in general, and potassium chromate in particular, have been attractive as compounds for hot-atom chemical study because of the favorable nuclear properties of chromium, the great thermal and radiation stability of the compounds, the apparent structural simplicity of the crystals and the presumed known and simple chemistry of the expected recoil products. A wealth of information has been accumulated over the past 25 years, from which the anticipation of a straightforward chemistry has given way to an expanding realization that these systems are actually quite complex.

More solid-state hot-atom chemical studies have dealt with potassium chromate than with any other compound. Thus, a major fraction of this review is given to this compound. Our emphasis is on recent literature and on the present views of phenomena which affect the chemical fate of recoil chromium atoms in chromates. Many other data are tabulated so that the interested reader can speculate independently on the results of a wide variety of experiments.

1. Nuclear Properties of Chromium

1.1. RADIATIVE NEUTRON CAPTURE TO PRODUCE ^{51}Cr

Most of the hot-atom chemical studies with chromates have utilized the ^{50}Cr(n,γ)^{51}Cr reaction with thermal neutrons. Nearly all (> 98%) of the recoil ^{51}Cr atoms receive recoil energies greater than 50 eV from the gamma-ray deexcitation cascade following neutron capture (Andersen 1968) and a majority (55%) of the ^{51}Cr atoms undergo gamma deexcitation transitions of 8.5 MeV or more (Bartholomew et al. 1966) producing nuclear recoil energies greater than 700 eV. Thus, nearly every recoil ^{51}Cr atom has sufficient initial kinetic energy to break its chemical bonds and most have sufficient kinetic energy to cause much local damage and disorder in the neighbourhood of a recoil event.

In addition to the initial recoil impulse experienced by nearly all recoil atoms, within perhaps 10^{-15} sec following neutron capture, and the small but finite probability of internal conversion–Auger charging events occurring on a similar time scale (von Egidy 1969), there are at least two energy levels of ^{51}Cr [at 776 keV and 748 keV above ground (Bartholomew et al. 1966) of sufficiently long lifetime [5.53 and 7.35 nsec, respectively (Haar and Richter 1970)] to affect the chemical fate of the recoil atoms. The resulting delayed nuclear deexcitation, by either gamma-ray emission or by an internally converted transition, takes place long after the initially deposited 'hot-zone' energy has been dissipated (Jones 1970a) and can deposit significant recoil energy and/or electronic displacement energy at the site to affect the ultimate fate of the recoil atom.

1.2. OTHER NUCLEAR TRANSFORMATIONS OF CHROMIUM

The $^{54}Cr(n,\gamma)^{55}Cr$ activation is readily accessible with thermal neutrons, but the short half-life (3.5 min) has discouraged its use. Both ^{49}Cr (42 min) and ^{51}Cr (27.8 d) are produced, from ^{50}Cr and ^{52}Cr, respectively, by (n,2n) and (γ,n) activations. Results from studies using these activations will be discussed in a later section.

2. Physical and Chemical Aspects of Irradiated Chromates

2.1. THERMAL PROPERTIES

The thermal stability of potassium chromate is sufficiently great to allow preparation of single crystals (Andersen and Olesen 1965) or of doped crystals (Andersen and Maddock 1963d, Costea and Podeanu 1967, Milenković and Veljković 1967, Milenković and Maddock 1967) from the molten (960° C) salt. Potassium dichromate is stable to at least 800° C (Spitsyn et al. 1960). Other chromate compounds (Spitsyn et al. 1960, Flood and Muan 1950, Skorik et al. 1967, Udupa et al. 1970) decompose at somewhat lower temperatures, with ammonium chromate, ammonium dichromate and ammonium trichromate decomposing at less than 200° C (Fischbeck and Spingler 1938, Simpson et al. 1958, Mahieu et al. 1971).

The principal atmospheric pressure solid state phase transition in potassium chromate occurs at 663° C (Pistorius 1962) while that in sodium chromate occurs between 392° C and 424° C (Marchart and Grass 1965a). At temperatures above 600° C, oxygen atoms from the surrounding atmosphere readily exchange with oxygen of the chromates of lithium, sodium, potassium, rubidium and cesium (Shakhashiri and Gordon 1965). However, both the exchanges and the phase transitions occur at temperatures above those commonly used in hot-atom chemical studies.

2.2. RADIOLYTIC PROPERTIES OF CHROMATES

Radiolysis studies on potassium chromate at ambient temperature, using ^{60}Co gamma radiation or ambient reactor radiation, have shown that the chemical damage which results in measurable amounts of Cr(III) being present after dissolution of the sample gives a *G*[total Cr(III)] value of about 10^{-2} for relatively low (ca. 10^7 rad) absorbed dose (Yang et al. 1973) and about 10^{-4} at high ($> 10^{10}$ rad) absorbed dose (Shibata et al. 1969, Gütlich et al. 1971b). Radiolysis, using reactor radiation, of other chromates gives greater amounts of Cr(III) (Harbottle and Maddock 1958). Most heavily irradiated chromates give an insoluble product which has been tentatively identified, for the case of ammonium dichromate, as $[Cr(OH)_2]_2CrO_4$ (Nawojska 1967). The radiolytic yield of Cr(III) from potassium chromate appears to decrease at both higher and lower temperatures than the ambient region and is greater for irradiations in vacuo than in air (Yang et al. 1973).

Several recent investigations have utilized IR (Bancroft et al. 1970) and ESR (Constantinescu et al. 1968, Schara et al. 1970, Lister and Symons 1970, Debuyst 1971, Debuyst et al. 1972a, b, c) spectroscopy to identify, in situ, radiation-produced species in solid potassium chromate. Several observations have significance to hot-atom chemical studies. If the irradiation is carried out at low temperature (−196° C), there is evidence only for the production of species in which *no* Cr–O bonds are broken; CrO_4^- and CrO_4^{3-} (Lister and Symons 1970, Debuyst et al. 1972a). Gamma irradiation at ambient temperature gives as many as six different species, including some identified as fragmented CrO_4^{2-} [CrO_3 (Bancroft et al. 1970), CrO_3^- (Constantinescu et al. 1968, Debuyst et al. 1972c) and CrO_2^- (Debuyst et al. 1972a)]. Some of these species survive heating at greater than 100° C (Bancroft et al. 1970, Schara et al. 1970, Debuyst et al. 1972a). Ambient temperature irradiations to 10^8 rad result in a *G* (total radicals) of 10^{-4} (Debuyst 1971). Significant quantities of Cr(III) species, although indicated by dissolution studies, are not seen by these in situ methods.

3. Analysis of ^{51}Cr Recoil Species

Recoil chromium atoms very probably exist in a set of different states within the crystalline chromate lattice. Detecting, identifying and measuring these species directly by physical means is not yet possible. Hence, the natures of the ^{51}Cr-tagged species within the irradiated crystals have been inferred, and their quantities measured, by chemical analysis after dissolution of the crystals in aqueous solution.

For many years, the ^{51}Cr species in the crystals were differentiated only into $^{51}Cr(VI)$ and $^{51}Cr(III)$ by precipitation of Cr(VI) (Green et al. 1953, Harbottle

1954, Veljković and Harbottle 1962) or Cr(III) (Popov 1959, Ikeda 1963, Yeh et al. 1969, 1970, Collins et al. 1972), extraction (Marchart and Grass 1965a, b), electrophoresis (Andersen and Maddock 1963b), alumina column separation (Mahieu et al. 1971, Apers et al. 1964, Collins et al. 1965, Veljković et al. 1965a, Khorana and Wiles 1971) and anionic (Dimotakis 1968, Dimotakis and Stamouli 1968, Baumgärtner and Maddock 1968, Brune 1967) or cationic (Costea and Podeanu 1967, Dimotakis and Stamouli 1964) exchange column separations. Of these, the most widely used separation procedure has been that of Cr(VI) precipitation as $PbCrO_4$ (Green et al. 1953, Harbottle 1954). Recent experiments (Ackerhalt et al. 1970) suggest that quantitative errors may have been made in at least some of these measurements, although most qualitative observations were probably valid.

The similarities between the measured quantities of Cr(VI) observed with different aqueous phase analytical procedures or observed following dissolution in non-aqueous solvents (Harbottle 1954, Veljković and Harbottle 1962, Andersen and Maddock 1963b) or in aqueous solutions containing added foreign ions (Green et al. 1953, Veljković and Harbottle 1961b, 1962) or reducing agents (Harbottle 1954) support the supposition that the ^{51}Cr(VI) in the solution of the irradiated crystals does, indeed, represent Cr(VI) species present within the neutron-irradiated crystal (Harbottle 1965) before dissolution. Thus, analyses which measure only the Cr(VI) are principally of value only in studying changes in the initial or annealed yield values.

Several cation analysis procedures have been described (Gütlich and Harbottle 1966, Andersen and Sørensen 1966a, Ackerhalt et al. 1969, Gütlich et al. 1971a, Collins et al. 1971) which measure the ^{51}Cr(VI) yield and the relative amounts of several aqueous ^{51}Cr(III) species. Dissolution is often carried out in solutions of low bulk pH (Ackerhalt et al. 1969, Collins et al. 1971) to minimize the hydrolysis reactions of the several Cr(III) species (Laswick and Plane 1959, Kolaczkowski and Plane 1964). However, the actual pH at the interface of the dissolving crystals is undoubtedly higher. This factor, as well as effects of changes in other dissolution parameters, on the measured distribution of the Cr(III) species, needs further study.

Several procedures have been used to search for the presence of ^{51}Cr(II) in air-free solutions of neutron-irradiated chromates (Andersen and Olesen 1965, Gütlich and Harbottle 1966, Andersen and Sørensen 1966a). The presence of ^{51}Cr(II) as a recoil product is reported for ammonium chromate and for several chromic salts (Gütlich and Harbottle 1966). It has also been reported in non-chromate targets subjected to implantation by $^{51}Cr^+$ ions (Andersen and Sørensen 1966a). No ^{51}Cr(II) has been seen in neutron-irradiated potassium chromate (Gütlich and Harbottle 1966) or in potassium chromate exposed to high energy $^{51}Cr^+$ ions (Andersen and Sørensen 1966a). These results must be viewed with some caution, however, since the rapid interaction between Cr(II) and Cr(VI) (Hegedus and Haim 1967) would tend to produce ^{51}Cr(III)-

monomer from any ^{51}Cr(II) present in chromates, due to the high Cr(VI) concentration in the immediate environment of any dissolving Cr(II) species.

The relationships between the solid-state ^{51}Cr-labelled species and those found in solution are not known. Nevertheless, it is useful to assume that each measured solution phase species [^{51}Cr(VI), ^{51}Cr(III)-monomer, ^{51}Cr(III)-dimer, ^{51}Cr(III)-trimer and ^{51}Cr(III)-higher polymer*] derives from one or more corresponding solid state precursors. This assumption is supported by the recrystallization studies (Andersen and Sørensen 1966a, Ackerhalt et al. 1971) and by theoretical considerations (Maddock and Collins 1968).

Another useful assumption relates to the identification of the various solid state ^{51}Cr(III)-precursor species in terms of the differences seen in the $\underline{\mathrm{M}}$, $\underline{\mathrm{D}}$ or $\underline{\mathrm{P}}$ yields upon dissolution after samples are subjected to different treatments. Thus, a decrease in observed solution phase $\underline{\mathrm{M}}$ is ascribed to a similar decrease in solid state precursor, *M*. Also useful is the present convention of ascribing increases or decreases in the relative amounts of a measured species, e.g. $\underline{\mathrm{M}}$, to discrete populations of different *M* species (e.g., *M1*, *M2*, etc.). Whether these assumptions are valid or not cannot be determined until new analytical procedures are available. Nevertheless, inferences about the behaviour of the relevant solid-state species can be obtained through various combinations of pre- and post-neutron irradiation treatments of the chromates, followed by dissolution analysis. Such data are the basis for much of this review.

4. Initial Yields of Neutron-Irradiated Potassium Chromate

The initial $\underline{\mathrm{VI}}$ yield, i.e., the $\underline{\mathrm{VI}}$ yield observed in the absence of post-irradiation treatment, is a quantity which serves as a reference value for post-irradiation annealing studies and as an index of the influence of various factors, such as pre-neutron irradiation treatments, which may influence the transient reactions of ^{51}Cr recoil atoms following neutron activation events.

The initial $\underline{\mathrm{VI}}$ yield depends on many parameters. Figure 8.1a summarizes much of the available reactor-dose dependence data from pure (non-doped, non-pretreated) potassium chromate crystals irradiated at ambient reactor temperature†. The observed scatter within the data is attributable to the actual

*These solution phase species will be subsequently represented by $\underline{\mathrm{VI}}$, $\underline{\mathrm{M}}$, $\underline{\mathrm{D}}$, $\underline{\mathrm{T}}$ and $\underline{\mathrm{P}}$, respectively. The solid-state precursors of these solution phase species will be represented by *VI*, *M*, *D*, *T* and *P*.

†Ambient reactor temperature has been measured in only a few cases (Veljković and Harbottle 1962, Stamouli 1971. In these cases, the temperature of the irradiation site (pneumatic conveyor tube, thermal column) was not the same as the nominal temperature of the cooling water (or air). In addition, the irradiation temperature can change during the course of an irradiation (Veljković and Harbottle 1962). Lacking definitive information, the term 'ambient reactor temperature' will be used to denote a temperature range of ca. 40–100° C, assumed to be present in most irradiation facilities (Andersen and Olesen 1965, Costea and Podeanu 1967, Milenković and Veljković

irradiation conditions in the different irradiation facilities used, such as different ambient temperature, different relative fluxes of fast and thermal neutrons and different levels of ambient gamma radiation.

The dependencies of M and DP yields on reactor dose for irradiations at ambient reactor temperature are shown in figs. 8.1b, c (Gütlich and Harbottle 1966, Andersen and Sørensen 1966a, Andersen and Baptista 1971a, Stamouli 1971, Collins and Collins, unpublished work). These dependencies are more pronounced than that of VI and may be related to a radiation-promoted *M* to *DP* reaction.

Whenever D and P have been determined separately, it has been shown that the D/P ratio is approximately constant (Ackerhalt et al. 1969, Stamouli 1971). Thus, these Cr(III) species are generally considered to behave as a single population, DP. The possibility of ^{51}Cr recoil species other than VI, M and DP cannot be excluded. Although direct experimental evidence is lacking (Gütlich and Harbottle 1966) $^{51}Cr(II)$-forming species (Andersen 1968, Matsuura and Matsuura 1968) and perhaps others (Batasheva et al. 1968) may comprise significant fractions of the non-$^{51}Cr(VI)$ recoil fragments. The solution phase behaviour of these initially non-hydrated species, like that of non-hydrated Cr(III) (Marchart and Grass 1965b), may be quite different from that of the hydrated species normally studied (Laswick and Plane 1959, Kolaczkowski and Plane 1964). Since available analytical procedures do not distinguish these species, this review will generally discuss only VI, M and DP as ^{51}Cr recoil products of neutron-irradiated potassium chromate and the other chromates.

The effect of lowering the irradiation temperature upon the initial VI, M and DP yields is pronounced. No reactor dose dependence is seen for crystals irradiated at −78° C or lower (Andersen and Baptista 1971a, Stamouli 1971, Gütlich and Harbottle 1967). The VI, M and DP yields from such low temperature irradiations are essentially constant, the average values being 47.1 ± 1.5%, 20.2 ± 1.2% and 32.2 ± 1.5%, respectively, over a reactor dose range from 1.7×10^{12} to 6.5×10^{15} $n \cdot cm^{-2}$ (Veljković and Harbottle 1961b, 1962, Yeh et al. 1970, Andersen and Baptista 1971a, Stamouli 1971, Gütlich and Harbottle 1967, Marqués and Wolschrijn 1969, Ackerhalt and Harbottle 1972, Collins et al. 1972). It is of interest that these initial values indicate a much higher DP content than M content. Short ambient temperature irradiations, followed by storage at low temperature, produce more M than DP (Andersen and Baptista 1971a, Stamouli 1971). The differences in all the yields caused by the change in irradiation temperatures reflect the rapid thermal-

1967, Shibata et al. 1969, Harbottle and Maddock 1958, Green et al. 1953, Harbottle 1954, Collins et al. 1965, Dimotakis 1968, Dimotakis and Stamouli 1964, Gütlich and Harbottle 1966, Andersen and Sørensen 1966a, Gütlich et al. 1971a, Dimotakis and Stamouli 1965, 1967, Costea 1969, Ackerhalt 1970, Costea et al. 1970, Andersen and Baptista 1971a).

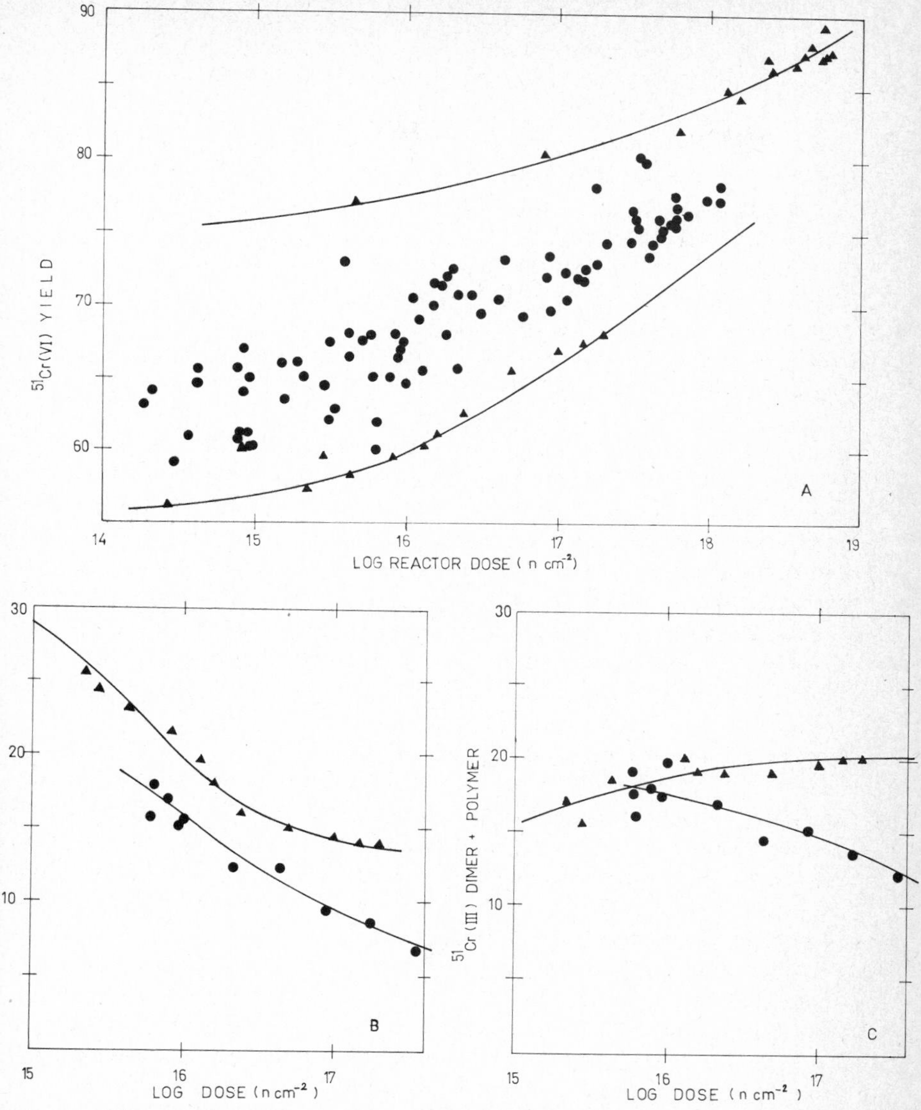

Fig. 8.1. Effect of reactor dose on (a) VI yield, (b) M yield and (c) DP yield from potassium chromate: ● samples irradiated at ambient reactor temperature, ▶ samples irradiated at relatively high reactor temperature, ▲ samples irradiated at ambient reactor temperature and stored, after irradiation, at low temperature. (From Andersen and Olesen 1965, Costea and Podeanu 1967, Milenković and Veljković 1967, Shibata et al. 1969, Harbottle and Maddock 1958, Green et al. 1953, Harbottle 1954, Veljković and Harbottle 1962, Collins et al. 1965, Dimotakis 1968, Dimotakis and Stamouli 1964, 1965, 1967, Gütlich and Harbottle 1966, Andersen and Sørensen 1966a, Gütlich et al. 1971a, Costea 1969, Ackerhalt 1970, Costea et al. 1970, Andersen and Baptista 1971a, Stamouli 1971, Collins and Collins, unpublished work.)

and/or radiation-promoted processes which occur when samples are irradiated at ambient reactor temperature. These concurrent processes result in higher VI yields than are obtained from room temperature storage (annealing) in the absence of radiation, although a rapid thermal annealing process does account for a ca. 10% difference in the observed VI yield for samples irradiated and stored at −196° C, then annealed at room temperature (Stamouli 1971).

The radioactive decay events within irradiated potassium chromate crystals may also promote changes in ^{51}Cr distribution during storage. This phenomenon, observed when irradiated crystals are heated at 130° C during ^{42}K decay (Maddock et al. 1963), presumably also accounts for the observation that room temperature storage after an ambient temperature neutron irradiation results in a higher VI yield than is observed when ambient temperature-irradiated samples are stored at −196° C for 8 days prior to 8 days storage at room temperature (Stamouli 1971). The effect that this generally overlooked factor, post-irradiation storage temperature, has on both initial values and those observed after subsequent treatment emphasizes the importance of specifying the detailed history of each experimental sample.

Several other factors affecting initial VI yields are listed, and their effects summarized, in table 8.1. None of the pre-treatments thus far tested has any measurable effect on the initial yields when the neutron irradiation is carried out at low temperature (Andersen and Baptista 1971a, Stamouli 1971, Collins et al. 1972a). On the other hand, experiments which have been designed to increase (or decrease) concentrations of lattice defect species, prior to ambient temperature neutron irradiation, do cause observable changes, although the results are qualitatively quite mixed. An investigation of the apparent inconsistencies observed in cation-pre-doped crystal studies indicates that concentration of the dopant, the influence of reactor irradiation conditions and the reactor dose received all affect the observed VI yield in a non-additive manner (Costea and Podeanu 1967, 1968). Thus, reactor irradiation parameters may tend to obscure the contributions attributable to the crystal pre-treatment parameters.

The relatively high values of initial VI yields obtained from neutron-irradiated potassium chromate (and most other chromates) have long been of interest. At first, these high values were attributed to the influence of solution phase redox reactions on fragment species upon dissolution of the irradiated crystals in water (Green et al. 1953). However, the fact that the VI yield is insensitive to wide differences in the composition of the dissolution media (Green et al. 1953, Harbottle 1954, Veljković and Harbottle 1962, 1961b, Andersen and Maddock 1963b) strongly suggests that this interpretation is not adequate (Harbottle 1965). As an alternative possibility, the relatively high VI yield may measure very efficient, low-temperature thermal annealing processes (Maddock and Vargas 1961a, b), those which give slow annealing even at −196° C (Ackerhalt 1970, Marqués and Wolschrijn 1969). The rapid thermal

Table 8.1
Effects of irradiation parameters and crystal pre-treatments on initial and annealed yields from neutron-irradiated potassium chromate

Variable	Effect on initial VI yield[a]	Ref.*	Annealing treatment[b]	Effect on annealing to VI[a]	Ref.*
Ambient-temperature neutron irradiations					
Lower irradiation temperature	–	[1,9,36, 37,54]	Δ< 150 γ	– 0	[1] [1]
Increase irradiation temperature	+	[8,55]			
Evacuate before irradiation	0	[2,3]	Δ> 150	0	[2,3]
Irradiation to higher reactor dose	+	[1,2,8,9, 30,55]	Δ> 150	+	[4–6]
Pre-mix with α-emitter:					
H_3BO_3	+	[7,56]			
$NH_4HB_4O_7 \cdot 4H_2O$	+	[7]	Δ> 150	+	[7]
$(Li)_2CO_3$	+	[8,43]	Δ> 150	+	[8]
Increase crystal size	+	[5,9,10, 11]	Δ> 150 e	+ +	[5] [9]
Pre-crush crystals	–	[3,8,10, 11,16,19, 56]	Δ> 150	0[c]	[3,8,10, 11]
	+[d]	[8]	Δ> 150	0[c]	[8]
Pre-compression to 2000 atm	0	[12]	Δ> 150	0	[12]
Pre-heat (with slow cooling) to:					
350° C	0 –	[14] [13]	 Δ> 150	 0	 [13,14]
600° C	+ –[e]	[9] [9]	 e	 –	 [9]
Pre-heat with rapid cooling (quenching) from:					
500° C, 700° C, 900° C	0	[1,15,16]	Δ< 150 Δ> 150 γ	0 + –	[1] [15,16] [1]
Pre-heat to and quench from melt (960° C)	+	[4]	Δ> 150	+	[15]

*Refs. on p. 173.

Pre-irradiate with 1.8 MeV electrons to:					
10 Mrad	–	[9]	$\Delta > 150$	+	[9]
150 Mrad	–	[2,10,11]	$\Delta > 150$	–	[10]
Pre-irradiate with ^{60}Co-γ to 200 Mrad	0	[1,2]	γ	–	[1]
Pre-dope with cations by co-melting with chromates of:					
Na^+ or Rb^+ at 10^{-4}–5×10^{-3} MF[f]	0	[15,16]	$\Delta > 150$	0	[15,16]
Tl^+ at 10^{-4}–5×10^{-3} MF	–	[15,16]	$\Delta < 150$	–	[15]
			$\Delta > 150$	+	[15,16]
Ca^{2+} at 10^{-4}–10^{-2} MF	0	[16]	$\Delta > 150$	+	[16]
			e	–	[16]
	–	[15]	$\Delta < 150$	–	[15]
			$\Delta > 150$	+	[15]
	+	[1]	$\Delta < 150$	–	[1]
			γ	–	[1]
Sr^{2+} at 5×10^{-5}–10^{-3} MF	0	[15]			
5×10^{-3} MF	+	[15]			
Ba^{2+} at					
5×10^{-5}–5×10^{-4} MF	0	[15]			
10^{-3} MF	+	[15]			
10^{-5}–10^{-3} MF	–	[4,6]	$\Delta > 150$	–	[4,6]
10^{-5} MF	+[e]	[4,6]	$\Delta > 150$	+[e]	[4,6]
10^{-4}–10^{-3} MF	–[e]	[4,6]			
10^{-5}–10^{-3} MF	–[d]	[4]			
La^{3+} at					
5×10^{-5}–5×10^{-3} MF	+	[16,17]	$\Delta > 150$	+	[16,17]
10^{-4}–10^{-3} MF	–	[15]			
Cr^{3+} at 10^{-5} MF	–	[18]	$\Delta > 150$	–	[18]
Pre-dope with anions by co-melting with potassium salts of:					
Cl^-, Br^- or I^- at 10^{-4}–10^{-1} MF[f]	–	[19]	$\Delta < 150$	+	[19,20]
			$\Delta > 150$	–	[19,20]
PO_4^{3-} at 10^{-6}–10^{-1} Mf	+	[13]	$\Delta < 150$	0	[13]
			$\Delta > 150$	–	[13]
BeF_4^{2-} at 0.13–0.99 MF	–	[21,22,23, 24,25,30]	$\Delta > 150$	–[g]	[21,22, 23,24]
			P	–	[25]
Co-crystallize with:					
K_2SO_4 at 0.75–0.99 MF[f]	+	[2,24,30]	$\Delta < 150$	+	[24]
			$\Delta > 150$	–	[24]
			P	–[h]	[25]
	–	[21]			
$(NH_4)_2SO_4$ at 0.99 MF	–	[24]			
$Co(NH_3)_6Cl_3$ at 0.4 MF	0	[45,57]			

Pre-coated on surface of:					
MgO at 4×10^{-4}–4×10^{-3} MF[i]	–[d]	[26,27]	Δ> 150	–[d]	[26,27]
4×10^{-2} MF	0[d]	[26,27]	Δ> 150	–[d]	[26,27]
Al_2O_3 at 4×10^{-4} MF	–[d]	[27]	Δ> 150	–[d]	[27]
SiO_2 at 4×10^{-4} MF	–[d]	[27]	Δ> 150	–[d]	[27]
Low-temperature neutron irradiation					
Evacuate before irradiation	0	[9]	Δ> 150	0	[9]
Irradiate to higher reactor dose	0	[1,5,9]	Δ< 150	0	[1]
Increase crystal size	0	[9]			
Pre-heat (with slow cooling) to: 400° C, 600° C or 900° C	0	[9,28]			
Pre-heat and quench from: 500° C, 700° C or 900° C	0	[1,28]	Δ< 150	0	[1]
			Δ> 150	0	[28]
Pre-irradiate with ^{60}Co-γ to 200 Mrad	0	[1]	Δ< 150	0	[1]
Pre-dope with cations by co-melting with chromates of: Ca^{2+} at 10^{-4}–5×10^{-3} MF[f]	0	[1]	Δ < 150	0	[1]

[a] Effects: + = increase in observed VI yield, – = decrease, 0 = no change.
[b] Treatments: Δ = thermal treatment, e = 1.8 MeV high energy electron irradiation, γ = ^{60}Co-gamma irradiation, P = compression to 2000 atm.
[c] Rate of reaching plateau enhanced; no change in plateau value.
[d] Irradiated at relatively high reactor temperature.
[e] Irradiated to higher total reactor dose than previous entry.
[f] MF = mole fraction of non-potassium chromate component in all doping and co-crystallizing entries.
[g] Plateau value is lower although degree of annealing is enhanced.
[h] No compression annealing observed.
[i] MF = mole fraction of potassium chromate deposited on oxide surface for all oxide-coating entries.

[1] Stamouli 1971. [2] Green et al. 1953. [3] Podeanu and Costea 1971. [4] Costea and Podeanu 1967. [5] Gütlich and Harbottle 1967. [6] Costea and Podeanu 1968. [7] Dimotakis and Stamouli 1964. [8] Costea 1969. [9] Andersen and Baptista 1971a. [10] Maddock et al 1963. [11] Maddock and Vargas 1959. [12] Andersen and Maddock 1963c. [13] Milenković and Maddock 1967. [14] Pertessis and Henry 1963. [15] Andersen and Olesen 1965. [16] Andersen and Maddock 1963d. [17] Costea et al. 1970. [18] Collins et al. 1965. [19] Milenković and Veljković 1967. [20] Milenkovic et al. 1971. [21] Andersen and Sørensen 1966a. [22] Maddock and de Maine 1956b. [23] de Maine et al. 1957. [24] Maddock and Vargas 1961a, b. [25] Andersen 1963. [26] Veljković et al. 1965a. [27] Veljković et al. 1965b. [28] Collins unpublished. [29] Baumgärtner and Maddock 1968. [30] Harbottle 1954. [31] Costea et al. 1963. [32] Ackerhalt and Harbottle 1972. [33] Marchart and Grass 1965a. [34] Marques and Wolschrijn 1969. [35] Yeh et al. 1970. [36] Veljković and Harbottle 1962. [37] Veljković and Harbottle 1961b. [38] Yeh et al.

1969. [39] Andersen and Maddock 1963a. [40] Ikeda 1963. [41] Gorla and Lazzarini 1967. [42] Gütlich and Harbottle 1966. [43] Dimotakis and Stamouli 1965. [44] Marchart 1965b. [45] Gainar and Gainar 1962. [46] Muxart et al. 1947. [47] Ottar 1953. [48] Green and Maddock 1949. [49] Getoff and Maddock 1963. [50] Andersen and Maddock 1963b. [51] Harbottle and Maddock 1958. [52] Babeshkin 1969. [53] Gili Trujillo and Canwell Morcuenda 1968. [54] Ackerhalt 1970. [55] Shibata et al. 1969. [56] Dimotakis and Stamouli 1968. [57] Ikeda and Kujirai 1971. [58] Milenković and Veljković 1974. [59] Costea et al. 1971a. [60] Sherif et al. 1974. [61] Ladrielle et al. 1974. [62] Stamouli 1974. [63] Stamouli 1975. [64] Costea et al. 1971b. [65] Batasheva et al. 1970.

Table 8.1a

Effects of irradiation parameters and crystal pre-treatments on initial and annealed yields from neutron-irradiated potassium chromate

Variable	Effect on initial VI yield	Ref.*	Annealing treatment	Effect on annealing to VI	Ref.*
Ambient temperature neutron irradiations					
Pre-dope with anions by co-melting with potassium salts of:					
Br^- at 10^{-2}MF		[58]	$\Delta < 150$	+	[58]
			$\Delta > 150$	+	[58]
PO_4^{3-} at 10^{-2}MF			$\Delta < 150$	+	[58]
			$\Delta > 150$	–	[58]
PO_4^{3-} at $10^{-4} - 10^{-2}$MF	–	[59]	$\Delta < 150$	–	[59]
			$\Delta > 150$	–	[59]
Pre-dope with anions by co-melting with sodium salts of:					
PO_4^{3-} at $10^{-4} - 10^{-2}$MF	–	[59]	$\Delta < 150$	0	[59]
			$\Delta > 150$	–	[59]
Co-crystallize with:					
K_2SO_4 at 0.05–0.90MF	+	[60]			

*See footnote to table 8.1.

annealing which occurs when samples are brought to room temperature from −196° C (Ackerhalt 1970) further supports the possibility that low-temperature thermal processes may cause some portion of the high initial VI yield.

Another possible explanation to account for the relatively high VI yields from most neutron irradiated chromates is that of localized chemical activation of fragmented ^{51}Cr-recoil atom sites by secondary recoils which accompany decay of the long-lived levels populated in deexcitation cascades from the (n,γ) activation of ^{50}Cr. Eighty percent or more of the ^{51}Cr recoil atoms pass through the 5.53 nsec–776 keV state, through the 7.35 nsec–748 keV state or through both states (Bartholomew et al. 1966, Haar and Richter 1970). Delayed transitions of these energies can give 5–6 eV of recoil energy to the ^{51}Cr atom long after the initial thermalization (Jones 1970). Such a low energy recoil will, on average, not greatly affect the population of ^{51}Cr atoms which have already been stabilized as $^{51}Cr(VI)$-producing states upon thermalization following the initial high-energy recoil. However, fragmented ^{51}Cr-containing species (those which would tend to give one of the several $^{51}Cr(III)$ species upon dissolution) probably exist in metastable chemical states in the lattice which may be activated for 'reentry' reactions (Müller 1967a) as a result of a 5–6 eV recoil event. Thus, the relatively high VI yield in chromates and dichromates might result from promotion of *III* → *VI* processes by delayed-state recoil-producing transitions.

5. Thermal Annealing of Neutron-Irradiated Potassium Chromate

Potassium chromate was the first solid in which thermal annealing of recoil atoms was demonstrated (Green and Maddock 1949). Since that first report, considerable effort has gone into the study of the processes which form *VI* in heated crystals. These studies have shaped most of the current views on what happens to recoil chromium atoms in crystalline chromates.

The effect on the annealed VI yield of changing irradiation parameters (for ambient temperature irradiation) is quite pronounced. Increasing the reactor dose increases the proportion of VI obtained from a given thermal pulse (Costea 1969). Irradiation and storage at low temperatures, which markedly decreases the initial VI yield, also somewhat lowers the absolute VI yield from a given thermal pulse (Stamouli 1971) although the 'degree of annealing' (the fraction of potentially annealable sites which are actually involved in a thermal reaction to produce *VI*) is substantially greater for samples irradiated at low temperature (−78° C, −196° C) than for samples irradiated and stored at ambient temperature (Stamouli 1971).

Irradiation and annealing of samples of potassium chromate in different atmospheres have generally shown little or no effect on the VI yield (Green et al. 1953, Andersen and Baptista 1971a, Maddock and de Maine 1956b, de Maine et al. 1957). However, the presence of 2 to 5 atm of oxygen does depress the VI yield from longer (4 h and 22 h) although not from short term (60 min) annealing at 245° C (Ackerhalt et al. 1971). A similar depression due to air is reported at shorter annealing times in potassium chromate crystals which have been crushed, heated and recrushed prior to irradiation (Podeanu and Costea 1971).

The effects of various pre-treatments on the annealed VI yield are summarized in table 8.1. In each case, the value of the isothermal 'pseudo-plateau' has been used to determine the effect of the given pre-treatment on the thermal annealing process. Irradiation at low temperature prevents the appearance of any significant effect from the pre-treatments (Andersen and

Baptista 1971a, Stamouli 1971, Collins et al. 1972b). For samples irradiated at ambient reactor temperature, the qualitative influence on the thermal annealing is not necessarily the same as the effect on the initial yields. The direction of the effect often appears to be dependent on the temperature of annealing used, with the effect being different at temperatures greater than about 150° C than it is below this temperature. The lack of a predictable pattern to the data is probably due to a combination of several factors, with the influence of the irradiation and storage parameters possibly being the most important.

Isochronal and linear tempering annealing studies show that there are several temperature regions in which annealing processes readily give rise to *VI* while in other temperature intervals the change in VI yield is less pronounced (Andersen and Olesen 1965, Milenković and Veljković 1967, Milenković and Maddock 1967, Andersen and Sørensen 1966a, Ackerhalt et al. 1969, Ackerhalt et al. 1971, Andersen and Baptista 1971a, Gütlich and Harbottle 1967). This is illustrated in fig. 8.2 which gives typical 60 min isochronal annealing curves for a sample of potassium chromate irradiated at −78° C (Ackerhalt et al. 1971). Comparison of a number of isochronal curves from samples having different histories prior to heating suggests that three distinct *VI*-forming processes can be distinguished; a low-temperature process completed by a 60 min pulse at less than 20° C, an intermediate process which becomes significant at about 100° C for a 60 min pulse and a high-temperature process which becomes important with a 60 min pulse at a temperature of about 200° C. This latter process can be seen at temperatures nearer to 150° C in isochronal annealing curves from 4 to 20 h of heating.

Typical isochronal curves for M and DP are also shown in fig. 8.2. In addition to changes in M and/or DP which occur in temperature regions where *VI* is formed, significant changes in M and DP yields occur in temperature regions where little formation of *VI* is observed (Ackerhalt et al. 1969, 1971, Andersen and Baptista 1971a). Additionally, the M and DP yields appear quite sensitive to the annealing atmosphere. Thus, from fig. 8.2 it appears that, in the presence of air, the ^{51}Cr(III) species which produces *VI* in the temperature region from 100–180 is *DP* with no apparent change in *M*, while the reverse is true when the 60 min annealing pulse is carried out in vacuo. However, despite this apparent drastic change in the *VI*-producing precursor, the rate at which *VI* is formed is independent of the annealing atmosphere.

The quandary presented by these conflicting observations can be better examined using isothermal annealing curves. Figure 8.3 shows the results of four different annealing programmes. The notation on each curve describes the histories of the individual samples*. The dotted lines in fig. 8.3, at 60 min

*This notation (Collins and Willard 1962, Collins and Harbottle 1964) serves to give an abbreviated description of solid-state chemical experiments. It is particularly useful where sequences of operations (sample treatments) are involved. Each independent operation is separated by a diagonal slash (/). Thus, the sequence vac/$(n,\gamma)^{10'}_{-78}$/s^{10d}_{-196}/Δ^{t}_{245}, describes a

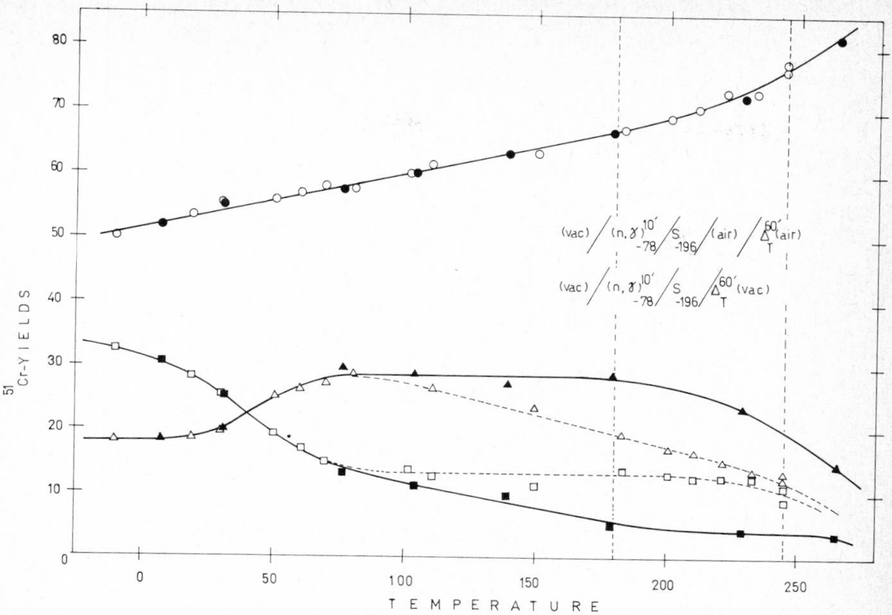

Fig. 8.2. Isochronal annealing curves for potassium chromate heated for 60 min in vacuum or in air: ● VI yield (air); O VI yield (vac); ▲ M yield (air) △ M yield (vac); ■ DP yield (air); □ DP yield (vac). (From Ackerhalt et al. 1971.)

illustrate the VI, M and DP values seen at the respective temperatures, also indicated by dotted lines in fig. 8.2.

Figure 8.3a shows that the 'crossover' reaction (Ackerhalt et al. 1969), wherein some *DP* is converted to *M*, is complete in about two days at room temperature (Stamouli 1971). The rate of this reaction is faster as the temperature is increased; the crossover reaction being complete in 60 min at 70° C in most samples. The temperature at which the crossover reaction occurs appears to be sensitive to crystal preparation; in some crystals a 140° C pulse of 60 min is required for its completion (Andersen and Baptista 1971a), although others report the crossover to be complete in 60 min at temperatures between 35° C and 60° C (Ackerhalt et al. 1969, 1971, Andersen and Baptista 1971a, Collins and Collins 1971). Figure 8.3b shows thermal

---

sample's history as: pre-neutron-irradiation evacuation of sample, neutron irradiation for 10 min at −78° C, subsequent storage (S) for 10 days at −196° C followed by thermal annealing (still under vacuum) at 245° C for time *t*. Other symbols used in this review are: γ for gamma irradiation, e for high energy electron irradiation, P for pressure treatment and UV for ultraviolet radiation. The subscript AT is used for ambient temperature (irradiations); RT for room temperature (storage).

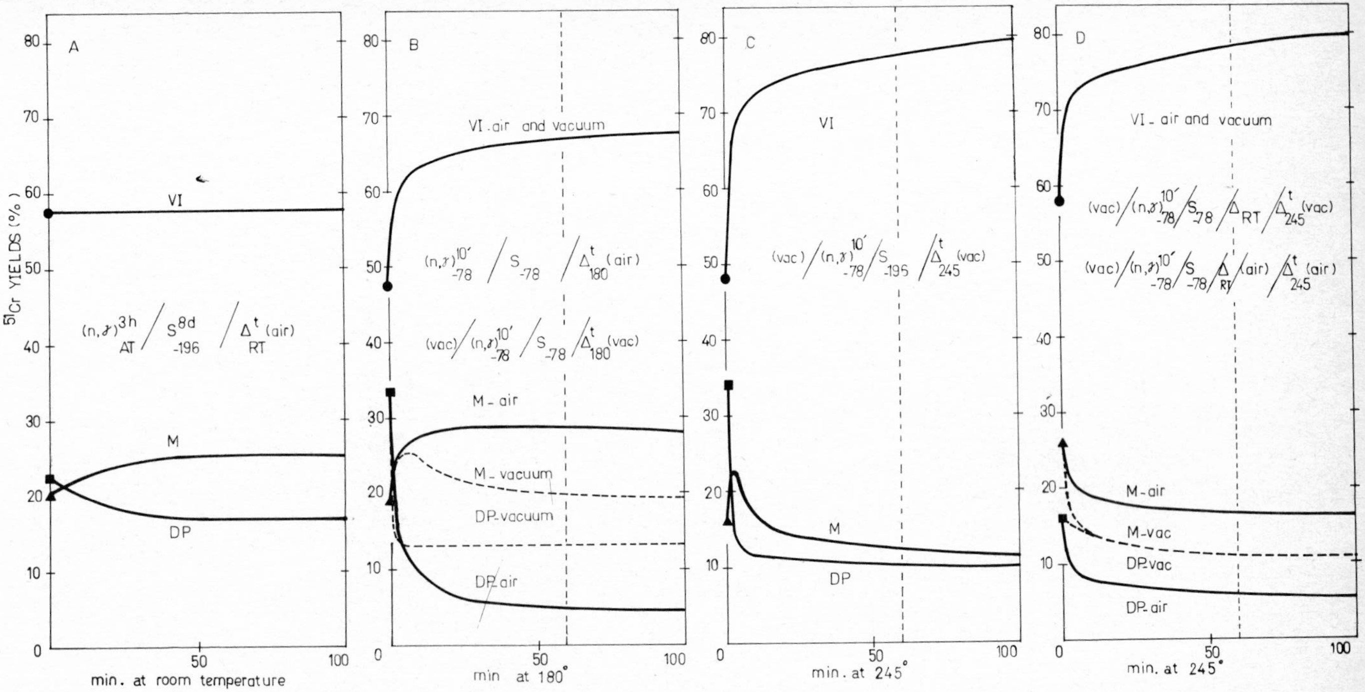

Fig. 8.3. Typical isothermal annealing curves for potassium chromate with different sample histories. (From Ackerhalt et al. 1969, Stamouli 1971, Collins and Collins, unpublished work.)

annealing treatments at 180° C (Collins and Collins, unpublished work). In both air and vacuum, the rate of production of *VI* is the same. In both cases, an increase in M reflects the *DP* to *M* crossover reaction. However, the extent of this reaction is considerably less than that observed at, say, 100° C and, in vacuo, the M yield goes through a maximum, suggesting that some of the newly formed *M* may be annealing to *VI*. This is confirmed by fig. 8.3c, thermal treatment in vacuo at 245° C (Ackerhalt et al. 1969), where rapid increase in M is also followed by a fall as *M* produces *VI*. In this case, the increase–decrease quantitatively represents the *M* formed from *DP* in the crossover reaction since the slower portion of the M curve extrapolates back to the amount of M initially present. Thus, the reaction which rapidly produces *VI* at 180° C and at 245° C is of the type $M2 \rightarrow VI$, with *M2* being produced from *DP2* by the crossover reaction*, $DP2 \rightarrow M2$.

Figures 8.3b, d indicate that the extent of the *DP* to *M* reaction is sensitive to atmosphere (Collins unpublished) although the isochronal data (fig. 8.2) indicate that the crossover reaction occurring below 100° C is not atmosphere-sensitive (Ackerhalt et al. 1971, Andersen and Baptista 1971a). It thus appears that a second, oxygen-promoted *DP* to *M* reaction occurs (Ackerhalt et al. 1971) and this results in the higher M yields seen in the presence of air in figs. 8.2, 8.3b and 8.3d. This reaction, $DP3 \xrightarrow{O_2} M3$, occurs only at temperatures above 120° C.

Additionally, in contrast to the *M2* species formed by the lower temperature crossover reaction, *M3* does not anneal rapidly to *VI* but instead enters the 'pool' of ^{51}Cr(III) species which anneal slowly to *VI*.

Annealing from the 'pool' of ^{51}Cr(III), which includes *M*, *D*, *P* and (possibly) other non-^{51}Cr(VI) species (Andersen 1968, Matsuura and Matsuura 1968, Batasheva et al. 1968) occurs at temperatures greater than 150° C. The rate of these *VI*-forming reactions may be controlled by diffusing bulk-crystal entities in such a way that these reactions occur essentially independently of the *initial* identity of the ^{51}Cr(III) species involved (Collins et al. 1965, Ackerhalt et al. 1969, 1971, Collins et al. 1972b). The constancy of the M/DP ratio (Andersen and Sørensen 1966a, Ackerhalt et al. 1969) in this temperature range, after the initial fast reactions, supports this interpretation.

In contrast to earlier reports (de Maine et al. 1957), extended time isotherms indicate that, even at 245° C (Ackerhalt 1970), the production of *VI* from ^{51}Cr(III) occurs at a slow but measurable rate with VI not approaching 100% in 100 days at 200° C (Andersen 1968).

Sequences of thermal treatments at increasingly higher temperature (for samples having the same pre-thermal annealing histories) reveal that the same VI, M and DP values are observed as are obtained from comparable

*The numerical notation given here to the several *M* and *DP* species assists in distinguishing the several (presumably) discrete species and their reactions.

experiments employing only the final treatment (Ackerhalt 1970, Maddock et al. 1963, Maddock and de Maine 1956b). This indicates that the several thermal annealing processes which produce *VI* involve distinct populations of crystal phase species which are not appreciably altered by heating at a lower temperature.

The kinetics of thermal annealing processes forming *VI* involve several components. No studies have been made of the low-temperature processes suggested by the isochronal annealing experiments. The intermediate-temperature processes shown by the isochronal data have been described in terms of two first-order processes of activation energies 0.8 eV (for a 60–90° C process) and 1.1 eV (for a 120–140° C process) (Andersen and Olesen 1965). Other studies suggest activation energy spectra for the intermediate temperature processes centered at 0.71 eV (Marqués and Wolschrijn 1969), 0.4–0.7 eV (Yeh et al. 1969) or 0.8–1.2 eV (Veljković and Harbottle 1962). The kinetics of processes taking place at temperatures above 150° C are complex (Andersen and Olesen 1965, Maddock 1965c). Several analyses of these kinetics have been attempted (Milenković et al. 1971, Maddock and De Maine 1956, Harbottle and Sutin 1959, Veljković 1970, Costea et al. 1963) with varying degrees of success. No discrete activation energies can be attributed to these high temperature processes, but determination of activation energy spectra indicate bands centered at 1.1–1.4 eV (Harbottle and Sutin 1959, Maddock and Wolfgang 1968), 1.36 eV (Andersen and Olesen 1965) 1.76 eV (Costea et al. 1970), 1.83 eV (Marqués and Wolschrijn 1969), 1.87 eV (Costea 1969, Costea and Podeanu 1968) or 2.0–2.1 eV (Milenković et al. 1971). The activation energies for the high-temperature processes decrease at higher reactor doses or in the presence of crystal dopant species (Costea 1969, Costea and Podeanu 1968), further supporting the interpretation that annealing at temperatures above 150° C is controlled by properties of the bulk crystal.

6. Radiation Annealing of Neutron-Irradiated Potassium Chromate

Exposure of neutron-irradiated potassium chromate to either high-energy electrons or gamma radiation produces changes in VI, M and DP yields similar to those produced by increasing the reactor dose (Andersen and Baptista 1971a, Stamouli 1971). A typical radiation annealing experiment, fig. 8.4 (Stamouli 1971), shows a tendency toward 'saturation' which commonly occurs at about 100 Mrad (Andersen and Maddock 1963, Green et al. 1953, Harbottle 1954, Stamouli 1971, Maddock et al. 1963d). However, the degree of radiation annealing to this point can vary widely, apparently depending on the sample history, as is illustrated in table 8.2. In contrast to the thermal annealing case, the degree of annealing to *VI* during ambient temperature irradiation is essentially the same, whether the sample is neutron irradiated at −196° C or at ambient reactor temperature (Stamouli 1971). Some data suggest that the degree of radiation annealing to *VI* is sensitive to changes in dose rate and in the 'quality' of the radiation (Harbottle 1954, Baumgärtner and Maddock 1968). However, these conclusions are uncertain since similar changes in the degree of annealing are also found for samples having (presumably) the same prior history at (presumably) the same dose rate (Andersen and Maddock 1963d, Baumgärtner and Maddock 1968, Stamouli 1971 and Maddock et al. 1963).

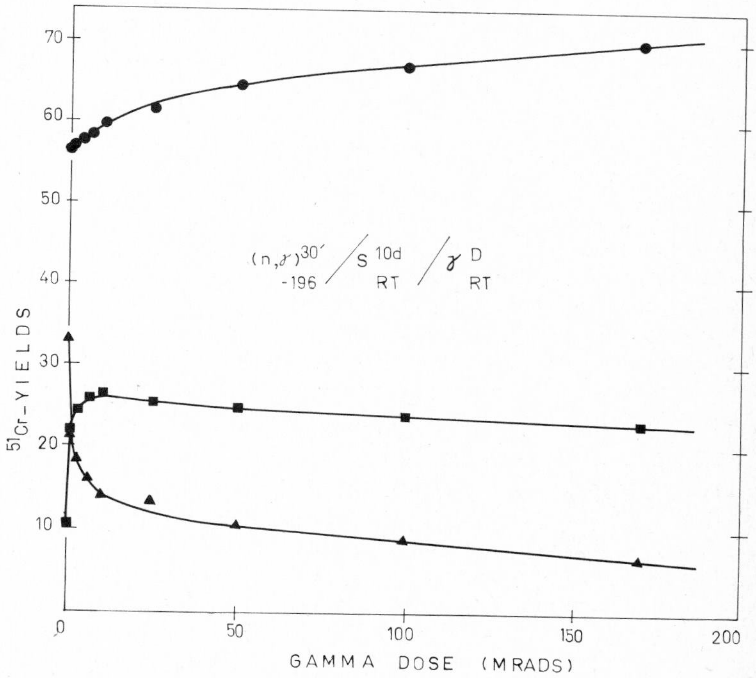

Fig. 8.4. Typical radiation annealing curves for potassium chromate: ● VI yield; ▲ M yield; ■ DP yield. (From Stamouli 1971.)

Essentially no radiation annealing to *VI* occurs with either gamma (Stamouli 1971) or high-energy electron (Baumgärtner and Maddock 1968) irradiation at −196° C, in agreement with the lack of significant reactor dose effect during low-temperature neutron irradiation. Thermal annealing at room temperature of samples which have been both neutron irradiated and exposed to gamma irradiation at −196° C shows no effect attributable to the gamma dose received; rather it reflects only the thermal processes observed in the absence of gamma radiation (Stamouli 1971). Samples which are neutron irradiated at

Table 8.2
Summary of radiation annealing effects in neutron-irradiated potassium chromate

Radiation source	Dose rate (rad/h)	Temperature of irradiation (°C)	% annealed to VI in 20 Mrad	% annealed to VI in 100 Mrad	Ref.*
Neutron irradiation at ambient reactor temperature					
1.8 MeV e^-	–	ambient	10.2	44.6	[10]
	–	ambient	16.2	57.6	[10]
	–	ambient	8.7	30.7	[16]
	–	ambient	5.0	20.6	[16]
	–	ambient	9.2	40.2	[29]
10 MeV e^-	–	ambient	9.2	–	[9]
^{60}Co-γ	2.5×10^5	ambient	19.2	65.4	[2]
	4×10^5	ambient	9.1	26.9	[1]
	4×10^5	ambient	–	15.2	[1]
	7.8×10^5	ambient	23.4	–	[30]
^{182}Ta-γ	5.8×10^4	ambient	49.0	–	[30]
Fission products	–	ambient	36.5	52.3	[29]
	–	ambient	38.0	78.3	[29]
^{60}Co-γ	2.5×10^5	−28° C	17.1	67.3	[30]
1.8 MeV e^-	–	−196° C	0.8	1.6	[29]
Neutron irradiation at −196° C					
^{60}Co-γ	4×10^5	ambient	9.6	25.9	[1]
		0° C	6.25[a]	–	[1]
		−196° C	1.1[b]	–	[1]

[a]Irradiated to 16 Mrad only. [b] Irradiated to 10 Mrad only.
* See footnote to table 8.1.

ambient reactor temperature and subsequently exposed to high-energy electron bombardment at −196° C do show some subsequent thermal annealing to *VI* upon being warmed to room temperature or higher (Baumgärtner and Maddock 1968). However, in contrast to an earlier report (Maddock 1965c), the sequence $(n,\gamma)_{AT}/e_{-196}/\Delta_T$ produces considerably lower VI yields than does the sequence $(n,\gamma)_{AT}/e_T$ (Baumgärtner and Maddock 1968).

Although the reactions which produce *VI* in most radiation annealing situations are relatively slow [typically, 9–10% of the available ^{51}Cr(III) sites anneal in 20 Mrad (Andersen and Maddock 1963d, Baumgärtner and Maddock 1968, Andersen and Baptista 1971a, Stamouli 1971, Maddock et al. 1963)] a very rapid reaction, complete in 2 Mrad at ambient temperature, occurs in which some *M* is converted to *DP*. Samples neutron irradiated at −196° C and subsequently gamma irradiated at this temperature show none of the *M* to *DP* reaction if the samples are also stored at −196° C (Stamouli 1971). If the samples are held at room temperature for one week (to permit the low-temperature thermal annealing processes which produce *VI* and cause the *DP2* → *M2* reaction to occur) subsequent gamma irradiation at −196° C reveals a significant *M* to *DP* reaction (Stamouli 1971), suggesting that the gamma-sensitive *M* species is that formed by the thermal crossover reaction. Since gamma irradiation at −196° C does not appear to promote significant bond disruption (Lister and Symons 1970, Debuyst 1971), the radiation-promoted *M2* → *DP2* reaction which occurs even at −196° C may require little or no mobility of the species involved.

Thermal annealing at temperatures of 130° C or higher, of samples receiving 50–100 Mrad of gamma or high-energy electron irradiation at ambient temperature prior to the thermal treatment, does not cause significant change in VI (Andersen and Maddock 1963d, Baumgärtner and Maddock 1968, Stamouli 1971, Maddock et al. 1963) but does change the M and DP distribution (Stamouli 1971), presumably due to *reinitiation* of the *DP2* → *M2* thermal crossover reaction. Thus, a subsequent (second) radiation treatment produces a second radiation-promoted *M2* → *DP2* reaction. In addition, although little change in *VI* occurs due to the thermal treatments, a new, rapid (< 5 Mrad), radiation-promoted reaction of the type *M* → *VI* occurs. Since a thermal pulse of 130° C also causes the oxygen-promoted $DP3 \xrightarrow{O_2} M_3$ reaction to occur, the *M* species which rapidly give *VI* during the second irradiation is probably the one produced by this oxygen-promoted reaction. Support for this view comes from experiments in which neutron-irradiated samples are heated at temperatures of 130° C or higher prior to the gamma (or high-energy electron) irradiation, causing considerable enhancement of the VI yield in a fast (< 5 Mrad) reaction (Andersen and Maddock 1963d, Baumgärtner and Maddock 1968, Stamouli 1971, Maddock et al. 1963).

For neutron-irradiated samples which have been heated and then exposed to radiation, the effect of a subsequent (second) thermal treatment appears to depend on the temperature used. Thermal treatment at 80° C produces no change in VI and only a slight change in M and DP (Stamouli 1971), while thermal treatment at 150° C shows a small (< 1.5%) though rapid increase in the VI yield (Stamouli 1971) (probably from some of the *M2* formed in the *DP2* → *M2* thermal reaction). A final heating for 16 h at 190° C is reported to cause a significant change in the VI yield (Andersen and Maddock 1963d). No annealing curve is given for this, however, so it is not known if this latter *VI*-forming reaction is primarily a slow process from the ^{51}Cr(III) pool, possibly assisted by some radiation-produced entities, or whether it involves a rapid increase from an unidentified precursor.

The effect of treatments carried out prior to neutron irradiation on the subsequent radiation annealing are summarized in table 8.1. All pre-treatments for which data are available appear to suppress the radiation-promoted *M* to *DP* reaction (Andersen and Baptista 1971a, Stamouli 1971), while most also depress the radiation annealing processes which produce *VI* (Stamouli 1971).

The kinetics of radiation annealing, as well as the degree of annealing, appear to be very dependent on sample pre-history. Some data indicate that the initial reaction is first order, giving a linear plot of log(100 −VI) vs absorbed dose (Green et al. 1953, Harbottle 1954, Maddock et al. 1963d) while other data reveal no such linear relationship (Andersen and Maddock 1963, Baumgärtner and Maddock 1968, Andersen and Baptista 1971a, Stamouli 1971). More probably, most of the annealing to *VI* results from crystal (and radiation) dependent reactions from the 'pool' of $^{51}Cr(III)$ species. Thus, the apparent kinetics would not be expected to be first order, although they might appear to be so after certain crystal pre-treatments which would level the $^{51}Cr(III)$-pool to entities of similar reactivity.

7. Pressure Annealing of Neutron-Irradiated Potassium Chromate

Crushing of neutron-irradiated potassium chromate does not alter the observed VI yield (Maddock et al. 1963, Maddock and Vargas 1959) but does enhance subsequent thermal annealing (Andersen and Maddock 1962). Similarly crushing a sample which has reached the 'pseudo-plateau' by thermal treatment does not change the VI yield but does promote additional thermal annealing to a higher plateau (Andersen and Maddock 1962, 1963c).

Rapid processes initiated by application of high pressure increase the observed VI yield and enhance the yields observed in subsequent thermal annealing, when this pressure is applied in the presence or absence of carbon tetrachloride (as an inert agent). Surprisingly, no pressure effects are observed when the high pressure is applied in the presence of paraffin. Repetitive compressions have no cumulative effect unless the later compressions are at a higher pressure. Compression during thermal annealing gives a higher VI yield than is observed by first applying pressure and then heating the sample, while this latter sequence [(n,γ)/P/Δ] gives a higher VI yield than the reverse sequence [(n,γ)/Δ/P], indicating that some of the same species which anneal rapidly during thermal treatment are also involved in the compression annealing processes (Andersen and Maddock 1963c, Andersen 1963).

A small increase in VI yield is observed with 'explosion annealing' [using a silver azide detonation to achieve a high instantaneous compression (Andersen and Maddock 1963c)], demonstrating the rapidity with which such annealing can occur when sufficient activation energy is available.

8. Other Annealing Experiments with Neutron-Irradiated Potassium Chromate

The VI yield in neutron-irradiated potassium chromate is not affected by

Table 8.3
Initial yields for other chromates and dichromates

Compound	Reactor dose ($n.cm^{-2}$)	Irrad. temp.[a]	^{51}Cr present (%) VI	M	DP	Annealing treatment[b]	Ref.*
Li_2CrO_4	3.7×10^{14}	AT	66	–	–	–	[30]
	6×10^{15}	AT	76.2	–	–	Δ	[24]
	6×10^{15}	AT	68	–	–	Δ	[10,11]
	3.5×10^{16}	AT	83.5	–	–	Δ	[31]
	3×10^{15}	– 78° C	70.8	13.4	15.8	–	[32]
$^7Li_2CrO_4$	6×10^{15}	AT	64	–	–	Δ	[10,11]
Na_2CrO_4	3.7×10^{14}	AT	74	–	–	–	[30]
	6×10^{15}	AT	72.2	–	–	Δ	[33]
	2.5×10^{14}	−196° C	44	–	–	Δ	[34]
	3×10^{15}	– 78° C	65.0	20.2	14.8	–	[32]
	6.5×10^{15}	– 78° C	61.5	–	–	Δ	[35]
$Na_2CrO_4 \cdot 4H_2O$	3.7×10^{14}	AT	88	–	–	–	[30]
	8.8×10^{14}	AT	89.3	–	–	–	[36]
	4.2×10^{13}	−196° C	35.3	–	–	–	[36,37]
	2.5×10^{14}	−196° C	38	–	–	Δ	[34]
	8.8×10^{14}	– 78° C	39.8	–	–	–	[36,37]
	6.5×10^{15}	– 78° C	44	–	–	Δ	[35]
	1.1×10^{17}	−196° C	55.6	–	–	Δ	[38]
Rb_2CrO_4	6×10^{15}	AT	58.1	–	–	Δ,e	[24]
	6×10^{15}	AT	68.5	–	–	–	[33]
CS_2CrO_4	6×10^{15}	AT	62.8	–	–	–	[33]
Tl_2CrO_4	6.5×10^{15}	– 78° C	74	–	–	Δ	[35]
$MgCrO_4$	3.7×10^{14}	AT	55	–	–	–	[30]
	6×10^{15}	AT	55.8	–	–	Δ	[39]
	6×10^{15}	AT	53.5	–	–	–	[33]
	1.5×10^{17}	−196° C	25	–	–	Δ	[38]
	1.0×10^{18}	−196° C	31	–	–	Δ	[38]
$^{24}MgCrO_4$	6×10^{15}	AT	56.7	–	–	Δ	[39]
$^{26}MgCrO_4$	6×10^{15}	AT	52.7	–	–	Δ	[39]
$MgCrO_4 \cdot H_2O$	1.5×10^{17}	−196° C	23	–	–	Δ	[38]
	1.0×10^{18}	−196° C	24	–	–	Δ	[38]
$CaCrO_4$	6×10^{15}	AT	74.9	–	–	–	[33]
	2×10^{17}	AT	62	–	–	Δ	[40]
$SrCrO_4$	6×10^{15}	AT	72.3	–	–	–	[33]
$BaCrO_4$	6×10^{15}	AT	77.2	–	–	–	[33]
$MnCrO_4$	6.5×10^{15}	– 78° C	93	–	–	Δ	[35]

$MnCrO_4 \cdot H_2O$	6.5×10^{15}	– 78° C	90	–	–	Δ	[35]
$ZnCrO_4$	3.7×10^{14}	AT	35	–	–	–	[30]
	3×10^{15}	– 78° C	31.0	57.6	11.5	–	[32]
$PbCrO_4$	6×10^{15}	AT	73.5	–	–	–	[33]
$(NH_4)_2CrO_4$	2.6×10^{14}	AT	16.7	–	–	–	[36]
	3.7×10^{14}	AT	17.5	–	–	–	[30]
	8.8×10^{14}	AT	17.3	–	–	–	[36]
	5.8×10^{15}	AT	17.5	–	–	Δ	[7]
	6×10^{15}	AT	24.1	–	–	e	[24]
	6×10^{15}	AT	15.6	–	–	Δ	[33]
	6×10^{15}	AT	15.1	–	–	Δ	[41]
	7.9×10^{15}	AT	20	36	44	–	[42]
	1.8×10^{16}	AT	16.5	–	–	–	[43]
	8.8×10^{14}	– 78° C	10.1	–	–	–	[36]
	3×10^{15}	– 78° C	10.9	48.3	40.8	–	[32]
$[C(NH_2)_3]_2CrO_4$	6×10^{15}	AT	10.6	–	–	Δ	[33,44]
$[Co(NH_3)_6]_2[CrO_4]_3$	6.5×10^{14}	AT	10.3	–	–	–	[45]
$Li_2Cr_2O_7$	3.7×10^{14}	AT	54.5	–	–	–	[30]
	3×10^{15}	– 78° C	54.9	31.5	13.9	–	[32]
$Na_2Cr_2O_7$	1.3×10^{16}	AT	85.4	–	–	–	[45]
	3×10^{15}	– 78° C	62.8	26.3	10.9	–	[32]
$Na_2Cr_2O_7 \cdot 2H_2O$	3.7×10^{14}	AT	73	–	–	–	[30]
$K_2Cr_2O_7$	–	AT	99	–	–	–	[46]
	–	AT	98.5	–	–	–	[47]
	3.6×10^{14}	AT	87.3	6.1	6.4	–	[1]
	3.7×10^{14}	AT	90	–	–	–	[30]
	6×10^{14}	AT	86–89	–	–	–	[48]
	2.2×10^{15}	AT	87.8	5.9	6.0	–	[1]
	4.3×10^{15}	AT	88.2	5.5	5.6	–	[1]
	5.2×10^{15}	AT	89–91	–	–	Δ,e,UV	[49]
	5.8×10^{15}	AT	89	–	–	–	[7]
	6×10^{15}	AT	91.3	–	–	e	[24]
	6×10^{15}	AT	89.5	–	–	Δ	[50]
	1.3×10^{16}	AT	90.4	–	–	–	[45]
	1.3×10^{16}	AT	90.2	4.9	4.5	Δ,γ	[1]
	1.8×10^{16}	AT	86	–	–	–	[43]
	2.6×10^{16}	AT	90.5	4.5	4.6	–	[1]
	4×10^{17}	AT	87.1	–	–	–	[51]
	1.3×10^{18}	AT	96.8	–	–	–	[51]
	3.6×10^{14}	–196° C	65.5	22.4	12.1	–	[1]
	8.8×10^{14}	– 78° C	65.3	–	–	–	[36]
	1.4×10^{15}	–196° C	65.5	22.7	11.8	–	[1]
	2.9×10^{15}	–196° C	65.1	22.9	12.0	–	[1]
	3×10^{15}	– 78° C	74.8	19.4	5.8	–	[32]
	4.3×10^{15}	–196° C	65.3	22.5	12.2	Δ	[1]
	7.2×10^{16}	–196° C	65.6	22.5	11.9	–	[1]

$K_2Cr_2O_7$ (monoclinic form)	6×10^{15}	AT	82.4	–	–	Δ	[50]
$Tl_2Cr_2O_7$	6.5×10^{15}	– 78° C	73	–	–	Δ	[35]
$(NH_4)_2Cr_2O_7$	3.7×10^{14}	AT	32	–	–	–	[30]
	5.2×10^{15}	AT	34	–	–	Δ,e,UV	[49]
	6×10^{15}	AT	31.6	–	–	Δ,e	[24]
	6×10^{15}	AT	38	–	–	Δ,UV	[50]
	7.9×10^{15}	AT	36	23	40	–	[42]
	1.3×10^{16}	AT	31.5	–	–	Δ	[45]
	3×10^{15}	– 78° C	25.7	37.4	46.9	–	[32]
$[N(CH_3)_4]_2Cr_2O_7$	6×10^{15}	AT	49	–	–	UV	[50]
$K_3Cr_3O_{10}$	4×10^{17}	AT	69.0	–	–	–	[51]
	1.3×10^{18}	AT	72.7	–	–	–	[51]
CrO_3	–	AT	87	–	–	–	[47]
	8.8×10^{14}	AT	68.8	–	–	–	[36]
	6×10^{16}	AT	65	–	–	Δ	[8]
	4×10^{17}	AT	79.1	–	–	–	[51]
	4.2×10^{17}	AT	50	–	–	–	[52]
	1.3×10^{18}	AT	80.8	–	–	–	[51]
	8.8×10^{14}	– 78° C	69.1	–	–	–	[36]
$Cs_2(CrOCl_5)$	1.8×10^{16}	AT	18.5[c]	–	–	–	[53]

[a] Reactor irradiation temperature used; AT is ambient reactor temperature.
[b] Annealing treatment reported in reference; Δ is thermal annealing, γ is gamma irradiation annealing, e is high energy electron irradiation annealing, UV is ultraviolet radiation treatment.
[c] Cr(VI) fraction is 18.5%, Cr(V) fraction (parent yield) is 37.5% and Cr(III) fraction is 44.0%.
*See footnote to table 8.1.

Table 8.3a
Initial yields for other chromates and dichromates

Compound	Reactor dose (n cm^{-2})	Irrad. temp.	^{51}Cr present (%)			Annealing treatment	Ref.*
			VI	M	DP		
$^7Li_2CrO_4$	–	AT	63.5	–	–	Δ	[61]
$^nLi_2CrO_4$	–	AT	72.5	–	–	Δ	[61]
$^6Li_2CrO_4$	–	AT	78.7	–	–	Δ	[61]
$^7Li_2CrO_4 \cdot H_2O$	–	AT	58	–	–	Δ	[61]
$^nLi_2CrO_4 \cdot H_2O$	–	AT	66	–	–	Δ	[61]
$^6Li_2CrO_4 \cdot H_2O$	–	AT	76.5	–	–	Δ	[61]
$^7Li_2CrO_4 \cdot D_2O$	–	AT	56	–	–	Δ	[61]
$^nLi_2CrO_4 \cdot D_2O$	–	AT	67	–	–	Δ	[61]
$^6Li_2CrO_4 \cdot D_2O$	–	AT	71	–	–	Δ	[61]

Na_2CrO_4	2.25×10^{14}	−196° C	47.9	24.5	27.6	Δ	[62]
	2.25×10^{14}	AT	63	25.5	11.5	–	[62]
	6.75×10^{14}	AT	64.6	22	13	–	[62]
	1.35×10^{15}	AT	68.5	19.5	12	–	[62]
	2.7×10^{15}	AT	70.5	17.5	12.6	γ	[62]
	5.4×10^{15}	AT	72.2	16.8	11	–	[62]
	8.1×10^{15}	AT	73.6	15.4	11	–	[62]
	1.1×10^{16}	AT	74.8	15	10.2	–	[62]
$Na_2CrO_4 \cdot 4H_2O$	2.25×10^{14}	−196° C	41.8	18.4	39.8	Δ	[62]
	2.25×10^{14}	AT	57.5	9.5	33	–	[62]
	4.5×10^{14}	AT	66.5	7.5	26	γ	[62]
	6.75×10^{14}	AT	72	6.5	–	–	[62]
	1.35×10^{15}	AT	75.5	6	18.5	–	[62]
	2.7×10^{15}	AT	86.5	4	9.5	–	[62]
	5.4×10^{15}	AT	91.5	2.5	6	–	[62]
	8.1×10^{15}	AT	93.5	4	2.5	–	[62]
	1.1×10^{16}	AT	94	2.5	3.5	–	[62]
$(NH_4)_2CrO_4$	2.25×10^{14}	−196° C	10.2	43.3	46.5	Δ	[63]
	2.25×10^{14}	AT	13.0	33.3	53.7	γ	[63]
	1.35×10^{15}	AT	13.9	30.2	55.9	–	[63]
	6.75×10^{15}	AT	15.0	27.5	57.5	–	[63]
$La_2(CrO_4)_3$	8.6×10^{14}	AT	70	–	–	Δ	[64]
$Na_2Cr_2O_7$	2.25×10^{14}	−196° C	53.6	30.0	16.4	Δ	[62]
	2.25×10^{14}	AT	70.6	22.2	7.2	–	[62]
	6.75×10^{14}	AT	73.5	18.5	8	–	[62]
	1.35×10^{15}	AT	75	18.5	8.5	–	[62]
	2.7×10^{15}	AT	77	15.5	7.5	γ	[62]
	5.4×10^{15}	AT	78	14.5	7.5	–	[62]
	8.1×10^{15}	AT	78	14	8	–	[62]
	1.1×10^{16}	AT	80	13.5	6.5	–	[62]
$Na_2Cr_2O_7 \cdot 2H_2O$	2.25×10^{14}	−196° C	45.0	24.7	30.3	Δ	[62]
	2.25×10^{14}	AT	65	8	27	–	[62]
	6.75×10^{14}	AT	66	6.5	27.5	–	[62]
	1.35×10^{15}	AT	66	6.5	27.5	–	[62]
	2.7×10^{15}	AT	66.5	6	27.5	γ	[62]
	5.4×10^{15}	AT	66.5	6	27.5	–	[62]
	8.1×10^{15}	AT	66.5	6	27.5	–	[62]
	1.1×10^{16}	AT	68	5.5	27.5	–	[62]
$K_2Cr_2O_7$	1.8×10^{17}	AT	92.7	–	–	–	[65]
	1.3×10^{18}	AT	98.8	–	–	–	[65]
$(NH_4)_2Cr_2O_7$	2.25×10^{14}	−196° C	24.0	32.0	44.0	Δ	[63]
	2.25×10^{14}	AT	27.5	22.2	50.3	γ	[63]
	1.35×10^{15}	AT	27.2	20.9	51.9	–	[63]
	6.75×10^{15}	AT	26.4	20.3	53.3	–	[63]
CrO_3	1.8×10^{17}	AT	62.1–69.3	–	–	–	[63]
	1.3×10^{18}	AT	87.2	–	–	–	[63]

*See footnote to table 8.1.

exposure to sunlight (Harbottle 1954) or to low-powered (50 W) ultrasonic treatment at ambient temperature (Maddock et al. 1963). However, higher-powered (250 W) ultrasonic treatment accompanying thermal treatment increases the VI yield more than thermal treatment alone (Andersen and Maddock 1963c).

9. Chromates other than Potassium Chromate

Data on initial yields (and annealing) for a number of different chromates and dichromates are given in table 8.3. No simple trend appears to result from changing the cation. As with potassium chromate, low-temperature neutron irradiation reduces the initial VI yield of most of these compounds, sometimes quite dramatically (Veljković and Harbottle 1961a, 1962, Yeh et al. 1969, 1970, Marqués and Wolschrijn 1969, Ackerhalt and Harbottle 1972), indicating that significant thermal- and/or radiation-promoted processes occur during neutron irradiation at ambient temperature. Some of the chromates appear to have a reactor dose dependence even with low temperature irradiation.

Thermal annealing studies of potassium dichromate (Stamouli 1971), sodium chromate (Yeh et al. 1970, Marqués and Wolschrijn 1969), magnesium chromate (Yeh et al. 1969) and other compounds irradiated at low temperature show significant thermal annealing at room temperature and below. As with potassium chromate, isochronal curves for these compounds suggest several different reaction regions (Yeh et al. 1969, 1970, Stamouli 1971, Marques and Wolschrijn 1969). In contrast to potassium chromate, the thermal and radiation annealing behaviours of *M* and *DP* species in potassium dichromate appear to involve only one (or possibly two) dominant reactions and the M/DP ratio remains sufficiently constant through these annealing programmes to suggest that both M and DP may be derived from a single precursor (Stamouli 1971).

Thermal treatment of neutron-irradiated calcium chromate causes annealing to produce *VI* and diffusion of ^{37}Ar (produced by the $^{40}Ca(n,\alpha)^{37}Ar$ reaction) from the crystals. Samples of calcium chromate which are pre-heated before neutron irradiation show the same amount of annealing to produce *VI* but greatly reduced ^{37}Ar diffusion (Ikeda 1963); an observation consistent with the idea that escape of ^{37}Ar from the bulk crystal requires dislocation channels or some other diffusion-promoting defects while the reactions producing *VI* do not.

Further detailed study of VI, M and DP yields from thermal and radiation annealing in other chromates and dichromates is desirable, although low-temperature neutron irradiation and storage may be required.

10. Results from other Nuclear Transformations

Differences in initial yields and annealing reactions of chromium recoil atoms produced in chromate target crystals by different nuclear processes may be interpreted in terms of differences in the details of the nuclear activation events and in the decay schemes of the resulting nuclei (Ackerhalt and Harbottle 1972). The VI yields from potassium chromate have been studied for ^{51}Cr atoms resulting from (n,γ), (n,2n) (Andersen and Sørensen 1966a, Ackerhalt and Harbottle 1972, Maddock and Treloar 1962, Maddock 1965a) and (γ,n) (Popov 1959, Omori et al. 1969) activations. The similarities in initial VI, M and DP yields (and in their thermal annealing) for ^{51}Cr atoms from the (n,γ) and (n,2n) activations carried out at low temperature show that the chemical fate of these recoil atoms is not strongly dependent on the magnitudes of their recoil energies (Ackerhalt and Harbottle 1972). The deexcitation cascades from the ^{51}Cr nuclei produced by both activations may be quite similar, insofar as such yield-affecting factors as nuclear recoil momenta, internal conversion probability and delayed states are concerned. This view is further supported by the similarities in the initial VI, M and DP yields of ^{51}Cr following (n,γ) and (n,2n) activations at low temperature of lithium chromate, sodium chromate, ammonium chromate and zinc chromate. The small observed differences in initial yields (about 6% for potassium chromate, less than 1% for sodium chromate and ammonium chromate) which remain essentially as a constant difference throughout isochronal annealing (Ackerhalt and Harbottle 1972) may reflect either (minor) differences in the deexcitation cascades or a small recoil energy effect.

In contrast to the similar yields from (n,γ) and (n,2n) activations producing ^{51}Cr in potassium chromate, [Cr(VI)] yields of 47.5% and 53.2%, respectively, the $^{49}Cr(VI)$ yield from the (n,2n) activation of ^{50}Cr at −196° C is markedly different, 80.8% (Ackerhalt and Harbottle 1972). Since the recoil energies of both ^{51}Cr and ^{49}Cr produced by (n,2n) activation are similarly high, the great difference in yield demonstrates that isotopic effects, such as the possible presence of delayed and/or internally converted transitions, are more important than recoil energy in determining initial yields.

The initial VI yield from the $^{50}Cr(\gamma,n)^{49}Cr$ reaction is 84–86%, depending on irradiation conditions (Omori et al. 1969). From these irradiations, carried out at ambient temperature, the simultaneously produced $^{51}Cr(VI)$ yield [from the $^{52}Cr(\gamma,n)^{51}Cr$ reaction] is also very high (82–84%). In the light of the much lower $^{51}Cr(VI)$ yield observed from the (n,2n) activation at −196° C (Ackerhalt and Harbottle 1972), the high $^{51}Cr(VI)$ yield reported for the (γ,n) activation at ambient temperature probably measures thermal- and/or radiation-promoted processes occurring during the irradiation, rather than direct effects of the $^{52}Cr(\gamma,n)^{51}Cr$ activation itself.

Initial $^{51}Cr(VI)$ and $^{55}Cr(VI)$ yields from ambient temperature (n,γ) activation in potassium chromate are 54% and 48%, respectively (Collins and Amin Singhh, unpublished). Without low-temperature irradiation to minimize thermal- and/or radiation-promoted processes, however, it cannot be concluded that the small difference between these yields represents a meaningful difference between the corresponding nuclear processes.

11. Annealing of Dopant Atoms in Chromates

Annealing reactions of lower-valency ^{51}Cr atoms introduced either as dopants or as ion-implants have been studied in potassium chromate (Mahieu et al. 1971, Apers et al. 1964, Collins et al. 1965, 1972a, b, Andersen and Sørensen 1965, 1966a, Ackerhalt et al. 1971, Collins and Collins 1971), ammonium chromate (Mahieu et al. 1971) and chromic chromate (Maddock 1965a, b) as well as in non-chromate crystals (Mahieu et al. 1971, Khorana and Wiles 1971, Andersen and Sørensen 1966a). High kinetic energy $^{51}Cr^{+}$ ions implanted into potassium chromate might be expected to produce *VI*, *M* and *DP* species which would show annealing reactions similar to those of recoil ^{51}Cr species, as is observed (Andersen and Sørensen 1965, 1966a), as these high energy ions (70 keV) formally resemble high kinetic energy recoil atoms. On the other hand, dopant ^{51}Cr ions introduced at thermal energies from co-crystallization of $^{51}Cr(III)$ with potassium chromate from aqueous solution (Mahieu et al. 1971, Apers et al. 1964, Collins et al. 1965, 1972, Collins and Collins 1971) or by recrystallization of neutron-irradiated potassium chromate (Andersen and Sørensen 1966a, Ackerhalt et al. 1971, Collins and Collins 1971) might be expected to be present as different species in sites of different structure within the chromate lattice and, hence, show little if any similarity in annealing reactivity to recoil ^{51}Cr species. However, in annealing experiments carried out at temperatures of 150° C or higher, the rates of production of *VI* from ^{51}Cr-dopant species and from ^{51}Cr-recoil species are indistinguishable although marked differences do exist when annealing is carried out at lower temperatures (Collins et al. 1965, 1972b, Collins and Collins 1971). These results lead to the important conclusion that information about 'hot-atom' (i.e. recoil-related) reactions of chromium recoil atoms may be obtained from thermal annealing studies below 150° C, but that thermal annealing at higher temperatures cannot give information about the chemical state of the recoil ^{51}Cr atom or the neighbourhood in which it initially comes to rest. However, these higher temperature studies of ^{51}Cr recoil atoms may serve as a useful probe of bulk crystal species and their behaviour (Collins et al. 1965, Harbottle and Sutin 1959, Collins 1965).

Annealing of dopant $^{51}Cr(III)$ atoms in ammonium chromate crystals gives data that are similar to that of ^{51}Cr-recoil atoms at temperatures above 80° C (Mahieu et al. 1971). Thus, the temperature range in which recoil-related processes can be investigated in ammonium chromate may be even lower than in potassium chromate.

Recent experiments on annealing reactions in doped crystals of potassium chromate indicate that the gross Cr(III) content of the crystals remains constant, even though ^{51}Cr(III) is being transformed to ^{51}Cr(VI). These results are apparently consistent with an 'exchange' mechanism (Mahieu et al. 1971). However, the fact that a similar process (the change of chromium oxidation state from III to VI) occurs upon heating or gamma-irradiating a *non-chromate* lattice doped with ^{51}Cr(III) (Khorana and Wiles 1971) suggests that the actual mechanism of transfer annealing and, by implication, the annealing of recoil atoms at higher temperatures, is probably not a true exchange reaction.

Heating $^{51}Cr_2O_3$ in a powdered mixture with potassium chromate at 350° C and 500° C produces some ^{51}Cr(VI) (Marchart 1965a, c), demonstrating that a heterogenous transfer process with, perhaps, long-range diffusion of ^{51}Cr to the crystal–crystal interface, can occur at these temperatures.

12. Species and Mechanism: Summary

The chemistry of recoil chromium atoms in neutron-irradiated potassium chromate is quite complex; many contributing species and individual reactions are required to describe the available data. One description of the species involved is summarized in tables 8.4 and 8.5. No convincing chemical identification of individual species has been made nor, indeed, have quantitative measurements of such species been possible, due to the great sensitivity of some species to details of crystal preparation and sample treatment. However, appropriate choice of treatment sequence (i.e. the detailed histories of experimental samples) may ultimately permit selective study of characteristics of individual processes and species.

Although *chemical* identifications of the various crystal phase species which form ^{51}Cr(III) and ^{51}Cr(VI) upon dissolution of irradiated crystals have not been made, various mechanisms to account for the *III→VI* annealing reactions have been proposed. Fragment recombination (Milenković and Maddock 1967, Ikeda 1963, Baumgärtner and Maddock 1968, Dimotakis and Stamouli 1967, Maddock and de Maine 1956b, Maddock and Vargas 1961a, b), ligand oxygen transfer (Mahieu et al. 1971, Maddock and Collins 1968), or electronic rearrangement triggered by an ion vacancy (Andersen and Maddock 1963d, Maddock et al. 1963), a mobile electron (Andersen and Olesen 1965, Andersen and Sørensen 1966a, Nath 1964, Nath et al. 1966, Nath and Klein 1969, Andersen et al. 1968), a mobile electronic hole (Andersen 1968, Andersen and Olesen 1965, Milenković and Veljković 1967, Collins et al. 1965, Khorana and Wiles 1971, Andersen and Sørensen 1966a, Costea 1969, Costea et al. 1970, Podeanu and Costea 1971, Costea and Podeanu 1968, Nath and Klein 1969) or an exciton (Dimotakis and Stamouli 1968, Stamouli 1971, Babeshkin 1970) have been proposed as principal mechanistic steps for thermal and/or

Table 8.4
Reactions of species in neutron-irradiated potassium chromate

Threshold temperature	Dose required	Reaction			Relative rate[a]
Thermally-initiated reactions					
$<-10°$ C	–	*M1*	$\longrightarrow$	*VI*	very fast
$<-10°$ C	–	*DP1*	$\longrightarrow$	*VI*	very fast
20° C	–	*DP2*	$\longrightarrow$	*M2*	fast
100° C	–	*M2*	$\longrightarrow$	*VI*	moderate
120° C	–	*DP3*	$\xrightarrow{O_2}$	*M3*[b]	fast
$>150°$ C	–	*III+x*	$\longrightarrow$	*VI*[c]	slow
Radiation-initiated reactions					
$-196°$ C	$<$ 5 Mrad	*M2*	$\xrightarrow{\gamma}$	*DP2*	fast
0° C	$<$ 5 Mrad	*M3*	$\xrightarrow{\gamma}$	*VI*[d]	fast
0° C	$>$100 Mrad	*III+x′*	$\xrightarrow{\gamma}$	*VI*[c]	slow

[a] Rate relative to other processes taking place at the same temperature.
[b] Reaction occurs only in air (oxygen).
[c] Annealing from the 'pool of ^{51}Cr(III) species; x and x' refer to diffusing trigger species from the bulk crystal.
[d] Reaction requires prior formation of *M3* by thermal processes.

Table 8.5
Summary of species present in neutron-irradiated potassium chromate

Treatment sequence	^{51}Cr species present
$(n,\gamma)_{-196}/S_{-196}$	*VI, M1, DP1, DP2, DP3, III*
$(n,\gamma)_{-196}/S^{8d}_{RT}$	*VI, M2, DP3, III*
$(n,\gamma)_{AT}/S_{RT}$	*VI, M2, DP2, DP3, III (M3)*[a]
$(n,\gamma)/S/\Delta^{60'}_{100}$	*VI, M2, DP3, III*
$(n,\gamma)/S/\Delta^{60'}_{180}$(air)	*VI, M3, III*, some *M2*
$(n,\gamma)/S/\gamma^{50Mrad}_{AT}$	*VI, DP2, DP3, III*
$(n,\gamma)/S/\gamma^{50Mrad}_{AT}/\Delta^{60'}_{150}$(air)	*VI, M3, III*, some *M2*
$(n,\gamma)/S/\gamma^{50Mrad}_{AT}/\Delta^{60'}_{150}(air)/\gamma^{D}_{AT}$	*VI, III*, some *DP2*

[a] *M3* appears in long-term irradiations in air at relatively high ambient reactor temperature.

radiation annealing. However, such is the complexity of the overall system that more than one such process may contribute importantly to the overall annealing observed.

Many interesting experiments can be devised – such as those based on selected sequences of operations – which can help to clarify the overall picture of what kinds of recoil species are produced and what reactions these recoil species undergo in potassium chromate. Other chromates, having (possibly) less complicated recoil atom chemistries, may yield more readily to intensive investigation. The need for better analytical procedures as well as input from other lines of investigation is clear. Nevertheless, the rich variety of unexplained phenomena known to occur in hot-atom chemical studies of potassium chromate and other chromates should continue to make such investigations attractive.

Addendum (December 1976)

Recent thermal annealing and radiation annealing studies of the hot-atom chemistry of several chromates and dichromates, summarized in table 8.3a (Batasheva et al. 1970, Costea et al. 1971b, Ladrielle et al. 1974, Stamouli 1974, 1975), have revealed the complex annealing behaviour of M, DP and VI precursor species in these systems. $^{51}Cr(VI)$ yield data from ambient temperature neutron irradiation of K_2CrO_4 (Batasheva et al. 1970, Dimotakis and Yavas 1970, Karim 1973, Ke et al. 1976, Sherif et al. 1974) fall well within the yield/dose distribution of fig. 8.1. Thermolysis (Mahieu et al. 1971) and radiolysis (Collins et al. 1978, Debuyst et al. 1972a, c, Krishnamurthy 1974, Milenković and Veljković 1974) of chromate systems are also consistent with earlier interpretations. Other recent studies include development of a new analytical procedure for separating $^{51}Cr(III)$ and $^{51}Cr(VI)$ (Mahieu et al. 1976), use of $^{52}Cr(p,d)^{51}Cr$, $^{50}Cr(d,p)^{51}Cr$ and $^{50}Cr(p,d)^{49}Cr$ reactions to introduce high-energy recoil chromium atoms into K_2CrO_4 (Ladrielle et al. 1975a, b) and a clarification of the effect of introducing dopant phosphate ions into K_2CrO_4 (Costea et al. 1971a).

Use of ^{51}Mn as a relatively clean source of non-chromate dopant ^{51}Cr species in K_2CrO_4 (Ladrielle et al. 1975a, b) has given clear support to the view that the thermally activated $^{51}Cr(III)$ to $^{51}Cr(VI)$ conversion occurring at temperatures above 150° C, as seen in both the doping experiments and experiments with ^{51}Cr recoil atoms, proceeds by transfer of oxygen from surrounding oxyanions or from molecules of hydration (Collins et al. 1972b, Maddock and Collins 1968). This overall process continues to be consistent with a mechanism in which electronic holes are mobilized at such temperatures to promote the overall process (Collins et al. 1972b, Costea et al. 1970, 1971a, b). Mechanistic interpretations of the solid state thermal annealing processes involving recoil ^{51}Cr atoms at temperatures below 150° C remain to be developed and may require more definitive experimental programmes.

Acknowledgement

The authors express sincere appreciation to Professors A.G. Maddock and D.J. Apers for making available recent results from their laboratories prior to publication.

CHAPTER 9

CHEMICAL BEHAVIOUR OF RECOIL PHOSPHORUS PRODUCED BY $^{31}P(n,\gamma)^{32}P$ IN SALTS OF ANIONS OF PHOSPHORUS

L. LINDNER

Instituut voor Kernphysisch Onderzoek, Oosterringdijk, Amsterdam, The Netherlands

Chemical Effects of Nuclear Transformations in Inorganic Systems
Edited by G. Harbottle and A.G. Maddock

Contents

1. Introduction

Phosphorus has attracted the attention of hot-atom chemists, in both its organic and inorganic forms and in different states of aggregation, for a long time (Libby 1940, Halmann 1964). The accessibility of the isotope ^{32}P by reactor irradiations with thermal neutrons [$^{31}P(n,\gamma)^{32}P$] and with fast neutrons [$^{31}S(n,p)^{32}P$ and $^{37}Cl(n,\alpha)^{32}P$] – all three with favorable cross sections – undoubtedly has contributed to this. In addition the half-life of 14 days and the relatively hard beta rays emitted ($E_{max} = 1.72$ MeV) ease the radioassay of the isotope.

Equally important, however, are the intrinsic chemical properties of the element itself; both the inorganic (Van Wazer 1958) and organic chemistry of phosphorus have shown rapid growth in scope and understanding since World War II.

Several chemical considerations conspire to render phosphorus extremely interesting for investigations in the hot-atom field. First, there are four different oxidation states of the *single* atom, known respectively as orthophosphate ($PO_4^{\equiv}$), phosphite ($HPO_3^{=}$), hypophosphite ($H_2PO_2^{-}$) and phosphine (PH_3), which are chemically stable. Covalently bonded hydrogen is characteristic of both phosphite and hypophosphite ions which have tetrahedral configurations similar to that of the orthophosphate ion (Van Wazer 1958). A single phosphorus atom in the neutral state – to be distinguished from elementary phosphorus which in the red, white and black modifications has a polymer-like structure (Van Wazer 1958) – by contrast should be extremely reactive and therefore can only exist as a transient species.

Particularly interesting, however, is phosphorus' ability to form highly polymerized structures in which phosphorus atoms are linked to each other, either directly, in P–P bonds, or by means of oxygen bridges in P–O–P bonding. Combinations of the two possibilities leading to skeletons like P–P–O–P are also known. It should be pointed out here that the different phosphorus atoms in these condensation products are not necessarily in the highest oxidation state; configurations with one or more P atoms in a reduced state also exist. Branching, leading to two-dimensional structures like

P–O–P–O–P, known as ultraphosphates can only exist in a solid matrix. On
|
O
|
P

dissolution in an aqueous medium one branch is split off by rapid hydrolysis. This implies that in aqueous solutions only straight chain or ring structures can exist. No rings larger than eight-membered are known (tetrametaphosphate with four P-atoms bridged in a ring by four O-atoms). In this chapter, the various oxyanions of phosphorus will be represented by a symbolic description proposed by Blaser and Worms (1959c), now commonly used (see table 9.1); polymers will occasionally be denoted as P_n, the index giving the number of P atoms. For the chemical properties of these anions, the reader is referred to the literature (Van Wazer 1958, Baudler 1959, Blaser and Worms 1959a, b, 1961 b). In general, both acid and alkaline hydrolysis of the condensed structures lead to complete degradation yielding the corresponding monomers; rate constants, however, differ widely from one ion to another. Oxidation of reduced forms generally occurs with bromine water; in neutral media the oxidation often proceeds faster. Oxidation with iodine is much more selective than with bromine.

The rather fortunate property that exchange of phosphorus does not occur in aqueous media between the various anions mentioned (Van Wazer 1958), has played a key role in the development of experimental techniques to determine the distribution of recoil phosphorus among these anions. Although this chapter is strictly confined to the chemical behaviour of ^{32}P recoils produced by slow neutron capture in solid salts of phosphorus oxyanions, mention will be made of a small number of investigations dealing with anions of phosphorus with ligands different from oxygen and hydrogen, such as fluorine (Claridge and Maddock 1963b) and sulphur atoms (Ujimoto et al. 1970).

Of peripheral interest to our main subject are the hot atom reactions of outer (ligand) atoms of complex ions: for instance, work has been reported on recoiling tritium atoms in reactions with phosphate, phosphite and hypophosphite ions (Lindner 1958, Lindner et al. 1965, Van Urk 1970).

It seems certain that our understanding of the chemical behaviour of energetic phosphorus atoms has been substantially increased by the studies reviewed here and in other chapters. Apart from purely academic interest, hot atom investigations have played their role in applied studies, for example, of fertilizer (Thomas and Nicholas 1949, Fried and MacKenzie 1950, Fiskell et al. 1952, MacKenzie and Borland 1952, Scheffer and Ludwig 1957), and in the production of ^{32}P in high specific activity but free from ^{33}P (Tanaka 1964c, Matsuura et al. 1967). Possible uses of a recoil process in the synthesis of ^{32}P-labelled inorganic compounds have also been reviewed (Lindner 1962a).

Table 9.1
Oxyanions of phosphorus

Blaser–Worms notation (Blaser & Worms 1959a)	Name	Formula
P^5	orthophosphate	PO_4^{3-}
P^3	(ortho)phosphite	HPO_3^{2-}
P^1	hypophosphite	$H_2PO_2^-$
P^5OP^5	pyrophosphate	$O_3POPO_3^{4-}$
P^5OP^3	isohypophosphate	$O_3POP(H)O_2^{3-}$
P^3OP^3	pyrophosphite	$O_2(H)POP(H)O_2^{2-}$
P^4P^4	hypophosphate	$O_3P{-}PO_3^{4-}$
P^4P^2	diphosphite	$O_3PP(H)O_2^{3-}$
P^4P^2		$O_2(H)PP(H)O_2^{2-}$
$P^5OP^5OP^5$	tripolyphosphate	$O_3POP(O_2)OPO_3^{5-}$
$P^4P^4OP^5$		
$P^4P^4OP^3$		
$P^4P^3P^4$		
$(P^5O)_3$-ring	trimetaphosphate	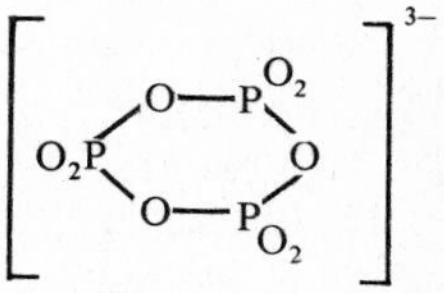
$(P^3)_6$-ring		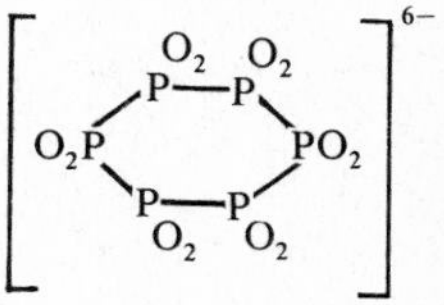
$P^5OP^5OP^5OP^5$	tetrapolyphosphate	$O_3POP(O_2)OP(O_2)OPO_3^{6-}$
$(P^5O_3)_4$-ring	tetrametaphosphate	$(-P(O_2)O-)^{4-}$–ring
$P^5O(P^5O)_nP^5$	(long chain) polyphosphates	$O_3PO(PO_3)_nPO_3^{[n+4]-}$

2. Experimental Aspects

2.1. ANALYSIS OF NEUTRON-IRRADIATED PHOSPHORUS COMPOUNDS

Chemical analytical techniques have played so crucial a role in the investigation of the hot atom reactions of phosphorus that they deserve special mention.

In the early days classical precipitation methods were the only ones available and hence used throughout. In 1940 Libby (1940) reported on the chemistry of ^{32}P produced in a number of orthophosphates. Two fractions were distinguished; one consisting of ^{32}P which precipitated with magnesium ammonium phosphate in an alkaline medium, the other the residuum which oxidized with bromine and then similarly precipitated was thought to contain ^{32}P originally present as phosphite. About a decade later several investigations in which more sophisticated precipitation techniques were applied drew attention to the formation of labelled non-orthophosphates of a polymeric nature in neutron-irradiated solid phosphates (Fiskell 1951, Fiskell et al. 1952, MacKenzie and Borland 1952, Borland et al. 1952, Aten et al. 1952). Fiskell et al. reported the formation of hypo-, pyro- and condensed phosphates, while Aten et al. showed that precipitates of both zinc and cadmium pyrophosphate yielded on subsequent hydrolysis ^{32}P in lower oxidation states. They arrived at the important conclusion that part of the ^{32}P activity in irradiated orthophosphates was probably present as pyrophosphite, isohypophosphate and other species.

Strain and coworkers (Sato et al. 1953, Sellers et al. 1957) made an invaluable contribution when they showed in principle that the separation of monomer- and polymer-type anions of phosphorus by means of paper electrophoresis could very well be applied in hot atom chemistry studies with ^{32}P. Unfortunately, because of severe conditions during the irradiation of their samples, their spectrum of product species appeared to be essentially limited to the parent compound. Only later the full potentialities of the method came to light through the work of Lindner and Harbottle (1960). By rigid control of conditions it was demonstrated that by *high*-voltage (80 V/cm) electrophoresis on paper wetted with 0.1 M lactic acid, it was possible to resolve within approximately one hour, more than ten different species (assumed to represent most of the known monomeric, dimeric and trimeric anions of phosphorus) labelled with ^{32}P, produced in crystalline disodium hydrogen orthophosphate and other similar salts (Lindner and Harbottle 1961). The extensive spectrum of ^{32}P labelled species was obtained by irradiation with neutrons under relatively mild conditions. Subsequently Claridge and Maddock (1963a) showed that group separations by classical precipitation methods ($^{32}P^5$ as phosphomolybdate, the activity carried down by magnesium ammonium phosphate representing *total* phosphate-^{32}P) yielded results which, within certain limits, were consistent with the figures obtained by an electrophoretic separation. Electrophoresis also appeared applicable in the case of fluorophosphates (Claridge and Maddock 1963b).

In the same period the technique of one-dimensional ascending paper chromatography was applied for the first time (Yoshihara and Yokoshima 1961, Shima and Utsumi 1961). Depending on the developing solvent mixture, species movement amounted to 20–25 cm in 10 hours. Yoshihara and Yokoshima (1961) separated various monomeric and polymeric anions of phosphorus from each other, but the resolution obtained was not as good as with paper electrophoresis. Shima and Utsumi (1961) claim the separation of chain phosphates as long as heptapolyphosphates from each other as well as from the two known ring phosphates. However, no research was done on chains with phosphorus in lower oxidation states. A disadvantage of paper chromatography is that (radioactive) cations travel along with the phosphorus compounds, whereas in electromigration they move from the starting point in the opposite direction. Recently Kitaoka et al. (1970a, b) have reported on a more detailed study of the influence of various parameters in ascending paper chromatography, resulting in improved resolution. They also applied reversed-phase paper chromatography and a two-dimensional technique. A disadvantage paper electrophoresis and paper chromatography have in *common*, is the rather limited capacity of the paper: only small amounts of irradiated material can be treated, conflicting with the important goal of keeping the irradiation dose received by the sample in the reactor as low as possible. Ion exchange does not have this limitation; the diameter of the column can be adjusted to meet the capacity required. Also radioactive cationic impurities are easily eliminated as they will be eluted before the anions of phosphorus.

The first application of ion-exchange for the separation of ^{32}P-labelled recoil products was reported by Anselmo (1965). Elution in a discontinuous mode, with sodium chloride solutions of increasing concentration, resulted in group separations between monomers and various groups of labelled polymers. Elution with a continuous gradient resulted in improved separations for the monomeric and dimeric species (Anselmo 1967, Anselmo and Sanchez 1967, 1969). Even better results were obtained by Ujimoto et al. (1970) by means of gradient elution with ammonium acetate solutions (pH 6.9) in a column of Dowex-X8 resin in the acetate form. Separations, however, required two days! The same technique also proved to be useful in the case of ^{32}P produced in sodium monothiophosphate. Ujimoto et al. also tested a gel chromatographic method with Sephadex G-25 and demonstrated that group separations between monomers, dimers and higher polymers are possible.

In conclusion the following can be said: With respect to resolution both paper electrophoresis and ion-exchange chromatography seem superior to paper- and gel-chromatography and precipitation methods. The two first techniques lead to a satisfactory-to-excellent separation of the individual monomers and most of the dimers. The trimeric anions, however, and higher polymers, cause a more serious problem. By virtue of its greater flexibility ion-exchange seems the method of choice for the separation of more condensed species. A word of caution is in order concerning the influence on the separation process of the presence of certain cations on the one hand, and the presence (or absence) of carriers for the various labelled products on the other. For instance in the case of paper electrophoresis Sellers et al. (1957) observed that addition of zinc ions retarded the movement of P^5OP^5 at the origin, but did not affect other migrating species. Fenger, among others, applied this variation on the method successfully in the course of his extensive investigations (Fenger 1964, Fenger and Nielsen 1965, Fenger 1968, 1969), but he also realized that this technique carried the danger of coprecipitation of other polyphosphates at the origin (Fenger 1971). In order to improve the resolution complex formation of phosphorus anions with both zinc and cadmium ions as a function of pH and concentration has been utilized (Kitaoka et al. 1970b). The concentration of both carrier and zinc ions, as well as pH, appeared to have a profound influence on migration distance, the order of mobility of various phosphorus ions, and the apparent distribution of the ^{32}P due to (partial) precipitation of species. Fenger and Nielsen (1965) have reported similar findings. 'Carrier effects' have also been observed in the case of ion-exchange separations and gel-chromatography; according to Ujimoto et al. (1970) certain carriers must be present in order to obtain the true ^{32}P distribution. For polymeric anions this effect was most pronounced: the addition of carrier greatly facilitated desorption from the column. One reasonable speculation is that the behaviour of carrier-free solutions is due to radiocolloid formation by ^{32}P species absorbed on the surface of dispersed particles. The same authors claim to have observed similar effects in paper electrophoresis. Whatever their explanation, these disturbing phenomena, especially for the more complex labelled ions of phosphorus emphasize the need for more reliable and adequate separation techniques. This is particularly true when we realize the important role these ions play in the present-day conflicting theoretical discussions on their existence and mechanism of formation as recoil products.

2.2. IDENTIFICATION

Identification of the labelled monomeric, dimeric and trimeric oxyanions of phosphorus is no longer an insurmountable problem as their corresponding carriers are well-characterized and available by purchase or synthesis. However, more complex ions, especially those with phosphorus atoms in lower oxidation states still offer a serious problem. Detection is generally achieved by degradation and oxidation to orthophosphate which is then determined by classical methods. Also specific chemical properties can be tested (Lindner and Harbottle 1961) as an additional check.

2.3. COUNTING

Paper strips resulting from chromatographic or electrophoretic separations of recoil species labelled with ^{32}P have been radioassayed on a qualitative basis by autoradiography. Quantitative counting is almost invariably done with end-window G.M. or proportional counters. The strips were either scanned (Fenger 1968) or cut up in pieces to be counted individually (Lindner and Harbottle 1961), often with automatic sample changers. The effluent of ion exchange columns was either radioassayed on-line with thin-window G.M. flow counters (Anselmo 1965) or afterwards on fractions collected (Matsuura et al. 1967,

Ujimoto et al. 1970). Scanning and flow-counting are less desirable than counting of individual samples if activities are low, as the latter procedure allows one to apportion counting times in an optimal way. It should be mentioned that in the author's laboratory batch-wise counting of ^{32}P samples from electrophoresis or thin layer chromatography on the one hand and liquid samples resulting from ion exchange on the other, is quite satisfactorily accomplished either with liquid scintillators (with close to 100% efficiency) or, without scintillator added, directly on the Cerenkov radiation produced by the hard beta-rays (this procedure has higher efficiency than the so-called beta-liquid G.M. counters) in commercially available automatic liquid-scintillation counting equipment.

3. Experimental Results

3.1. GENERAL

From a historic point of view one can split the research into two periods, the year 1960 approximately indicating the dividing point. Prior to this date, information had been gained primarily by means of wet chemical precipitation techniques. Thanks to results obtained with more powerful separation procedures it became clear in the early sixties that only sophisticated analytical techniques could unravel the very complex spectrum of ^{32}P-labelled recoil products formed in crystalline phosphates. We know now that a magnesium ammonium phosphate precipitate carries, in addition to P^5, most of the ^{32}P (Claridge and Maddock 1963a) present in the form of polymeric species. Conditions under which phosphomolybdate is precipitated also unfortunately facilitate acid hydrolysis of less stable condensed products; as a result this precipitate represents to some extent the fraction of ^{32}P present in oxidized form. One immediate success of the 'chromatographic' techniques was their revelation of the profound influence of the irradiation and storage (annealing) conditions on the experimental results. At the same time these techniques provided the more detailed information needed to formulate conclusions concerning mechanisms.

3.2. INITIAL ^{32}P DISTRIBUTIONS

Libby (1940) observed that about half of the total ^{32}P activity produced in sodium orthophosphates could not be precipitated by magnesium ammonium phosphate and assigned this fraction to phosphite. In essence this was confirmed by others in the period before 1960 (Thomas and Nicholas 1949, Fried and MacKenzie 1950). Clear evidence was soon given for the existence in the product spectrum of labelled species of a condensed nature (Scheffer and

Ludwig 1957), such as pyrophosphate (MacKenzie and Borland 1952, Borland et al. 1952, Aten et al. 1952), hypophosphate and products more condensed than pyrophosphate. Aten et al. made the important observation that some of the ^{32}P produced in orthophosphates ended up in pyro-type structures but with the labelling atoms in a reduced state. Although the investigations just mentioned yielded significant qualitative information, a quantitative evaluation is difficult because information on irradiation conditions is incomplete and the real distribution of the ^{32}P is rather uncertain due to the non-specificity of the precipitation techniques used. Strain and coworkers in a series of papers (Sellers et al. 1957, Sato et al. 1959, Sato and Strain 1961) described experiments whose outcome (retentions close to 100%) led them to believe that in nearly all instances the original chemical bonds of the phosphorus (in various phosphates, phosphites and hypophosphites) were not disrupted in the nuclear event. This conclusion was rather unexpected in view of earlier observations on similar systems and also on the basis of an extensive investigation of the chemical behaviour of ^{32}P recoils produced by fast neutron reactions in a large number of inorganic compounds of sulphur and chlorine (Lindner 1958). Accordingly the validity of the arguments put forward by Strain and co-workers was tested by Lindner and Harbottle (1960) by irradiating orthophosphates at dry-ice temperatures with a more moderate neutron flux. It was demonstrated, by employing paper electrophoresis separations like those of Strain, that the ^{32}P generated in Na_2HPO_4 was distributed on dissolution among at least thirteen different labelled products. The majority of these species could be tentatively identified as known monomeric, dimeric and trimeric oxyanions of phosphorus. Moreover, the retention amounted to only 9.3%, giving a much smaller upper limit for the probability of non-bond-rupture in crystalline phosphates. In a subsequent detailed paper (Lindner and Harbottle 1961) it was shown, by varying the irradiation conditions, that retention values are raised drastically if those conditions become more severe, i.e. higher neutron flux, higher accompanying radiation field, high temperature. Thus the very high retentions observed by Strain could be attributed to very extensive annealing during irradiation. Actually this phenomenon had been pointed out earlier (Borland et al. 1952). Experimental results of other workers in the same period (Claridge and Maddock 1959, 1961a, 1961b, Yoshihara and Yokoshima 1961, Shima and Utsumi 1961, Anselmo 1961, Baba et al. 1961) similarly conflicted with the non-rupture hypothesis of Strain and co-workers. More recent evidence also confirms the extensive rupture of the parent phosphorus ion (and in polymeric anions as well) on neutron capture by the phosphorus atom; at present there is general agreement that the probability of complete failure of initial rupture of bonds in solids amounts to only a few percent. These experimental observations agree with calculations of the distribution of recoil energies imparted to ^{32}P atoms after slow neutron capture in ^{31}P (Cifka 1963, Fenger 1964). If the minimum recoil energy necessary to cause bond rupture in an orthophosphate ion is assumed to be 50 eV, retentions of only 1–2% can be explained. A classical trajectory study (Bunker and Van Volkenburgh 1970) also predicts that most ^{32}P recoils leave their original site.

Comparison of the experimental results obtained by different investigators studying the same system is difficult not only because of differences in irradiation and storage conditions (as mentioned above), but also because of the resolution obtained by the analytical procedure and the ambiguity which still exists with regard to the identification of more condensed products, such as trimeric species. As an illustration, table 9.2 shows some recoil yield data on alkali orthophosphates from experiments performed under similar conditions with good resolution in the labelled product spectrum. Examination of this table reveals that differences in yields are sometimes considerable and difficult to understand. For instance, consistently higher P^5 yields are reported by Nakamura et al. (1970). Fenger's (1969) P^5OP^5 yields are clearly higher than the others and one wonders whether this might be due to coprecipitation of other condensed species at the origin where the pyrophosphate fraction is deliberately trapped by the addition of zinc ions. However, considering the complexity of the product spectrum and the difficulties generally experienced in this kind of work, there is reasonable and reassuring agreement. It is especially noteworthy that none of these results (with a single exception) indicates high yields of highly-condensed labelled products. The presence or absence of these products plays a role in current theories of the hot atom process in phosphates; therefore it seems important to examine this particular aspect in some detail. Nakamura et al. (1970, Ujimoto et al. 1970) claim that labelled tetra- to hexa-polyphosphates occur in irradiated orthophosphates to some extent. One should realize, however, that identification of these supposed P_4 to P_6 compounds is difficult inasmuch as a complete set of potential candidates (for example, condensed anions with phosphorus in a reduced state) in the form of 'known' compounds is not as yet available. It is further of interest to recall the observation of Dahl and Birkelund (1962) that, on standing, carrier-free $^{32}P^5$ solutions produced ^{32}P labelled chain polyphosphates. This seems relevant in that Nakamura et al. (1970, Ujimoto et al. 1970) state that after dissolution the irradiated samples were stored for a period at a temperature $\leqslant 5°$ C. A further complication arises from recoil ^{32}P experiments with ammonium salts: the results suggest the existence of a new class of ^{32}P-labelled products containing nitrogen. Anselmo and Sanchez (1969) have presented evidence that this may be the case when ammonium hypophosphite is the target material. On the other hand, the observation of Anselmo (1965) that comparatively high yields (up to 47%) of highly condensed labelled products are formed in orthophosphates is clearly inconsistent with most recent observations by others: this result may be an artifact of the ion-exchange separation method employed.

Apart from orthophosphates, also pyrophosphates, hypophosphates, tripolyphosphates, trimetaphosphates, tetrametaphosphates and long-chain

Table 9.2
Distribution of ^{32}P in different orthophosphates of alkali metals

	Ref.	P^1	P^3	P^5	P^3OP^3	P^5OP^3	P^5OP^5	P^2P^2
Li_3PO_4	a	3.2	5.3	30	0.3	19.6	5.4	–
anh.	b	0.6	11.1	15.5	–	4.6	50.8	–
$Na_3PO_4 \cdot$	a	10.3	22.7	38.1	0.3	9.3	13.0	
$12H_2O$	b	21.3	30.2	15.8	0.7	14.3	11.3	–
	c	29.1	31.6	21.2		11.6		–
Na_2HPO_4	a	8.6	9.1	9.5	0.2	26.3	11.2	–
anh.	b	9.0	6.7	6.6	1.3	32.6	31.5	–
	c	5.9	8.0	13.6		36.6		
$Na_2HPO_4 \cdot$	a	14.7	13.2	16.2	–	22.8	16.6	–
$2H_2O$	b	13.5	10.6	14.6	1.1	14.6	28.0	–
$Na_2HPO_4 \cdot$	a	12.5	21.9	31.9	1.1	11.9	11.4	–
$12H_2O$	b	13.5	15.9	28.8	1.8	12.8	19.3	–
	c	5.6	13.7	54.7		17.6		
$NaH_2PO_4 \cdot$	a	7.9	15.7	17.9	1.0	19.1	13.8	–
$1H_2O$	d	3.7	8.8	15.1		13.2	22.0	–
	b	10.9	7.8	9.9	1.7	32.7	26.3	–
$NaH_2PO_4 \cdot$	a	6.6	12.5	23.1	0.4	14.1	21.0	–
$2H_2O$	b	10.5	10.8	16.6	1.5	20.3	27.8	–
	c	4.9	12.6	32.1		26.7		–
K_3PO_4	a	6.2	11.4	18.5	0.1	22.0	8.8	–
anh.	d	7.9	7.5	5.3		14.8	19.8	–
	c	4.1	11.5	24.9		30.1		–
K_2HPO_4	a	7.0	10.4	9.5	1.0	26.7	10.1	–
anh.	d	10.3	13.1	3.8			22.1	–
	b	7.1	8.3	7.4	1.1	21.7	33.7	–
	c	4.6	10.0	17.1		37.4		–
KH_2PO_4	a	8.0	11.4	10.9	0.5	21.6	13.8	–
anh.	d	9.2	14.2	9.8		15.4	14.0	–
	b	10.1	4.4	7.4	0.8	24.1	32.5	–
	c	6.1	3.9	18.3		18.3		–

[a] Lindner and Harbottle (1961): separations by electrophoresis; irradiations with 2×10^{12} n/cm^2·sec during 30 min at dry ice temperature.
[b] Fenger (1969): separations by electrophoresis in the presence of Zn^{2+} ions; irradiation with 4×10^{12} n/cm^2·sec during 15 min at liq. N_2 temperature.

<table>
<tr><th>P^4P^2</th><th>P^4P^4</th><th>$P^5OP^5OP^5$</th><th>$P^4P^4OP^5$</th><th>$P^4P^4OP^3$</th><th>P_4</th><th>P_5</th><th>P_6</th><th>P_n</th></tr>
<tr><td>9.4</td><td>~3.0</td><td>15.0</td><td colspan="2">7.8</td><td></td><td></td><td></td><td></td></tr>
<tr><td>1.1</td><td>11.1</td><td>2.3</td><td>0.9</td><td></td><td></td><td></td><td></td><td>0.4</td></tr>
<tr><td>2.0</td><td>2.1</td><td>1.0</td><td colspan="2">1.1</td><td></td><td></td><td></td><td></td></tr>
<tr><td>0.7</td><td>1.3</td><td>0.7</td><td>0.7</td><td></td><td></td><td></td><td></td><td>1.1</td></tr>
<tr><td>2.8</td><td></td><td colspan="3">0.4</td><td></td><td></td><td></td><td></td></tr>
<tr><td>9.3</td><td>2.6</td><td>11.3</td><td colspan="2">12.5</td><td></td><td></td><td></td><td></td></tr>
<tr><td>3.7</td><td>3.1</td><td colspan="2">2.2</td><td></td><td></td><td></td><td></td><td></td></tr>
<tr><td>4.2</td><td></td><td colspan="5">26.5</td><td></td><td>1.6</td></tr>
<tr><td>4.9</td><td>1.1</td><td>5.3</td><td colspan="2">3.9</td><td></td><td></td><td></td><td></td></tr>
<tr><td>1.8</td><td>1.5</td><td>1.1</td><td>0.9</td><td></td><td></td><td></td><td></td><td>1.5</td></tr>
<tr><td>3.0</td><td>2.5</td><td>1.3</td><td colspan="2">0.9</td><td></td><td></td><td></td><td></td></tr>
<tr><td>1.4</td><td>0.8</td><td colspan="2">1.1</td><td></td><td></td><td></td><td></td><td></td></tr>
<tr><td>1.9</td><td></td><td colspan="3">0.8</td><td></td><td></td><td></td><td>0.9</td></tr>
<tr><td></td><td></td><td></td><td></td><td></td><td></td><td></td><td></td><td>2.5</td></tr>
<tr><td>5.2</td><td>1.5</td><td>8.7</td><td colspan="2">8.0</td><td></td><td></td><td></td><td></td></tr>
<tr><td>1.3</td><td>14.7</td><td>12.8</td><td colspan="2">5.5</td><td></td><td></td><td></td><td></td></tr>
<tr><td>3.5</td><td>3.3</td><td>1.1</td><td>1.2</td><td></td><td></td><td></td><td></td><td>1.1</td></tr>
<tr><td>2.2</td><td>0.7</td><td>13.0</td><td colspan="2">5.4</td><td></td><td></td><td></td><td></td></tr>
<tr><td>2.9</td><td>2.0</td><td>0.9</td><td>1.4</td><td></td><td></td><td></td><td></td><td>0.7</td></tr>
<tr><td>1.8</td><td></td><td colspan="5">17.6</td><td></td><td></td></tr>
<tr><td>13.0</td><td>6.2</td><td>5.0</td><td colspan="2">8.5</td><td></td><td></td><td></td><td></td></tr>
<tr><td>2.3</td><td>13.4</td><td>14.0</td><td colspan="2">11.6</td><td></td><td></td><td></td><td></td></tr>
<tr><td>9.9</td><td></td><td colspan="3">14.5</td><td></td><td></td><td></td><td></td></tr>
<tr><td>9.6</td><td>2.1</td><td>7.7</td><td colspan="2">13.8</td><td></td><td></td><td></td><td></td></tr>
<tr><td>4.7</td><td></td><td>2.9</td><td colspan="2">4.5</td><td></td><td></td><td></td><td></td></tr>
<tr><td>6.0</td><td>3.1</td><td>2.1</td><td>1.4</td><td></td><td></td><td></td><td></td><td>2.2</td></tr>
<tr><td>2.5</td><td></td><td colspan="3">14.5</td><td></td><td></td><td></td><td></td></tr>
<tr><td>2.8</td><td>1.0</td><td>13.4</td><td colspan="2">13.4</td><td></td><td></td><td></td><td></td></tr>
<tr><td>2.2</td><td>10.6</td><td>10.1</td><td colspan="2">12.0</td><td></td><td></td><td></td><td></td></tr>
<tr><td>5.4</td><td>6.3</td><td>1.8</td><td>1.6</td><td></td><td></td><td></td><td></td><td>0.3</td></tr>
<tr><td>0.7</td><td></td><td colspan="3">22.3</td><td>18.9</td><td>5.9</td><td>1.4</td><td></td></tr>
</table>

[c] Nakamura et al. (1970): separations by ion exchange with gradient elution; irradiations with 5×10^{12} n/cm²·sec during 10 min at dry ice temperature.
[d] Claridge and Maddock (1963a): separations by electrophoresis; irradiations with 1×10^{12} n/cm²·sec during 120 min at liq. N_2 temperature.

polyphosphates have been used as targets. General trends can be summarized as follows. Retentions are low for ring-phosphates in comparison with chain phosphates (Shima and Utsumi 1961) and surprisingly high in pyrophosphates, amounting to almost 50% for $K_4P_2O_7$ (Fenger 1969) compared to $\leqslant$ 10% for other dimeric and trimeric target compounds (Lindner and Harbottle 1961). Product spectra are at the least as complex as in orthophosphates, although as would be expected, some new components are observed. These spectra are characterized by relatively high yields of degradation products compared to those of the more complex built-up products. Most species observed have been identified satisfactorily but for a few the structure is still unknown.

There is general experimental agreement that for virtually all types of target compounds investigated the yield of labelled monomers (P^5, P^3 and P^1) increases with the number of crystal water molecules present in the lattice. In phosphates (both monomers and polymers) one can say, as a rather crude rule of thumb, that half or more of the labelled monomer fraction consists of P^3 and P^1. The formation of labelled phosphine has generally been assumed not to occur in phosphate lattices: Fiskell et al. (1952) however reported its formation in calcium hypophosphite. Reduced forms of ^{32}P can also appear in substantial quantities in the polymeric fraction. For example isohypophosphate (P^5OP^3) generally is an important constituent: the observation (Lindner and Harbottle 1961) that this product is specifically labelled as $P^5O^{32}P^3$ has been confirmed by others (Claridge and Maddock 1963a, Fenger 1964) and sheds light on the mechanism of the interaction of the thermalized ^{32}P recoil atom with its crystal environment. A search for specific labelling has also been undertaken in tripolyphosphate formed in orthophosphates and pyrophosphates (Lindner et al. 1965), in tripolyphosphates as targets (Lindner et al. 1965, Matsuura et al. 1969), and in polyphosphates (Matsuura and Lin 1970). Lindner et al. (1965) found that in recoil reactions in orthophosphates the tripolyphosphate formed is preferentially labelled in the center position, $P^5O^{32}P^5OP^5$ as predicted (Lindner and Harbottle 1961), whereas for pyrophosphate and tripolyphosphate targets the label is by preference in the end position, $P^5OP^5O^{32}P^5$. This is consistent with a part of the work of Matsuura and Lin (1970), who conclude that with polyphosphates as targets, labelling occurs at the end atom of the chain. In contrast, for tripolyphosphates Matsuura et al. (1969) report a uniform labelling. Matsuura's results, however, were obtained through partial alkaline hydrolysis of the irradiated sample as a whole, and not of a *separated* labelled tripolyphosphate fraction. There is yet another matter of concern: the yields of tripolyphosphate-^{32}P reported in one study by Lindner et al. (1965) are higher than those published elsewhere (Lindner and Harbottle 1961, Fenger 1971). One possible explanation is that these high tripolyphosphate yields, which were obtained by repeated 'selective' precipitation, were caused by a contamination of other labelled species. Later unpublished work of Lindner and co-workers in which the tripolyphosphate fraction was purified by carefully performed anion exchange separations, however, supported the earlier conclusion (Lindner et al. 1965) with respect to the position of the label. Also, the yields were in better agreement with other work.

Crystals containing the lower oxidation states, phosphite and hypophosphite, have also been used as targets in hot atom studies (Libby 1940, Aten et al. 1952, Sellers et al. 1957, Lindner and Harbottle 1961, Yoshihara and Yokoshima 1961, Baba et al. 1963, Tanaka 1964b, Yoshihara et al. 1963, Anselmo 1965, Anselmo and Sanchez 1969, Fenger 1969, Kitaoka et al. 1970a, b). Retentions and the combined yield of $P^3 + P^1$ are high to very high, depending on irradiation conditions. The most detailed recent information is given by Fenger (1969). As in his results for orthophosphate targets, Fenger reports substantially higher yields of labelled pyrophosphates from phosphites than do some others.

3.3. ANNEALING

MacKenzie and Borland (1952, Borland et al. 1952) were the first to demonstrate not only that a higher neutron flux and a higher temperature *during* irradiation resulted in higher retentions in orthophosphates but also that prolonged *post*-irradiation thermal treatment resulted in an almost complete conversion of labelled non-orthophosphates into the parent compound. Lindner and Harbottle (1960) also noted that when their irradiation conditions (moderate flux, dry-ice temperature) were varied by increasing the ambient temperature of bombardment, retentions in orthophosphates rose from less than 10% to very high values. Yields of other labelled products were also affected by changes in irradiation and storage conditions (Lindner and Harbottle 1961). Since then a large body of experimental annealing data has become available, giving us some information on the nature of the primary products of the recoil event. In these studies annealing was by thermal treatment, ultraviolet light and electron-irradiation, gamma radiation, and compression (Lindner and Harbottle 1961, Claridge and Maddock 1959, 1961a, b, 1963a, Fenger 1964, 1968, 1971, Fenger and Nielsen 1965, Yoshihara and Yokoshima 1961, Anselmo 1961, Baba et al. 1961, 1963, Jovanovic-Kovacevic 1969, Matsuura et al. 1969, Claridge 1965, Tanaka 1964a, b, Andersen 1963).

There is general consensus on a number of aspects of the annealing phenomenon. Apart from oxidation and/or recombination of fragments on thermal annealing as is observed in many other systems, in phosphates there is also quite obviously rupture of bonds involved. The curves cannot be fitted with kinetic expressions of a definite order (Baba et al. 1963), in contrast to the first-order kinetics for the initial rapid annealing followed by second-order processes in the slow part, observed in some other systems. Some of the important conclusions drawn by Claridge and Maddock (1961a, b, 1963b) and by Fenger (1964, 1968, Fenger and Nielsen 1965) respectively from their extensive studies of annealing are summarized below.

At low temperatures the main process appears to be bond rupture, schematically $P^5O^{32}P^3 \rightarrow {}^{32}P^3$. At higher temperatures *polymeric* ${}^{32}P^5$ is similarly converted into monomeric ${}^{32}P^5$. Oxidation proceeds in parallel to thermally induced bond rupture: at low temperatures primarily ${}^{32}P^1 \rightarrow {}^{32}P^3$, while at higher temperatures oxidation of ${}^{32}P^3$ to ${}^{32}P^5$ takes place simultaneously leading to ${}^{32}P$ incorporation in oxygen-bridged structures. Radiation and photo (ultra violet) annealing are fundamentally different and presumably arise from electronic excitation by transference of energy through the lattice, leading to oxidation at the 'defect' site, i.e. the trapped ${}^{32}P$ recoil atom. The most prominent processes are the conversion of ${}^{32}P^1$ into ${}^{32}P^3$ and stabilization by oxidation of ${}^{32}P$ leading to preservation of existing skeletons as in the process $P^5O^{32}P^3 \rightarrow P^5O^{32}P^5$. As they are normally found, compounds of phosphorus in lower oxidation states (P^1, P^2, P^3) contain at least one P–H bond; however, the exact configuration of reduced ${}^{32}P$ recoil atoms, especially in hydrogen-free lattices remains questionable. Whatever the precise nature of their precursors, there is the interesting observation that non-radioactive phosphite radical ions $\cdot{}^{31}PO_3^{2-}$, observed by means of ESR measurements in irradiated ammonium dihydrogen phosphate anneal out thermally long before the ${}^{32}P^3$ fraction starts to disappear (Fenger 1964, Fenger and Nielsen 1965). Baba et al. (1963) have determined distributions of activation energies for thermal annealing processes (applying the Vand–Primak analysis) in neutron-irradiated orthophosphates. They found 0.60–1.0 eV for the polyphosphate fraction and for the reduced species 1.1–1.4 eV. Claridge and Maddock (1963a) found that the maximum wavelength causing photo annealing lies at about the photon energy that would be predicted for promotion of an electron to the conduction band in these phosphates.

Annealing studies of non-orthophosphates show that in potassium pyrophosphate, for example, as in the orthophosphates, there is a clear tendency for thermal restoration of the original lattice leading to an increase in retention (Fenger 1971).

4. Discussion

Libby (1940) concluded from his experiments that the oxidation state of a ${}^{32}P$ atom recoiling from an (n,γ) process was determined in the nuclear event, and that in at least fifty percent of the cases an oxygen *atom* was expelled, leading to reduction. Present thinking assumes that in a solid matrix geometrical conditions at the site of entrapment as well as considerations of a chemical nature play decisive roles in determining the ultimate chemical fate of the recoiling particle. If one compares the chemical forms of ^{32}P produced in aqueous solutions by the $^{31}P(n,\gamma)^{32}P$ (Libby 1940) and the more energetic $^{32}S(n,p)^{32}P$ processes, (Lindner 1958) one finds rather similar results, showing that neither the initial form of the target, nor the nuclear event itself determines the oxidation number of the recoiling ^{32}P. However, the presence of rather small amounts of various additives in these solutions greatly influences the distribution of the ^{32}P among the oxidized and reduced states (Lindner 1958, Kobayashi et al. 1971, Matsuura and Lin 1971, Fenger and Pagsberg 1973). This behaviour is typical of a scavenged system and is strong evidence that the recoiling ^{32}P atom is subject to thermal or near-thermal chemical reactions after being slowed down. In the case of a solid matrix it is still unknown at what level of energy the events occur which lead to chemical incorporation of a recoiling ^{32}P. However, this kind of uncertainty is not unique for ^{32}P recoils, but a matter of basic concern in the whole field of solid-state hot-atom chemistry. In this regard the observation of Claridge and Maddock (1963a, b, Claridge 1965), that quite small amounts (10^{-3} mole %) of arsenate present in mixed crystals of KH_2PO_4 with KH_2AsO_4, substantially reduce the yield of ^{32}P in the orthophosphate fraction, is of great interest in that it clearly indicates the operation of a long-range effect. This result is reminiscent of the similar long-range effect exerted by a number of cations on the oxidation state of ^{32}P generated by (n,p) reactions in mixed crystals of isomorphous sulphates (Lindner 1971, unpublished). Formally, at least, these solid-state effects resemble the well-known scavenger reactions observed in the hot atom chemistry of liquid and gaseous systems: they may reflect the trapping of electrons or holes resulting in the 'protection' of reduced ^{32}P species.

Claridge and Maddock argue that following the nuclear event it is difficult to distinguish the process of oxide *ion* loss followed by rapid electron capture, from simple oxygen *atom* ejection. They visualize successive loss of oxide ions yielding sequentially the primary recoil fragments PO_3^-, PO_2^+, PO^{3+} and P^{5+}. The latter two fragments are thought to capture electrons, thus forming the precursors for P^3 and P^1. The first two fragments are supposed to be responsible for the formation of P–O–P bridges. The P–P bond formation is considered by them to be the best evidence for electron capture from inactive phosphate groups in the crystals. A pattern similar to the rupture of phosphorus–oxygen bonds is assumed in the case of fluorophosphate, PF_6^-, to explain the formation of various labelled anions including those containing fluoride ligands.

There is general agreement based upon a growing body of evidence (Fenger 1969) that the oxidation state and also the skeleton in the case of polymeric labelled products, are essentially determined in situ in the lattice and that subsequent dissolution steps are unable to change them. The presence of hydrogen in the lattice in various chemical forms has been related to the formation or stabilization of lower oxidation states of phosphorus characterized by P–H bonds (Lindner 1958), but the relationship is of a qualitative nature. The absence of hydrogen in a lattice, for instance, does not prevent the formation of species which are presumably precursors to P^3 and/or P^1. Even NH_4^+ ions, which have been cited for their reducing properties in some recoil atom reactions, do not display this behaviour in any definite way in the case of orthophosphate targets (Lindner and Harbottle 1961, Claridge and Maddock 1963a, Fenger 1964, 1969, Fenger and Nielsen 1965). Water of crystallization tends to enhance the $^{32}P^5$ yield relative to that for $^{32}P^3 + {}^{32}P^1$. A comparison of recoil yields in crystals containing NH_4^+ with those having ND_4^+ reveals an isotope effect which has been taken to indicate opposing tendencies concerned with oxidation of the recoil ^{32}P.

The specific labelling of certain recoil products – for instance in orthophosphate: isohypophosphate as $P^5O^{32}P^3$ and tripolyphosphate as $P^5O^{32}P^5OP^5$ – seems to give us clues to the synthetic mechanisms. These observations and others: the relatively low yields of labelled P^3OP^3 in comparison with P^5OP^3 and P^5OP^5 (cf. table 9.2) and the preferential labelling as $^{32}P^5OPOP$ for tripolyphosphate recovered from both irradiated pyrophosphates and tripolyphosphates all seem to imply that lattice groups like PO_4^{3-} are not much affected by a ^{32}P recoil particle being trapped at the site. This appears to give further support for a model of 'little disorder' suggested by Lindner (1958). The main feature of this model is the assumption that mean free paths for collisions toward the end of the track of the recoiling ^{32}P particle are sufficiently long and relaxation times sufficiently short that only limited amounts of energy are deposited locally, giving moderately high (equivalent) temperatures; thus the site of entrapment rapidly cools down and is characterized by limited fragmentation and little disorder. As a consequence the chemical processes which take place are strongly influenced by configurational considerations. One such process would be the linking of the ^{32}P to a free electron pair of one of the oxygen atoms of, for example, a phosphate lattice group, leading to the formation of structures with $PO^{32}P$ bridges labelled by preference at certain positions. This linking is probably not a high energy process as has been suggested (Fenger 1964, 1969), but may proceed after the recoil ^{32}P has reached an interstitial position, without having caused many atomic displacements and/or excessive rupture of bonds. The formation of a $P{-}^{32}P$ bond on the other hand may well be the result of a 'hot' process as it probably requires the rupture of a P–O bond and also the displacement of the oxygen (atom or ion). The inability of the relatively light-weight tritium recoils to form P–T bonds in an orthophosphate lattice (Lindner et al. 1965) supports this point of view. Cífka (1963), in a study dealing with recoil ^{32}P produced in the white allotropic modification of elementary phosphorus, concludes that his results cannot be considered as unequivocal evidence for the existence of a 'hot zone' [hot spot or 'thermal spike' with a relatively high temperature (Harbottle and Sutin 1958, 1959)] and leaves open the possibility that heat dissipation "can occur much faster than is frequently supposed". The calculations of Bunker and Van Volkenburgh (1970) indicating rather *short* ranges for the (n,γ) produced ^{32}P also agree with the model of little disorder. The moderately low yields of labelled products with P–P bonds, the 'shielding' effect (Lindner and Harbottle 1961, Yoshihara and Yokoshima 1961) exerted by water of crystallization on both the product spectrum and the ^{32}P distribution [leading to less polymeric and more monomeric products (Fenger 1969)] also fit a model of little disorder. If large yields of highly condensed labelled products were found, it would indicate a large zone with high temperatures, and that would not agree with this model. There are contradictory reports on this subject in the literature and the ambiguities need to be resolved. To accomplish this important matter would require improvement in the analytical techniques applied, especially with regard to better separation and identification of all the polymeric species of potential importance. Linkage of a ^{32}P particle to an oxygen atom of a long-chain phosphate in the manner already described, would be governed by a statistical probability for the formation of a branched ultraphosphate, which would increase with chain length. This class of products instantaneously hydrolize on dissolution (Van Wazer 1958) and consequently always yield end-labelled 'fragmentation' products (Lindner et al. 1965). The occurrence of such end-labelled products after dissolution of irradiated long-chain target material therefore is not necessarily proof of selectivity in end-labelling by recoil as suggested by Matsuura and Lin (1970).

In general one hopes that the model proposed has implications beyond the systems under discussion. Therefore it is reassuring that Müller (1965a, 1969c) and others (Rössler and Otterbach 1971a, b) dealing with recoil process in quite different inorganic crystalline systems have arrived at similar conclusions. We take this to mean that the combined effort of all investigators involved in the chemistry of recoil ^{32}P in phosphates has contributed to a better understanding of the field of hot atom chemistry in solid inorganic systems as a whole.

Addendum (1978): Recent Work

Akaboshi, Kawai and collaborators have shown that when silica containing ^{31}Si is mixed with adenosine, labelled with ^{14}C, to facilitate subsequent paper chromatographic analysis, small quantities of adenosine monophosphate (1%, 5′ AMP; 0.5%, 2′ AMP and 0.5%, 3′ AMP) are produced by the β decay of the ^{31}Si (Akaboshi et al. 1971). Mixtures with adenosine mono and diphosphates gave tiny yields (~0.2%) of the di and triphosphates respectively (Akaboshi and Kawai 1971a). They suggest that such processes might have been involved in primordial synthesis and speculate on the possibilities of asymmetric synthesis in this way (Akaboshi and Kawai 1971b). Kitaoka et al. (1973a, b) have investigated the products of the (n,γ) reaction in adenosine triphosphate (Kitaoka 1973). They confirm that the products are greatly influenced by the conditions of irradiation. The products include most of the oxyanions found with irradiated phosphates (Kitaoka et al. 1973a, b, Nakamura et al. 1970). Various improvements to the detailed procedures for chromatographic or electrophoretic analysis have appeared (Kiso et al. 1972; Jovanovic-Kovacevic 1973, 1976). The technique of cross-electrophoresis has been recommended. In this procedure phosphorus anions move towards the anodic end of the paper, zinc ions move to meet and pass them, fixing insoluble species (Kitaoka 1974).

Fenger and Pagsberg have found an interesting example of the interconnection of recoil and radiolytic processes with irradiation of orthophosphate solutions. The ^{32}P product distribution is not so very different from that in a solution of the irradiated solid orthophosphate. Some pyro and polyphosphates are found, the yields increasing with the phosphate concentration in the solution. Further, the yield is dependent on an oxidation by OH radicals and is greatly reduced in the presence of hydroxyl scavengers. It is also dependent on the dose rate of ionising radiation concurrent with the neutron irradiation (Fenger and Pagsberg 1973, 1975). They suggest the oxidation by OH of ^{32}P recoil species leads to a ligand deficient entity that can attach an O–P containing species to form a ^{32}P–O–P species. These observations may be related to those of Dahl and Birkelund (1962 v. supra).

Lin and Matsuura (1975) have extended their studies of polyphosphates using $(NaPO_3)_n$ with $n = 50$–100. About 90% of the ^{32}P is reported to appear as terminal phosphorus atoms in polyphosphates. Most of the reduced ^{32}P forms mono-nuclear anions. Thermal annealing increases the amount of $^{32}P^5$. Gamma annealing is less effective.

Anselmo (1973) finds that the proportion of $^{32}P^5$ in neutron irradiated phosphates, and of $^{32}P^3$ in phosphites, decreases as the mass of the cation present increases. He interprets this in terms of escape of the ^{32}P from its lattice site – the "Exploded lattice" model. But it must be noted that the data all refer to ambient temperature irradiations and must include a substantial contribution from annealing.

Finally more data have appeared on hypophosphites. Kobayashi et al. (1972) find $^{32}P^2P^4$, $^{32}P^3{\cdot}O{\cdot}P^5$, $^{32}P^5$ and $^{32}P^4{\cdot}P^4$ on dissolving irradiated sodium hypophosphite. They also find good evidence, based on its disproportion kinetics, for $^{32}P^2{\cdot}P^2$. This latter, together with $^{32}P^1$ and $^{32}P^3$ are the major products. Application of non-aqueous methods of analysis to neutron irradiated ammonium hypophosphite by Chandratillake et al. (1976) and comparison with aqueous procedures shows that all the ^{32}P species of oxidation state greater than 3 are formed during solution of the irradiated salt. The non-aqueous analytical procedure was similar to that developed by

Mahmood (1972) and involved solution in alcohol. Appreciable amounts of a PH_3 forming, or exchanging, species were found, as well as traces of zero valent phosphorus.

Acknowledgement

This work is part of the research program of the Institute for Nuclear Physics Research (IKO), made possible by financial support from the Foundation for Fundamental Research on Matter (FOM) and the Netherlands Organization for the Advancement of Pure Research (ZWO).

The author is much indebted to Miss J.C. Kapteyn, Miss N. Kuyl and Mrs. M. Oskam-Tamboezer, who all contributed substantially to the finalizing of the manuscript.

CHAPTER 10

MAIN GROUP ELEMENTS

N. SAITO* and K. YOSHIHARA**

**Department of Chemistry, The University of Tokyo,*
Bunkyo-ku, Tokyo, Japan
and
***Department of Chemistry, Tohoku University,*
Sendai, Japan

Chemical Effects of Nuclear Transformations in Inorganic Systems
Edited by G. Harbottle and A.G. Maddock

Contents

Introduction

Research in the field of hot-atom chemistry of complexes of the main group elements may be divided into two categories:

(1) The enrichment of radioisotopes or the production of radioisotopes of high specific activity by means of the Szilard–Chalmers process.
(2) Studies whose objective is to understand the mechanisms of the hot-atom reactions. However, research in this particular area of hot-atom chemistry is scanty. This is simply because few main group element complexes are suitable for hot-atom studies.

1. Enrichment of Radioisotopes

There have been some interesting studies whose aim was to produce radioisotopes of high specific activity for various purposes. The results of these studies are summarized in table 10.1.

Early work along this line until 1960 includes work by Herr (1952c) to enrich gallium and indium isotopes; work by Payne et al. (1958) to concentrate radiogallium; work dealing with the enrichment of radioantimony by Williams (1948a, b); and work by Sharp et al. (1959) regarding Szilard–Chalmers separations of several radioisotopes.

One of the interesting results obtained by recent investigations was the Szilard–Chalmers separation of radiocalcium which is of particular biological interest. Bruno and Belluco (1956a) made the first unsuccessful attempt to enrich radiocalcium using calcium EDTA complex as a target. Recently, Ebihara (1965) and Ebihara and Yoshihara (1960) employed several other calcium compounds in order to obtain radiocalcium of high specific activity. For instance, high enrichment was observed with a target of calcium oxinate. In one of the experiments, the neutron-irradiated oxinate was dissolved in n-butylamine. The resulting solution was poured onto a cation exchange resin (Diaion SK-1) column. Ca-45 adsorbed on the resin was then eluted with 6M HCl. The radiocalcium appeared in the effluent with an enrichment factor of 120. However, the yield of extractable activity was only 8.3%. In a

Table 10.1
Enrichment of radionuclides by Szilard–Chalmers process

Nuclide	Target[a]	Enrichment factor	Ref.
^{24}Na	NaPc[b]	—	Pertessi and Henry (1963)
^{45}Ca	Ca oxinate	120 (max)	Ebihara and Yoshihara (1960)
^{47}Ca	Ca oxinate	40 (max)	Ebihara (1965)
^{47}Ca	Ca oxalate	140 (max)	Ebihara (1965)
^{47}Ca	Ca tartrate	24 (max)	Ebihara (1965)
^{47}Ca	Ca citrate	88 (max)	Ebihara (1965)
^{87m}Sr	$SrCO_3$(n;2n)	6 (max)	Shiokawa et al. (1967)
^{72}Ga	Ga oxinate	2880 (max)	Ebihara (1962)
*Ga	GaPc	574 (max)	Payne et al. (1958)
$^{70,72}Ga$	GaPc	$> 5 \times 10^3$	Herr (1952b)
^{114m}In	(InCl)Pc	6	Mühl and Grosse-Ruyken (1967)
^{114m}In	(InCl)PcCl	5.2 (max)	Mühl and Grosse-Ruyken (1967)
^{114}In	(InCl)Pc + (InCl)PcCl (mixture)	12.3 (max)	Mühl and Grosse-Ruyken (1967)
^{116m}In	In oxinate	$\sim 10^6$ (max)	Ebihara and Yoshihara (1961)
^{116m}In	InPc	$\sim 10^3$	Herr (1952b)
^{124}Sb	NH_4SbF_6	large	Williams (1948a and b)

[a] Unless otherwise stated, nuclear processes involved are (n,γ) reactions.
[b] Pc = phthalocyanine.

continuation of previous work, Ebihara (1965) studied a method of extracting Ca-47 from the irradiated solid oxinate with a ethanol(10%)–water mixture, which gave a better yield of 44%.

The Japanese group, Ebihara and Yoshihara (1961) carried out the extraction of indium-116m by NaOH solution from a chloroform solution of the irradiated indium oxinate with an enrichment factor of $\sim 10^6$. Ebihara (1962) also successfully used gallium oxinate as a target for the (n,γ) reaction to concentrate Ga-72 with an enrichment factor of $\sim$ 2600. Mühl and Grosse-Ruyken (1967) conducted an investigation into the production of indium-114m using indium phthalocyanine complexes. They carefully studied the effects of irradiation conditions, pretreatments of the targets, storage conditions after the irradiation, and methods of separation upon the specific activity of the radioindium product. However, the range of enrichment factors observed in their experiments were not very promising.

Since the preliminary study of Na-24 by Sharp et al. (1959), one paper has been published on the separation of Na-24 from a target of sodium phthalocyanine. In this paper, Pertessis and Henry (1963) reported the yields of Na-24 in the Szilard–Chalmers separation for different extraction processes and also the effect of the neutron dose and thermal annealing before and after the irradiation upon the extraction yield. No data were given, however, for the enrichment factors.

Apart from the (n,γ) reactions, Shiokawa and co-workers (1967) have attempted to separate Sr-87m by the recoil effect of (n,2n) reactions. A target of strontium carbonate powder suspended uniformly in a gelatine solution was irradiated with 14 MeV neutrons. A fraction of Sr-87m activity was found in the gelatine phase with a maximum enrichment factor of 6.

2. Studies on the Mechanisms of Hot-Atom Reactions

In this section, we shall consider only complexes of the main group elements. Numerous organometallic compounds will be excluded here as they are covered elsewhere (see ch. 13).

Only a few papers have dealt with phthalocyanines of the main group elements. One of the complexes which falls into this category is indium phthalocyanine. In recent studies on indium complexes, Kudo and Yoshihara (1971) investigated the neutron dose dependence of the apparent initial retention with the neutron-irradiated indium chlorophthalocyanine. This complex differs from other metal phthalocyanines in that chloride anion is attached to the central indium metal. The authors irradiated the compound at dry-ice temperature in three different reactors whose thermal neutron fluxes ranged from 5×10^{11}n/cm^2·sec to 5.2×10^{13}n/cm^2·sec. They calculated the apparent initial retention which was defined in their paper as the retention value for the target which has been irradiated at dry-ice temperature and subjected to chemical treatment at a temperature below 0° C, and the results were discussed in terms of neutron dose dependence.

The conclusions to be drawn from this experiment are summarized as follows:

(a) The apparent initial retention increases with the increasing neutron dose.
(b) In spite of the use of different reactors for irradiation, plots of the apparent initial retention vs the neutron dose (nvt) provide a single curve as shown in fig. 10.1.
(c) There is a linear relationship between the neutron dose and the reciprocal of the value $(R_p - R)$, where R and R_p stand for the apparent initial retention and its plateau value, respectively. This relationship is represented by a straight line shown in fig. 10.2.

From this, the value of 20% can be derived for the apparent initial retention at very low neutron fluences. This value may reflect the initial recoil reaction taking place in the solid excluding thermal and radiation effects during irradiation. It is interesting to note that, in contrast to the indium complex, the effect of neutron fluence on the apparent initial retention was hardly noticed with cadmium phthalocyanine.

Hot-atom studies of another series of indium complexes have also been

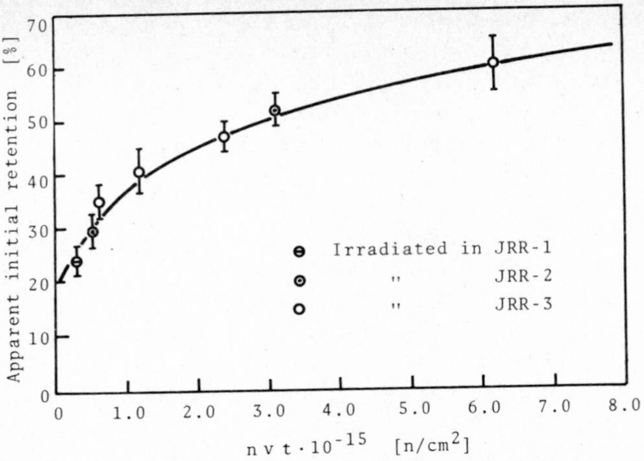

Fig. 10.1. Apparent initial retention of indium phthalocyanine as a function of the neutron dose (nvt).

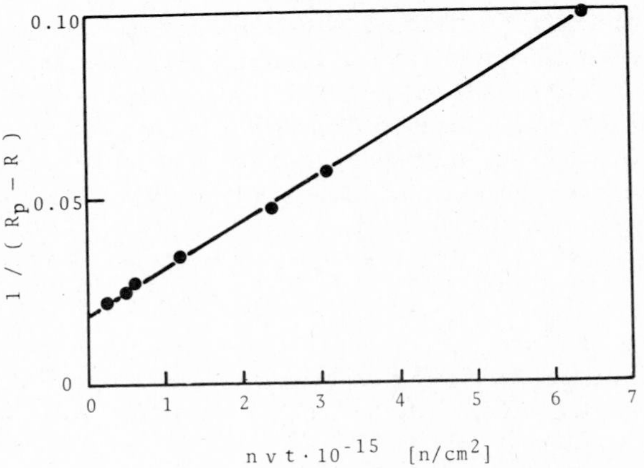

Fig. 10.2. Relationship between $1/(R_p - R)$ and nvt.

carried out by the same authors. Yoshihara and Kudo (1970) investigated recoil behavior of In-115m following the $^{115}In(\gamma,\gamma')^{115m}In$ reaction with an EDTA (= Y) complex, $NaInY \cdot 3H_2O$. The experiments involving the (γ,γ') reaction have an advantage that the energy of the recoil atom may be varied in the range of 1–300 eV. Irradiation of the complex with γ-rays was done at dry-ice temperature by means of an electron linear accelerator to which a platinum converter was attached. In this experiment, not only the (γ,γ') reaction but also the $^{115}In(n,\gamma)^{116m}In$ and the $^{115}In(\gamma,n)^{114m}In$ reactions were studied with the same target.

In fig. 10.3 is plotted the yield of radioindium ion following the (γ,γ′), (n,γ) and (γ,n) reactions as a function of the recoil energy of the radioindium atom. It is evident from fig. 10.3 that the yield of $^{115m}In^{3+}$ ions for the (γ,γ′) reaction was zero in the low-recoil-energy region but rose from zero when the recoil energy exceeded a certain value. The yield increased with increasing recoil energy and finally reached a saturation value in the higher energy region. It is worth mentioning that this sort of energy dependence was observed not only with indium recoils but with zinc recoil atoms from α- and β-zinc phthalocyanines (Yoshihara and Yang 1969b).

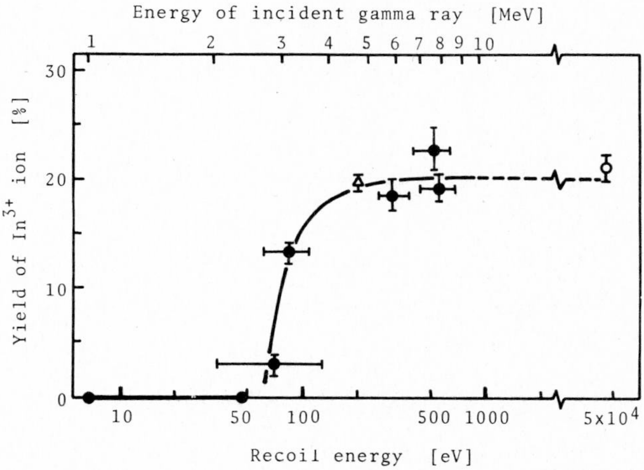

Fig. 10.3. Yield of radioactive In^{3+} ion following the nuclear reactions in the In-EDTA complex as a function of the recoil energy; ●, $^{115}In(\gamma,\gamma')^{115m}In$ reaction; △, $^{115}In(n,\gamma)^{116m}In$ reaction; ○, $^{115}In(\gamma,n)^{114m}In$ reaction.

The energy at which the yield rises from zero on the curve is called ‘appearance energy’, a new term in hot-atom chemistry proposed by Yoshihara and Kudo. The appearance energy of the $^{115m}In^{3+}$ ion in the indium EDTA complex was approximately 60 eV. The physical meaning of the appearance energy may be compared with that of the displacement energy in solid-state physics.

Indium isotopes produced by these reactions cannot be dealt with in the same manner because of the differences in the nuclear and physicochemical properties. Nevertheless, they were correlated by a single curve as shown in fig. 10.3. This study also dealt with the phase effect on the yield of $^{115m}In^{3+}$ ion following the (γ,γ′) reaction in the complex. It was observed that change of the phase of the complex from solid to liquid, namely, aqueous solutions of the complex, had a profound effect on the yield.

Yoshihara et al. (1970) carried out a series of experiments in which indium EDTA complexes, $Na[InY(OH_2)]\cdot 2H_2O$, $K[InY(OH_2)]H_2O$ and

H[InY(OH_2)], were irradiated with γ-rays of various mean energies at dry-ice temperature. Recoil energy dependence of the $^{115m}In^{3+}$ yield for these compounds is demonstrated in fig. 10.4. It is clear from this that almost no difference in yield was observed between the sodium and potassium complexes. On the other hand, the hydrogen compound gave a higher plateau value than the sodium and potassium compounds. With regard to the appearance energy, it decreased in the following order: K compound ≈ Na compound > H compound.

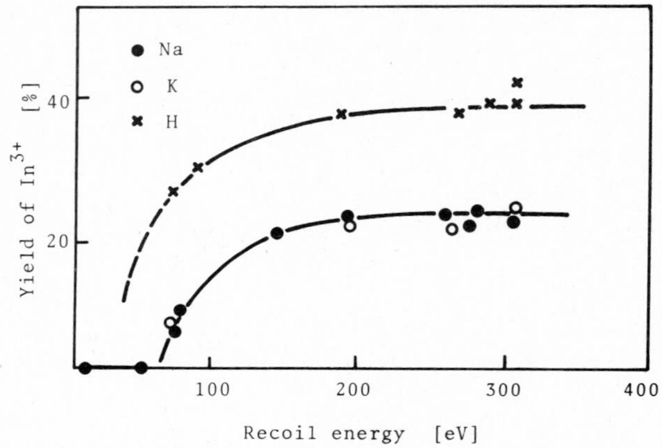

Fig. 10.4. Yield of $^{115}In^{3+}$ as a function of recoil energy.

In another experiment, Yoshihara and Mizusawa (1972) irradiated dehydrated indium EDTA complex, Na[InY(OH_2)], and lutetium EDTA complex, Na[LuY$(OH_2)_2$]·H_2O, with γ-rays at dry-ice temperature. The appearance energy obtained in this experiment for the lutetium complex was in the vicinity of 40 eV, whereas the corresponding value for the indium complex was 60–70 eV. The authors of the last two papers attempted to interpret the difference in the value of appearance energy in terms of the cordination number and the nature of cordinate bond of the EDTA complexes. It seems, however, that more data with other complexes are needed before a clear interpretation can be made.

Addendum (December 1976)

The appearance energy of aquo-N-2-hydroxyethylenediamine-N,N′,N′-triacetatoindate(III) monohydrate was found to be lower than that of sodium aquo-ethylenediamine-tetraacetatoindate(III) dihydrate, showing the same tendency in the plateau value of the recombination reaction after bond rupture. A chemical cage effect in the recoil atom reaction in solids was discussed in a recent paper by Yoshihara et al. (1976).

CHAPTER 11

FIRST TRANSITION FAMILY ELEMENTS

Takeshi TOMINAGA and Nobufusa SAITO

Department of Chemistry, Faculty of Science, The University of Tokyo, Bunkyo-ku, Tokyo, Japan

Chemical Effects of Nuclear Transformations in Inorganic Systems
Edited by G. Harbottle and A.G. Maddock

Contents

Introduction

The recoil and subsequent reactions of numerous stable complexes of the first transition family elements have been studied extensively. While the major object of recoil studies in the early stage was the production of highly enriched radioisotopes, recent works have been mainly designed to elucidate the mechanisms of the recoil and subsequent reactions. However, the mechanisms of recoil reactions in irradiated solid complexes are complicated by a number of factors which can hardly be controlled experimentally and none of the models proposed so far has been proved to be satisfactory for a general and detailed understanding of the reaction mechanisms.

The main problems we encounter in solid phase recoil experiments are annealing phenomena and chemical reactions taking place before, during and after dissolution of the irradiated solid complexes. Since the mechanisms of annealing phenomena will be discussed in ch. 26, we shall simply mention in this chapter some experimental data on post-irradiation effects observed in connection with transition metal complexes. Irradiations at lower temperatures and with lower radiation doses as well as some dynamic methods of separation (simultaneous with irradiation) have been used to try to approach the 'unannealed' primary distribution of recoil species.

The various recoil products formed in irradiated solid complexes are usually separated by chemical methods such as electrophoresis, ion exchange and solvent extraction, all of which inevitably involve a dissolution step. It is generally believed that the metastable recoil species present in the irradiated solid systems may undergo rapid and mostly unknown reactions on dissolution in the solvent. Hence, the primary distribution of the recoil species in the irradiated solid complexes may be disturbed significantly depending on the dissolving and analytical conditions, and the data obtained by chemical methods may not exactly reflect a true picture of the product distribution in the solid state. Although physical methods such as Mössbauer spectroscopy can in principle give more direct information on the primary stage of recoil reactions in the solid phase, they are not omnipotent for recoil studies in solid complexes. For example, Mössbauer spectroscopy can be applied to only a limited number of nuclides and systems and sometimes yields only ambiguous information as to the actual chemical forms of recoil products formed in the solid complexes.

Accordingly, chemical and physical methods are still complementary to each other for the complete understanding of recoil reactions in the solid complexes.

The typical reactions involving recoil species from complexes of first transition family reported so far are summarized in table 11.1.

1. Affected Atom in Ligand

While earlier studies on hot-atom chemistry of complexes have mainly dealt with the recoil behavior of central atoms, a simpler product distribution may be expected for the recoil reactions with ligands of complexes. The substitution of recoil halogen atoms for ligands in irradiated complexes was first reported by Saito et al. (1960a). Extensive studies have been carried out on the chemical effects of (n,γ), (n,2n) and IT reactions on bromine as outer anion in a series of pentamminecobalt(III) bromides, $[Co(NH_3)_5X]Br_2$ where X could be NO_2, NH_3, NCS, H_2O, NO_3, F, Cl, Br and I, and the substitution of radiobromine for X as ligand has been observed in all these complexes (Saito et al. 1961b, 1962b, 1965c, 1967b, Tominaga et al. 1971a). Since recoil radiobromine atoms arising from cobaltammine bromides are stabilized mostly as bromopentamminecobalt(III) and bromide ions (Yoshihara and Harbottle 1963), the recoil product spectra from these systems are simple enough for studying the recoil reactions. It is worth mentioning that the radiobromine-for-X substitution yield (defined as 'ligand yield' by Saito and his group) in these complexes correlates well with the stability of the target complexes as reflected in the frequency of the maximum in their first absorption band. The yield of $[Co(NH_3)_5Br^*]^{2+}$ is slightly increased by irradiation (or storage) at lower temperatures but the general correlation is not affected significantly. The above correlation indicates that the recoil bromine atoms can substitute with greater ease for less stably complexed ligands in the target complexes. Although the apparent dependence of the substitution yield (ligand yield) on a collision parameter (physical factor) observed in the earlier stage of their work cannot be totally neglected, the predominant role of chemical factors in recoil and subsequent reactions is shown by the fact that the above correlation appears to be essentially true for recoils from different types of nuclear reactions, e.g., (n,γ), (n,2n) and IT transformations. Recently, Lazzarini has attempted to explain the recoil ^{80}Br substitution yield-target stability correlation for the ^{80m}Br(IT)^{80}Br transformation in terms of a physical model originally proposed for interpreting the isomeric effects in ^{60m}Co and ^{60}Co recoils in cobalt complexes (Lazzarini and Fantola-Lazzarini 1971a). Analogous recoil bromine-for-ligand substitution reactions have been observed with bis(ethylenediamine)cobalt(III) bromides, $[Co(en)_2X_2]Br_{1,3}$, where X can be NH_3, NCS, Cl and Br, and the substitution yields are also related to the stability of the target complexes (Saito et al. 1961b, 1962b, 1967b). An

Table 11.1
Reactions involving recoil species from complexes of the first transition family

Subject of interest	Nuclear transformations investigated	Typical systems investigated	Typical modes of reaction observed and related topics
Recoil reactions with ligands	(n,γ) (n,2n) (d,n) (t,n) IT	Cobaltammines	Ligand substitution reaction: $[Co(NH_3)_5X]Br_{2,3} \longrightarrow [Co(NH_3)_5Br^*]^{2+}$ $[Co(en)_2X_2]Br_{1,3} \longrightarrow [Co(en)_2XBr^*]^{1,2+}$ $[Co(NH_3)_6]X_3 \rightarrow [Co(NH_3)_5X^*]^{2+}$
Recoil reactions with central atoms	(n,γ) (γ,n) (n,2n) IT ion implantation technique	Cobaltammines Nitroammines Anionic complexes Chelate complexes Phthalocyanines	Synthesis type reaction: $[Co(NH_3)_6]X_3 \rightarrow [Co^*(NH_3)_5X]^{2+}$, $[Co^*(NH_3)_4X_2]^+$, etc. Isomerization reaction: cis-$[Co(NH_3)_4(NO_2)_2]^+ \rightarrow$ trans-$[Co^*(NH_3)_4(NO)_2]^+$ Central atom substitution reaction: $[Co(NH_3)_6][Fe(CN)_6] \rightarrow [Co^*(CN)_6]^{3-}$ $[Co(NH_3)_6]_2(CrO_4)_3 \rightarrow [Cr^*(NH_3)_6]^{3+}$
Solid-phase Exchange reactions		Chelate complexes	Exchange reactions between doped ions and host complexes (similar to thermal annealing reactions)
Reactions of recoil species in solution and resin phase	(n,γ) IT	Chelate complexes	Thermal diffusive reactions in organic solvents (recombination of ligand deficient species with free ligand, etc.) ⟶ scavenging with metal salts
		Complex ions on ion exchange resin	Dynamic extraction of recoil species

interesting observation with this group of complexes is that recoil radiobromine arising from the isomeric transition can partially react with ethylenediamine molecules while the (n,γ) recoil radiobromine never reacts with them (Ambe et al. 1971). Similar substitution reactions of recoil anions for ligands are seen in recoil ^{38}Cl (and ^{60}Co) in hexamminecobalt(III) chloride (Ikeda et al. 1961, 1964) and recoil ^{18}F in hexamminecobalt(III) fluoride and nitrate (Saito et al. 1970). In the ^{18}F work recoil ^{18}F atoms were produced by three different types of nuclear transformations, i.e., ^{19}F(n,2n)^{18}F, ^{17}O(d,n)^{18}F and ^{16}O(t,n)^{18}F reactions.

The radiohalogen-labelled pentamminecobalt(III) ions formed as a recoil product in the hexamminecobalt(III) halides disappear gradually on annealing even at room temperature. In fig. 11.1 is shown the thermal annealing behavior of $[Co(NH_3)_5{}^{80m}Br]^{2+}$ in irradiated $[Co(NH_3)_6]Br_3$ (Yoshihara and Harbottle 1963). The disappearance of $[Co(NH_3)_5{}^{80m}Br]^{2+}$ on annealing has been explained by assuming an electronic mechanism producing a cobaltous complex ion which is unstable and decomposed on dissolution in water (Harbottle 1961, Yoshihara and Harbottle 1963), as follows:

$$[Co(NH_3)_5Br]^{2+} + e^- \longrightarrow [Co(NH_3)_5Br]^+ \xrightarrow{H_2O} Co^{2+} + Br^-$$

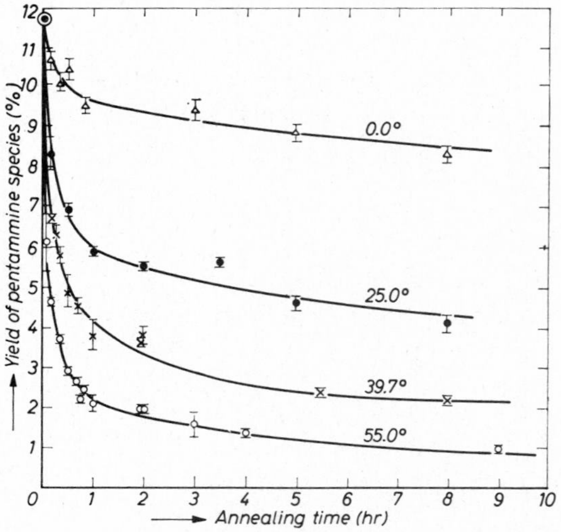

Fig. 11.1. Thermal annealing of ^{80m}Br-labelled bromopentamminecobalt species in neutron-irradiated hexamminecobalt(III) bromide (Yoshihara and Harbottle 1963).

Another mechanism has been proposed by Ikeda et al. (1961, 1964) to account for the disappearance of $[Co(NH_3)_5{}^{38}Cl]^{2+}$ and $[{}^{60}Co(NH_3)_5Cl]^{2+}$, formed in $[Co(NH_3)_6]Cl_3$. The activation energies obtained appear reasonable for the proposed reaction, i.e.,

$$[Co(NH_3)_5Cl]^{2+} + NH_3 \longrightarrow [Co(NH_3)_6]^{3+} + Cl^-.$$

Yasukawa (1966) has observed the formation of $[^{60}Co(NH_3)_6]^{3+}$ in neutron-irradiated halopentamminecobalt(III) complexes, $[Co(NH_3)_5Br]X_2$ and $[Co(NH_3)_5Cl]Br_2$($X = NO_3$, Cl, Br and I). The $[^{60}Co(NH_3)_6]^{3+}$ yield increases on annealing, and especially remarkably in a stream of ammonia. If hexamminenickel chloride $[Ni(NH_3)_6]Cl_2$ is irradiated with 20 MeV γ-rays, the ^{57}Co recoil atoms from the (γ,p) reaction are found in the forms of $[Co(NH_3)_5Cl]^{2+}$ and $[Co(NH_3)_6]^{3+}$ and their yields increase appreciably on annealing (Yoshihara 1964). If the same complex is labelled with ^{57}Ni, the ^{57}Co recoil atoms produced by the decay processes (β^+ and EC) behave differently. Thus the nature of these annealing reactions is complex and needs further investigation.

2. Metal Atom Affected

A number of labelled products are generally expected from recoil reactions of central atoms in transition metal complexes.

2.1. NATURE OF THE RECOIL PRODUCTS

Harbottle (1961) has found that seven ^{60}Co-labelled complex cations can be separated by paper electrophoresis from neutron-irradiated hexamminecobalt(III) nitrate and nitropentamminecobalt(III) nitrate. The ^{60}Co-labelled species found in the solutions of irradiated $[Co(NH_3)_6]$ $(NO_3)_3$ include $[^{60}Co(NH_3)_5NO_2]^{2+}$ and $[^{60}Co(NH_3)_5H_2O]^{3+}$, which may arise from reactions of the ligand deficient recoil species $[^{60}Co(NH_3)_5]^{3+}$ with NO_2^- ion or water, respectively.

Saito and coworkers (1960c, 1962c, d, 1963a, b, 1965d) have systematically investigated the recoil product distribution in a large number of irradiated cobalt(III) complexes such as nitroammine complexes $[Co(NH_3)_n(NO_2)_{6-n}]$ $(NO_{2,3})_{n-3}$, hexammine complexes $[Co(NH_3)_6]X_3$ ($X = NO_2$, NO_3, F, Cl, Br, I, etc.), aquopentammine complexes $[Co(NH_3)_5H_2O]X_3$, nitropentammine complexes $[Co(NH_3)_5NO_2]X_2$, and tris-ethylenediamine complexes $[Co(en)_3]X_3$. In their work, ^{60}Co-labelled species were separated and identified by electrophoresis, ion exchange, precipitation, recrystallization and paper chromatography. Figure 11.2 represents a typical electrophoresis histogram of irradiated hexamminecobalt(III) nitrite $[Co(NH_3)_6]$ $(NO_2)_3$. It is seen that the ^{60}Co-labelled complex species consist of the parent ion $[^{60}Co(NH_3)_6]^{3+}$ and products from the stepwise substitution of outer anions (NO_2^-) for ligands (NH_3) in the parent compound. While such substitution of ligands by outer anions is commonly observed in the recoil reactions of cationic cobalt complexes, complex recoil products containing less

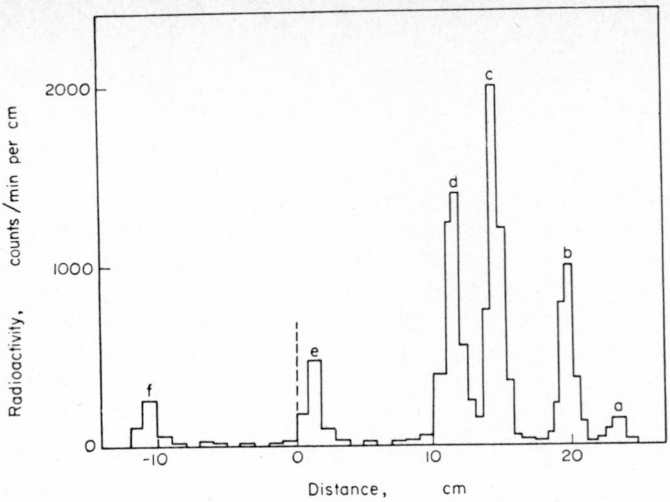

Fig. 11.2. Paper electrophoresis histogram of neutron-irradiated hexamminecobalt(III) nitrite. Cations move right. Dashed line indicated the starting point. Peak a, parent; b, $[Co(NH_3)_5NO_2]^{2+}$; c, Co^{2+}; d, cis- and trans-$[Co(NH_3)_4(NO_2)_2]^+$; e, $[Co(NH_3)_3(NO_2)_3]$; f, $[Co(NH_3)_2(NO_2)_4]^-$ (Saito et al. 1962d).

anionic ligands than the parent compound are seldom found in these systems. For instance, $[^{60}Co(NH_3)_6]^{3+}$ is not formed in nitroamminecobalt(III) complexes. Furthermore it is also found in the series of nitroamminecobalt(III) complexes $[Co(NH_3)_n(NO_2)_{6-n}]\,(NO_{2,3})_{n-3}$ that the outer anion-for-ligand substitution reactions are more favorable if the target complex ion is more positively charged or has more outer anions in its neighborhood (Saito et al. 1960c, 1962d). These observations indicate that the Coulombic interaction between the outer anions and the fragmented (ligand deficient) complex ions probably play an important role in the recoil or early annealing reactions.

The recoil product distribution in the irradiated complexes also depends on the nature of the outer anions. If the product spectrum is compared for salts of the same complex cation containing various anions such as $[Co(NH_3)_6]X_3$ where X can be NO_2, NO_3, F, Cl, Br, I, etc., the following behavior is generally observed: the outer anions which complex strongly with cobalt substitute most readily for ligands of the target complex and thus lower the retention in the parent form (Saito et al. 1962c, 1963a, b, 1965d). Costea and Dema (1962, 1963) also irradiated a number of cobaltammine complexes and studied the effect of chemical constitution on the percentage of the activity not separable as Co^{2+}. The reducing or oxidizing character of outer anions, and the radiation or thermal stability of the target compounds were considered to explain their results. The influence of chemical isomerism has also been studied in various cobalt ammine complexes (Costea and Dema 1961a, b, Costea et al. 1961, Costea 1961a, b).

2.2. COMPARISON OF (n,γ) AND ISOMERIC TRANSITION EFFECTS

Lazzarini has found that ^{60m}Co and ^{60g}Co recoil atoms produced in cobalt complexes behave differently and that the isomeric effect measured in terms of the ratio ρ of the ^{60m}Co retention to ^{60g}Co retention depends on the chemical constitution of the target complexes and on the irradiation and storage conditions. It is interesting to note that the ρ value is correlated with the d-orbital splitting Δ induced by the ligand field in a number of cobalt(III) complexes (Lazzarini 1967a, b, Lazzarini and Fantola-Lazzarini 1967a, b, c, 1969b) and in mixed crystals of some cobalt(III) and chromium(III) complexes (Lazzarini and Fantola-Lazzarini 1971a). The isomeric effect has been mainly ascribed to fast annealing reactions of ^{60g}Co recoil atoms (from the decay of ^{60m}Co) promoted by the energy released in the IT decay process. These annealing reactions could be the deexcitation of the excited crystal levels formed as a consequence of the IT process and the above correlation may be explained by assuming that the number of such excited levels is a function of Δ (Lazzarini and Fantola-Lazzarini 1971a).

2.3. ANNEALING PROCESSES IN THE IRRADIATED COMPLEXES

The annealing reactions in the irradiated cobalt(III) complexes generally convert the recoil species separable as Co^{2+} on dissolution to the parent chemical form, leading to ^{60}Co retention (Costea 1961a, Costea et al. 1961, Williams et al. 1961, Negoescu and Costea 1962, Dimotakis and Stamouli 1963, Dimotakis and Maddock 1964, Dimotakis et al. 1967, Dimotakis and Papadopoulos 1970). Although the kinetic analysis of annealing curves has become a common procedure for testing the proposed mechanism of reactions it is essential to obtain very reliable experimental data before any decisive conclusion can be reached. Harbottle (1961, 1965) has proposed that the process yielding $[^{60}Co(NH_3)_6]^{3+}$ on annealing may be

$$[Co(NH_3)_6]^{2+} \longrightarrow [Co(NH_3)_6]^{3+} + e^-$$

either through thermal ionization or electron transfer. However, experiments on the role of ambient gases or vapors on the annealing behavior indicate that the actual mechanism for parent reformation may not be as simple as shown above. Yasukawa has found that vapors of ammonia and ethylenediamine promote annealing in hexamminecobalt(III) and tris(ethylenediamine)-cobalt(III) complexes and assumed that the diffusion of ligand molecules (ammonia etc.) may play a more important role than electronic processes in annealing reactions in these systems (Yasukawa and Saito 1965, Yasukawa 1967). It has been shown by Venkateswarlu and Kishore (1968) that oxygen has an inhibiting effect on the yield of $[^{60}Co(NH_3)_6]^{3+}$ whereas it promotes the yield of another recoil species $[^{60}Co(NH_3)_5NO_2]^{2+}$, in the irradiated hexamminecobalt(III) nitrate. If the annealing consists in the conversion of $[Co(NH_3)_6]^{2+}$ to $[Co(NH_3)_6]^{3+}$ by a simple electron transfer process, the presence of oxygen should promote rather than inhibit the formation of $[^{60}Co(NH_3)_6]^{3+}$ in this system (Harbottle 1965). The formation of an oxygen carrier of the type $[^{60}Co(II)\ complex \cdot O_2 \cdot Co(III)\ complex]$ is proposed to explain the oxygen effect (Venkateswarlu and Kishore 1971). This oxygen carrier then dissociates leading to $[^{60}Co(NH_3)_5]^{3+}$ which may react with an ammonia ambient to form $[^{60}Co(NH_3)_6]^{3+}$, or to $[^{60}Co(NH_3)_5]^{2+}$ which may be decomposed to yield Co^{2+} on dissolution. Thus the mechanism of the annealing reactions is of fairly complex nature, which cannot be explained by a purely electronic donor–acceptor model as suggested earlier (Nath et al. 1964a, Rao and Nath 1966).

2.4. STEREOCHEMICAL ASPECTS OF ANNEALING PROCESSES

The recoil and annealing reactions in optical and geometrical isomers have been studied in some cobalt(III) and chromium(III) complexes. According to Zuber's work, only a small fraction of the parent activity was found in the l-isomer when d-tris(ethylenediamine)cobalt(III) nitrate was irradiated with neutrons (Zuber 1954a, b, Zuber et al. 1961). Although the ^{60}Co yields in both isomers increased on annealing, the reformation of the parent d-isomer was favored predominantly. Rauscher and coworkers (1960, 1961) have studied the thermal annealing of both the ^{60}Co and ^{38}Cl activities in the cis- and trans-isomers of $[Co(en)_2Cl_2]NO_3$. After one isomer was irradiated, neither ^{60}Co nor ^{38}Cl was found in any appreciable amount in the other isomer, nor did any appear as the other isomer on thermal annealing. The recoil and annealing behavior in these systems has been interpreted as an ordering process which is analogous to recrystallization, with molecules surrounding the recoil site acting as a template and the annealing proceeds therefrom, leading to the reformation of the parent isomer alone. Similar studies has been made by Saito et al. (1962a) in cis- and trans-$[Co(NH_3)_4(NO_2)_2]NO_3$. A considerable amount (one third to one half of the trans isomer yield) of the cis isomer was found in the neutron irradiated trans-tetrammine complex. The preservation of the parent configuration in the recoil and annealing reactions has been observed also with geometrical isomers of $[Co(en)_2Cl_2]NO_3$ and $[Co(en)_2Br_2]NO_3$ (Dimotakis and Maddock 1961), $[Cr(en)_2Cl_2]NO_3$ (Lin et al. 1967) and $[Cr(en)_2Cl_2]Cl$ (Jagannathan and Mathur 1968). When any of these complexes was irradiated, very little of the activity was found in the other isomeric form, with little or no increase on annealing. Wolf (1966) has studied the dependence of the stereospecificity of the recoil and annealing reactions on the recoil energy by comparing the behavior of ^{60}Co and ^{58}Co recoils from the (n,γ) and (n,2n) reactions respectively, in irradiated cis- and trans-$[Co(en)_2Cl_2]NO_3$. Only the parent isomer was formed stereospecifically in the recoil and annealing reactions of ^{60}Co (with lower recoiling energies), whereas a small yield of the other isomer was also found in the reactions of ^{58}Co (with higher recoiling energies). It has been suggested that in the ^{60}Co reaction the energy transferred locally to the lattice may not be sufficient to induce the necessary rearrangement of the ligands. Such processes are considered possible only in nuclear reactions involving highly energetic neutrons. In an extension of this work, Wolf and Fritsch (1969) have obtained very interesting results using ion implantation. Co^{2+} ions labelled with ^{57}Co were accelerated to 20–60 keV by means of an electromagnetic isotope separator and implanted into cis- and trans-$[Co(en)_2Cl_2]NO_3$ targets. It has been found that formation of the labelled parent isomer, preserving the original configuration, decreases with increasing energy transfer. The yield of the cis-isomer was very low in the irradiated trans-isomer, whereas the yield of trans-isomer was considerably higher and proportional to the energy transfer in the irradiated cis-isomer. These observations have been interpreted in terms of lattice defects and the different stability of the isomers against radiative lattice destruction.

3. Effects in Mixed Crystals

The Indian group has carried out a number of recoil and annealing studies on the cobalt complexes with organic ligands such as oxinate, glycinate and 'open-type' complexes (Shankar and Shankar 1960, Shankar et al. 1961a, b, 1964, Nath and Shankar 1961, Nath and Nesmeyanov 1962, Rao et al. 1962). One of the chelate complexes used successfully for such studies is tris(acetylacetonato)cobalt(III) (Shankar et al. 1961c, 1966, Nath 1961, Nath and Nesmeyanov 1963, Thomas 1969). Since the recoil and annealing behavior in these systems appear to be essentially the same for recoil cobalt atoms from the (n,γ), (n,2n) and (γ,n) reactions, the large differences in the recoil energies among these reactions do not influence their thermal annealing behavior. If tris(acetylacetonato)cobalt(III) dispersed in isomorphous acetylacetonates of aluminum(III), rhodium(III) or manganese(III), the mixed crystals show very interesting recoil and annealing behavior (Shankar et al. 1961d, 1965a). The initial and saturation retention values of $Co(acac)_3$ dispersed in $Al(acac)_3$ or $Rh(acac)_3$ are found to decrease sharply with increasing dilution and the thermal and radiation annealing is almost negligible at high dilutions. By contrast, the apparent retention of $Co(acac)_3$ dispersed in $Mn(acac)_3$ shows a sharp rise on dilution. This increase on dilution with the manganese complex is ascribed to a mechanism in which exchange reactions between the ^{60}Co recoil species and the host complex play an important role. Machado et al. (1965) have studied the influence of defects and of solid-state dilution on the radiation, photo and thermal annealing reactions of (n,γ) recoil fragments produced in Co(acac)$_3$ and Cr(acac)$_3$. They have observed a similar decrease in the ^{60}Co and ^{51}Cr retention values on dilution of Co(acac)$_3$ and Cr(acac)$_3$ with the corresponding isomorphous Al(III) complex. It is interesting to note that the annealing behavior of the mixed crystals is dependent on whether they are prepared by precipitation or vaporization.

4. Effects of Ambient Atmosphere

The effects of the ambient atmosphere on the annealing reactions in the cobalt chelate complexes have been investigated by the Indian groups. Nath and coworkers have reported that oxygen (an electron acceptor) inhibits annealing in tris(acetylacetonato)cobalt(III) and bis(salicylaldehyde)triethylenetetramine cobalt(III) chloride while electron donors, acetone and ethanol vapors, accelerate the annealing in the latter complex (Nath et al. 1964a, b, Rao and Nath 1966, Sarup and Nath 1967). They have explained the effect using the 'variable depth electron trapping model' with the adsorbed oxygen (or electron acceptor) molecules acting as traps for electrons. However, nitric oxide (electron acceptor) inhibits annealing in bis(salicylaldehyde)triethylenetetramine cobalt(III) chloride (Rao and Nath 1966) whereas it promotes annealing in tris(acetylacetonato)cobalt(III) (Thomas 1967). Furthermore, it has been shown recently by Venkateswarlu and Kishore 1971) that oxygen itself enhances the thermal annealing and solid state isotopic exchange reactions in potassium dinitrobis(glycinato)cobalt(III) and sodium EDTA cobalt(III). These contradictory observations have been interpreted by assuming the formation of adducts between the ambient gases and Co(II) chelates similar to synthetic oxygen carriers, as has been discussed earlier in this chapter. Similar studies have been made of the role of water of hydration in the annealing behavior of irradiated tris(dipyridyl)cobalt(III) perchlorate and its trihydrate (Shankar et al. 1965b, Nath and Vaish 1967, Rao 1969).

5. Relevance of Exchange Reactions

It has been pointed out that the reformation of the parent chemical form (corresponding to retention) during annealing may be partly accounted for by defect promoted solid phase isotopic exchange reactions. Solid phase isotopic exchange between the dopant metal ion and the central atom of complexes has been observed in $^{57}Co^{2+}$, $^{58}Co^{2+}$, or $^{60}Co^{2+}$-doped cobalt(III) complexes such as $Co(acac)_3$ and $[Co(dipy)_3](ClO_4)_3 \cdot 3H_2O$ (Nath et al. 1966, Nath and Khorana 1967, Nath and Vaish 1967, Khorana and Nath 1967, 1969, Shankar and Gupta 1969, Venkateswarlu et al. 1971), and in $^{51}Cr^{3+}$-doped tris(acetylacetonato)chromium(III) (Găinar and Ponta 1971a). A more direct verification of the isotopic exchange in the solid state has been obtained by studying the Mössbauer spectra of $^{57}Co^{2+}$-doped $[Co(dipy)_3](ClO_4)_3 \cdot 3H_2O$ (Nath and Klein 1969). Similar solid state exchange reactions between different kinds of metals have been reported in $^{51}Cr^{3+}$ or $^{57}Co^{2+}$-doped metal acetylacetonates (Ramshesh 1969, Vaish 1970), ^{103}Ru-doped tris(acetylacetonato)cobalt(III) (Meinhold and Reichold 1970) and $^{60}Co^{2+}$-doped tris(oxinato)chromium(III) (Lazzarini et al. 1970). The mechanisms of the solid-state exchange and of the thermal and radiation annealing of recoil damage in the corresponding systems are postulated as essentially the same since their kinetics appear remarkably alike.

6. Effects of Solution of Irradiated Solids

We may now consider the influence of the dissolving and analytical conditions on the apparent distribution of recoil species. The effects of pH on the chemical forms of recoil ^{60}Co have been studied in aqueous solutions of neutron-irradiated solid potassium hexacyanocobaltate(III) (Rauscher and Harbottle 1957) and sodium ethylenediaminetetraacetato cobaltate(III) (Shankar and Shankar 1960). Stucky and Kiser (1969) have dissolved neutron-irradiated cobalt complexes such as $[Co(NH_3)_6](NO_3)_3$ and $K_3[Co(C_2O_4)_3] \cdot 3H_2O$ in several polar and non-polar solvents and studied the correlation of the ^{60}Co-labelled product distribution obtained with the different analytical methods. One conclusion from their work is that variables such as solvent, pH and type of analytical procedure employed may generally influence the chemical behavior of recoil species.

Recently, several attempts have been made to study the behavior of recoil species after dissolution of the neutron-irradiated solid complexes in organic solvents. When irradiated solid tris(acetylacetonato)chromium(III) is dissolved in benzene or methanol, the ^{51}Cr retention as the parent form increases rapidly with the time after dissolution, approaching a temperature-dependent pseudo-plateau value (Găinar and Ponta 1968, 1971b). The yield of the acetylacetone-deficient ^{51}Cr species in benzene solutions is found to decrease with the time after dissolution. The increase in ^{51}Cr retention in benzene solutions is remarkably enhanced by the presence of acetylacetone (Omori and Shiokawa 1970, Tominaga and Nishi 1971).

These observations suggest that the acetylacetone-deficient ^{51}Cr recoil species may be recombined with acetylacetone to reform the parent species. It is pointed out that the curves for reformation of the parent complex in solutions are similar in shape to the annealing curves for the irradiated solid complex. Accordingly, the study of the behavior of recoil species in solutions will be useful for understanding the recoil and annealing phenomena in the solid phase. The behavior of recoil species in neutron-irradiated organic solutions (and frozen organic solutions) of metal complexes was first studied by Tominaga and coworkers (Tominaga and Fujiwara 1970, Tominaga and Nishi 1971, Tominaga et al. 1971b, Tominaga and Sakai 1972). They have found that the ^{60}Co retention in irradiated benzene (or ethanol, acetone, etc.) solutions of tris(acetylacetonato)-cobalt(III) decreases sharply with increase in the concentration of metal salts added as scavenger for acetylacetone prior to irradiation (fig. 11.3). It is interesting to note that the thermal reactions which contribute mostly to the apparent retention in the irradiated solutions can be suppressed more effectively by using salts of the metals which can form more stable complexes with acetylacetone (i.e., Fe(III), Cu(II), Al(III), etc.). Since the ^{60}Co retention in well-scavenged solutions gives a minimum value of $0.00 \pm 0.02\%$, this may be considered as an upper limit for the primary retention for $Co(acac)_3$ in the solution phase (Tominaga and Fujiwara 1970, Tominaga et al. 1971a, b). Similar

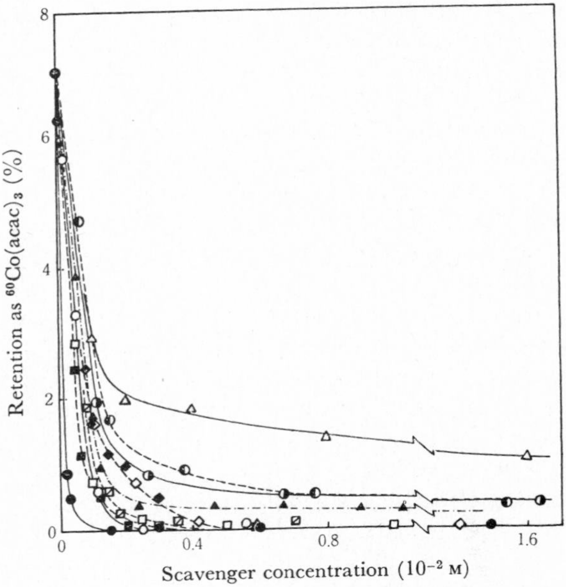

Fig. 11.3. Scavenger effect of various metal salts on ^{60}Co retention in the neutron-irradiated 0.2 M solutions of $Co(acac)_3$ in benzene containing 10% of ethanol. —●— $FeCl_3 \cdot 6H_2O$; —O— $FeCl_3$; ⧄ $AlCl_3$; --■-- $CuCl_2 \cdot 2H_2O$; --□-- $Cu(CH_3CO_2)_2 \cdot H_2O$; --▲-- $NiCl_2 \cdot 6H_2O$; —◑— $CoCl_2 \cdot 6H_2O$; —◐— $CoCl_2$; —△— $MgCl_2 \cdot 6H_2O$ (Tominaga et al. 1971b).

scavenger effect of metal salts has been observed in irradiated benzene solutions of tris(acetylacetonato)chromium(III) and of tris(nitrosonaphtholato)-cobalt(III) complexes (Tominaga and Nishi 1971, Tominaga and Sakai 1972). The thermal recombination reactions which proceed in solutions of the irradiated solid complexes can be scavenged similarly by using the metal salts. For instance, the apparent retention of irradiated solid tris(nitrosonaphtholato)-cobalt(III) complexes dissolved in benzene without scavenger is as high as 90%, whereas it decreases to 15–20% with cobalt or copper salt as scavenger. The latter values should be considered as the solid-phase retention, while the difference (~70%) may be ascribed to thermal reactions in solution. Some of the retention data reported for solid chelate complexes must be reexamined in this respect using the scavenger technique since they may include an appreciable contribution from thermal reactions in solution, which are influenced by analytical conditions such as temperature, time after dissolution, solvent and concentration of impurities (free ligand group, products from radiolysis of the target, etc.).

Unlike organic solvents, however, thermal recombination reactions after dissolution may not always be important in aqueous solutions of irradiated solid cobalt(III) complexes. Ligand-deficient recoil species such as $[^{60}Co(NH_3)_5]^{3+}$ may react with water instantaneously on dissolution to form stable species such as $[^{60}Co(NH_3)_5H_2O]^{3+}$ which will not undergo further reaction and reformation of the parent.

7. Dynamic Separation Technique

In connection with the behavior of recoil species in aqueous solutions we should mention of the dynamic extraction method with ion exchanges, proposed by Matsuura and coworkers (Matsuura 1965, 1966, Matsuura and Hashimoto 1966, Matsuura and Sasaki 1966b). The complex ion adsorbed on an ion exchange column is irradiated in a reactor while an eluant is passed through to elute the recoil species which is less readily adsorbed on the resin than the parent. Since the recombination of the recoil species is reduced by continuous elution, a high separable yield, or low retention, may be expected in this method. For the cobalt complex ions such as $[Co(NH_3)_6]^{3+}$, $[Co(en)_3]^{3+}$, $[Co(NO_2)_6]^{3-}$, $[Co(C_2O_4)_3]^{3-}$ and $[Co(CN)_6]^{3-}$, the separable yield, or the retention, is found to correlate with the stability of the target complex.

8. Other Complexes

Relatively few studies have been reported on the recoil reactions in complex anions such as cyano complexes of cobalt, nickel and iron (Rauscher and

Harbottle 1957, Shibata et al. 1966, Meriadec and Milman 1968, 1971, Fenger et al. 1970b) and oxalato complexes of cobalt, chromium and iron (Saito et al. 1961a, 1964, Siekierska and Fenger 1970, Fenger et al. 1970a). Recoil products found in solutions of the irradiated solid samples of these complexes are mainly aquo complexes such as $[Co(CN)_5H_2O]^{2-}$, and $[Co(CN)_4(H_2O)_2]^-$, which probably arise from reactions of ligand-deficient fragments with water. Saito and coworkers (1962e) have studied the behavior of ^{60}Co recoils in irradiated solid complex salts $[Co(NH_3)_6][Co(CN)_6]$ and $[Co(NH_3)_6][Fe(CN)_6]$ and identified at least nine ^{60}Co-labelled species. Appreciable amounts of $[^{60}Co(CN)_6]^{3-}$ ions are found in solutions of irradiated $[Co(NH_3)_6][Fe(CN)_6]$, indicating that ^{60}Co-for-Fe substitution, or interchange of central atoms, takes place in this system. It is worthwhile mentioning that the distribution of ^{60}Co-labelled species in $[Co(NH_3)_6][Fe(CN)_6]$ appears essentially similar to that in $[Co(NH_3)_6][Co(CN)_6]$. The $[^{60}Co(NH_3)_6]^{3+}$ yield is increased gradually on storage of the irradiated complexes at room temperature, while the $[^{60}Co(CN)_6]^{3-}$ yield remains almost unchanged. Analogous substitution reactions between outer sphere ^{51}Cr recoil atoms and central cobalt atom have been observed by Ikeda et al. (1970, 1971b) in solid $[Co(NH_3)_6]_2(CrO_4)_3 \cdot 5H_2O$ and solid mixture system of $[Co(NH_3)_6]Cl_3$ and K_2CrO_4 powders. About 1% of recoil ^{51}Cr is found in the form of $[^{51}Cr(NH_3)_6]^{3+}$ in both systems. In the mixtures, the $[^{51}Cr(NH_3)_6]^{3+}$ yield increased proportionally with respect to the $[Co(NH_3)_6]Cl_3$ mole fraction, while the ^{51}Cr retention as $^{51}CrO_4^{2-}$ decreases linearly with increasing $[Co(NH_3)_6]Cl_3$ mole fraction. A considerable fraction of the ^{51}Cr recoils appear to leave the mother particles and react outside. Since the grain size of the target is much larger than the recoil range of ^{51}Cr, surface processes involving cracks in the irradiated crystals must be considered important in this phenomenon. Fenger and coworkers (1970b) have compared the chemical consequences of (n,γ) and EC reactions in soluble Prussian Blue. When $KFe[Fe(CN)_6] \cdot H_2O$ samples with ^{58}Fe enriched in either the cation or the central atom of the complex are irradiated with neutrons, the ^{59}Fe retention in the hexacyano-complex is about 5% and can be increased only slightly by thermal annealing. However, parallel experiments with $K_4[Fe(CN)_6] \cdot 3H_2O$ indicate that the ^{59}Fe retention in hexacyanide is nearly 20% and increases to 30% on annealing. The low retention in soluble Prussian Blue is ascribed to competition between recoil ^{59}Fe and inactive Fe^{3+} for reformation of the parent complex. Mössbauer spectra of soluble Prussian Blue doped with ^{57}Co as cation or labelled with ^{57}Co in the complex ion reveal that ^{57}Fe arising from electron capture in the cationic ^{57}Co cannot substitute for the central atom in the complex, whereas ^{57}Fe from $[^{57}Co(CN)_6]$ is found mainly in a ligand deficient species, probably pentacyanide. In this system, the chemical effects of electron capture appears more similar to radiolysis than to thermal neutron capture.

9. Effects in Metal Phthalocyanines

In closing this section on transition metal complexes, we mention recoil studies on metal phthalocyanines. Because of their chemical and thermal stabilities, metal phthalocyanines have been used for production of various radioisotopes in high specific activities (Herr and Götte 1950, Herr 1952c, Ebihara 1966, Grossmann et al. 1968b). One interesting feature of this group of compounds is polymorphism. The different recoil and annealing behavior of different crystal modifications of the phthalocyanines have been reported by Cook (1960) and Apers et al. (1962). It is often observed that two different crystal modifications of the phthalocyanines have different retentions and anneal at different rates to different plateau values (Mathur 1969, Yoshihara and Ebihara 1964, 1966, Yoshihara and Yang 1969a, b, Yang et al. 1970a, b). However, Scanlon and Collins (1971) have recently observed a difference in annealing behavior between α- and β-cobalt phthalocyanines under various external atmospheres (oxygen, hydrogen, etc.) and suggested some alternative mechanism in which the observed difference in behavior results from the effect of oxygen on the annealing process. Apers and coworkers (1962) have found an isotope effect between ^{65}Zn and ^{69m}Zn recoils (in terms of retention) in zinc phthalocyanine and attributed the effect to the difference in recoil energy. Yoshihara and coworkers have studied the nature of the isotope effect systematically by using various nuclear processes and found that the recoil energy plays an important role in determining the fate of the recoil atoms in the metal phthalocyanines (Yoshihara and Yang 1969b, Yang et al. 1971). The correlation between the recoil energy and initial retention in zinc phthalocyanine is shown in table 11.2. Similar recoil energy dependence has also been observed in indium ethylenediaminetetraacetato complex (Yoshihara and Kudo 1970).

Table 11.2
Initial retention of zinc recoil atoms produced by various nuclear processes in α- and β-zinc phthalocyanines (Yang et al. 1971)

Nuclear process	Average recoil energy	Retention(%) α	β
$^{69m}Zn \xrightarrow{IT} {}^{69g}Zn$	1.5 eV	99.5	(~100)[a]
$^{68}Zn(n,\gamma)^{69m}Zn$	190 eV	17.7 ± 0.6	26.0 ± 1.2
$^{68}Zn(n,\gamma)^{69g}Zn$	255 eV	17.5 ± 0.9	24.5 ± 0.7
$^{64}Zn(n,\gamma)^{65}Zn$	480 eV	11.9 ± 1.0	14.3 ± 0.7
$^{64}Zn(\gamma,n)^{63}Zn$	74 keV	9.5 ± 0.2	11.8 + 0.5
$^{70}Zn(\gamma,n)^{69m}Zn$	75 keV	9.0 ± 0.6	12.6 ± 1.4
$^{64}Zn(n,2n)^{63}Zn$	520 keV	9.2 ± 0.9	12.1 ± 1.2

[a] Estimated value.

Recent Developments

In this additional paragraph, we describe briefly some recent developments mainly concerning the analytical techniques for the recoil products from the transition metal complexes. Sakanoue and his group have applied the vacuum sublimation technique to the radiochemical separation of volatile solid complexes. Carrier-free recoil products can be separated sublimatographically from the irradiated metal phthalocyanines (Endo and Sakanoue 1972) and metal acetylacetonates (Kawazu and Sakanoue 1974, Amano and Sakanoue 1974). Although the recoil species inevitably suffer thermal annealing on heating, this technique may still deserve attention since the irradiated solid systems can be analyzed directly without dissolution. In an extension of the liquid-phase reactions with the scavenger technique, Tominaga and coworkers have investigated the effects of the atmosphere and various additives such as labile metal acetylacetonates (which increase the apparent retention) on the reactions of recoil species after dissolution of the irradiated cobalt- and chromium acetylacetonates (Tominaga and Nishi 1972a, b, Tominaga et al. 1974). The overall mechanism of the behavior of such recoil species on dissolution as well as the effects of additives can be generally accounted for in terms of recombination, scavenging and ligand transfer reactions (Nishi and Tominaga 1974, 1976). The solution-solid mixtures and frozen solutions of $Co(acac)_3$ have also been studied with the similar technique (Tominaga 1973c). The comparison of the reactions in air with those in argon has revealed that at least 50% of the ^{60}Co recoil species on dissolution of the irradiated $Co(acac)_3$ consist of $^{60}Co(II)$ species including $^{60}Co(acac)_2$ (Tominaga et al. 1974). The formation of $^{60}Co(acac)_2$ in irradiated $Co(acac)_3$ has been verified later by means of the vacuum sublimation technique (Amano and Sakanoue 1974).

CHAPTER 12

EFFECTS IN COMPOUNDS OF THE SECOND AND THIRD TRANSITION ELEMENT FAMILIES

A.G. MADDOCK

University Chemical Laboratory, Cambridge, UK

Chemical Effects of Nuclear Transformations in Inorganic Systems
Edited by G. Harbottle and A.G. Maddock

Contents

Much of the earlier work with these elements was somewhat empirical and directed towards enrichment of the radioactive products. However these studies provided several important leads that have subsequently proved rewarding.

1. Application to Isotopic Enrichment

The early recognition of the high thermal neutron capture cross section of the mono-isotopic element gold led to it being chosen for the first studies. Majer (1937, 1939) obtained modest enrichments (ratio of specific activities of the element in the product and in the untreated irradiated material) of about 10 and reasonable yields using solid, or solutions of, $Na_3Au(S_2O_3)_2$ and $AuCl_3$. The radioactive product could be separated as gold amalgam or on coagulating a trace of gold sol in the solution. Herr (1948) obtained a much higher enrichment using Si $(PhCOCHCOPh)_3$ $AuCl_4$ and cathodically depositing the product on a platinum foil. A radioactive yield of 95% was obtained.

Steigman (1941) obtained enrichments of 56, 44 and 150 for Ir $en_3(NO_3)_3$, Pt $en_2(NO_3)_2$ and Rh $en_3(NO_3)_3$ (en = ethylene diamine) respectively, separating the products on a trace of the reduced metal. Recognising the necessity for the absence of exchange between the target material and the separable product he proposed that the possibility of resolution of the d and *l* enantiomorphs was a necessary, but perhaps not sufficient, condition for a successful enrichment.

Platinum and iridium isotopes produced by the (γ,n) reaction were enriched 10–20 fold with yields of 30–40% by Christian et al. (1952) using chloroplatinic acid, $Pt(NH_3)_4Cl_2$ and $Pt(NH_3)_4C_2O_4$ and similar iridium compounds. A more extensive study of the use of hydroxychloroplatinates IV for the same purpose was reported by Armento (1968a, b).

Haldar studied a wider range of platinum compounds, measuring always the 18 h ^{197}Pt neutron capture product. The analytical procedures used to separate the products are open to some criticism, but one particularly interesting feature of this study was the observation of the formation of a Pt IV species in a Pt II matrix. $[^{*}Pten_2Cl_2]^{2+}$ was found in irradiated $Pten_2Cl_2$: one of the rare examples of an oxidative recoil reaction. The retentions found are given in Table 12.1. Solutions of the compounds generally gave much lower retentions. In many cases there was evidence of several recoil products, (1954).

Table 12.1

Compound	cis $Pt(NH_3)_2Cl_4$	cis $Pt(NH_3)_2Cl_2$	$[Pt(NH_3)_5Cl]Cl_3$	$[Pt(NH_3)_4]Cl_2$
R for ^{197}Pt	11.4%	16.5%	20.7%	40.2%
Compound	$[Pt\ en_3]Cl_4 2\frac{1}{2}H_2O$	$[Pt\ en_3]Cl_4H_2O$	trans$[Pt\ en_2Cl_2]Cl_2$	$Pt\ en_2Cl_2$
R for ^{197}Pt	5.6%	4.2%	6.8%	23%

More recently Ebihara and Yoshihara (1960, 1961, 1962) have used the oxinate (8 hydroxyquinoline) complexes of molybdenum (1961), tungsten (1960, 1961) and palladium for this purpose. The irradiated oxinates were dissolved in chloroform and the product extracted with an aqueous buffer solution (see table 12.2). Much of the exploratory work on the compounds of these elements was carried out by Herr. He investigated the enrichment of ^{186}Re and ^{188}Re produced in K_2ReCl_6 and in rhenium trichloride by thermal neutron capture (1952b). With the chlororhenite IV he obtained a 30% yield of separable activity, but this rose to nearly 100% if a solution was irradiated.

Table 12.2

	^{99}Mo	^{185}W	^{109}Pd
pH of aqueous extractant	7	7	4
Enrichment	130	429	900
Yield	40%	34%	23%

The separable activity appeared as ReO_4^-, i.e. Re VII, the first instance of an oxidized recoil product (see below). The presence of iodide or sulphite in the irradiated solution did not affect the production of Re VII. With the solid trichloride the activity was distributed *Re III 30%; *Re IV 4%; *Re VII 66%, while in solution it all appeared as *Re VII. He applied the corresponding separation following isomeric transition to establish the mass number of the 18 min ^{188m}Re (1952d).

Herr discovered the very high yields and enrichments possible with the metal phthalocyanines. Yields of more than 90% and enrichments exceeding a thousand were found for the Mo, Pd, Os, Ir and Pt compounds (1952c). Payne et al. (1958) confirmed the yields but found that radiolytic decomposition of the complexes often led to lower enrichments in longer irradiations. The procedure was applied to establish the mass number of ^{140m}Rh (Herr 1954b). These complexes have also been studied by Merz (1966) who reported retentions for the Mo, W, Re, Ru, Pd, Ir, Os and Pt compounds. Merz found that the retentions were lower for irradiation in air or oxygen and were sensitive to impurities in the complex. The complexes showed thermal, photo and radiation annealing. Measurements of retentions were also made for the products following the (n,2n), (n,p) reactions as well as beta decay. He confirmed that these complexes do suffer radiolytic decomposition. Finally Herr has given a brief description of a separation of W using Zn_2 $W(CN)_8$ (1954b).

A recent investigation covers the main features of the recoil chemistry of radio-ruthenium produced by the (n,γ) reaction in ruthenocene (Jacob 1973, 1974).

2. Processes Essentially Dependent on Physical Recoil

Gold has been used to investigate the physical recoil following both the (n,γ) and fast neutron scattering processes (Magnusson 1951, Yosim and Davies 1952, Keller and Lee 1966). The former experiments were important because some Au^+ was found to recoil from a gold metal surface implying that the charging process occurred after the gold left the metal surface. This suggests internal conversion of a relatively long-lived excited state in the capture gamma de-excitation cascade.

Several separations, using heterogeneous systems, essentially dependent on the physical recoil have been described. Parker (1962) found that after neutron irradiation of a silver sol the radioactive silver is found in the very small particle size fraction not depositing on ultra contrifugation. This fraction gradually attaches itself to the silver particles in the sol and so the yield is time dependent.

Closely related processes can be used to enrich ^{99}Mo (Colonomos and Parker 1969) or ^{185}W (Parker and Perez-Alarcon 1970) by irradiating an initimate mixture of finely divided metal carbonyl and oxalic acid. With a carbonyl to oxalic acid ratio of 1 : 30 ^{99}Mo gave a yield of about 50% and an enrichment of 500 while the ^{185}W gave a yield of 69% and enrichment above 1000.

When the recoil energy is very high, for instance with the fission products or in high-energy reactions, one can arrange for the recoiling atom to pass into a reactant, such as a halogenating medium, and then utilize the physical characteristics of the halide produced to separate and or characterise the product. This technique, much used in the study of the highest atomic number elements was developed by Zvara et al. (1967) with fission product Zr and Mo. Its application to fission product niobium has been reported by Cavallini (1969).

3. Studies on Osmium Compounds

The complex halo-osmates IV, the cyano-osmates II and osmium tetroxide have been studied.

Mitchell and Martin (1955, 1956) obtained yields of 40–60% and enrichments of greater than 40 following neutron irradiation of $(NH_4)_2$ $OsCl_6$. The sodium salt gave a lower enrichment. The chemical separation was unusual in that it relied on the difference in the rate of conversion of the unidentified radioactive species and the $OsCl_6^=$ to OsO_4 by nitric acid. On solution of the irradiated salt and distillation with nitric acid the activity appeared in the first part of the OsO_4 distillate. Herr and Dreyer (1957) obtained similar results with the potassium salt and noticed that the irradiated salt underwent thermal annealing. Aten and Kapteyn (1968a), however, report much higher retentions following the (n,γ) reaction in K_2OsCl_6(100%) and $(NH_4)_2$ $OsCl_6$(98%). Osmium tetroxide showed a retention of 70% when dissolved in 0.1 M HCl and 81% when dissolved in 0.1 M KOH for analysis. Diefallah and Kay (Diefallah 1968, Diefallah and Kay 1972, 1973) have made an extensive study of $K_4Os(CN)_6$. They have used paper electrophoresis for the separation and found several products. The retention seems to be independent of the fast neutron and ionising radiation doses accompanying the thermal neutron irradiation at low doses. At higher doses the retention falls possibly due to radiolytic decomposition. In mixed crystals of $K_4Os(CN)_6/K_4Fe(CN)_6$ and $K_4Os(CN)_6/Cs_4Fe(CN)_6$, the retention falls from about 51% to 20% as the Cs/K ratio increases from 0 to 1.0. It was suggested that this might be due to the increased self-generated γ dose due to the caesium. (See below and ch. 25.)

4. Studies on Rhenium Compounds

The interesting pecularity of rhenium in giving oxidised recoil products has attracted a good deal of attention. The first detailed study was reported by Herr (1952a), who investigated the annealing in thermal neutron irradiated K_2ReCl_6. Annealing isotherms and, for the first time, isochronals were reported. The phenomenon of annealing by ultraviolet light was also discovered. Herr suggested that the production of $^*ReO_4^-$ might take place through the intermediary of $^*ReCl_6^+$. Schweitzer and Wilhelm (1956) found *Re VII production in a wider range of rhenium compounds (see table 12.3).

Table 12.3

Compound	$NaReO_4$	K_2ReCl_6	K_2ReBr_6	K_2ReI_6	ReO_2	ReO_2nH_2O
Retention	~ 100%	70%	81%	87%	74%	38%

Apers and Maddock (1960) used K_2ReCl_6 to explore isotope effects on retention (see below) and found traces of other radioactive rhenium species were formed, besides $^*ReCl_6^=$ and $^*ReO_4^-$, by paper electrophoresis.

An early success of the perturbed angular correlation technique was to show that although the retention of radioactive rhenium in neutron-irradiated potassium herrhenate appears to be substantially complete the osmium atoms formed by beta decay of the rhenium were not at normal lattice sites (Sato et al. 1966, see also ch. 24).

Aten and Kapteyn (1968a) found essentially complete retention in $(NH_4)ReO_4$ and Re_2O_7 as well as $KReO_4$.

More recent work has modified this picture considerably and shown that the formation of *Re VII takes place, at least to some extent, upon solution of the irradiated perrhenate for analysis. Facetti et al. (1969) have reinvestigated NH_4ReO_4 as well as ReO_2 and Re_2O_7 using solution of the irradiated compounds in hydrochloric acid containing Re III, Re IV and Re VII carriers followed by paper electrophoresis for analysis. Results are shown in table 12.4. The irradiations were at a thermal flux of 10^{12} n cm^{-2} s^{-1} for 200 s. No significant difference between the ^{186}Re and ^{188}Re was observed.

Table 12.4

Compound	*ReIII	*ReIV	*ReVII
ReO_2	12.5	64.9	22.3
Re_2O_7	1.2	77.5	21.0
NH_4ReO_4	6.6	8.1	85.0

The ammonium perrhenate showed a small thermal annealing ($\Delta R \sim 3\%$) at 210° C. But the distribution of radioactive rhenium in the ReO_2 and Re_2O_7 changed on storage at room temperature. In both compounds the proportion of *ReVII increased at the expense of *ReIV. ReO_2 thus resembles Na_2SeO_3 in that the annealing does not increase the proportion of the activity found in the target species. Irradiation of the Re_2O_7 with ionising radiation previous to the neutron irradiation increased the proportion of *ReVII, the retention reaching about 100% for doses in excess of 3.6 Mrad. Preheating the Re_2O_7 at 200° C before neutron irradiation also increased the retention. DeKimpe et al. (1969) also investigated NH_4ReO_4. The retention they observed after solution of the irradiated salt in water agreed with the value of Facetti et al. (1969). The irradiations were for 10 min at 2.1×10^{11} n cm^{-1} s^{-1}. Their analytical procedure simply involved absorption of the lower oxidation states on MnO_2. But if the irradiated salts were dissolved in liquid ammonia for the analysis the retention fell from 85% to 62%. Clearly some formation, at least, of *ReVII occurs by reaction with the water upon solution. They also found that hydrated Na_2ReCl_6 gave a much lower retention (6%) than unhydrated salt (49%). Any difference in behaviour of ^{186}Re and ^{188}Re was only a little greater than the experimental error (R for ^{188}Re was persistently a little higher than for ^{186}Re).

Further results on ammonium perrhenate have been reported by Defrance and Apers (1974a, b). Isochronal annealing of material irradiated at −100° C and ambient reactor temperature show an energy of activation for the thermal annealing of about 0.45 eV. About 5% of the radioactive rhenium cannot be annealed. Pre-irradiation of the ammonium perrhenate with gamma radiation has little effect on the low-temperature neutron-irradiated salt but leads to complex effects with the ambient temperature irradiated material. The initial retention for the −100° C irradiated material analysed after solution in liquid ammonia at − 55° C was as low as 68%.

Preliminary work on irradiated rhenium carbonyl shows that, in addition to the retention and "inorganic" radioactive rhenium, at least four chloroform soluble radioactive rhenium compounds are formed. Polynuclear carbonyls might be suspected. The system readily undergoes thermal annealing (Cavin et al. 1975).

Rhenium compounds have been used for several studies of high energy reactions using 660 MeV protons to initiate the nuclear reaction (see ch. 20). Zaitseva and Ianovici (1970) have found considerable similarities between the behaviour of radioactive rhenium atoms produced by the (p,p*x*n) reaction and that from the low recoil (n,γ) reaction in sodium, potassium and ammonium perrhenates and K_2ReCl_6. However, the thermal annealing behaviour seems to be more complicated both in NH_4ReO_4 (Zaitseva et al. 1969) and in K_2ReCl_6 (Zaitseva and Ianovici 1971).

When *Re isotopes are produced by 660 MeV proton bombardment of the iridium compounds, Na_2IrCl_6, $Na_2IrCl_6 6H_2O$, $(NH_4)_2IrCl_6$, Na_3IrCl_6, $Na_3IrCl_6 12H_2O$, $(NH_4)_3IrCl_6$; the proportion of *Re VII is greater in the hydrates and ammonium compounds. But hydration and dehydration has little effect after the proton bombardment.

Zaitseva et al. have explored several other aspects of the recoil reactions following 600 MeV proton bombardment of compounds of these elements. (Zaitseva et al. 1974a, b, Islamova et al. 1975) (see also ch. 20).

A comparison of the (n,γ) and (n,2n) reactions in K_2ReX_6 (X = F, Cl, Br and I) shows that all the radioactive halogen appears as X^- or $ReX_6^=$ and the yield of $^*X^-$ is independent of the mode of nucleogenesis (Bell et al. 1972). Bell and Stocklin (1970) have shown that the yield of $^*Cl^-$ from K_2ReCl_6 is independent of the length of irradiation; apparently no radiation annealing occurs.

The distribution of radioactive halogen in mixed crystals of K_2ReBr_6/K_2ReCl_6 is shown in fig. 12.1. No appreciable yields of species such as $[ReBr_nCl_{6-n}]^{2-}$ for $n \neq 0$, 1, 5 or 6 are found. The yields of the different forms of ^{38}Cl and ^{82}Br can be expressed as

$$^{82}Br^- = 25 - 5\,b \qquad ^{38}Cl^- = 95 - 56a - 181\,ab$$
$$Re^{82}BrBr_5^= = 14 + 66b \qquad Re^{38}ClCl_5^= = 5 + 56a + 87\,ab$$
$$Re^{82}BrCl_5^= = 61(1 - b) \qquad Re^{38}ClBr_5^= = 94\,ab,$$

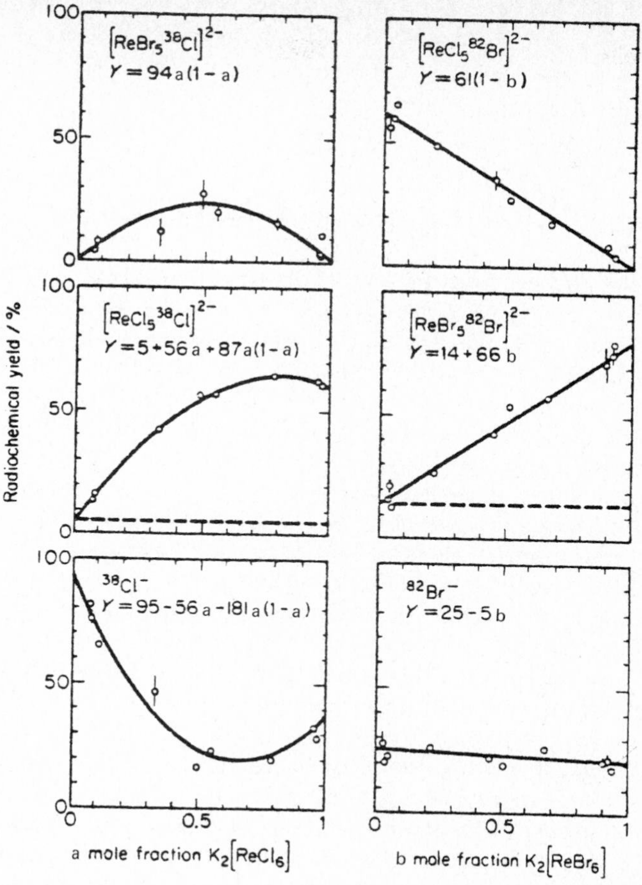

Fig. 12.1. Effect of mixed crystal composition on the ^{38}Cl and ^{82}Br product distribution in the $K_2[ReBr_6] - K_2[ReCl_6]$ system (from Rössler et al. 1972).

where a = mole fraction of $ReCl_6^=$ in the original mixed crystals and b = mole fraction of $ReBr_6^=$ in the original mixed crystals. They interpret terms not involving a or b to correlated recombination, i.e. all within the terminal recoil zone; terms involving a or b to substitution reactions (^{82}Br for Cl or ^{38}Cl for Br) and the terms in ab to exchange processes involving ligand deficient intermediaries (Rössler et al. 1972, Otterbach 1972).

A computer simulated study of these processes has been made by Robinson et al. (1974). Good agreement with the experimental data has been obtained assuming a volume containing about 50 halogen sites for the recoil affected zone. The distribution of the recoil halogen appears to be practically independent of the recoil energy above some threshold value. The atom generally comes to rest in a lightly damaged region of the lattice. Calculations have also been made for the recoil rhenium. In the latter case only a very short replacement cascade involving one or two rhenium atoms is concerned. It is still difficult to give a quantitative assessment to the role, if any, of direct, or hot, replacement reactions.

5. Effects in Cadmium and Mercury Compounds

A novel procedure has been applied to the determination of the retention in cadmium phthalocyanine. This complex is one of the 'ionic' type of phthalocyanines and looses its cadmium on solution in sulphuric acid so that the normal analytical procedure cannot be applied. Indium phthalocyanine does not behave in this way so that if the ^{115}Cd, formed by thermal neutron capture in ^{114}Cd, is allowed to decay to ^{115m}In the retention of this species can be measured in the usual way. A control experiment with ^{115}Cd labelled phthalocyanine shows that the beta decay step leads to complete retention, so that the ^{115m}In retention is equal to the ^{115}Cd retention (Yoshihara and Kudo 1969). A Vand–Primak analysis of the thermal annealing data on cadmium phthalocyanine shows two energy of activation peaks, centred at 0.12 and 0.40 eV. R_0 was about 34% for a $-80°$ C irradiation (Kudo and Yoshihara 1970). The initial retention of the complex has been shown to be rather sensitive to preheating and quenching of the crystals and less so to ionising irradiation before neutron irradiation. In each case some reduction in R_0 is observed (Kudo 1972a).

Some of the very few data on the physical effects of the Szilard–Chalmers effect relate to cadmium chalcogenides. Thermal neutron irradiation of CdTe (Barnes and Kikuchi 1970) and CdS (Kikuchi 1971) lead to reductions in their conductivity and modification of their luminescent properties. A similar fast neutron irradiation has a very much smaller effect. The effects are attributed to Frenkel defects produced by the recoiling Cd atoms.

Mercury compounds were not investigated until comparatively recently and there are still inconsistencies in the data. Bruno and Belluco (1956a) found that the ethylene diaminetra-acetic acid complex exchanged with mercury ions too rapidly for use, but that a yield of 33% and a high enrichment of ^{203}Hg could be achieved with 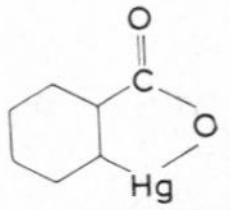

The irradiated compound was shaken with a dilute mercuric solution for the separation (1956b). Belmondi and Ansaloni (1958) found phenyl mercuric acetate could also be used, but the compound was quite sensitive to radiolysis. A similar sensitivity seems characteristic of the alkyls and aryls used in the remaining studies.

Nefedov et al. reported a 67% yield and the remarkably high enrichment of

2.5×10^5 using $Hg(C_2H_5)_2$. A similar separation accompanied isomeric transition in ^{195m}Hg and ^{197m}Hg and was used to separate the ground state species (Nefedov and Sinotova 1958, Murin et al. 1958).

Subsequent work has shown that high enrichments are not usually possible with these alkyls or aryls. Heitz and Adloff (1964) have measured retentions for both ^{197m}Hg and ^{203}Hg in $Hg(C_2H_5)_2$ and $HgPh_2$. The values for the two isotopes are practically the same and about 51% in the diethyl and 85% in the diphenyl compounds. In the liquid $Hg(C_2H_5)_2$ or in solution of the diphenyl compound they found evidence of exchange reactions which complicate the interpretation of the data. Grillet (1966) has also investigated these compounds.

Further work by Heitz (1967a, b, c) has included measurements of the G value for the radiolytic decomposition for the compound, the enrichment for short irradiations and the retentions (see table 12.5). The retentions were not very different if 0.1 M solutions in benzene were used. No isotopic effects were observed. Angenberger and Grass (1970) have obtained fairly similar results and have examined mercury dibenzyl. The identity of some, at least, of the recoil products still seems open to question. Collins and Collins (1968) found that if $HgPh_2$ were neutron irradiated in solution in tetrahydrofuran in a polythene tube the radiomercury adsorbed on the walls. An enrichment of about one thousand was possible for ^{203}Hg in short irradiations but on standing the retention increased quite quickly. Probably various organo-mercury compounds are formed and exchange reactions are certainly involved.

The high retention for some of these organo-mercury compounds is attractive for radiosynthesis. Nesmeyenov and Filatov (1961) irradiated solutions of $HgBr_2$ in benzene ($\sim 10^{-2}$M) and found 72% of the radiomercury as PhHgBr after standing for 30 days. Wheeler and McClin (1967a) found thermal neutron irradiation attractive for the direct synthesis of labelled $HgPh_2$. They looked more closely at the other radioactive products, dissolving their irradiated $HgPh_2$ in benzene chromatographing on neutral alumina and eluting, successively, with benzene, ether, ethyl alcohol, acetic acid, water, 10% hydrochloric acid and 20% nitric acid. Like Collins and Collins, they found a fraction that adsorbed strongly and yet reconverted to $HgPh_2$ readily. They could also obtain Ph*HgX from the solutions. Their results suggest a long-lived *HgPh species. Finally, Kronrad et al. (1970) have obtained a good yield

Table 12.5

Compound	Retention	Enrichment	G_{-Hg} organic
Liquid $Hg(CH_3)_2$	15	312	21
Liquid $Hg(C_2H_5)_2$	8	321	42
Solid $HgPh_2$	74	214	2.7

of labelled difluoresceinyl mercury by direct irradiation. de Jong and Wiles (1974, de Jong et al. 1974) have suggested that this high retention might be turned to advantage and provide a basis for "molecular" activation analysis.

Clearly if the retention is appreciable and reproducible, activation analysis can yield data on the amounts of different compounds of an element that are present in an irradiated sample, rather than only the amounts of the elements present. The sensitivity of activation analysis is often so high that even a very small retention would serve. However a great deal of work is necessary on the reproducibility of the retention and the importance, if any, of synthesis due to the recoil, before this technique can find wide application. It would also seem possible that the retention as ordinarily determined might not be the value applicable in this application. The preliminary work has been directed towards the assay of pollutant organo-mercurials.

6. Iridium Complexes

6.1. THE HALOIRIDATES

Alkali chloroiridate IV was first investigated in early studies of the carrier free separation of recoil species by paper electrophoresis (Croatto et al. 1951). The inert character of the iridium complexes and the high thermal neutron capture cross section of the iridium isotopes, both species yielding gamma active iridium isotopes of convenient half-life, encouraged one to believe that they might be well suited to the study of the details of the recoil fragmentation and annealing in such salts. The systems have now been studied extensively, by Herr in collaboration with Heine (1960, 1961), Heine and Schmidt (1962), Bell (1964, 1965, 1966) and in Heine (1961) and Bell's (1964) theses. These investigations were considerably elaborated by Cabral (1964, 1965, 1966b, Cabral and Maddock 1967) and more recently have been carried still further by van Ooij and Houtman (1967, 1973a, b, 1974a, b). (See also the theses of Cabral (1966a) and van Ooij (1971).)

An extremely complicated situation has been, at least partly, disentangled but the main impact of the results lies with the solution and complex chemistry of iridium rather than in the radiochemistry of the processes.

The first paper by Herr and Heine showed that paper electrophoresis could distinguish seven different radioactive products containing the iridium after solution of the neutron irradiated chloroiridate IV salts. Salts of different cations seemed to give widely different proportions of the various products. The proportions were also changed by thermal annealing, some products passing through a maximum. Herr et al. (1962) compared the behaviour of the radioactive iridium or osmium produced by the irradiation of K_2MBr_6 ($M = Ir$ or Os) with that of the radioactive bromine produced at the same time. The latter, appearing only as bromide or the complex ion, could be annealed almost completely at 210° C. The metals could not and indeed the annealing appeared best represented by a single first order process without temperature-dependent plateaux (Heine and Herr, 1961). Radiation annealing of the iridium species was also possible.

Subsequent work by Bell and Herr (1964, 1965, 1966) and especially by Cabral (1964, 1965, 1966b, Cabral and Maddock 1967) identified several of the species on the electrophoresis strips and revealed a number of the complicating features of iridium chemistry. By this time some two or three cationic, one or more neutral and eleven anionic species had been found.

The evidence and the conclusions are summarised by Cabral and Maddock (1967). It was found that chloroiridate IV solutions are normally in equilibrium with a small concentration of chloroiridate III. The proportion decreases as the hydrogen ion concentration increases. An electrophoretic separation disturbs this equilibrium and electrophoresis of even a pure chloroiridate IV solution always produces a trailing spot of faster moving chloroiridate III, the proportions decreasing with the pH of the supporting electrolyte.

Ageing solutions of chloroiridate III and solutions of the irradiated chloroiridate IV lead to the identification of some, and probable nature of other, species on the electrophoresis strips. Ageing chloroiridate III solution led to the separation of spots due to $Ir\,Cl_5\,H_2O^=$, $Ir\,Cl_4(H_2O)_2^-$, $Ir\,Cl_3(H_2O)_3$, $Ir\,Cl_2(H_2O)_4^+$ and $Ir\,Cl(H_2O)_5^{2+}$. Compositions of the spots were confirmed by activation analysis (Bell and Herr 1964). On irradiating solutions the proportions of the various products depend on the free chloride concentration in the solution, suggesting a competition between anation and aquation reactions of ligand deficient species.

There are qualitative differences between the distributions found for the same neutron irradiated crystals analysed with a supporting medium of 0.01 M HCl and 0.01 M NaCl. These are presumably due to deprotonation equilibria involving aquo species. For instance, $Ir(H_2O)_3Cl_3$, $Ir(H_2O)_2Cl_4$ and $Ir\,H_2O\,Cl_5^=$ might all be expected to deprotonate between pH 2 and pH 7. The experimental evidence suggests that a neutral species, present in 0.01 M NaCl, gives an anion of similar electrophoretic mobility to $IrCl_6^=$ in 0.01 M HCl.

Taking into account all these species, and allowing for the possibility of isomers, one cannot arrive at sufficient mononuclear iridium species to fit the data. The presence of polynuclear iridium complexes is further supported by the decrease in the proportion of some species and increase in the amounts of known mononuclear species on ageing dilute solutions of the irradiated salts and the formation of previously unidentified species on ageing concentrated solution of the unirradiated chloriridate III or IV.

A much clearer picture of the polynuclear complex products has been obtained by van Ooij and Houtman (1967, 1973a, b, c). They found that by combining electrophoretic and chromatographic separation techniques (for technique see van Ooij 1973) they could separate no less than thirty-five radioactive iridium species. Several of these they have shown to be di- or even trinuclear complexes. A tentative identification of others has been made by using an empirical additive relation for the $R_M[= \log 1/(R_f - 1)]$ values for the different species suggested by Martin (1944).

Houtman and Van Ooij conclude that the neutral species are not all mononuclear. They are not reduced by thiosulphate nor oxidised by chlorine and on ageing acid solution yield $IrCl_5H_2O^=$, $IrCl_4(H_2O)_2^-$, $IrCl_3(H_2O)_2$, $IrCl_2(H_2O)_4^+$ and $Ir_2Cl_8(H_2O)_2^=$. They also find evidence for protonation equilibria. They suggest that the most important polymeric product is H_5[*Ir V(Ir IV $Cl_6)_5$] but it cannot be said that this unusual species is very well substantiated. The stability of the polynuclear species certainly suggests ligation by double chloride bridges e.g. 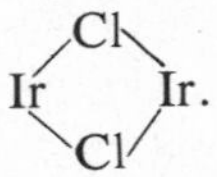

Changing the cation in the salt affects the distribution amongst the different species but more profound differences are found between hydrated and anhydrous salts and for the ammonium salts. With the latter at least three new radioactive iridium species can be separated, one of which forms on ageing an ammonium chloroiridate III solution. Anhydrous chloroiridate IV salts yield far less of the cationic species and conversely more of the neutral recoil products. Anhydrous salts preserve more *Ir IV than the hydrates or ammonium salts (Cabral and Maddock 1967).

Some of these conclusions have been modified by the results of van Ooij and Houtman (1974a) who have examined a large number of chloroiridate III and IV salts. They confirm that hydrated salts and salts of small cations usually give low retentions. The principal products are identified as the isomers of $IrCl_{6-x}(H_2O)_x^{x-3}$ and $Ir_2Cl_{10-x}(H_2O)_x^{x-4}$ and the metastable M_5[*Ir V(IrIV$Cl_6)_5$]. Their results suggest that the polynuclear products are not present as such in the irradiated crystals.

Finally the same authors (1974b) have returned to the question of the annealing reactions in this complex system. Unlike Cabral and Maddock they found that dehydration of irradiated Na_2IrCl_6 $6H_2O$ did not change the product spectrum. However hydration of irradiated Na_2IrCl_6 had a profound affect, eliminating the postulated H_5[*IrV (IrIV $Cl_6)_5$].

Jach and Chandra (1971) have shown that at high temperatures K_2IrCl_6 anneals more quickly than it decomposes.

6.2. THE IRIDIUM AMINE COMPLEXES

Some rather peculiar results have been reported for amine complexes of iridium of the type $[Ir(NH_3)_5Cl]^{2+}$ [Ir trien $Cl_2]^+$ (trien ≡ trisethylenediamine). A comparison was made between the recoil behaviour of both the metal and chlorine in the cobalt, rhodium and iridium complexes. Experiments were also made with labelled chlorine in the complex. (Gardner et al. 1970a, b). The data implied that each neutron capture event ruptured so many Ir–Cl bonds that a macroscopic decomposition of the complex occurred. Only small or negligible effects of this kind were found with the cobalt or rhodium complexes.

These papers were soon criticised in respect of various experimental details (Rössler and Otterbach 1971a). In particular, it was shown that the silver chloride precipitation separation used to isolate the radioactive chloride is not sufficiently selective and is particularly liable to adsorb complex iridium ions. The chlorine activity data certainly seems suspect. However, in a subsequent paper Gardner et al. (1972) found evidence of similar macroscopic effects using paper electrophoresis for the analysis and looking at the behaviour of the macroscopic iridium. The possibility of gross radiolytic effects was discounted but the control experiments do not seem to have simulated the reactor dose.

The present status of this topic is unsatisfactory; more work will be necessary before one can believe in a novel effect. The authors suggested their results implied a 'super hot spot', each recoil atom affecting a relatively large volume of the solid. Both the theoretical (see chs. 2 and 3) and experimental (see ch. 25) data make such an hypothesis improbable. An alternative explanation, if such an effect indeed exists, might be a recoil initiated chain reaction such as is involved in the radiolysis of choline hydrochloride.

7. Isotope Effects

One of the themes in which the compounds of this group of elements have played an important role has been the exploration of isotope effects. The possibility that different (n,γ) products produce in the one neutron irradiation of a compound could display quantitatively different behaviour was considered quite early (Green and Maddock 1951), and some evidence for such differences was found with sodium chloroiridate (Croatto et al. 1952).

A fairly detailed study was made on K_2ReCl_6 using paper electrophoresis for the separation and measuring the ^{186}Re and ^{188}Re produced by thermal neutron capture. Most, but not all, of the radioactive rhenium appeared as $*ReCl_6^{=}$, $*ReO_4^-$ or $*ReO_2$. A small, but reproducible, difference in retentions was found, the retention for the ^{188}Re being initially about 4% higher than the ^{186}Re. This difference gradually decreases on thermal annealing (Apers and Maddock 1960).

Much larger effects were found for the metal porphryn complexes (Rosenberg 1964, Rosenberg and Sugihara 1965). These are very radiation-resistant complexes. The ratio of the retention of ^{109}Pd to that of ^{103}Pd was 1.4. No thermal annealing was observed. For ^{191}Pt and ^{197}Pt the ratio was 2.2. The platinum complex does show thermal annealing. In neither complex was the retention very sensitive to the radiation or neutron dose.

It would be difficult to account for such substantial isotopic differences in terms of the differences in the recoil energy distributions (see chs. 2 and 3) and it seems probable that internal conversion of excited states in the photon cascade might account for these effects. There is not a great deal of direct physical evidence of such conversion. Hillmann, Weiss and collaborators have explored this hypothesis with the phthalocyanines and cyclopentadienyl derivatives of zirconium and hafnium.

With the phthalocyanines they have investigated (n,γ), (n,n′) and (n,2n) products both in the solid complexes and in solution in dimethylsulphoxide. For the (n,γ) products they find $R_0(^{95}\text{Zr})/R_0(^{97}\text{Zr}) = 1.60$ and $R_0(^{180\text{m}}\text{Hf})/R_0(^{181}\text{Hf}) = 2.11$, while $R_0(^{175}\text{Hf}) \approx R_0(^{181}\text{Hf})$. The retentions in solution were only slightly smaller than for the solids, and the ratios remained very different from one. They associated the greatest differences with large spin changes which might be expected to lead to more internal conversion.

The first measurements on the Cp_2MCl_2 compounds (M = Zr or Hf) were due to Wheeler and McClin (1967c), but their analytical procedure was not very satisfactory, Hillmann et al. (1969) again found substantial isotope effects both in the solids and in solution. The retentions in solution were much smaller than for the phthalocyanines or the solid complexes. The ratios for different isotopes were roughly parallel to the phthalocyanine data, but not simply quantitatively related.

Very substantial differences in retention have been found for the different platinum isotopes produced by the (n,γ) reaction in platinum acetylacetonate (see table 12.6).

Table 12.6

Isotope	191	197	199	193m	195m	197m
Retention%	8.6	7.9	4.0	17.9	11.5	15.8

(Belgrave 1971, 1974)

Harbottle and Zahn (1967) compared the retention of each of the elements in chromium and molybdenum carbonyls following 2 GeV proton and 50 MeV bremsstrahlung radiation (table 12.7). Henrich and Wolf (1969) also reported isotope effects in metal carbonyls.

Table 12.7

	^{48}Cr	^{49}Cr	^{51}Cr	^{11}C	^{15}O
2 GeV p	84		82	30	%
50 MeV γ		68		33	31 %

The most remarkable and significant result so far on this subject has been due to Groening and Harbottle (1970) who have found that the structures of the isochronals found for ^{99}Mo and ^{101}Mo produced by thermal neutron capture in molybdenum carbonyl are quite different. This implies different energy of activation spectra for the two isotopes which suggests either chemically different metastable fragment species for the two isotopes or different metastable sites.

8. Other Investigations

The complexes of these two families of transition elements have been used in several studies reported in other chapters. Investigations of the effects of isomeric transition in ^{80m}Br in labelled complexes of the type $[M(NH_3)_5Br]^{2+}$ or $MBr_6^{=}$ were begun by Adamson and Grunland (1951). Extensive further work by Herr, Schmidt and others is considered in ch. 15. Schmidt (1966) has also shown that following (n,γ) activation of the bromine in $[Rh(NH_3)_5Br](NO_3)_2$ the activity appears in five cationic species as well as free bromide.

Irradiation of K_2ReCl_6 or K_2RhCl_6 gives a reasonable yield of separable radioactive chloride. The salts are very radiation resistant and the retention is independent of the radiation dose received by the sample, although normal thermal annealing can occur (Bell and Stöcklin 1970).

Mixed crystals of the halide complexes, such as K_2ReBr_6 in K_2OsCl_6, have been extensively studied by Müller and other results are discussed in ch. 25.

Organometallic compounds such as $Mo(C_6H_6)_2$ and $RuCp_2$ have been investigated a number of times in connection with the effects of beta decay (see ch. 16). The state of the technetium and rhenium formed by beta decay in $^{99}Mo(CO)_6$ and $^{188}W(CO)_6$ (Nefedov and Toropova 1957, 1958a, b, Nefedov and Mikulaj 1973) has been applied to the separation of the daughter species. Labelled $^{99}Mo(CO)_6$ can be separated in reasonable yield after neutron irradiation of $Mo(CO)_6$; a second sublimation after decay of the ^{99}Mo, leaves most of the technetium as an involatile residue of high specific activity (Reichold and Anders 1966). Further details of results on carbonyls are to be found in ch. 13.

CHAPTER 13

RADIOCHEMICAL TRANSFORMATIONS IN ORGANOMETALLIC COMPOUNDS

D.R. WILES
Chemistry Department, Carleton University, Ottawa, Canada

and

F. BAUMGÄRTNER
Institut für Heisse Chemie, Gesellschaft für Kernforschung, Karlsruche, F.R. Germany

Chemical Effects of Nuclear Transformations in Inorganic Systems
Edited by G. Harbottle and A.G. Maddock

Contents

1. Introduction

The first paper describing the effect of nuclear transformation on an organometallic compound was that of Mortenson and Leighton (1934) who observed that a high proportion of $^{210}Bi(CH_3)_3$ was formed from the beta decay of $^{210}Pb(CH_3)_4$. Since that time a good deal of work has been published in this field; the subject has recently been reviewed (Wiles and Baumgärtner 1972, Wiles 1973) but the field seems to lack coherence. This general area is of special interest, since it represents a middle ground between the Szilard–Chalmers reactions in ionic compounds and those in organic compounds. The radiochemically interesting atom lies at the centre of the molecule, and hence these compounds are geometrically similar to many of the ionic compounds. Chemically, however, they are more like the organic compounds.

We are thus in this review concerned with compounds having metal-to-carbon covalent bonds – whether these be σ, π or any other form of covalent bonding. Not included are discussions of cyanide complexes or of the many complex ions whose bonding is through nitrogen, oxygen or sulphur, in which the processes of recombination seem to be more like those of ionic solids than like those of the molecular organometallic compounds (see chs. 10–12). The compounds of interest are mostly solids, with some liquids and a few gases, all of which are composed of discrete molecules having no strong electronic interaction with one another.

Generally speaking, radiochemical studies with organometallic compounds, as with most other compounds, have been undertaken with one or more of the following objectives:

(a) Isotope enrichment
(b) Molecular synthesis
(c) Mechanistic investigation

Examples of (a) and (b) are given in tables 13.1 and 13.2, respectively. It is perhaps an exaggeration to describe the studies to date as 'mechanistic'. The phenomena which occur are so complex, and vary so much from one situation to another that most of the work to date has been quite rightly directed towards learning general patterns of behaviour. In this sense, the mechanistic study of radiochemical transformations in metal–organic compounds began in 1953, with the papers by Edwards et al. (1953) and Maddock and Sutin (1953, 1955) on tetramethyl lead and triphenylarsine, respectively. Studies with this main objective form the subject of the present review. In most cases, the conclusions and explanations given are those of the original authors.

Table 13.1

A. Recoil separations for nuclear studies

Nuclide	Target	Ref.
^{76}As	$(CH_3)_2AsCOOH$ (n,γ)	(Starke 1940)
^{239}Np	Uranium benzoyl-acetylacetonate (n,γ)	(Starke 1942a)
^{104}Mo–^{104}Tc	$UO_3 + Cr(CO)_6$ (n,f)	(Baumgärtner et al. 1958)
^{103,105}(Mo,Tc,Ru)	$UO_3 + Cr(CO)_6$ (n,f)	(Baumgärtner et al. 1958)
$^{103,104,105}Tc$	$UO_2(CH_3COO)_2 + Cr(CO)_6$ (n,f)	(Baumgärtner et al. 1958)
^{134}Te	$UO_3 + Sn(C_6H_5)_4$ (n,f)	(Blachot and Carraz 1969)

B. Recoil Separations for isotope production or enrichment

Nuclide	Target[a]	Enrichment Factor	Ref.
$^{122,124}Sb$	$(C_6H_5)_3Sb$	100	(Melander 1948)
$^{122,124}Sb$	$(C_6H_5)_3Sb$	200	(Kahn 1951)
Zn,Ga,In,V,Mo, Pd,Os,Ir,Pt	Metal Phthalocyanines	1000	(Herr 1952)
Sn	$(C_6H_5)_4Sn$	2800	(Spano and Kahn 1952)
^{69}Ge	$(C_6H_5)_4Ge$ (γ,n)	10^3–10^4	(Murin et al. 1956)
^{122}Sb	$(C_6H_5)_3SbCl_2$ (γ,n)	10^3–10^4	(Murin et al. 1956)
^{74}As	$(C_6H_5)_3AsCOOH$ (γ,n)	10^3	(Murin et al. 1956)
	$(C_6H_5)_3As$ (γ,n)	10^3–10^4	(Murin et al. 1956)
^{51}Cr	$Cr(CO)_6$	10^4	(Toropova 1957)
$^{75,77}Ge$	$(C_2H_5)_4Ge$		(Nowak 1965)
^{51}Cr	$Cr(CO)_6$	104	(Riedel and Merz 1966)
^{99}Mo	$Mo(CO)_6$	260	(Riedel and Merz 1966)
^{187}W	$W(CO)_6$	130	(Riedel and Merz 1966)

[a] Bombardment is with neutrons to give (n,γ) unless otherwise noted.

2. Metal Alkyls and Aryls

Studies of metal alkyls and aryls deal largely with arsenic, germanium and antimony, although some work has also been done on tellurium, mercury, thallium, bismuth and lead. The major contributions can be neatly divided into four periods: early studies by Maddock, Sutin and Hall (Edwards et al. 1953, Maddock and Sutin 1955, Hall and Sutin 1956); studies by the Polish group (Halpern et al. 1964, Siekierska et al. 1959, 1961, Siekierska and Sokolowska

Table 13.2
Synthesis by recoil methods: a partial list of radioactive organometallic compounds, other than the starting material, synthesized by recoil methods

Molecule	Method	Ref.
$^{99m}Tc(C_6H_6)_2^+$	$^{99}Mo(C_6H_6)_2 \xrightarrow{\beta}$	(Baumgärtner and Reichold 1961c)
$^{103}RuCp_2$	$U_3O_8 + FeCp_2$ (n,f)	(Baumgärtner and Reichold 1961a)
$^{99}Mo(CO)_6$	$U_3O_8 + Cr(CO)_6$ (n,f)	(Baumgärtner and Reichold 1961d)
$Cp^{99}Tc(CO)_3$	$[Cp^{99}Mo(CO)_3]_2 \xrightarrow{\beta}$	(Baumgärtner and Reichold 1961a)
$^{105}RhCp_2$	$^{105}RuCp_2 \xrightarrow{\beta}$	(Baumgärtner and Reichold 1961b)
$Te\varphi_2$, $Sb\varphi_3$	$Sn(C_6H_5)_4 + UO_2$ (n,f)	(Blachot and Carraz 1969)
$CH_3{}^{56}Mn(CO)_5$	$CH_3CpMn(CO)_3$ (n,γ)	(Srinivasan et al. 1966)
$H^{56}Mn(CO)_5$	$CH_3CpMn(CO)_3$ (n,γ)	(de Jong et al. 1969)
$^{56}Mn_2(CO)_{10}$	$CpMn(CO)_3$ (n,γ)	(de Jong et al. 1969)
$(C_6H_6)_2{}^{51}Cr$, $^{51}Cr(CO)_6$	$C_6H_6Cr(CO)_3$	(Baumgärtner and Zahn 1963)
$As\varphi_3$	$AsCl_3$ + benzene	(Siekierska and Sokolowska 1962)
$^{103}RuCp_2$	U in Cp (n,f)	(Zahn et al. 1969)
φ_2Hg	$\varphi HgCl$ (n,γ)	(Heitz and Adloff 1964)
$\varphi HgCl$	φ_2Hg'(n,γ)	(Wheeler and McClin 1967a)

1962); work by Heitz and Adloff (1964) and by Riedel and Merz (Merz 1964, Merz and Riedel 1964, 1965, Claridge et al. 1965); and most recently a series of investigations by Duncan et al. (Duncan and Thomas 1967, Baulch et al. 1961), Grossmann (1968, 1969a, b, c, d, 1970) and Nowak 1965, Nowak and Akerman 1970). The early work suffered (as has so much subsequent work) from what Maddock and Sutin themselves recognized to be a poor separation method. Despite this handicap, Maddock and Sutin were able to establish (1953, 1955) that a large fraction of the radioarsenic atoms were bonded to at least one phenyl group. Thermal annealing effects were also evident. It was speculated that many of the arsenic atoms may form $As\varphi$ or $As\varphi_2$ radicals through hot processes, which then could combine by thermal reactions with phenyl radicals produced by the recoil or by radiation damage.

Irradiation of $AsCl_3$ or $As\varphi_3$ in benzene solution in the presence and absence of scavengers led Siekierska and Sokolowska (1962) to a much clearer picture of what is actually going on. The results showed that the presence of oxygen caused a decrease in the yield of $As\varphi_3$ and $-As\varphi_2$, and an increase (not equal to the sum of the other decreases) in $-As\varphi$. The decrease is clearly due to thermal scavenging effects, while the increase in $As\varphi$ was ascribed to scavenging reactions such as

$$\varphi As = + O_2 + H_2O \rightarrow \varphi AsO(OH)_2 \qquad (13.1)$$

which would preserve a φAs species otherwise susceptible to further decomposition or further reaction to form $φ_2$As species.

These results had led to the conclusion (Siekierska et al. 1961) that the observed products are formed through a series of consecutive reactions, occurring at different stages in the slowing down of the recoil atom: the first Asφ bond is formed in the high-energy region, the second during the cooling, and the third after the cooling.

The study of the radiochemical reactions of arsenic atoms in benzene solution was carried further (Heitz and Adloff 1964) by comparing the product spectra of neutron irradiated $AsCl_3$ solutions and $^{77}GeCl_4$ solutions which have undergone beta decay. The product spectra were found to be remarkably similar, especially when considered only as to the number of As–φ bonds reformed. Only the Asφ yield from ^{77}Ge seemed to be sensitive to the application of an external ^{60}Co γ field (up to 3.5 Mrad h^{-1}) changing from 9% to 18% for a total dose of 25 Mrad. This result, different from that found for $Asφ_3$ solutions, must reflect a slow reaction of some inorganic arsenic compound with radiation-produced phenyl radicals.

A series of papers by Merz and Riedel (see Bibliography) describe work designed to compare radiochemical behaviour following n,γ; EC and $β^-$ decay. Gallium isotopes are produced in most of the cases studied, but isotopes of Sn, Pb, Ge and Sb were also involved. Unfortunately, the various chromatographic fractions were not well identified, so that it is not easy to draw definite conclusions from this work. The results were found not to be very sensitive to the presence of scavengers such as atmospheric oxygen, iodine and $FeCl_3$, although the effect was sufficient to indicate clearly the involvement of radical reactions, at least in the later stages. It is interesting to note here the very high ionic yield following electron capture. Riedel and Merz take the 2.5% retention following electron capture as representing failure of bond rupture, since this decay produced no gamma ray and thus no recoil is likely.

A detailed study of the radiochemical reactions of phenylarsenic compounds has been published by Grossmann (see Bibliography). He was able to get reliable values for the sums of all compounds with one, two and three phenyl–arsenic bonds, respectively, as well as ionic arsenic and a further organic-soluble fraction which appeared to be a group of polymeric phenylarsenic compounds.

Studying in addition the effect of irradiation temperature, presence or absence of air, gamma irradiation, and thermal annealing, Grossmann found results which can be summarized as follows:

–the reformation of φAs bonds is stepwise, much as was described by Halpern et al.

–both phenyl radicals and arsenic-containing radicals seem to be involved. Both can be scavenged by oxygen present either in the atmosphere or in the compound itself.

–On thermal annealing, φAs usually produces $\varphi_2 As$ or $\varphi_3 As$, but sometimes gives ionic arsenic, depending on how much absorbed or constituent oxygen is present in the target. The φ groups may come from neighbouring molecules through exchange-like processes.

A recent study of neutron-irradiated germanium tetraethyl was reported by Nowak and Akerman (1970). The most important result of this work was the observation of a large number of different products, separated and identified by gas chromatography. These products involve methyl and vinyl groups bonded to germanium, as well as hydrogen and ethyl. The yields were mostly fairly low, and varied somewhat with the gamma dose received. Virtual disappearance of the parent compound at high radiation doses is attributable to the interception of the stepwise reformation by competing radical reactions.

Processes following β^- decay in RaE have been used for the synthesis of organic polonium compounds (Murin et al. 1961e, Nefedov et al. 1961a, b).

To summarize the foregoing brief report on alkyl and aryl metal compounds, we can cite the following points.

–The reactions seem to be primarily radical combinations, with some abstractions being evident. Bond reformation occurs at each of the several possible stages: by hot processes, by fast, but scavenger-sensitive processes, and during thermal treatment.

–The reformation, or formation, of molecules appears to be a step-wise process – that is, it occurs by sequential reactions which may be interrupted at certain stages:

$$\begin{array}{l} As \xrightarrow{\varphi} As\varphi \xrightarrow{\varphi} As\varphi_2 \xrightarrow{\varphi} As\varphi_3 \\ \qquad\;\; \llcorner\!\xrightarrow{\quad O \quad} \varphi AsO \end{array} \qquad (13.2)$$

No information is available as to the nature of the primary fragment, however, and it must not be concluded that all As–φ bonds are broken.

–It is interesting to note that scavenging of radicals can be done by oxygen atoms from the target compound itself (Grossmann 1969b).

–The radicals are produced by various processes, although little is known for certain: by the initial bond rupture (Merz and Reidel 1964, 1965); by recoil impact (Blachot and Carraz 1969); by Auger ionization (Adloff-Bacher and Adloff 1964); by gamma radiation damage (Nowak 1965); and by normal exchange (Duncan and Thomas 1967).

–The initial ligands are not essential if the solvent molecules suffice. On the other hand, solvent participation is unnecessary where the target's own ligands are also involved.

3. Pi-bonded Ligands

3.1. π-RING METAL COMPOUNDS

Ferrocene was the earliest π-bonded metallorganic compound whose Szilard–Chalmers reactions were studied. Sutin and Dodson (1958) did a very convincing study which showed the retention to be 12%. Thermal treatment was found to increase this to about 21%. Room temperature annealing was also evident: a sample irradiated at −196° C rose from $R = 8.9\%$ to $R = 14.3\%$ on standing at room temperature for 6 days. The distribution of activity was also dependent on the fast neutron and the gamma ray doses. The data were interpreted as resulting from radical reactions involving $\cdot C_5H_5$ and $\cdot Fe(C_5H_5)$ radicals formed by the recoil or by gamma radiolysis. Recent work by Hillman (unpublished) has provided close confirmation of these results. By comparison with ferrocene, ferricinium picrate was found to have a low retention (Jach and Sutin 1958).

Subsequent studies have dealt with ruthenocene, dibenzenechromium and other transition metal π-ring compounds. In the course of this work it has been found that appreciable yields of the radioactive target compounds is obtained from such diverse initiating reactions as n,γ; fission (Baumgärtner and Schön 1963, 1964) and β decay. The results are summarized in table 13.3.

Table 13.3
Summary of experimental studies on metal 'sandwich' compounds

Compound	Conditions	Retention	Ref.
$RuCp_2$	(n,γ)	20–25%	(Baumgärtner et al. 1958)
$Cr(C_6H_6)_2$	(n,γ)	11.8 (19.4[a])%	(Baumgärtner et al 1960)
$RuCp_2$	$UO_2 + FeCp_2$ (fission)	40–60%	(Baumgärtner and Reichold 1961a, b)
$RuCp_2$	$FeCp_2 + RuO_2$ (n,γ)	0.01%	(Baumgärtner et al. 1958, Baumgärtner and Reichold 1961a, c)
$RuCp_2$	(n,γ)	~10%[b]	(Harbottle and Zahn 1965)
$RuCp_2$	UO_2 + Cp (fission)	~90%	(Zahn and Harbottle 1966)
	$UO_2 + Cp_2$ (fission)	2.5%	(Zahn and Harbottle 1966)
$NiCp_2$	(n,γ)	65%	(Wheeler and McClin 1967b)
$CoCp_2$	(n,γ)	25%	(Wheeler and McClin 1967b)
$CoCp_2$	various	30–85%[c]	(Ross 1969)
$HfCp_2Cl_2$	(n,γ)	15–60%	(Hillman et al. 1969)
$ZrCp_2Cl_2$	(n,γ)	15–60%[d]	(Hillman et al. 1969, Wheeler and McClin 1967c)

[a] The higher value was found on sublimation at 120° and probably reflects annealing.
[b] An isotope effect was found which was particularly strong in solution.
[c] By studying transmutation by radioactive decay, Ross found what he claimed to be a marked destruction of the molecule following internal conversion.
[d] A very strong isotope effect was found here.

Significant conclusions to be drawn from this work are the following:
–The molecules can be efficiently formed from a metal atom interaction with molecules of the matrix.
–Molecular bonding is evidently preserved through β decay.
–Monomeric Cp, but neither dimeric Cp nor benzene, is able to form sandwich compounds by thermal reaction in liquid solutions.
–Strong isotope effects are observed with ruthenium, hafnium and zirconium.
–Annealing effects have been frequently observed, and increase the yield of the radioactive parent.
–Internal conversion seems to lead to molecular destruction.

3.2. CARBONYLS

The first study of metal carbonyls was that of Toropova (Toropova 1957, Nefedov and Toropova 1958) whose objective was isotope enrichment using $Cr(CO)_6$. After dissolving the target compound in chloroform, she found nearly 90% of the ^{51}Cr to be extracted further into 0.1 M HCl, with isotopic enrichment factors greater than 10^4. This implies retention values of the order of 10% or less. Baumgärtner and Reichold (1961d) prepared carrier-free $Mo(CO)_6$ in high yield by neutron irradiation of powdered mixtures of U_3O_8 and $Cr(CO)_6$.

Harbottle and Zahn (1967) studied $Cr(CO)_6$ and $Mo(CO)_6$ irradiated in various ways so as to produce nuclear reactions in oxygen and carbon, as well as both high- and low-energy reactions in Cr and Mo. In all cases, reasonably high yields – up to 85% – were obtained (see table 13.4). It appears from this work that the final step is the addition of a carbonyl group to complete the molecule. The high retention and absence of isotope effect for the chromium isotopes suggests that the critical stages in the reformation are thermal reactions, rather than reactions dependent on the initial nuclear event.

Table 13.4
Recoil yields from metal carbonyl targets

Target	Parent yield	Other yields	Ref.
$Ni(CO)_4$	98.7%		(Wheeler et al. 1970b)
$Fe(CO)_5$	41	$Fe_3(CO)_{12}$ 26%	(Narayan and Wiles 1969)
$Cr(CO)_6$	40–80		(Harbottle and Zahn 1967, Zahn et al. 1969)
$Mo(CO)_6$[a]	60–80		(Groening and Harbottle 1970, Henrich and Wolf 1969, Zahn et al. 1969)
$W(CO)_6$[a]	50–60		(Narayan and Wiles 1969, Zahn et al. 1969)
$Mn_2(CO)_{10}$	12	$\cdot Mn(CO)_5$ 4.5%	(de Jong et al. 1969)
$Fe_3(CO)_{12}$	26	$Fe(CO)_5$ 17%	(Narayan and Wiles 1969)
$Ru_3(CO)_{12}$	41		(Narayan and Wiles 1969)

[a] Yield increased on thermal annealing.

Further insight into the reactions of metal carbonyls is provided by a study (Zahn et al. 1969) of the thermal annealing of $Cr(CO)_6$, $Mo(CO)_6$ and $W(CO)_6$ in the presence of carbon monoxide atmospheres at various pressures. The CO pressure, up to 100 atm, greatly increased the effect of annealing at higher temperatures. A maximum retention of about 80% was observed for annealing at 120° C for 2 h under 100 atm of CO. The authors suggest that the product molecule is reformed through a series of thermal reactions.

$$M(CO)_x + CO \longrightarrow M(CO)_{x+1} \tag{13.3}$$

where x may be between 3 and 5, after the cooling of the hot zone.

Nickel carbonyl has been shown (Wheeler et al. 1970b) to have a very high retention – 98.7% – both in the pure liquid and as 10% solution in *n*-heptane. It is apparent that this represents the yield following essentially complete isotopic exchange. It thus seems unlikely that much can be learned about hot or epithermal reactions in $Ni(CO)_4$.

Henrich and Wolf (1969) have studied the formation of $Mo(CO)_6$ by catching ^{90}Mo and ^{93}Mo recoils in $Cr(CO)_6$. It was found that the yields of $Mo(CO)_6$ with the two isotopes differed from each other, but varied only slightly as a function of initial recoil energy (table 13.4).

Groening and Harbottle (1970) have found a similar isotopic difference in $Mo(CO)_6$ between ^{99}Mo and ^{101}Mo. The retention of the latter isotope is a few percent lower than that of the former.

In $IMn(CO)_5$, both ^{128}I and ^{56}Mn retentions have been studied (de Jong et al. 1971, de Jong and Wiles 1972). The retention of ^{128}I is quite high (~30%). The retention of ^{56}Mn, on the other hand, is 11%. The retention values increase somewhat (^{128}I much more strongly) on heating of the samples at temperatures up to 80° C, while the $Mn_2(CO)_{10}$ activity (2.8%) remains unchanged. $BrMn(CO)_5$ has been similarly studied (de Jong and Wiles 1971). Yields in $RMn(CO)_5$ ($R = H, D, CH_3, C_6H_5$) are reported. No isotopic difference is shown in $HMn(CO)_5$ (Jakubinek et al. 1971).

3.3 POLYNUCLEAR CARBONYLS

$Mn_2(CO)_{10}$ was studied (de Jong et al. 1969, de Jong and Wiles 1968) by dissolving the irradiated target in dilute solutions of I_2 or $IMn(CO)_5$ in petroleum ether, so as to scavenge any manganese-containing radicals. By this means it was shown that in addition to the retention of 12% in $Mn_2(CO)_{10}$, a considerable fraction (4.5%) of the activity appeared as $IMn(CO)_5$. This was interpreted as indicating the presence of the radioactive $\cdot Mn(CO)_5$ radical trapped in the crystal.

Zahn (1967a) found that $Mn_2(CO)_{10}$ irradiated in dilute (~10%) solution showed very low retention – as low as 0.004%. This value, she contended, must represent the true failure of bond rupture. When these same solutions were irradiated as frozen solutions, the retention was 11.4%, quite similar to the value found for the solid.

Trinuclear carbonyls have been studied (Narayan and Wiles 1969) with the anticipation that the retention would prove to be in some way inversely related to the molecular complexity. The values obtained were surprisingly high, despite careful chemical purification, as is shown in table 13.4.

Conclusions to be derived from these studies tend largely to corroborate those mentioned earlier regarding the radical nature of the reactions. The occurrence of quasi exchange reactions seems to be more pronounced. One point is the appearance of polynuclear molecules more complex than the parent compound, which must reasonably be due to such exchange-like reactions.

3.4. ARENEMETAL CARBONYLS

These compounds provide interesting subjects for study, since there arises the possibility of two quite different types of product molecule, in addition to the target compound: the carbonyl and the pure diarene. These possibilities are well illustrated in the study of benzenechromium (Baumgärtner and Zahn 1963) tricarbonyl, as is shown in table 13.5.

Table 13.5
Recoil yields from π-ring metal carbonyls

Target	Parent yield	Other yields	Ref.
$(\varphi H)Cr(CO)_3$	10.0	$(\varphi H)_2Cr$ 0.2 $Cr(CO)_6$ 13.5	(Baumgärtner and Zahn 1963)
$CpMn(CO)_3$	12–20		(Costea et al. 1966, Zahn 1967)
	7.0	$HMn(CO)_5$ 10–12 $Mn_2(CO)_{10}$ 0.4	(de Jong et al. 1969)
$CH_3CpMn(CO)_3$	8	$CH_3Mn(CO)_5$ 2	(de Jong et al. 1969, Yang and Wiles 1967)
		$HMn(CO)_5$ 10	(de Jong et al. 1969)
Fulvalene $[Mn(CO)_3]_2$	9.1	$CpMn(CO)_3$ 0.2 $HMn(CO)_5$ 4. $Mn_2(CO)_{10}$ 0.3	(de Jong and Wiles 1970)
$[CpFe(CO)_2]_2$[a]	16.	$FeCp_2$ 1.4	(Kanellakopulos-Drossopulos and Wiles 1971)
		$Fe(CO)_5$[a] 3.2	

[a] Yield increased on thermal annealing.

A series of studies (Costea et al. 1966, Yang and Wiles 1967, de Jong et al. 1969) of cyclopentadienylmanganese tricarbonyl and related compounds has provided interesting results. As with the chromium compound mentioned above, it was found that carbonyl-rich compounds are formed in yields comparable to the retention. In these compounds, however, it is not the bi-nuclear $Mn_2(CO)_{10}$ but rather the mononuclear $Mn(CO)_5$ compounds which are prominent. Some results are given in table 13.5.

It is noted in table 13.5 that CH_3 is not produced in $CpMn(CO)_3$, showing that the methyl group involved in $CH_3CpMn(CO)_3$ is the ring substituent, evidently pyrolysed or knocked off of a nearby molecule. A strange annealing behaviour in $CpMn(CO)_3$ was reported (Costea et al. 1966) in which the activity in the parent fraction [evidently containing an impurity (Zahn 1967b)] was observed to increase markedly during the first minutes of annealing, and then to decrease again.

Radical diffusion processes were shown to be involved in the case of $CH_3CpMn(CO)_3$ irradiated in benzene solution (Yang and Wiles 1967).

Data obtained (de Jong and Wiles 1970) for hexacarbonylfulvalene-dimanganese are a bit surprising, in that one would expect the yield of monomeric $CpMn(CO)_3$ to be higher. Similarly, the retention in $CpFe(CO)_2O_2$ is high (Kanellakopulos-Drossopulos and Wiles 1971). It is noted here that the yield of $Fe(CO)_5$ is higher than that of $FeCp_2$.

Again supporting previous conclusions, these works provide two further points of interest:

–The competition between CO and the π-ring seems to lie distinctly in favour of the carbonyl. This statement is at present an extrapolation from limited data. If such a preference turns out to be more widely valid, we are still left with the question of whether it is a kinetic or a thermodynamic effect. Suggestions have been made that it is kinetic in nature.

–The survival of large molecules (fulvalene $Mn_2(CO)_6$ and $CpFe(CO)_2O_2$) is not easily explained in terms of either a kinetic or a thermodynamic model. One is left, then, with failure of bond rupture somehow causing these molecules to be preserved. It may thus be that the above-mentioned preference for carbonyl groups is rather a preferential polarization of the metal–carbonyl bond.

Addendum (December 1976)

Attention is drawn to two reviews (Wiles and Baumgärtner 1972, Wiles 1973) of this subject which discuss fundamental questions and possible answers. Several significant developments have been reported recently:

–Studies with implanted ions (Jenkins and Wiles 1972, Kanellakopulos-Drossopulos and Wiles 1976) and with ions formed by high-energy nuclear reactions (Wiles 1974) have shown that, at least in systems with non-labile ligands, molecules are formed only to a very small extent following simple kinetic recoil. This leaves electronic excitation as a possible cause of reactions which may occur.

–A strong correlation of the retention of several rhenium isotopes in $Re_2(CO)_{10}$ with the spin of the ground state has been found (Henrich et al. 1973). The authors conjecture that this may be related to the overall spin change in the capture γ cascade and hence to the likelihood of the occurrence of longer-lived internally-converted isomers.

–Studies have been made of the effects of β^- decay in ^{99}Mo compounds, and new technetium compounds have been made (Mikulaj et al. 1973, de Jong and Wiles 1973, 1976). The yields of $\cdot Tc(CO)_5$ and $CpTc(CO)_3$ (72–75%) were compared with the technetium atom's recoil spectrum and it is found (de Jong and Wiles 1976) that a minimum recoil energy of 10 eV is required to remove the Tc atom permanently from the molecule.

–Retention of ^{59}Fe in bridged ferrocenes shows (Hillman et al. 1975) a marked decrease with increased numbers of ($-CH_2-CH_2-CH_2-$) bridges, which was interpreted as showing the decreased likelihood of recombination following recoil. An 'intrinsic' retention of roughly 0.2% was inferred from the behavior of these compounds in DMSO solutions.

Two possible applications of hot atom chemistry to other fields have recently emerged.

–The potential use of constant retention in species-specific activation analysis (de Jong et al. 1974) was not successful when applied to methylmercury in fish (Webber and Wiles 1976), because a retention of 72% in CH_3HgCl became a variable 3–8% for CH_3Hg- in fish protein. Nevertheless, the idea is worth pursuing, and may in favourable cases lead to very sensitive and selective molecular activation analysis.

–A further possible use of organometallic hot atom chemistry lies in the identification of super heavy elements. The model for this comes from the formation, by fission product recoil (Baumgärtner and Schön 1964), of $^{99}Mo(CO)_6$ (Baumgärtner and Reichold 1961d) and $^{105}Ru\ Cp_2$ (Baumgärtner and Reichold 1961a). Thus, if formed in appropriate matrices, elements 106 and 160 should give $C_6H_6(M)(CO)_3$ while 107 and 161 should give $C_5H_5(M)(CO)_3$. Formation of these volatile compounds should be selective enough to provide unambiguous identification.

CHAPTER 14

RECOIL REACTIONS INVOLVING OXIDATION AND HOT ATOM CHEMISTRY OF THE RARE GASES

Garman HARBOTTLE

Department of Chemistry, Brookhaven National Laboratory,
Upton, New York 11973, USA

Chemical Effects of Nuclear Transformations in Inorganic Systems
Edited by G. Harbottle and A.G. Maddock

Contents

1. Oxidative Recoil Reactions

The process of internal conversion of a capture gamma ray, or alternatively, of inner-shell electron capture, creates an inner-shell vacancy that leads quickly to the production of an atom bearing a positive charge. If the atom is in collision-free space, the charge is preserved and can be measured, as described in ch. 2 above. If however the atom is in the liquid or solid phase the immediate physical consequences of the charging–discharging process are not easy to visualize, although the chemical reactions have been repeatedly observed (ch. 3 and 15). Since the permanent loss of electrons constitutes an oxidation in the chemical sense, research workers have attempted to find evidence in inorganic systems for oxidative recoil reactions which could be related to positive charging following internal conversion. Of course, in crystals containing oxygen-bearing ions, or other potentially oxidative species, such reactions could equally well be attributed to 'hot-zone' or diffusion-controlled processes involving the recoil atom and any other electron acceptor, for example a hole.

When alkali halides are irradiated with neutrons either in the solid state or in solution, virtually all the activity is found in the halide state (Libby 1940, Maddock and del Val Cob 1959). In the case of iodine recoils formed by high-energy proton bombardment of cesium chloride crystals, Dema and Zaitseva (1966a) also found that, at the lowest integrated fluxes, where radiation damage was least, almost all the activity was present as iodide. Maddock and del Val Cob did, however, note that a fraction of a percent of recoil bromine in KBr could be identified as atomic bromine, and separated from the bromide bulk.

In the case of metallic cations, the search for oxidative recoil has been, in at least a few cases, rewarded. In 1952, in a paper which foreshadowed several ideas later developed in research on hot atom chemistry in crystals, Herr (1952a) found that irradiated rhenium III and IV salts produced substantial amounts of Re(VII) (perrhenate) on dissolution. Similarly, it was shown that neutron capture in chromic salts led to small amounts (ca 10%) of chromate (Harbottle 1954). However, in two ethylenediamine complexes of Cr(III), a much greater tendency to produce Cr(VI) was seen (Turco and Scatena 1955). No additional work seems to have been reported along this interesting line. The results with rhenium are especially interesting in that the crystals bombarded did not contain oxygen. The corresponding experiments with manganous compounds, both solid and in solution, were carried out very early by Libby (1940): no MnO_4^- activity was found.

In further experiments Maddock and de Maine (1956) searched for oxidative recoil in a number of inorganic compounds: Sb_2O_3, $KSbC_4H_4O_7 \cdot \frac{1}{2}H_2O$, As_2O_3, $Ce_2(SO_4)_3$ and various thallous salts. Within experimental error, no yields of oxidized species were found in any of these salts, when neutron capture had taken place. Kawahara and Harbottle (1959), however, bombarding crystals at lower temperature, were able to detect about 4% of As(V) in neutron-irradiated arsenious oxide and sodium arsenite. These experiments were confirmed by Baró (1961).

Low-temperature bombardment and storage conditions also proved to be the key to observation of oxidative recoil in the thallium system. Both $TlNO_3$ and $TlClO_4$ crystals undergoing $^{203}Tl(n,\gamma)^{204}Tl$ reactions show substantial percentages of Tl(III) which, however, anneals out at moderate temperatures. Implantation of ^{208}Tl following α-decay of ^{212}Bi, or generation of ^{202}Tl recoils by n,2n reactions all give similar results (Ackerhalt and Harbottle 1972, Ackerhalt et al. 1972, Butterworth and Harbottle 1966).

After all of this work, it is not at all clear whether the presence of oxidized recoil atoms reflects in any way the loss of electrons, i.e. positive charging, taking place as a result of internal conversion. More direct evidence could be given by the observation of oxidized species in Mössbauer spectra, for example, Fe(III) or Fe(IV) peaks in decay of cobalt-57 in labelled crystals (ch.24). There are so many opportunities for physical loss of electrons during the slowing-down process, for electron loss through trapping of one or two holes on the interstitial recoil atom, or through simple oxidation in the crystal or in solution by oxidizing entities such as halogen atoms or molecules or oxygen atoms, formed by radiolysis, that we can never be sure that there is any correlation between inner-shell vacancy production and the observed oxidized recoil species.

2. Noble Gas Hot Atom Chemistry

The extraordinary discovery that xenon could be made to form chemical compounds (Bartlett 1962, Claasen et al. 1962) was followed, of course, by numerous investigations of the properties of these compounds, for example xenon fluorides and oxide and ions such as perxenate. Hot atom chemists have participated in this work in various ways, through Mössbauer studies (Perlow and Perlow 1964a, b, 1965) of the consequences of nuclear decay leading to xenon, and through conventional, wet-chemistry techniques. Two principal lines of approach have opened: the recoil chemistry of the noble gases in their compounds, following n,γ activation, and the study of radiosynthesis of noble gas compounds following decay processes of atoms such as ^{131}I in IF_5 decaying to ^{131m}Xe. The field of noble gas radiochemistry including hot atom chemistry was reviewed by Adloff (1966).

Studies of the hot atom [(n,γ) initiated] recoil chemistry of xenon compounds were begun by Starke and Gunther (1964) and have been pursued in the laboratory of Adloff (Schroth and Adloff 1964a, b, Margraff and Adloff 1966). The results are interesting: surprisingly high retentions are seen in compounds like XeF_2, XeF_4 (in which both labelled XeF_2 and XeF_4 are found after irradiation) and $Ba_2XeO_6 \cdot 2H_2O$. The isomeric transition $^{133m}Xe \rightarrow {}^{133}Xe$ occurring in labelled XeF_4 does not lead to significant release of xenon. Such results suggest very short-range recoil effects, since one would expect that, if the xenon were once free of its original bonds and lodged interstitially, thermal diffusion and/or radiation defect trapping would hardly be able to alter its chemistry, i.e. return it to chemical combination.

The work of Heitz and Cassou (1971) on the neutron irradiation of xenon trioxide in solution, in which the retentions of no less than seven xenon nuclides were measured, is interesting for the same reason. Retentions of ca. 2% were observed, with substantial isotopic effects. The retentions can almost unequivocally be assigned to cancellation of recoil momenta and the differences between ground-state retentions following isomeric transition, to differing probabilities of internal conversion (Cassou 1970, Heitz and Cassou 1970).

The second avenue of hot atom research with noble gases, i.e. the observation of radiosynthetic processes, has been the subject of very extensive work at Leningrad. Much of this work has been directed to the synthesis of xenon–oxygen compounds by nuclear decay of iodine incorporated in the corresponding iodates and periodates (Kirin and Gusev 1966, Kirin et al. 1966, Murin et al. 1965a, c, d) and even to the exploration of organic molecules containing xenon (Toropova et al. 1968a, b). Further discussion will be found in ch. 16.

It is understandable that compounds of krypton will be far less stable than those of xenon (Pasternak and Sonnino 1967, Murin et al. 1965c), but on the other hand, it should be relatively easy to synthesize compounds of radon for study via recoil chemistry (Harbottle 1971). Radon is known to unite with fluorine (Fields et al. 1962) to form compounds less volatile than the xenon fluorides. But the only experiments involving recoil and/or solution chemistry of radon are somewhat equivocal (Haseltine and Moser 1967, Haseltine 1967, Flohr and Appleman 1968). Clearly, this is a field that deserves additional research, and one in which the hot atom chemist could make a substantial contribution to inorganic chemistry.

CHAPTER 15

EFFECTS OF ISOMERIC TRANSITION

C.H.W. JONES

The Chemistry Department, Simon Fraser University, British Columbia, Canada VSA IS6

Chemical Effects of Nuclear Transformations in Inorganic Systems
Edited by G. Harbottle and A.G. Maddock

Contents

1. Introduction

In comparison with the very extensive work which has been carried out on the results of the (n,γ) reaction in inorganic solids, surprisingly little has been reported on the effects of radioactive decay in labelled compounds. Such studies of radioactive decay are extremely important because they allow the investigation of the chemical effects resulting from very specific, well-defined physical phenomena associated with the decay event. This is in contrast with the study of the effects of (n,γ) reactions, where the details of the de-excitation γ-cascade are rarely well-characterised, and where, as a consequence, there is considerable uncertainty as to the relative importance of recoil energy and electronic excitation in leading to the observed effects. Moreover in (n,γ) and related studies, the temperature during the irradiation and the concomitant γ-irradiation have very marked and extremely complex effects on the chemical form of the recoil atom in the solid. In contrast, in studying the chemical effects of radioactive decay in labelled molecules these complicating factors may be more readily controlled, if not eliminated.

2. The Isomeric Transition Process

Metastable nuclear isomers may de-excite by the emission of a γ-ray or by the ejection of an internal conversion electron from an inner shell of the atom. A third mode of de-excitation is that in which the nucleus emits a positron–electron pair. The last mode is of little interest in the present discussion since it can only occur if the excitation energy exceeds 1.02 MeV and even then the process competes only very unfavourably with γ-ray emission.

Isomeric transitions which occur predominantly by γ-ray emission are generally only observed for relatively high-energy transitions (> 100 keV) of low multipole order, and in such instances the parent isomeric state is short-lived. As a result very few studies of the chemical effects of such decay processes have been made, since the half-life of the metastable state is usually so short as to preclude the labelling and study of a molecule of interest. One case which does allow investigation, however, is $^{69m}Zn \xrightarrow{13.8\ h} {}^{69}Zn \xrightarrow{69\ min}$, where the decay is predominantly by emission of a 0.439 MeV γ-ray, which imparts a recoil energy of 3.8 eV to the ^{69}Zn atom. An early study by Seaborg et al. (1940a, b) of gaseous $^{69m}Zn(C_2H_5)_2$ and a much more recent study of labelled zinc phthalocyanine in the solid state by Yang et al. (1969, 1970a) have shown that there is failure to bond rupture in $> 95\%$ of events in this decay. But for these two investigations, all other studies have been concerned with isomeric transitions which are highly internally converted.

Decay by internal conversion is generally observed for relatively low-energy transitions (< 100 keV) of high multipole order, and in such cases the metastable isomer is often long-lived with a half-life of minutes, days or longer. The details of the internal conversion process have been well reviewed by Wexler (1965b) and the important points will now be outlined. Where it is energetically allowed, internal conversion occurs predominantly in the K-shell of the atom, the internal conversion coefficients decreasing markedly for each of the successive outer atomic shells. The ejected internal conversion electron will generally impart a recoil energy to the daughter atom much smaller than chemical bond energies. Of far greater importance for the fate of the daughter atom is the hole created in the inner atomic core. This hole is unstable and rapidly moves out through the successive electron shells to the outer, valence shell of the atom. These electronic transitions may occur with the emission of X-rays or by the Auger process, the latter leading to hole multiplication with the production of a high positive charge. The two processes compete, resulting in a spectrum of charges on the daughter atom, and the electronic transitions are complete in ca. 10^{-16} to 10^{-14} seconds.

Extensive studies have been made of the physical and chemical effects of the Auger charging process on isolated atoms and molecules in the gas phase at low pressures, as reviewed by Wexler (1965b) and reported in recent articles by Carlson and White (1966, 1968). Such studies have used both radioactive decay and X-ray bombardment to create the initial hole in the K-shell of the atom of interest. These investigations have shown that the Auger process generates a very large positive charge in most events. For example, following the $^{131m}Xe \longrightarrow {}^{131}Xe$ decay in xenon gas, it was found by Pleasanton and Snell (1957) that the resulting ^{131}Xe has a charge distribution peaking at +8 with a maximum charge of +22, the mean charge being +7.9. Thus in a large fraction of events the xenon atoms lose their entire valence shell following the nuclear transformation.

Where the parent isotope is initially incorporated in a molecule, extensive molecular decomposition is found to accompany the decay in the gas phase. Bond rupture may occur through direct loss of valence or bonding electrons in the Auger cascade. However, Wexler (1965b) and Carlson and White (1968) have proposed another mechanism to explain the decomposition process. Since the Auger cascade reaches the valence shell in 10^{-16} to 10^{-14} sec, i.e. in a time much shorter than bond vibration times, the atoms will remain fixed in their relative equilibrium positions in the molecule. A redistribution of electrons in the molecule then occurs in 10^{-15} sec and this is immediately followed by a molecular explosion as a result of Coulomb repulsion between the various positively charged centres in the molecule. The decomposition process is then characterised by the ejection of a number of positively charged fragments with kinetic energies up to 100 eV depending on the mass and charge of the fragments. Carlson and White (1968) have experimentally studied the decompositions of CH_3I in the gas phase following the production of a hole in the K-shell of the iodine by X-ray bombardment. Their experimental findings are in excellent agreement with the features of the model described above.

3. The Effects of the Isomeric Transition Process in the Solid State

The gas phase studies mentioned above have provided a very clear picture of the molecular decomposition which occurs following a highly internally converted transition in an isolated molecule. In examining the effects of such a decay in the solid state it is immediately necessary to ask what effects the close proximity of neighbouring molecules and ions will have on the primary physical processes which occur.

In the first instance, the high positive charge generated on the daughter atom in the Auger cascade will be dissipated not only by transfer of electrons from other atoms in the molecule but also by transfer of electrons from adjacent ions and molecules in the lattice. Thus the daughter molecule may never acquire a number of highly charged centres, but instead the positive charge may be shared by a large number of atoms in the immediately surrounding lattice. The electronic nature of the lattice and its constituent atoms will determine to what extent this occurs. Thus a metallic lattice will totally dissipate the charge in 10^{-15} seconds by neutralization through the conduction band. In an ionic or insulating solid, the relative electron affinities of the various ions and that of the positively charged atoms in the molecule will be a determining factor.

The ability of a solid lattice to dissipate electronic excitation energy without producing attendant molecular decomposition has been extensively studied in the general field of radiolysis, as reviewed by Johnson (1970). It is found that on irradiating a solid with ionising radiation such as γ-rays or electrons, in some cases the *G*-value of the solid (the number of molecules decomposed per 100 eV of energy absorbed) is smaller, the smaller the crystal free-volume. This is evidence that tightly packed lattices more readily delocalise excitation energy into vibrational modes of the crystal without producing molecular decomposition. However, a simple relationship between lattice parameters and *G*-values is not always observed and Johnson (1970) has concluded that the significant parameter affecting the radiolytic decomposition of molecular ions in solids is the so-called 'total' crystal environment. Factors such as crystal free-volume, cation size, and lattice energies are all variables of the total crystal environment, and are all probably involved in some complex manner in the radiolytic decomposition process. Moreover the ability of a lattice to transfer energy through electron-hole pairs (excitons), annealing which can occur following the primary decomposition, and the thermal stability of the primary products, must all play a role in determining the mode of radiolytic decomposition. Many, if not all, of these factors may also play a part in determining the stability towards fragmentation of molecular ions in solids following Auger charging.

Even if bond rupture of the parent molecule does occur, either through loss of valence electrons or through Coulomb repulsion of positive charges, the solid lattice will still play an important role through the classical cage-effect. The fragments formed in the decomposition will be constrained by the surrounding lattice and will not be free to escape far from the parent site. (This cage effect may again be dependent on lattice parameters such as the crystal free-volume.) The primary decomposition products may then react with one another or with surrounding substrate molecules, and these reactions may mask the identity of the primary products formed in the decay. Further chemical reactions of the daughter fragments may also occur as a consequence of thermal annealing either during or following the time that the parent isotope is decaying in the solid. As in other inorganic systems, a systematic study of such annealing reactions may provide information about the chemical environment in which the isomeric transition daughter atoms find themselves.

A final point concerns the phenomenon of autoradiolysis, i.e. the radiolytic decomposition of the surrounding environment produced by the decay radiations themselves. Geissler and Willard (1963) have pointed out that low-energy conversion and Auger electrons will have relatively short ranges in liquids and solids, and that extensive radiolytic decomposition of molecules immediately surrounding the decaying atom will occur. In liquids, the radiolytic products are free to diffuse to, and react with, the daughter atom and this leads to a marked similarity between the product distribution observed following nuclear transformations accompanied by internal conversion, and that for simple radiolysis. [However, de Fonseca et al. (1969) have recently observed that this model is not generally applicable to all liquid systems.] In solids the radiolytic products will not be quite so free to diffuse to, and react with, the recoil atom. However, the atom born in the decay will find itself surrounded by a region of the lattice containing a much larger concentration of radiolytic decomposition products than that found in the bulk lattice. Moreover, the transitory free electrons and positive holes generated in the radiolysis may be trapped nearby in the crystal at defect lattice sites or at the radiolytic products themselves. On subsequently heating the crystal, the thermal release of these electrons and positive holes may then induce annealing reactions involving the recoil atom. It may therefore be anticipated that isomeric transition recoil products will be very susceptible to thermal annealing.

4. Studies in Labelled Inorganic Compounds

4.1. BROMINE

The $^{80m}Br \rightarrow {}^{80}Br$ decay has been the most extensively studied isomeric transition processes. The relevant portion of the decay scheme is shown in fig. 15.1. The K-shell internal conversion coefficients for the two steps in the cascade, as measured by Schmidt-Ott et al. (1960a, b), are ca. 298 for the 0.049 MeV transition and 1.6 for the 0.037 MeV transition. One of the most important features of the $^{80m}Br \rightarrow {}^{80}Br$ decay seems to have attracted little if any attention from the investigators who have studied this system. The intermediate 0.037 MeV state has a half-life of 7.4 ns which is considerably longer than the time it takes for an Auger cascade to reach the outer, valence shell of an atom (10^{14} sec). Thus the decay in essence can generate two distinct and separate rounds of charging, and whereas the first cascade will occur for a bromine bound in the parent chemical form, the second may occur for atoms which are now in the chemical form of decomposition products formed following the first event. It is obviously not possible experimentally to separate the effects due to the two rounds of ionisation and excitation. However, the existence of this phenomenon should be recognized and some caution exercised in interpreting experimental observations in terms of a detailed physical model.

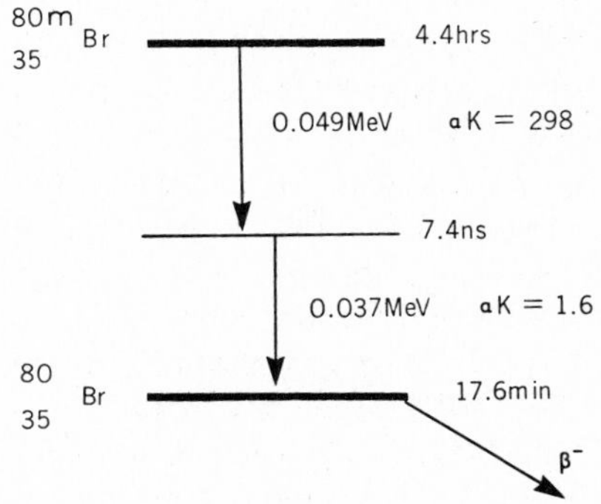

Fig. 15.1. The decay scheme for ^{80m}Br, showing the 7.4 ns intermediate state in the decay, and the K-shell internal conversion coefficients, α_K, for the two transitions.

4.1.1. *Metal complexes*

A very detailed study has been carried out by Herr et al. on the hexobromo complexes of tetravalent rhenium, osmium and iridium, and on the brompentammine complexes of cobalt, rhodium and iridium. For the solid ^{80m}Br-labelled potassium hexabromo complexes Herr and Schmidt (1962) found the ^{80}Br to be exclusively in the parent form, i.e. no bond rupture was observed. However, on replacing the cation successively with ammonium, tetraethyl ammonium, and tetra n-butylammonium Schmidt and Herr (1965) found that an increasing fraction of the ^{80}Br activity appeared as free Br^- ion following the decay. Thus as the size of the cation was increased the percent decomposition increased. However, lack of crystal structure parameters precluded the study of a possible relationship between the yield of $^{80}Br^-$ and the geometry of the crystal environment.

More recent experiments by Müller and Cramer (1970) have shown that in mixed crystals of $K_2Re^{80m}Br_6/K_2SnCl_6$, varying in composition from 11 to 85 mole % of K_2ReBr_6, the ^{80}Br in all cases was found to be 100% in the parent $ReBr_6^{2-}$ form. A detailed analysis of the structures of these mixed crystals suggests that the densely packed lattices present in all instances must be the explanation for the apparent lack of bond rupture. The tightly packed lattice may serve to dissipate the electronic excitation energy effectively and must also provide a rigid cage around the excited molecule which would ensure a low probability of bond rupture. This finding is certainly in agreement with the work of Herr et al.

In studying the solid brompentammine complexes of Co, Rh and Ir, $[Me(NH_3)_5{}^{80m}Br]X_2$, Schmidt and Herr (1963, 1965) found evidence of extensive molecular decomposition, the yield of free $^{80}Br^-$ being strongly dependent on the chemical nature of the anion, X, and/or the physical properties of the crystals. In general, the complexes with an anion with oxidising properties (NO_3^-, ClO_4^-, NO_2^-) showed higher $^{80}Br^-$ yields than those with reducing properties ($S_2O_6^{2-}$, $C_2O_4^{2-}$). This may be evidence that intramolecular electron transfer between the inner and outer coordination spheres of the complex may be an important factor in stabilising the electronically excited and ionised complex following the Auger cascade. Where the anion of the complex may readily lose electrons to the ionised parent molecule, bond rupture and molecular dissociation may be inhibited, and the daughter ^{80}Br atoms will then remain bonded in the inner coordination sphere of the complex. Saito (1965) hypothesised that the stability of cobalt complexes to molecular fragmentation in decay might parallel the stability of the lattice to radiolytic decomposition. However, a study by Saito et al. (1965c) of the radiolytic stability of these complexes did not indicate any simple relationship between the yields of products formed in radioactive decay and radiolytic *G*-values.

The thermal annealing of the recoil fragments was also investigated in the nitrate complexes of cobalt, rhodium and iridium by Schmidt and Herr (1961, 1963, 1965) and it was concluded that an equilibrium was reached between the annealing reaction and the radioactive decay of the daughter ^{80}Br atoms, which are being continually replaced in the transient equilibrium process in the solid. In contrast, for the halide complexes $[Co(NH_3)_5{}^{80m}Br]Br_2$ or Cl_2, of rhodium and iridium, the $^{80}Br^-$ yield was found to increase on heating, reaching limiting values of ca. 66% at 210–150° C. This is evidence for a thermally induced exchange between the complex-bound bromide and the anionic bromide or chloride. However, it is important to note that the excitation energy resulting from the isomeric transition must play an important part in facilitating the exchange. In ^{82}Br labelled $[Ir(NH_3)_5{}^{82}Br]Br_2$, thermally induced exchange occurs only slowly (8% exchange in 24 h at 210° C). Thus the exchange observed following the isomeric transition occurs far more readily than normal isotopic exchange, and thus must be traceable to a lowering of the activation energy for exchange in the environment locally altered by the isomeric transition.

Several investigators have studied the effects of the isomeric transition in anionically labelled complexes. Schmidt and Herr (1963, 1965) found that in the rhodium and iridium complexes $[Me(NH_3)X]^{80m}Br_2$, the yield of $[Me(NH_3)_5{}^{80}Br]^{2+}$ varied between 9 and 19%. This yield varied with X but more importantly did not vary with temperature or the mass of the metal atom for any one ligand. This was interpreted as evidence for a geniune hot-atom reaction between the ionised and excited ^{80}Br and the ligands of the complex. Ambe et al. (1971) have systematically investigated the complexes $[Co(NH_3)_5X]Br_{2,3}$ and $[Co\ en_2X_2]Br_{1,3}$, where X = NH_3, NCS, NO_2, OH_2, ONO, ONO_2, F, Cl, Br, and I. For crystals allowed to decay at −78° C or −196° C, they observed that the yield of $[Co(NH_3)_5{}^{80}Br]^{2+,3+}$ or $[Co\ en_2\ X\ Br]^{1+,3+}$ appeared to be related to the frequency of the maximum in the first absorption band of the parent complex. Thus, the less stable ligands, X, were replaced more readily by radiobromine as a consequence of the isomeric transition. A similar relationship was observed for the bromine-labelled products of the (n,γ) and (n,2n) reactions in these complexes by Saito et al. (1967a, Saito and Tominaga 1965). The approach taken by Ambe et al. is a very interesting one and an analysis of the data of Schmidt and Herr, using the same approach, may show similar trends.

In earlier experiments Yoshihara and Harbottle (1963) had studied the chemical effects associated with five different nuclear transformations in $[Co(NH_3)_6]Br_3$. Of particular interest for the present discussion was their study of the ^{80m}Br isomeric transition. They found that following decay at dry-ice temperature, $5.6 \pm 0.2\%$ of the ^{80}Br activity was present as $[Co(NH_3)_5Br]^{2+}$, dropping to only 1% for decay at room temperature. It was proposed that on heating electrons may be thermally ionised from traps in the crystal and that an electron reduction of the cobaltic ion to cobaltous ion could then occur.

$$[Co(NH_3)_5{}^{80}Br]^{2+} + e^- \longrightarrow [Co(NH_3)_5{}^{80}Br]^{1+}.$$

On dissolving the crystals for analysis, the cobaltic ion would survive as a stable ion in solution, while the unstable cobaltous ion would decompose, yielding free $^{80}Br^-$. This model was also used to explain the annealing results for the other recoil atoms in this system. The Auger cascade and attendant local radiolysis of the crystal in the isomeric transition would produce an appreciable number of electrons trapped near the recoil site by lattice defects or radiolysis products, and these would be able to participate in the thermal annealing reactions. Certainly the annealing reactions observed for the isomeric transition products were far more rapid than those observed for the bromine or cobalt (n,γ) recoil atoms produced in the same crystal.

4.1.2. Bromates

The first studies of the $^{80m}Br \rightarrow {}^{80}Br$ isomeric transition in crystalline bromates were made by Campbell (1959a, b) and Harbottle (1960). The focus of these investigations was the identification of the fragments formed following decay in solution and in the solid state. While following nuclear transformation in solution, an unstable fragment, $^{80}BrO_2^{2+}$, was identified, this fragment was not observed following the transformation in the solid state. In the latter case the fragment must be stabilised in the crystal by interaction with the substrate molecules. The retention as $^{80}BrO_3^-$ was found to be ca. 30% both in solution and the solid at room temperature. In these investigations only a cursory investigation was made of the thermal annealing reactions of the products of the decay in the solid. Campbell (1960) observed only a small change in the $^{80}BrO_3^-$ retention in $Na^{80m}BrO_3$ for crystals allowed to reach radiochemical equilibrium at 225° C. On the other hand Harbottle (1960) noted an increase from 35% at room temperature to 50% at 98° C in $K^{80m}BrO_3$.

More recently, further investigations of the thermal annealing reactions of the ^{80}Br labelled products in bromates have revealed an interesting and complex picture. In discussing the results of these experiments it must be noted that the $^{80m}Br \rightarrow {}^{80}Br \rightarrow$ decay is a transient equilibrium system, the mean lifetime of the ^{80}Br atoms being ca. 25 min. In an annealing experiment which lasts for a time long in comparison with the mean life of the ^{80}Br atoms, any change in the rate of annealing with time will be directly reflected in the distribution of ^{80}Br activity. Jones (1967, 1970b, Campbell and Jones 1968a, b) has observed that in $Na^{80m}BrO_3$, on heating above 150° C the retention as $^{80}BrO_3^-$ initially rapidly increases, reaches a maximum value, and then decreases to a plateau value at an annealing time of ca. 2 h (fig. 15.2). This annealing behaviour was explained in terms of the presence of two discrete annealing reactions producing $^{80}BrO_3^-$, the first arising from the release of electrons and positive holes in the bulk of the lattice and giving rise to the initial rapid increase in retention; the second arising from defects generated near to the ^{80}Br atom in the autoradiolysis, this resulting in the final plateau or equilibrium value of the retention. This proposal was supported by the observation that the first part of the annealing reaction was dependent on the thermal history of the sample and could be removed by heating the crystals before the annealing study, whereas the latter was totally independent of such treatment.

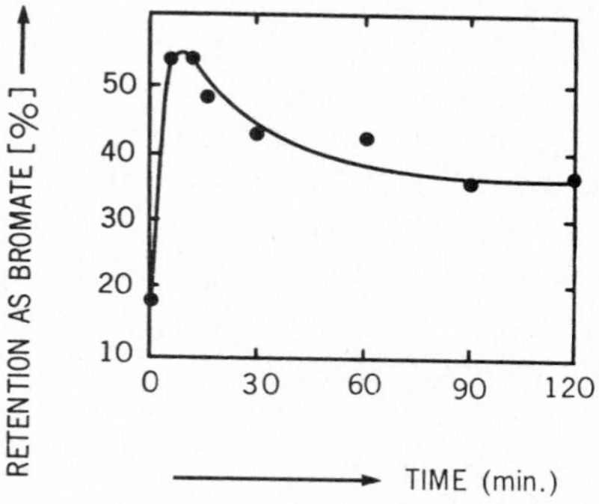

Fig. 15.2. The isothermal annealing curve at 216° C for the ^{80}Br recoil atoms in Na $^{80m}BrO_3$. The crystals were allowed to reach transient equilibrium at room temperature prior to the heat treatment.

Shiokawa et al. (1969b) and Sasaki and Shiokawa (1970) have reported that if the annealing is extended to 3 or 4 h, the retention as $^{80}BrO_3^-$ begins to increase again. This result is most unusual, since if the retention continues to increase over times longer than that required to reach transient equilibrium in the crystal, then the rate of annealing must in fact be speeding up with time. A possible explanation may lie in the presence of competitive oxidative and reductive processes which occur at different rates. The possible existence of such processes has been pointed out by Andersen et al. (1966) who have compared and contrasted the thermal annealing of radiolytic decomposition products and (n,γ) recoil atoms in $KBrO_3$. They found that while the annealing reactions of the two processes are different, they are closely related, and that thermal annealing of the radiolytic products may be the determining factor for the annealing of the recoil fragments. Thus in the thermal annealing of ^{82}Br(n,γ) recoil atoms in $KBrO_3$ they observed maxima and minima in the annealing curves, and concluded that the annealing of trapped hole centres led to the oxidation of ^{82}Br fragments while the annealing of trapped electron centres correspondingly caused a reduction of ^{82}Br fragments, the hole and electron centres being associated with the radiolytic decomposition products. The detailed and elegant analysis of Andersen obviously has considerable significance for the isomeric transition experiments. The pocket of radiolytic decomposition products surrounding the ^{80}Br atoms may govern the thermal annealing of the isomeric transition fragments in much the same way as described for the (n,γ) recoil products.

Müller and Cramer (1970) have commented on the extensive molecular decomposition which accompanies the isomeric transition in $K^{80m}BrO_3$, and contrasted it with the failure to bond rupture observed in $K_2Re^{80m}Br_6$. They estimated the recoil energy acquired by the oxygen atoms through Coulombic repulsion as between 21 and 58 eV, and suggested that extensive bond rupture occurs because of the large crystal free-volume in the $KBrO_3$ lattice. However, Sasaki and Shiokawa (1970) have studied the isomeric transition decay in all the alkali metal bromates at $-196°$ C and report that, while these bromates show radiolytic G-values which are simply related to the crystal free-volumes as reported by Boyd et al. (1962), no such relationship is observed for the retention as $^{80}BrO_3^-$. However, the change in retention observed for a standard annealing treatment was found to show a trend, being largest for Li or Na and smallest for Cs. The authors proposed that this trend in some way reflected the effective electron density on the BrO_3^- ions in the crystal. Another explanation may be that the decomposition fragments formed in the decay would not move far in the tightly packed $LiBrO_3$ crystal and that fragment recombination reactions may then occur more readily on annealing.

4.2. TELLURIUM

The chemical effects of both the $^{127m}Te \rightarrow {}^{127}Te$ and $^{129m}Te \rightarrow {}^{129}Te$ isomeric transitions have been studied. Both decays are highly internally converted, although there is some question concerning the precise details of the decay scheme for ^{129m}Te. Hahn (1963, 1964) observed that the percent molecular decomposition accompanying the isomeric transition for $H_6{}^{129m}TeO_6$ in aqueous solution appeared to differ depending on which γ-ray of ^{129}I was used in assaying the activities of the daughter fractions containing ^{129}Te. The data was used to determine the relative probabilities of populating the 1.12 MeV state in ^{129}I in the decay of ^{129m}Te and ^{129}Te respectively. Work by Carillo and Nassif (1967) appeared to be in agreement with the conclusions arrived at by Hahn. More recently Hillman and Weiss (1971) studying the chemical effects of the decay of both ^{129m}Te and ^{127m}Te in aqueous solution have refuted Hahn's findings for the decay scheme of ^{129m}Te. However, in this latter work no attempt was made to study the effects of the isomeric transition by monitoring the 1.12 MeV γ-ray of ^{129}Te and thus the basis of Hillman and Weiss's argument appears to be invalid.

The chemical effects of the tellurium isomeric transition were originally studied by Seaborg et al. (Seaborg and Kennedy 1939, Seaborg et al. 1940a, b) who observed molecular decomposition in H_6TeO_6 (solution) and $Te(ethyl)_2$ (gas), and by Williams (1948a) who in addition to studies in aqueous solution also investigated the effects of the ^{129m}Te decay in TeF_6 (gas). Many studies have subsequently been made both in aqueous solutions and the solid state.

4.2.1. Telluric acid and the tellurates

Kirin et al. (1968) have studied the decomposition observed following isomeric transition in a number of tellurates, but observed no systematic trend in the decomposition yield with variation of the cation. Stevovic and Muxart reported effects in telluric acid (1968). All other studies have been essentially concerned with the thermal annealing reactions of the labelled recoil products formed following the isomeric transition in tellurates or telluric acid. Dancewicz and Halpern (1964) carried out the first study of this type, and found that in solid $Na_2{}^{127m}TeO_4$, the daughter ^{127}Te was present as Te(VI) and Te(IV), and that on heating following the decay the yield of Te(IV) decreased. Halpern and Dancewicz (1969) subsequently published a more extensive study of this system.

Andersen et al. (1967) found that the ^{127m}Te and ^{129m}Te isomeric transitions in crystals of telluric acid, H_6TeO_6, both led to 34% to 36% of the daughter activity in a Te(IV) state. They studied the thermal annealing of the ^{127}Te labelled Te(IV) fragment in the crystals in detail, employing annealing times short in comparison with the half-life of the daughter isotope, thus avoiding any complications from the equilibrium nature of the decay. A quantitative fit to the annealing data was obtained, assuming a simple kinetic model. It was proposed that electrons and positive holes which had originated in autoradiolysis and were trapped close to the ^{129}Te atoms at lattice defects or radiolytic decomposition products, are liberated from their traps on heating, and may then (i) be retrapped, (ii) interact with ^{127}Te atoms producing an electronic oxidation to Te(VI) or (iii) recombine with holes or electrons in the bulk of the material. Consistent with this proposal, it was found that defects introduced into the crystal by crushing or electron irradiation had no effect on subsequent thermal annealing reactions. Jones and Warren (1968) have also observed that the annealing reactions of ^{129}Te atoms in $H_6{}^{129m}TeO_6$ are independent of the previous thermal history of the sample, and concluded that the annealing reactions must involve only defects populated or generated in the decay event itself. The work of Bulbulian and Maddock (1971) is in good agreement with these observations.

While discussing studies of the isomeric transition process in tellurium compounds, it is also interesting to note the work carried out on the $^{132}Te \xrightarrow{\beta^-} {}^{132}I$ decay, for which the 53 keV transition in ^{132}I has a large internal conversion coefficient. It has been shown by Jones and Warren (1970b) that in telluric acid the effects of this decay have some characteristics similar to those of the ^{129m}Te decay, evidence of the importance of the Auger charging process in this system. Similarly, studies by Llabador and Adloff (1967a, b, Llabador 1970) in $^{132}TeCl_4$, $^{132}Te(phenyl)_2$ and in ^{132}Te (butyl)$_2$ have also shown the importance of autoradiolysis accompanying the decay in determining the product distributions.

4.3. XENON

Studies of the $^{131m}Xe \rightarrow {}^{131}Xe \rightarrow$ isomeric transition which has a K-shell internal conversion coefficient of 4, are of some interest, since they present the possibility of observing the formation of novel xenon compounds, although their identification by radiochemical techniques may pose considerable problems. Margraff and Adloff (1966) observed that in $^{133m}XeF_4$(solid) no free xenon was liberated following the decay, a rather surprising finding which may be interpreted as evidence that no molecular fragmentation occurs in the decay. They also found that in the $^{125}Xe \xrightarrow{\text{E.C.}} {}^{125}I$ decay in $^{125}XeF_4$ solid, the daughter ^{125}I was evidently in the form of $^{125}IF_5$. The electron capture decay is again accompanied by Auger charging, and while some rearrangement must occur to stabilise the daughter atom as IF_5, it is again apparent that extensive fragmentation does not occur. In contrast, in solutions of $^{133m}XeO_3$ Nefedev et al. (1966b) and Heitz and Cassou (1969, 1970) observed an 85% rupture, in good agreement with the known internal conversion coefficient.

5. Isomeric Transition Following the (n,γ) Reaction

It has been recognised for some time that in the γ-cascade following neutron capture, delayed or metastable states of the daughter may be populated and that de-excitation of these states may proceed by internal conversion. In this case the Auger cascade must be expected to have some influence on the chemical form in which the recoil atom is finally stabilised in the solid. This will particularly be the case if the internal conversion occurs in the de-excitation of a state with a half-life longer than 10^{-11} sec, since the recoil atom produced in the (n,γ) reaction will lose its kinetic energy in less than this time and thus the Auger charging will occur in an atom chemically in the form of a recoil product produced in the (n,γ) reaction.

It is possible to study the importance of internal conversion following an (n,γ) reaction by using radiochemical techniques, as will be described below. In these cases, the metastable states populated in the (n,γ) reactions have half-lives of minutes or hours, and rapid separation techniques allow the study of both the metastable isomer and its isomeric transition daughter. Studies involving short-lived states in the (n,γ) cascade with half-lives of nano to microseconds can only be made using special techniques.

5.1. COBALT

The neutron irradiation of cobalt produces both ^{60m}Co and ^{60}Co directly by the (n,γ) reaction on ^{59}Co. The thermal neutron capture cross sections for the production of these two isotopes are 100 b for ^{60m}Co and 6 b for ^{60}Co. The half-lives are 10.5 min for ^{60m}Co and 5.2 y for ^{60}Co. The K-shell internal conversion coefficient as measured by Schmidt-Ott (1963) is 41.

Lazzarini (1967a, b, Lazzarini and Fantola-Lazzarini 1967b, c, 1969) has studied the retention in the parent chemical form for both ^{60m}Co and ^{60}Co produced by neutron irradiation in a very large number of cobalt (III) complexes. The ratio of the retentions for the two isotopes, $\rho = R(^{60m}\text{Co})/R(^{60}\text{Co})$, was found to vary between ca. 0.9 and 1.5 depending on the target. Following a series of very detailed experiments, Lazzarini and Fantola-Lazzarini (1967b, c, 1969) arrived at the following conclusions. (1) The ratio, ρ, is strongly dependent on the nature of the inner co-ordination sphere of the complex and in particular on the splitting of the Co(III) 3d orbitals. The ratio ρ was found to increase with increasing ligand-field splitting, and this effect was found to be an additive property of the ligands. (2) The outer co-ordination sphere of the complex, and the presence of cis-trans isomerism, have only a second order influence on ρ.

This isomeric effect was explained in the following way. The ^{60m}Co and ^{60}Co recoil atoms produced directly in the (n,γ) reactions will probably acquire the same kinetic energies and will then undergo identical reactions on slowing down in the crystal, and will exhibit the same thermal annealing characteristics. However, during the irradiation, and prior to analysis, some of the ^{60m}Co atoms will decay to ^{60}Co, the internal conversion coefficient for this transition being 41. In this decay, bond rupture may occur and the daughter ^{60}Co atoms may be formed as free interstitial atoms in the lattice. This would then give rise to the observed isomeric effect. The dependence of ρ on the inner co-ordination sphere of the complex, and the electronic environment of the Co(III) atom, may be explained by differences in the positive charge acquired by the cobalt atom following the isomeric transition decay in the different complexes. The X-ray fluorescence yields (the fraction of events which occur by X-ray emission, rather than by Auger processes) for the electronic transitions triggered by the internal conversion event may differ in the various complexes, because of the perturbations in electron energy levels arising from the ligand-field effects. The resulting different charges would then lead to differences in the production of free interstitial ^{60}Co atoms through bond rupture.

Lazzarini and Fantola-Lazzarini (1967c, 1969) have also studied the dependence of ρ on the energy of the neutrons used in producing the two isotopes, and also on the time and temperature of irradiation and of storage prior to analysis. The results of these experiments were very complex and suffice it to say that radiation-induced annealing and thermal annealing both play an important part in determining ρ for any one compound. These factors tend to mask the dependence of ρ on such crystal parameters as d-orbital splitting

5.2. BROMINE

The neutron irradiation of ^{81}Br produces ^{82}Br through the transformation sequence ^{81}Br(n,γ)^{82m}Br $\xrightarrow{6.1\text{ min}}$ ^{82}Br $\xrightarrow{35\text{ h}}$ in 90% of events. The ^{82m}Br → ^{82}Br isomeric transition is highly internally converted. Several studies have been made in organic liquids of the ^{82m}Br → ^{82}Br isomeric transition although no analogous studies have been made in inorganic systems. Many experiments have been reported in inorganic solids in which comparative studies have been made of the product distributions of ^{80m}Br, formed directly in the (n,γ) reaction on ^{79}Br, and ^{82}Br which is formed through the ^{82m}Br isomer. [See Campbell and Jones (1968a, b) for a review of the references.] Prior to the discovery Emery (1965) of 6.1 min ^{82m}Br, it was assumed that ^{82}Br was produced directly in the (n,γ) reaction on ^{81}Br. In general, similar product distributions are observed for the two isotopes ^{80m}Br and ^{82}Br, and they are also found to undergo quite similar annealing reactions. In terms of Lazzarini's nomenclature, the isomer ratio $\rho = R(^{80m}\text{Br})/R(^{82}\text{Br})$, is found to be about 0.9 for the initial retention, the retention of ^{80m}Br as bromate being 12–20% depending on the irradiation conditions and the particular bromate studied. An exception is $LiBrO_3$, where $\rho \approx 1$. The extensive ionisation and radiolysis of the crystal produced by the neutron-induced fission of ^{6}Li appears to mask the isotope effect. This shows that the isotope effect does not derive exclusively from the localised physical effects of the ^{82m}Br → ^{82}Br isomeric transition, but must also involve annealing effects which can be influenced by the production of free electrons and positive holes in the solids. The fact that $\rho = 0.9$, i.e. $R(^{82}\text{Br}) > R(^{80m}\text{Br})$, also appears to imply that annealing effects must be present, since if the isomeric effect derived simply from rupture in the $^{82m}BrO_3^-$ product through Auger charging, then $R(^{82}\text{Br})$ should be less than $R(^{80m}\text{Br})$.

A more direct study of the effects of an isomeric transition following an (n,γ) reaction may be made by studying ^{80m}Br and its daughter ^{80}Br: ^{79}Br(n,γ)^{80m}Br → ^{80}Br →. Here it is also possible to study the effects of the isomeric transition occurring in situ in an (n,2n) recoil site: ^{81}Br(n,2n)^{80m}Br → ^{80}Br →. Moreover the transient equilibrium in the ^{80m}Br → ^{80}Br → decay allows a far more detailed study of the isomer effect than is possible using the isomers of ^{82}Br. Thus it is possible to study the effect of the isomeric transition in a thermally annealed (n,γ) or (n,2n) recoil site, as well as investigating the thermal annealing of the ^{80}Br isomeric transition daughter atoms in these different lattice environments. Experiments of this kind have been reported by Jones (1970b) and the results may be briefly reviewed as follows. The isomeric transition occurring in an ^{80m}Br (n,γ) recoil site produces an ^{80m}Br/^{80}Br isomer effect on the initial retention and on subsequent thermal annealing which is analogous to that previously reported for ^{80m}Br/^{82}Br. The effects of the isomeric transition occurring in a thermally annealed (n,γ) recoil site are the same as those observed in an unannealed site. On the other hand, essentially no isomeric effect is observed for ^{80m}Br and its ^{80}Br daughter in an (n,2n) recoil site. A full explanation of the results of this work was not proposed and it is evident that further detailed experiments will be necessary in order to unravel this complex picture. It was pointed out that thermal annealing reactions must be present even at room temperature in all of the different recoil environments investigated. In this sense the results of this work are similar to those of Lazzarini et al. in that the presence of annealing reactions tend to confuse and complicate the picture.

5.3. SHORT-LIVED STATES IN THE (n,γ) CASCADE

In the early 1950's Wexler and Davies (1952) and Yosim and Davies (1952) demonstrated that I, Br, In, and Au (n,γ) recoil atoms acquire a positive charge in a certain fraction of events as a consequence of internal conversion in the γ-cascade. These experiments were performed by using charged electrodes to study the charge on atoms ejected by recoil in the (n,γ) reaction from the surfaces of irradiated foils or films. The experiments were subsequently extended by Thomson and Miller (1963) and by Tumosa and Ache (1970), and it has been shown that the positive charge acquired by In, Dy, and Mn (n,γ) recoils must originate from transitions involving states with half-lives in the range of 10^{-12} to 10^{-5} sec, since the charge is generated sometime after the recoil leaves the target surface and before it is collected at the catcher-electrode.

While these studies provide some evidence for the existence of delayed states and the number of atoms which are charged as a result of internal conversion, it would be desirable to measure directly the half-lives of these states, the number of events populating them, and the relevant internal conversion coefficients. Such measurements involve the study of the γ-cascade in the neutron capture reaction, and while the data for most (n,γ) reactions is still incomplete, there is direct evidence for delayed states in several cases of interest to recoil chemistry, as reviewed by Jones (1970a). In particular ^{32}P, ^{51}Cr, ^{56}Mn and ^{128}I formed in (n,γ) reactions are all produced in an appreciable fraction of events through excited states with half-lives of nanoseconds, while ^{65}Zn has a state with a half-life of 1.6 μsec. The internal conversion coefficients for the transitions from these states may be estimated from the deduced multipolarities, and for Mn, Zn and I these will be appreciable. Thus, in these instances at least, Auger charging will occur in a given fraction of events after the (n,γ) recoil atom has lost its energy in the crystal and come to rest. As more spectroscopic data becomes available for the prompt cascade in (n,γ) reactions, it must be anticipated that other delayed states of importance will come to light.

6. Summary

It is apparent that attempts to relate the product distribution following isomeric transition to lattice parameters such as crystal free-volume, to chemical bond strengths or to the oxidizing or reducing properties of ligands, have met with only limited success. It must be emphasized that such approaches may have to take into account more complex factors than have previously been considered. While molecular decomposition observed following Auger charging need not parallel in every way that following radiolytic decomposition, it is possible that similar factors will be involved in both cases, and Johnson (1970) has concluded that radiolytic decomposition of molecular ions is dependent upon the so-called 'total crystal environment'. The 'total crystal environment' is of course a concept and not a parameter: as such it is probably dependent upon a number of crystal parameters and interactions at best poorly known.

The thermal annealing studies of isomeric transition recoil products have demonstrated the importance of trapped electrons and positive holes generated in autoradiolysis. However, further studies should be made to correlate the thermal annealing of recoil fragments with the annealing of radiolytic decomposition products in these systems, since, as pointed out by Andersen, the two may be intimately related.

The possible occurrence of internal conversion following neutron capture remains a question of some importance, which will hopefully be answered for specific nuclei as more nuclear spectroscopic data becomes available. Further radiochemical studies of suitably long-lived isomeric pairs may help unravel the chemical effects of Auger charging occurring in situ in (n,γ) and (n,2n) recoil sites. However, the presence of annealing reactions in the solid obviously complicates the interpretation of the data considerably.

The study of the isomeric transition process in labelled molecules, or in situ in recoil sites, may be carried out directly in the solid state for specific nuclei by Mössbauer spectroscopy, and experiments conducted at liquid nitrogen or liquid helium temperature may then minimize (or conversely, reveal) the effects of thermal annealing. Such studies will hopefully lead to a better understanding of the mode of decomposition accompanying isomeric transition in the solid state, the nature of the primary products formed, and the mechanisms of annealing reactions.

Addendum (December 1976)

Relatively few papers have appeared in recent years on the chemical effects associated with isomeric transition in inorganic solids. Of interest to the present discussion, electronic excitation and ionisation as a result of internal conversion of long-lived states in the (n,γ) cascade, rather than concomitant γ-irradiation, has been demonstrated to be an important factor in determining the chemical distribution of the products in liquid ethyl iodide; De Halter and Cruset (1974). Further studies on the behaviour of ^{60}Co (n,γ) recoil atoms in solid cobalt complexes have been reported, although the emphasis has shifted from a study of the ^{60m}Co, ^{60}Co isotope effect to a study of the annealing of ^{60}Co recoils in both cationic and anionic sites in cis-[Cr en_2Cl_2][Co $glyc_2(NO_3)_2$] nH_2O for example, and the solid state isotope exchange between cationic and anionic sites in $[Co(H_2O)_6][CoEDTA]_2 4H_2O$. Such studies have implications for the interpretation of the ^{60m}Co, ^{60}Co isotope effects previously reported; Lazzarini and Fantola-Lazzarini (1974a, b, 1975). Isotope effects for ^{80}Br, ^{80m}Br and ^{82}Br have been reported for the n,γ reaction in alkali metal and ammonium perbromates and reference is made to the importance of the $^{82m}Br \rightarrow {}^{82}Br$ isomeric transition in this system; Hassan and Heitz (1975). A number of investigations have been reported in which the $^{129m}Te \rightarrow {}^{129}Te$ isomeric transition has been studied using the Mössbauer effect as a solid state probe; Lebedev et al. (1970b, 1971b), Jones et al. (Jones and Warren 1970a, Warren and Jones 1971, Warren et al. 1971, Johnstone et al. 1972).

Acknowledgement

A grant from the National Research Council of Canada is gratefully acknowledged by the author.

CHAPTER 16

CHEMICAL EFFECTS OF β-DECAY IN INORGANIC SOLID SYSTEMS

A. HALPERN

Institut für Chemie der Kernforschungsanlage Jülich, GmbH, Jülich, F.R. Germany

Chemical Effects of Nuclear Transformations in Inorganic Systems
Edited by G. Harbottle and A.G. Maddock

Contents

1. Introduction

In this chapter we shall deal with the chemical consequences of β-decay and electron capture (excluding those in the naturally radioactive heavy elements, which are discussed in ch. 21) in inorganic and metal–organic solid compounds. A few experiments dealing with gaseous and liquid systems will also be mentioned.

2. Basic Background

The three types of nuclear decay which we will consider (β^-, β^+ and electron capture) may be schematically represented as follows:

$$n \rightarrow p + e^- + \nu^* + Q, \tag{1a}$$
$$p \rightarrow n + e^+ + \nu + Q, \tag{1b}$$
$$p + e^- \rightarrow n + \nu_\nu + Q, \tag{1c}$$

where n and p are the neutron or proton present in a decaying nucleus, and e^-, e^+, ν and ν^* symbolize the emitted electron, positron, neutrino and antineutrino, respectively. Q is the energy of the nuclear decay. According to the law of conservation of momentum the daughter nucleus in which n has changed to p or vice versa recoils with a momentum counter-balancing the vector sum of the electron and neutrino momenta:

$$\boldsymbol{P}_R + \boldsymbol{P}_\beta + \boldsymbol{P}_\nu = 0, \tag{2}$$

where $\boldsymbol{P}_R$, $\boldsymbol{P}_\beta$ and $\boldsymbol{P}_\nu$ are the momenta of the recoil atom, β-particle and neutrino, respectively. The maximum recoil energy is given by

$$E_R = \frac{E_\beta}{2mc^2}[E_\beta 2m_e c^2]^{1/2}, \tag{3a}$$

more conveniently

$$E_R = \frac{536}{M} E_\beta \; [E_\beta + 1.02], \tag{3b}$$

where E_β and E_R are maximum kinetic energies of the β-particle and recoil atom in MeV and eV, respectively and M is the mass of the recoil atom in amu.

The maximum recoil energy after K-capture is similarly:

$$E_R = 536 \frac{E_\nu^2}{M}, \tag{3c}$$

where E_ν is the energy of the neutrino in MeV.

Calculation of the recoil energy spectrum after β-decay requires the knowledge of the electron-neutrino angular correlation. Unfortunately, this is often lacking. Edwards and Davies (1948) have derived equations for three particular cases (θ is the angle between the electron and neutrino emission). (1) Emission in the same direction (θ = 0°):

$$E_R = \frac{536}{M} [(E_\beta{}^2 + 1.02\, E_\beta)^{1/2} + E_\nu]^2 \text{ eV}. \tag{3d}$$

(2) Emission in opposite direction (θ = 180°):

$$E_R = \frac{536}{M} [(E_\beta{}^2 + 1.02\, E_\beta)^{1/2} - E_\nu]^2 \text{ eV}. \tag{3e}$$

(3) Isotropic emission:

$$E_R = \frac{536}{M} [E_\beta^2 + 1.02\, E_\beta + E_\nu^2 + 2E_\nu (E_\beta^2 + 1.02\, E_\beta)^{1/2} \cos\theta] \text{ eV}. \tag{3f}$$

Figure 16.1 shows the general features of the recoil spectra for these three cases. Thus, the highest kinetic energy is transferred to the daughter atom when the directions of the β-particle and neutrino are the same.

All the above relations apply to pure β-decays. When β-decay is followed by γ-emission the actual recoil spectrum of the daughter atom also depends on the β–γ correlation. But the mean recoil energy is independent of the angular correlation; it is governed by the following equation:

$$E_{R(av)} = (p^2{}_{R(av)} + \sum p^2{}_{casc})/2M,$$

where $p_{R(av)}$ is the average momentum of the β-recoil nucleus, and p_{casc} are momenta of γ-quanta emitted in cascade following β-decay.

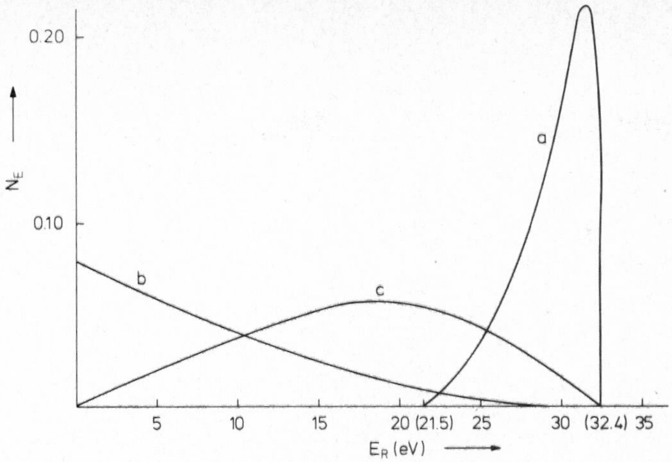

Fig. 16.1. Recoil spectra of various cases of β–ν angular correlation; $M = 100$ amu, $E_{max} = 2.0$ MeV. a: $\theta = 0°$; b: $\theta = 180°$; c: isotropic (approximated from peak values of spectra at eight angles, weighted by sin θ) (from Edwards and Davies 1948).

It should be emphasized that eqs. (3a–f) are true for free atoms only. If the decaying atom is bonded chemically only a fraction of the recoil energy is available for bond rupture. The available energy is $E_R \cdot R/(R + m_N)$, where m_N and R are atomic weights of the recoiling atom and the rest of the molecule, respectively, as was shown by Steinwedel and Jensen (1947) for diatomic molecules. In the decay of tritium, a pure β^--emitter, the contribution of the antineutrino to the recoil of the daughter ^3He ion can be neglected due to the low value of E_{max} (18.6 keV), and the calculated maximum recoil energy is 3.6 eV. But more than 90% of the daughter ions receive a recoil energy below 0.18 eV (Cacace 1970). Using the above relationship it can easily be shown that in the decay of T_2 or TH the maximum energy available for bond rupture is 1.8 eV and 0.9 eV. The problem was discussed more recently by Hsiung and Gordus (1962, 1965) who employed classical mechanics. Both internal and external excitation will occur when a recoil energy E_R is imparted to an atom in a molecule. In table 16.1 (Hsiung and Gordus 1965) equations are given for these various forms of energy, where m_N is the atomic weight of the atom that is receiving the momentum impulse, M is the molecular weight, and F is the fraction of the internal excitation energy deposited in the bond joining the atom to the remainder of the molecule.

On average, the recoil energy that must be acquired by the daughter atom in order to rupture from the molecule is about 25% greater than that calculated assuming a pseudodiatomic molecule. Summarizing, the kinetic energy of the daughter atom is often insufficient to cause the rupture of the

Table 6.1
Distribution of imparted energy E_R

Species	External energy	Internal excitation energy	
		Bond joining activated atom	Other bonds
Single atom	E_R	0	0
Diatomic molecule	$\frac{m_N}{M} E_R$	$\frac{M - m_N}{M} E_R$	0
Polyatomic molecule	$\frac{m_N}{M} E_R$	$F \frac{M - m_N}{M} E_R$	$(1-F) \frac{M - m_N}{M} E_R$

chemical bond if the decaying nucleus is of medium or high mass and if the emission of strongly converted photons does not accompany the β-decay.

The other sources of chemical excitation following β-decay have been discussed in ch. 2 and shall be omitted here. Papers by Wexler (1965b, 1967) and Baulch and Duncan (1958a, b) may also be consulted, and for tritium the review by Cacace (1970). Considering all the factors which determine the bond rupture and formation processes after β-decay, i.e. (a) the change of chemical identity, (b) mechanical recoil, (c) electronic, vibrational and rotational excitation and (d) the positive charge imparted to the daughter atom, it is reasonable to expect a rather high probability of secondary molecular consequences. This has been elegantly confirmed for the isolated gaseous molecules HT, T_2 and $^{14}CO_2$ (as well as for a number of organic molecules in the gas phase) in classical experiments carried out independently by Snell and Pleasonton (1958) and Wexler (1959), who analysed the positively charged fragments resulting from β-decay using the technique of charge mass spectrometry (details are to be found in ch. 2). On the other hand, all the literature suggests that even in condensed phases the nonbreaking of bonds is negligible, or at least smaller than could be expected from the comparison of the recoil energy and the bond energy. There is only one exception reported, the β-decay of ^{144}Ce-acetylacetonate (Edwards and Coryell 1948), where no evidence of bond rupture was obtained either in the solid state or in CCl_4 or CS_2 solutions. This result is difficult to interpret on the basis of decay scheme data. It would be highly desirable to repeat this work, which is 30 years old, using modern radiochemical techniques. More recently studies have been reported by Glentworth et al. (Glentworth and Betts 1961, Glentworth and Wiseall 1965, Glentworth and Wright 1969), on the β-decay of ^{143}Ce and ^{144}Ce in various aminopolycarboxylate complex systems, which demonstrated a rather significant proportion of bond rupture.

Special caution is required in interpreting the chemical effects of nuclear decay in the solid state. In the previous section on isomeric transition Jones has considered some problems specific for the solid state. Much of this is also valid for β-decay. However, a great difference between these two modes of radioactive decay is that in β-decay daughter atoms are produced which are non-isotopic with the parent. Daughter atoms in the parent solid should thus be considered as a very dilute impurity in the nonisotopic matrix, i.e., these atoms do not enjoy the 'protective' effect of *isotopic*, non-radioactive atoms normally formed by radiolytic processes in the lattice. Yields in such systems have been found to be extremely sensitive to even very small numbers of lattice defects formed during the preparation and purification of the parent crystalline compound. Consequently, the experimenter is often faced with a troublesome scatter of results despite his effort to maintain the experimental conditions unchanged. For example, the annealing pattern of ^{77}As was found to vary significantly for $^{77}GeO_2$ samples from different syntheses (Halpern and Sawlewicz 1968). Evidence was also obtained that in ^{131}Te-labelled telluric acid the oxidative annealing of ^{131}I was promoted by crystal defects inherently present in the parent lattice (Jones and Warren 1968). This 'sensitivity to structure', as well as possible chemical changes during the dissolution of solid crystals prior to analysis, were not always taken into account in older work. The desire to investigate the state of a daughter atom immediately after β-decay has lead to the application of new non-destructive (in situ) experimental techniques. In particular, measurements of perturbed angular correlation in ^{188}Re-containing potassium perrhenate (the decay $^{188}\mathrm{Re} \xrightarrow{\beta^-} {}^{188}\mathrm{Os}$ is followed by coincident gamma transitions) demonstrated that a considerable fraction of the ^{188}Re atoms which had been generated by ^{187}Re (n,γ) were not present as perrhenate ions in the original solid before the latter was treated by wet chemistry (Sato et al. 1966), but in fact seemed to reform perrhenate when the sample was dissolved. Another non-destructive technique which has found wide application is Mössbauer spectroscopy (see ch. 24).

3. Experimental Methods

A typical experiment involves three steps: (1) labelling of the parent crystalline compound, (2) storage under controlled conditions to allow decay of the β-emitter, and (3) separation and determination of different chemical forms of the daughter atoms. In early experiments the labelling was usually done by irradiation of the parent compound with neutrons in the reactor; thus the isotope ${}^{A}_{Z}\mathrm{X}\,(n,\gamma) \rightarrow {}^{A+1}_{Z}\mathrm{X}$ which then underwent β-decay to the daughter ${}^{A+1}_{Z+1}\mathrm{Y}$. Currently the synthesis of a labelled compound from the pure radioisotope ${}^{A+1}_{Z}\mathrm{X}$ is preferred, thus avoiding radiation damage caused by ionizing radiation which accompanies the neutron flux.

Investigation of β-decay induced reactions may be based upon (a) the investigation of the intermediate species, or (b) analysis of the end-products. Charge mass spectrometry, which has contributed so much to our understanding of these processes in the gas phase (see ch. 2), perturbed angular correlation (Sato et al. 1966, Bădică et al. 1971) and Mössbauer spectroscopy illustrate the first of these methods. The latter technique provides, for several specific nuclei, a direct way to observe transient molecular fragments formed in a solid by β-decay in a time interval of 10^{-7}–10^{-9} sec after the nuclear event. An extensive discussion of Mössbauer studies on the chemical consequences of nuclear decay is given in ch. 24, and also by Bondarevskii et al. (1971) and Wertheim (1971). The advantage of the above three methods is that wet chemistry is avoided and thus the possible reactions of intermediate products during and after dissolution of the parent compound are eliminated. This is unfortunately not the case with the radiochemical technique, which was commonly and necessarily used in most studies in the past. In this technique, after dissolution of solid, the radioactive products which incorporate daughter atoms are separated by a selective chemical method such as paper electrophoresis, solvent extraction, or ion exchange, with or without carriers. One has to be cautious in interpreting the results of such experiments, since many factors, which often elude one's control, may obscure the actual product distribution. The following review of the work on the forms of ^{77}As found in $^{77}GeO_2$ after β-decay well illustrates this point.

4. Typical Experimental Results for Selected Systems

In order to give the reader some concrete examples of investigations of the chemical forms of daughter atoms and their fate during annealing, we now focus our attention on the following nuclear decay processes:

$$^{77}\mathrm{Ge} \xrightarrow{\beta^-} {}^{77}\mathrm{As}$$

$$^{68}\mathrm{Ge} \xrightarrow{\mathrm{E.C.}} {}^{68}\mathrm{Ga}$$

$$^{125}\mathrm{Sn} \xrightarrow{\beta^-} {}^{125}\mathrm{Sb}$$

$$^{125}\mathrm{Sb} \xrightarrow{\beta^-} {}^{125m}\mathrm{Te}$$

$$^{131,132}\mathrm{Te} \xrightarrow{\beta^-} {}^{131,132}\mathrm{I}$$

$$^{125}\mathrm{Xe} \xrightarrow{\mathrm{E.C.}} {}^{125}\mathrm{I}$$

$$^{143}\mathrm{La} \xrightarrow{\beta^-} {}^{143}\mathrm{Ce} \text{ and other rare earth elements,}$$

$$^{129,131,133,135}\mathrm{I} \xrightarrow{\beta^-} {}^{129m,131m,133,135}\mathrm{Xe}.$$

Both simple inorganic compounds and metal–organic complexes have been investigated. Beta decay should be formally an oxidizing process (i.e. $^{143}La^{3+} \xrightarrow{\beta^-} {}^{143}Ce^{4+}$), but in fact there are relatively few examples of simple oxidation traceable in the daughter. Chemically speaking, a positive charge, acquired by daughter atoms in β-decay, should imply a high degree of reactivity.

4.1. GERMANIUM–ARSENIC SYSTEM

^{77}Ge decays to ^{77}As with a half-life of 11.3 h. The three β transitions have maximum energies of 710 keV (23%), 1379 keV (35%) and 2196 keV (42%). From this the calculated maximum recoil energies are 1.7, 4.5 and 10.2 eV, respectively.

4.1.1. Germanium dioxide

This system was first studied by Baró and Aten (1961), who reported the distribution of ^{77}As between the arsenite and arsenate fractions (69.2% and 28.8%, respectively, at room temperature) in neutron-irradiated GeO_2. At temperatures appreciably higher even than those which usually cause a strong annealing in neutron-irradiated solids, they found only slow annealing. This first work was followed by studies carried out independently by two groups (Halpern and Sawlewicz 1968, Genet 1969, Genet and Ferradini 1969a, b). In two of these, experimental conditions differed from those of Baró and Aten in that crystals of GeO_2 were synthesized after the reactor irradiation (to produce ^{77}Ge) was completed, i.e. practically all the ^{77}As was that of the second generation. The valence distribution of ^{77}As was found to depend critically on the concentration of the arsenite carrier in the solvent (KOH solution) prior to dissolution of the germanium compound (Genet and Ferradini 1964, 1969a, Halpern and Sawlewicz 1968).

Figure 16.2 gives some results for $^{77}GeO_2$ and $^{77}GeI_4$, illustrating typical behavior. The daughter ^{77}As is produced in the matrix of the ^{77}Ge-labelled compound as a very dilute impurity and is therefore sensitive to different oxidizing agents present in the solvent in macro-amounts compared with the carrier-free arsenic. Arsenite carrier thus protects the daughter ^{77}As from this kind of incidental oxidation; for both germanium dioxide and tetraiodide a concentration of arsenite carrier of about 3×10^{-2}M is required to eliminate this side-effect. In reactor-irradiated GeO_2 Genet and Ferradini assume that Ge^{2+} had also been formed during the irradiation and, on dissolving the oxide, the Ge^{2+} reacted with oxygen in the solution to form H_2O_2. The oxidation of As(III) by H_2O_2, a reaction dependent on pH, may also account for the effect of carrier.

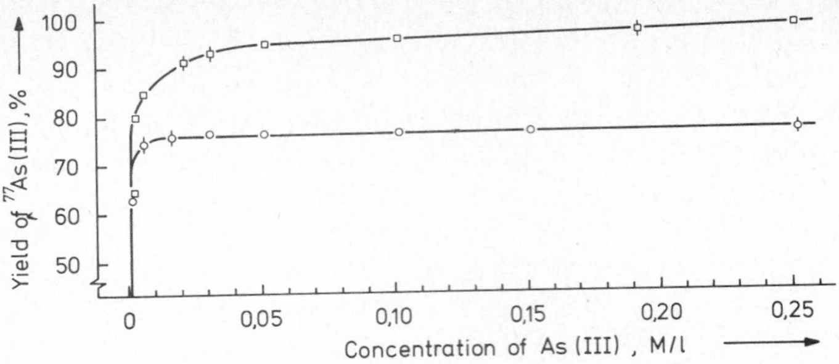

Fig. 16.2. Yield of trivalent ^{77}As from β-decay of ^{77}Ge in GeO_2 (o) and GeI_4 (□) versus the concentration of the arsenite carrier (adopted from Halpern and Sawlewicz 1968).

Divergent results concerning both initial valence distribution and thermal annealing in synthesized $^{77}GeO_2$ were reported by Halpern and Sawlewicz (1968) and Genet and Ferradini (1969b). In particular the initial valence distribution in untreated $^{77}GeO_2$, stored in liquid nitrogen, was found by Genet and Ferradini to be 40% As(III) and 60% As(V), as compared with 76% As(III) and 24% As(V) reported by Halpern and Sawlewicz. This disagreement illustrates another special problem associated with such studies; the fate of the daughter atom formed in β-decay in a crystal lattice seems to depend strongly on the crystal preparation method, i.e. on the concentration of defects of chemical origin. If the concentration of crystal defects was further increased (for example, by doping $^{77}GeI_4$ crystals with Al^{3+} prior to β-decay of ^{77}Ge, or by pre-decay powdering or quenching of $^{77}GeO_2$), an increase in the pentavalent arsenic fraction was observed. It is interesting that ^{77}As(III) appears at all, inasmuch as the radioarsenic should be expected to be initially pentavalent. However, a high yield of trivalent ^{72}As from electron capture in $H_2{}^{77}SeO_3$ was also observed (Halpern 1959), even though this nuclear process should produce even more highly initially oxidized radioarsenic.

Thermal annealing of ^{77}As daughter atoms was also investigated by both groups, although only qualitatively. Genet and Ferradini found that when ^{77}Ge β-decay takes place at room temperature, no thermal annealing is observable. On the other hand, when labelled $^{77}GeO_2$ was heated during the entire period of β-decay, oxidation of ^{77}As(III) was observed: at 450° C a 6% As(III) and 94% As(V) distribution was obtained. Irradiation of the $^{77}GeO_2$ with gamma rays reduces the As(V) fraction to 90%. The results were interpreted by assuming the formation of Ge^{2+} in GeO_2. It was shown that 50% of the radioarsenic can be oxidized by thermal annealing and reduced by γ irradiation. Genet (1969) has also correlated the oxidation of radioarsenic on thermal annealing with macroscopic changes on the $^{77}GeO_2$ caused by activation and by different methods of treatment of the solid phase, using thermogravimetry, gas analysis, EPR and determination of H_2O_2 to establish the correlation.

4.1.2. Germano-organic compounds

Riedel and Merz (1965, Merz and Riedel 1965) studied the β-decay of ^{77}Ge-labelled germanium tetraphenyl and compared the radioarsenic daughter products with those produced by neutron irradiation of $AsPh_3$ and $GePh_4$. These products were unfortunately characterized only by their elution properties from alumina with various solvents. The authors observed that β-decay apparently produces practically the same chemical products observed in (n,γ) reactions. However, the remarkably increased yield of the labelled parent is typical for β transitions, indicating a high percentage of non-bond-rupture in addition to the formation of the parent form by the secondary recombination of molecular fragments. The yield of $^{77}AsPh_3$ was found to be about 18.4% in the solid state and 21.9% in benzene solution. The authors estimated that the 'true retention' arising from non-bond-rupture could be as high as 14%, so that the fraction of $^{77}AsPh_3$ yield resulting from the recombination of molecular fragments was about 5%, approximately the same as in the (n,γ) reaction on $AsPh_3$. In benzene solution, the authors observed a 3% increase in this fraction, due to a higher phenyl radical concentration and faster diffusion, in good agreement with the results of Halpern et al. (1964), who found a 2.8% yield of $^{77}AsPh_3$ after the decay of $^{77}GeCl_4$ in benzene. Merz (1966) also studied the β-decay of ^{77}Ge in several phthalocyanines.

4.2. GERMANIUM–GALLIUM SYSTEM

The chemical effects of $^{68}\text{Ge} \xrightarrow[275\,\text{d}]{\text{E.C.}} {}^{68}\text{Ga}$ decay were investigated by Merz and Riedel (1965, 1964) in ^{68}Ge-tetraphenyl. The calculated recoil energy of the ^{68}Ga atom is 3.9 eV, just about equalling the Ge–Ph bond energy. Bond rupture could also result from the large positive charge caused by the Auger cascade and electronic readjustment. The low observed retention (2–3%) represents a failure of bond rupture rather than a secondary recombination process. Whatever the mechanism, the fragments formed in the decay of $^{68}GePh_4$ apparently do not posses high kinetic energy, and the authors have hypothesized that no hot reactions take place after electron capture.

4.3. TIN–ANTIMONY SYSTEM

The chemical effects of $^{125}\text{Sn} \xrightarrow{\beta^-} {}^{125}\text{Sb}$ and of $^{121m}\text{Sn} \xrightarrow{\beta^-} {}^{121}\text{Sb}$ decay have also been investigated (Andersen et al. 1966, Facetti 1965, Lebedev et al. 1970). Andersen and Knutsen reported (1961) that a significant fraction of the radioantimony is formed in the pentavalent state in potassium and ammonium hexachlorostannate, this fraction being only slightly affected by isothermal annealing at 100° C. In a later, more extensive study (Andersen et al. 1966) it was shown that the Sb valence distribution in $K_2{}^{125}SnCl_6$, like that of ^{77}As in

$^{77}GeO_2$, depends critically on the preparation and purification procedure. The yield of Sb(V) in crude crystals was only 10%, but in highly purified material went above 90%. This dramatic change is obviously related to the differing contents of impurity atoms and crystal defects introduced in synthesis. The radiation damage caused by the β-particles emitted by the labelling isotope may also lead to a higher concentration of trapped electron centres in the crude crystals than in the highly purified ones. If the number of defects was further increased in the crude crystals, either by doping with Ca^{2+} ions or by pre-irradiation with electrons, an increase in the pentavalent antimony was observed. The influence of crystal defects was also studied through annealing and the chemical changes of the recoil species compared with the electronic changes in the crystals during post-decay treatment, as revealed by their thermoluminiscence. The annealing pattern was also found to be dependent on the crystal preparation (fig. 16.3), and was markedly different from that usually found in thermal neutron capture studies. The customary rapid increase in the yield of a higher valence state with temperature was found to be combined with a series of reducing reactions. The authors have proposed a mechanism for the isothermal annealing reaction. The values of the energy of activation obtained from thermoluminiscence measurements and from Arrhenius plots of the rate constants of the reactions responsible for annealing were identical. The findings reported by Andersen thus show that there exists a close relationship between the electronic changes in K_2SnCl_6 and the chemical changes observed after post-decay heat treatment. The electronic centres act as precursors for both thermoluminiscence and chemical changes. The data seem to support the hypothesis that the 'extrinsic' annealing reactions, i.e. those involving interaction between crystal defects and the daughter atoms, reflect the

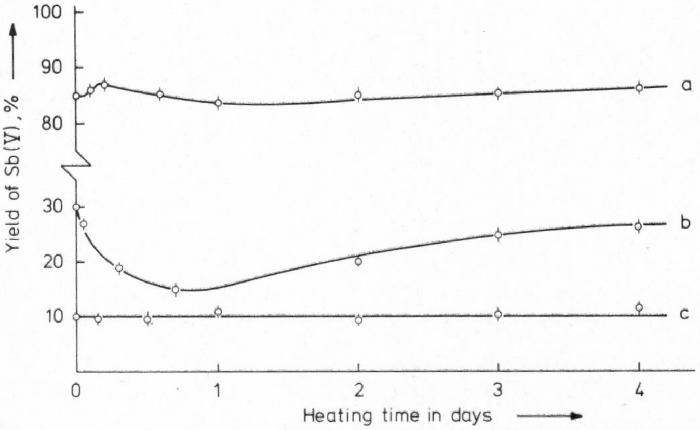

Fig. 16.3. Isothermal annealing curves for β-decayed $K_2{}^{125}SnCl_6$ at 108° C. a: recrystallized; b: grown slowly from solution; c: crude crystals (from Andersen et al. 1966).

liberation of charge carriers, electrons or holes, from centres produced by ionizing β or γ radiation or from other types of defect centres.

Lebedev et al. (1970) studied the $^{121m}Sn \xrightarrow{\beta^-} {}^{121}Sb$ decay in $(NH_4)_2$ $SnCl_6$, SnO_2, $K_2Sn(OH)_6$, $Sn(OH)_2$ and SnC_2O_4 using Mössbauer spectroscopy. Their evidence suggests that the daughter ^{121}Sb atoms appear in the matrices in a certain unusual, unstable state, which can undergo different transformations during post-decay chemical treatments. Recently Ambe and Ambe (1975) reported that the β^--decay of ^{121m}Sn in SnS_2 leads to preferential stabilization of ^{121}Sb in the isoelectronic Sb(V) state. After the electron capture decay of $^{119m}Sb(III)$ in SbS_3 the higher valence state $^{119}Sn(IV)$, rather than the isoelectronic $^{119}Sn(II)$, was formed predominantly (Ambe et al. 1974).

4.4. ANTIMONY–TELLURIUM SYSTEM

Nefedov and coworkers (1962, 1964b) studied the decay $^{125}Sb \xrightarrow{\beta^-} {}^{125m}Te$ in the solid phenyl- and tolyl-derivatives of tri- and pentavalent antimony. The half-life is 2.7 y; the eight β transitions have maximum energies of 96 keV (2.2%), 112 keV (0.8%), 124 keV (5.9%), 131 keV (30.5%), 240 keV (1.0%), 302 keV (40.4%), 433 keV (5.9%) and 619 keV (13.4%). Conversion coefficients of the accompanying gamma rays are low. Calculation of the maximum recoil energy shows that it is insufficient for bond rupture even in the case of emission of the β-particle and neutrino in the same direction (E_R is 4.3 eV for $E_{max} = 619$ keV, but only 1.7 eV and 0.6 eV for the most abundant transitions at 302 keV and 131 keV). Nevertheless, at least two ^{125}Te-labelled products are often observed. For example, in labelled $^{125}SbR_3$ up to 60% of the daughter ^{125m}Te was found as TeR_2 and up to 28% as TeR_3Cl, while 90% of the ^{125m}Te was found as TeR_2Cl_2 in $^{125}SbR_3Cl_2$ (R = phenyl or tolyl radical). The authors interpret these results in terms of the thermodynamic stability of the corresponding primary molecular ions formed after β-decay and their subsequent fragmentation. In the above cases these are $(^{125m}TeR_3)^+$ and $(^{125m}TeR_3Cl_2)^+$, respectively. The first is thermodynamically stable, since the tellurium atom has the noble gas (xenon) electronic structure, while the second is thermodynamically unstable. Two competitive pathways are open for the excited molecular ion $(^{125m}TeR_3)^+$, (1) deactivation, leading to the formation of $^{125m}TeR_3Cl_2$ (28%), and (2) the splitting off of one positively charged radical, leading to the 60% yield of $^{125}TeR_2$. The excitation of the molecular ion $(^{125m}TeR_3)^+$ is high enough to supply the activation energy required for splitting off a radical. In contrast, the decay of the thermodynamically unstable $(^{125m}TeR_3Cl_2)^+$ ion occurs entirely by the exothermic splitting off of a positively charged radical. Similar interpretations have been offered for other systems. In the decay sequence

$$^{125}SbPh_2Cl_3 \xrightarrow{\beta^-} [^{125m}TePh_2Cl_3]^+ \begin{cases} \rightarrow Ph^+ + {}^{125m}TePhCl_3 \text{ (Path A)} \\ \rightarrow Cl^+ + {}^{125m}TePh_2Cl_2 \text{ (Path B)} \end{cases}$$

the effect of added carriers on the observed yields was studied. It was found that, in the absence of carriers, the yields in Paths A and B were 35% and 56% respectively, while in the presence of carriers the corresponding yields were 20% and 72%.

In ^{125}Sb-labelled ammonium halo-antimonates(V) the daughter ^{125m}Te atoms appear predominantly in tetravalent species (Ambe and Saito 1971). The yield of ^{125}Te(VI) was 33% in hexafluoroantimonate, and only 3% in chloro- and 0.2% in bromo-derivatives, i.e. decreased with the electronegativity of the halogen.

4.5. TELLURIUM–IODINE SYSTEM

The $^{131}\text{Te} \xrightarrow{\beta^-} {}^{131}\text{I}$ and $^{132}\text{Te} \xrightarrow{\beta^-} {}^{132}\text{I}$ decays were very often studied in the early period of hot-atom chemistry. The β-particles emitted by ^{131}Te have the following maximum energies: 2150 keV (42.3%), 1680 keV (16.2%) and 1360 keV (41.4%), while for ^{132}Te, E_{max} are 280 keV (60%) and 100 keV (40%). The conversion of the gamma transitions from ^{131}Te is thought to be small, although according to some measurements one could reckon up to 20% conversion (Lederer et al. 1969). The gamma radiations from ^{132}Te are 83% converted.

4.5.1. TeO_2 and other oxygen-containing compounds

In 1961 Cummiskey et al. (1961) reported on the separation of labelled ^{132}I products resulting from the decay of ^{132}Te-tellurate in aqueous solutions, using the carrier method. The products observed were reduced forms of iodine, iodate and periodate (66–77%, 16–26% and 7–8%, respectively, depending on pH). Gordon (1967) again studied the same system, this time using high voltage electrophoresis without addition of carriers as the analytical method. Whereas Cummiskey et al. had observed a considerable amount of periodate, Gordon found this fraction to be quite small. Because of the low recoil energy following β-decay one would expect the daughter iodine to appear almost completely as isoelectronic IO_4^-; however, the Auger charging following the internal conversion of the 53 keV γ-photon (see ch. 2) should give rise to bond breaking and thus also to the formation of the reduced forms of radioiodine.

Bertet et al. (1964) studied ^{131}Te β-decay in crystals and in aqueous solutions of TeO_2 and telluric acid. In the solid state most of the investigations (Stevović et al. 1965, Kronrád 1964, Ortega 1966, Teofilovski 1966, Hashimoto et al. 1970) have been concerned with the distributions of valence states and the thermal annealing of ^{131}I formed by the β-decay products of the (n,γ) reaction. More recently Jones and Warren (1968, 1970b) published an extensive study on decay products in telluric acid labelled with ^{131m}Te, ^{131}Te and ^{132}Te. Thermal annealing was found to cause a redistribution of ^{131}I between I^- and IO_3^- forms. However, heating of the labelled crystals before decay blocks the thermal annealing entirely. It was concluded that the annealing reactions were here dependent on the inherent crystal defects present throughout the lattice, the preheating altering irreversibly the defect population and thus inhibiting the subsequent annealing reactions. An interesting feature of the Jones and Warren study was that the relatively 'soft' ^{131}Te decay was compared with the ^{132}Te decay, which, due to the high degree of internal conversion of the accompanying gamma rays, must give rise to a much more profound excitation and ionization of the radioiodine species. While, somewhat surprisingly, the initial product distribution was almost the same, the course of the thermal annealing reactions was different. ^{132}I was found to participate in two separate annealing reactions. The first, A, involved some bulk property of the lattice and gave rise to competitive reductive–oxidative annealing reactions. The second process, B, appeared to involve only fragments generated locally in the recoil zone during the decay process itself, and this annealing reaction was found to give rise to the reduction process. On heating the crystals for a substantial time before studying the annealing, reaction A was presumably taken irreversibly to completion. In this process crystal defects may be participating; on heating, the defect population of the crystal is irreversibly altered. On allowing the crystals to stand for 24 h the recoil products present during the initial heat treatment would have decayed away radioactively to be replaced by a new recoil product population with reduced forms of iodine amounting to 27% of the total. On heating the crystals for the second time, only reaction B was observed, leading now to a time-independent plateau value for ^{132}I in the reduced form of about 48%. By contrast, ^{131}I was found to be involved only in annealing reaction A, and on pre-heating labelled crystals little or no annealing was observed, indicating the absence of the reaction B.

Recently the chemical forms of the daughter ^{83}Br atoms from β^--decay of ^{83}Se in potassium selenate stored in aqueous solution or as a solid has been investigated (Tenorio et al. 1976). Like radioiodine in tellurates, ^{83}Br was found distributed among reduced forms, bromate and perbromate. In potassium selenite no BrO_4^- was found.

Interesting results concerning the chemical state of ^{129}I atoms in ^{129}Te-labelled TeO_2, H_6TeO_6, $(NH_4)_2H_4TeO_6$ and $Na_2H_4TeO_6$ were reported by Lebedev et al. (1971). By Mössbauer spectroscopy the authors could see only one chemical form of ^{129}I, which corresponded to a formal valence increase of one over that of the parent tellurium atom. This is in contrast to the work of Pasternak (1967), who found ^{129}I in $H_6{}^{129}TeO_6$ to exist in two valence forms. However, Pasternak had generated ^{129}Te in his Mössbauer sources by irradiating them in the reactor, so that it could have already existed in two forms due to the Szilard–Chalmers process. According to Lebedev et al., the daughter ^{129}I atom 'inherits' the valence shell of the parent ^{129}Te intact. The authors concluded that neither the recoil (maximum $E_R = 15$ eV) nor the ionization due to 'shake-off' or internal conversion influence the stabilization of ^{129}I in the solid state, the only factor determining the final chemical state being the inconsistency of the chemical nature of the daughter atom and the matrix. When the solid is dissolved to perform standard radiochemical procedures, various side reactions obscure the picture. In this context it is interesting to note that Mössbauer and radiochemical studies have provided different information on the effects of the β^--decay of ^{129}Te or ^{131m}Te in telluric acid (Warren et al. 1971, Lebedev et al. 1969, 1971b, 1972). The Mössbauer experiment demonstrated that at 80 K ^{129}I atoms remain bonded to the ligand –OH groups over a time long by comparison with the 15 nsec half-life of the Mössbauer transition. Probably the primary IO_6^{5-}-ion decomposes in the H_6TeO_6 lattice upon heating or during dissolution prior to radiochemical analysis.

Johnstone et al. (1972) studied the ^{129}I Mössbauer emission spectra of the compounds $(NH_4)_2{}^{129m}TeX_6$ and $^{129m}TeX_4$ (X = Cl, Br, I) at 4 K. The spectra gave evidence in the first case for the formation of the octahedral ions IX_6^-. In the TeX_4 compounds, the spectra indicated that the ^{129}I atoms are not found in an environment iso-structural and iso-electronic with that of the parent.

4.5.2. Telluro-organic compounds

Halpern and Sochacka (1961) studied the decay of ^{131}Te in dibenzyltelluride. In the solid state the yield of daughter ^{131}I atoms in benzyl iodide was found to be about 2%, the remainder being involved in fragmentation processes:

$$(C_6H_5CH_2)_2{}^{131}Te \xrightarrow{\beta^-} (C_6H_5CH_2)_2{}^{131}I^+ \begin{cases} \xrightarrow{98\%} \text{Fragmentation} \\ \xrightarrow{2\%} C_6H_5CH_2{}^{131}I \end{cases}$$

More than 50% of the ^{131}I activity was found in unidentified high-boiling iodo-organic compounds; these perhaps were formed in secondary reactions involving species generated by bond rupture. It is now well established (Stöcklin and Tornau 1966, Berei and Stöcklin 1971, Siekierska et al. 1970, Halpern 1971) that in both liquid and solid aromatic systems significant fractions of the daughter halogen atoms from (n,γ) activation are associated with high-boiling products; the same seems to be valid for radioiodine after β-decay (Halpern 1963). One cannot explain the very small 'retention' as benzyl iodide by considering only the recoil energy spectrum; apparently the electronic excitation from the change in Z and possibly also the internal conversion of γ-transitions play a decisive rôle.

One would expect even more profound molecular consequences from the ^{132}Te β-decay. This was investigated by Llabador and Adloff (1966, 1967a, b, 1968, Adloff and Adloff 1966) for diphenyltelluride and dibutyltelluride in the solid and dissolved states. The yield of $C_6H_5{}^{132}I$ from diphenyltelluride was found to be 32%. This comparatively high value of the 'retention', however, does not necessarily indicate a non-bond-rupture effect, but rather a 'parent' rebuilding by the recombination reactions of radicals arising through high primary charge. The authors emphasized the rôle of the accompanying radiation-chemical effects, arising from the Auger electrons.

4.6. IODINE–XENON SYSTEM AND SYNTHESIS OF NOBLE GAS COMPOUNDS

Perlow and Perlow (1964a, b, 1965, 1968) exploited the Mössbauer effect to study the chemical forms of ^{129m}Xe atoms resulting from the decay of ^{129}I in iodine compounds. Using $Na^{129}IO_3$ as Mössbauer source and xenon clathrate as an absorber they detected a quadrupole-split spectrum, as would be expected from a pyramidal $^{129m}XeO_3$ product. Similarly, $K^{129}IO_4$ and $K^{129}ICl_4 \cdot H_2O$ as Mössbauer sources allowed them to establish the existence of $^{129m}XeO_4$ and $^{129m}XeCl_4$. In general, the β-decay of ^{129}I was found to lead to the xenon compound most like the source compound. The remarkable achievement of the Perlows was that the structures of the previously unknown compounds $XeCl_2$ and $XeCl_4$ could be elucidated.

The Russian group of Nefedov and his coworkers (Murin et al. 1965a, b, Kirin et al. 1965, 1966, Mosewich et al. 1965, Gusev et al. 1967a, c) found that the β-decay of ^{129}I, ^{131}I, ^{133}I and ^{135}I, incorporated in various oxygen- and fluorine-containing iodocompounds yielded the corresponding oxygen or fluorine compounds of xenon labelled with ^{129m}Xe, ^{131m}Xe, ^{133}Xe or ^{135}Xe, respectively, which were separated chromatographically. The decay of radioiodine was found to produce partially chemically bound, and partially elementary xenon. For example, for heptavalent iodine the following reaction schemes were found:

$$IO_4^- \xrightarrow{\beta^-} [XeO_4]^* \left\{ \begin{array}{l} \longrightarrow XeO_4 \\ \longrightarrow XeO_3 \\ \longrightarrow Xe \end{array} \right.$$

or

$$H_2IO_6^{3-} \xrightarrow{\beta^-} [H_2XeO_6^{2-}]^* \left\{ \begin{array}{l} \longrightarrow H_2XeO_6^{2-} \\ \longrightarrow XeO_3 \\ \longrightarrow Xe \end{array} \right.$$

The important advantage of the radiochemical method in this case was that radioxenon compounds were not only identified but also isolated in carrier-free amounts. The possibility of synthesizing organic xenon derivatives in a similar way (Toropova et al. 1968a, b), the xenonium cation being an intermediate, should also be mentioned. For further details the reader is referred to a recent review (Nefedov et al. 1969). Other nuclear decays can also be utilized for hot-atom syntheses of molecules labelled with noble gases. Adloff (1966) has listed some nuclides appropriate for this purpose (not all the daughter nuclides are radioactive) (table 16.2). For example, the possible radiosynthesis of $^3HeF_2$ in a lattice of KHF_2 was suggested by Pimentel and Spratley (1963). About 20 Ci of tritium might yield 10 μmoles of HeF_2 in a period of 4–5 months, an amount that would probably be ample for detection by spectroscopic methods (Pimentel et al. 1964).

Table 16.2

Parent nuclide	Half-life	Mode of decay	Daughter nuclide	Nuclear reaction used to produce the parent nuclide
T	12.28 y	β^-	^3He	$^6Li(n,\alpha)$ or $^3He(N,p)$
^{20}F	10.7 s	β^-	^{20}Ne	$^{19}F(n,\gamma)$
^{22}Na	2.6 y	β^+	^{22}Ne	$^{24}Mg(d,\alpha)$
^{36}Cl	3.08×10^5y	β^-	^{36}Ar	$^{35}Cl(n,\alpha)$
^{38}Cl	37.3 min	β^-	^{38}Ar	$^{37}Cl(n,\gamma)$
^{40}K	1.25×10^9y	EC(11%)	^{40}Ar	natural
^{80}Br	18 min	β^-(92%)	^{80}Kr	$^{79}Br(n,\gamma)$
^{82}Br	35.85 h	β^-	^{82}Kr	$^{81}Br(n,\gamma)$
^{126}I	13.3 min	β^-(44%)	^{126}Xe	$^{127}I(n,2n)$
^{128}I	25 min	β^-	^{128}Xe	$^{127}I(n,\gamma)$
^{131}I	8.08 d	β^-	^{131}Xe	fission
^{132}I	2.33 h	β^-	^{132}Xe	fission
^{132}Cs	6.2 d	EC	^{132}Xe	$^{132}Xe(p,n)$
^{131}Cs	9.6 d	EC	^{131}Xe	$^{131}Ba(11,5d) \xrightarrow{E.C.}$
^{212}Fr	19.3 min	EC(56%)	^{212}Rn	Th + p

4.7. SOME OTHER SYSTEMS

A number of systems were studied by early workers. Some of the results must, however, be looked at with caution. All the ^{51}Cr from the decay of ^{51}Mn in solid $CsMnO_4$ was found after dissolution in water as chromate (Burgus and Kennedy 1950). But even if other chemical forms of chromium had been present, they must have been converted to chromate during dissolution! Another early study dealt with the β-decay of lanthanides, for example, $^{143}La \xrightarrow{\beta^-} {}^{143}Ce$ in La^{3+} solutions (Davies 1948) or $^{144}Ce \xrightarrow{\beta} {}^{144}Pr$ in the acetylacetonate complex (Edwards and Coryell 1948). However, the first work must now be considered unsatisfactory because of possible Ce^{3+}–Ce^{4+} exchange with Ce^{3+} impurities in lanthanum and because of the strong oxidizing properties of Ce^{4+}. In the cerium acetylacetonate complex, surprisingly, no bond rupture was observed. More recently studies were reported independently by Glentworth et al. [op. cit.] and Shiokawa et al. (1965, 1969a, Shiokawa and Omori 1965) on β-decay of ^{143}Ce and ^{144}Ce in ethylenediaminetetraacetate (EDTA), diethylenetetraminepentaacetate (DTPA), and trans 1,2-diaminocyclohexanetetraacetate (DCTA) complexes: here the authors demonstrated a much higher proportion of bond rupture in the complex than can be attributed to shake-off and recoil. The maximum recoil energies of the daughter atoms are 12.7 eV for ^{143}Pr and 1.6 eV for ^{144}Pr; these values are too low to cause the rupture of several strong bonds holding the metal in the complexes. Nevertheless, 30–85% of the praseodymium was found as uncomplexed Pr^{3+}. This, however, can be attributed not only to the intrinsic properties of the daughter atom, but also to a pH-dependent decomposition of a charged Pr^{3+}-complex formed after the β-decay of cerium. Accordingly, the experimental results can be explained qualitatively by a mechanism which includes the decomposition of an intermediate complex formed by intramolecular electron transfer in addition to a substitution reaction between the ^{144}Pr and the Ce^{3+}-complex. In the more recent studies of the chemical effects of β-decay or electron capture in lanthanides ^{151}Gd, ^{151}Sm (Glentworth et al. 1973), ^{171}Er and ^{177}Yb (Asano et al. 1974) were employed.

Some attention was paid to the decay $^{99}Mo \xrightarrow{\beta^-} {}^{99m}Tc$. Baumgärtner et al. (1961a, b, c) reported in 1961 the synthesis of the diphenyl technetium (I) cation from the decay of $^{99}Mo(C_6H_6)_2$ and in 1958 Mo-hexacarbonyl was utilized to produce carrier-free ^{99m}Te-carbonyl (Nefedov and Toropova 1958, a). The decay of ^{99}Mo in $[C_5H_5Mo(CO)_3]_2$ was found to yield the previously unknown compound $C_5H_5{}^{99m}Tc(CO)_3$. Ferradini et al. (1969) found only the highest oxidation state, Tc^{7+}, whatever the conditions of the decay of $^{99}Mo^{6+}$. This result contrasts with the earlier work of Munze (1961) according to whom small amounts of Tc^{6+} and Tc^{5+} are also possibly present in alkaline solution. Recent studies of the β-decay induced reactions of Tc in ^{99}Mo-labelled hexacarbonyl (Nefedov and Mikulai 1973) revealed the presence of a volatile compound labelled with ^{99m}Tc (yield: 30%). The complex cation $[^{99m}Tc(CO)_6]^+$ (yield: 65%) and pertechnate were also formed. de Jong and Wiles (1973) found the $^{99m}Tc(CO)_5$ radical to be formed with a high yield in the solid $^{99}Mo(CO)_6$. Cifka and Vesely (1971) studied once more the ^{99}Mo decay with the practical aim of determining the optimal parameters for ^{99m}Tc generators.

Two isotopes of nickel have been studied: ^{56}Ni and ^{57}Ni. ^{56}Ni decays ($T_{\frac{1}{2}} = 6.1$ d) 100% by electron capture, while ^{57}Ni ($T_{\frac{1}{2}} = 36.0$ h) undergoes 32% β^+ emission ($E_{max} = 0.85$ MeV) and 68% electron capture. The ensuing gamma transitions are only slightly converted, a significant fact for the following chemical consequences. So far only nickelocene has been studied (Ross 1969). The essential result is that in the decay of these two nuclides no volatile cobaltocene ($CoCp_2$) is observed unless inactive $CoCp_2$ is present in the sealed capsule during the decay. With this carrier present, however, the yield rose to 80–90%. Ferrocene was not able to perform the carrier function. The comparison with results from the decay of $^{58m}CoCp_2$, where the decay radiation is 100% converted, is noteworthy. In that case, 70–90% of the ^{58g}Co is found as Co^{2+}, 10–30% as $CoCp_2^+$ and none as volatile $CoCp_2$. The decay of

$^{58m}CoCp_2^+$ gives no $^{58g}CoCp_2^+$ and only inorganic Co^{2+}. Omori et al. (1970a, b) reported on the distribution of ^{57}Co-labelled species between the oxidation states (II) and (III) in the ^{57}Ni-hexamminenickel (II) complexes. Heating was found to increase the yield of tervalent species, which contrasts with a generally observed effect of thermal annealing – rebuilding the original parent form. It was thus concluded that the chemical state of ^{57}Co produced from β-decay in a non-isotopic nickel compound is not necessarily determined by the crystal matrix, but rather by the chemical properties of the daughter complex.

Finally we mention the decay $^{125}Xe \xrightarrow[18\ h]{100\%\ E.C.} {}^{125}I$. This offers an excellent example for studying the chemical consequences of electron capture. The recoil energy of ^{125}I is very low (< 4.4 eV), but the Auger charging should give rise to substantial molecular effects. It can be deduced from the work of Schroth and Adloff (1964a) on the reactions of ^{125}I with CH_4 that 40% of the ^{125}I atoms from E.C. are in the state $I^+(^1S_2)$ and 18% are highly excited atoms and positively charged ions, $I^+(^3P_0)$ or $I^+(^3P_1)$. Among inorganic compounds, the chemical state of ^{125}I was studied in $^{125}XeF_2$ and $^{125}XeF_4$. ^{125}I in XeF_4 is stabilized in oxidation states +5 and +7, indicating the formation of compounds such as $^{125}IF_5$ and $^{125}IF_7$. In XeF_2, hydrolysed after the decay, 95% of the ^{125}I was found as $^{125}IO_4^-$, the rest being $^{125}IO_3^-$. The oxidation of pentavalent iodine by XeF_2 during hydrolysis, which may increase the content of IO_4^-, cannot be excluded. However, the data currently available indicate that in both the xenon fluorides the highly reactive ^{125}I atoms undergo bonding in the lattice, the oxidation states +5 and +7 being characteristic for the daughter iodine. Kuzin et al. (1970a, b) studied the following systems: (1) ^{125}Xe-xenon at a pressure of 2–3 atm; (2) ^{125}Xe–atmosphere of saturated water vapor; (3) $^{125}Xe–H_2$; (4) $^{125}Xe–O_2$. The major form of ^{125}I in these systems is I^+. This follows from the fact that ionization potentials of the molecules of the medium in which the ^{125}Xe decayed (Xe 12.13 eV, H_2O 12.60 eV, O_2 14.10 eV, H_2 15.43 eV) are higher than the first but lower than the second ionization potential of iodine (10.45 eV and 19 eV, respectively). The I^+ ions interact with Xe atoms according to the reactions (Schroth and Adloff 1964a):

$$I^+(^1S_2) + Xe \longrightarrow I(^2P_{3/2}) + Xe^+ + 56.3\ \text{kcal},$$

or

$$I^+(^1S_2) + Xe \longrightarrow I(^2P_{1/2}) + Xe^+ + 32.9\ \text{kcal}.$$

Further changes in the chemical state of iodine are associated with the interactions of I^+ and elementary iodine with the surrounding medium, free electrons and/or the walls of the reaction vessel. In the $^{125}Xe–H_2$ and $^{125}Xe–H_2O$ systems practically all the ^{125}I was found in the form of I^-: in the atmosphere of xenon oxidized forms of iodine were not observed, and the yield of reduced iodine was found to depend on the solvent used to wash the walls; in the atmosphere of oxygen about 90% of ^{125}I was found to be reduced and 10% as IO_3^-. The most probable process leading to the formation of iodate is the addition of I^+ to one of the unshared electron pairs of oxygen, forming IO_2^+:

$$I^+ + O_2 \longrightarrow [I{-}\ddot{O}{=}\underset{\cdot\cdot}{\ddot{O}}]^+ \longrightarrow IO_2^+,$$

followed by the reaction of IO_2^+ with traces of water absorbed on the walls: $IO_2^+ + H_2O \rightarrow IO_3^- + 2H^+$.

It is clear that a better understanding of reactions involving excited $^{125}I^+$ ions from electron capture requires further experimental work. ^{119}Te, which decays by electron capture, can also be utilized for the investigation of the after-effects of the Auger process (Ambe et al. 1973). Less convenient is ^{73}Se (Dulpatre and Herment 1976) which undergoes E.C. (35%) and β^+-decay (65%).

4.8. MOLECULE SYNTHESES BY β-DECAY

We have mentioned the preparative possibilities of the chemical changes initiated by β-decay. This practical approach has already led to the synthesis of a number of new labelled carrier-free compounds. The interesting molecules $C_5H_5{}^{99m}Tc(CO)_3$, $^{105}Rh(C_5H_5)_2$, and $^{99m}Tc(C_6H_6)_2^+$ were first synthesized in this way by Baumgärtner (1965) and co-workers; Murin and Nefedov and their collaborators have also contributed substantially to the field (Murin et al. 1965a, b, Nefedov et al. 1969). Baumgärtner and Schön (1964) described a novel technique of molecular synthesis in the solid phase, utilizing nuclear fission followed by β-decay of a particular fission product. The target compound (ferrocene, ruthenocene or osmocene) mixed with uranium oxide was bombarded with neutrons and the newly formed labelled compound $^{105}Ru(C_5H_5)_2$ separated from the uranium oxide and fission products. The yield of ^{105}Ru-labelled ruthenocene from different targets was found to be related to the bond energy of the central atom. The authors thus concluded that the molecular synthesis originates in the β-decay process, which supplies a kinetic energy comparable to the bond energies, rather than from nuclear fission itself. The rôle of the primary fission recoil is only to transport the newly formed atoms to favorable sites in the lattice of the target compound. Further work along this line would be welcome. Adloff (1975) has recently reviewed β-decay induced inorganic syntheses.

5. Conclusions

The studies of the chemical effects of β-decay have enriched our knowledge of recoil chemistry. Important achievements have also been made in new-molecule synthesis and the chemistry of noble gases. However, it must be admitted that the overall scientific gain is more limited than was expected initially. In particular, the motivation behind many of the studies over the past 10 or 15 years was the desire to investigate the rôle of lattice defects in post-recoil reactions in the solid phase, since β-decay allows one to generate the recoil species largely free from defects produced by external ionizing radiation. However, evidence was obtained that the fate of daughter atoms from β-decay in solids depends critically on defects formed during the preparation and purification of the parent material, which are difficult to control. Thus our understanding of the link between primary physical events and final chemical consequences of β-decay is still shadowy. The need for further studies with the use of modern non-destructive (in-situ) experimental techniques is quite evident.

Studies on metal–organic compounds have perhaps yielded more valuable results than those on simple inorganic compounds. Further studies of annealing processes in metal–organic compounds might eventually reveal new aspects concerning the mechanisms and kinetics of annealing reactions, reflecting the role of free radicals. Some studies have demonstrated that thermal annealing in pure organic crystals (e.g. solid haloaromatics) differs substantially from the well-known annealing pattern for inorganic solids (Siekierska and Halpern 1966, Siekierska et al. 1968).

The most important studies of effects related to the β-decay of incorporated radioisotopes are, undoubtedly, of cases where that decay occurs in biologically significant molecules. Some information on the biological consequences of the decay of tritium, ^{14}C, ^{32}P, ^{33}P, ^{35}S and ^{125}I already available in the literature support this view (Krisch and Zelle 1969, Halpern and Stöcklin 1977). In particular, effects in genetic or mutagenic material should be extensively further explored, as even small numbers of nuclear decays can have very pronounced biological consequences. The hitherto accumulated knowledge on the chemical consequences of β-decay in more simple systems will be important to the theoretical and experimental interpretation of these more complex situations.

CHAPTER 17

^{35}S IN CRYSTALLINE POLAR CHLORIDES

J. CIFKA

Nuclear Research Institute, Czechoslovakian Academy of Sciences, Prague, Czechoslovakia

Chemical Effects of Nuclear Transformations in Inorganic Systems
Edited by G. Harbottle and A.G. Maddock

Contents

1. Introduction

The behaviour of ^{35}S in inorganic chlorides has been studied for nearly twenty-five years. In more than twenty papers it has been shown that ^{35}S produced in a target by the $^{35}Cl(n,p)^{35}S$ reaction is present in several chemical states and that many factors influence the distribution amongst these states. Although inorganic chlorides seemed to be fairly simple systems, the data concerning ^{35}S from different laboratories are not in quantitative agreement and the comparison of published results is not easy. Nevertheless some general conclusions can be reached.

The reasons for these difficulties are the rather low neutron cross section for the above-mentioned reaction and the long half-life of ^{35}S; many investigators therefore irradiated the targets for a long time and the initial distribution of ^{35}S was changed.

2. Chemical Analysis and Handling of Irradiated Targets

2.1. ANALYTICAL PROBLEMS

It is well known that chloro-compounds of sulphur are generally hydrolysed to sulphite while the most stable oxyanion of sulphur is sulphate. For purely formal reasons one would expect the ^{35}S produced by the (n,p) reaction in chlorides to be present as sulphide. However, the early experiments of Levey et al. (1948) have shown that after dissolution of the irradiated crystals in water nearly all the ^{35}S was present as sulphate. In these early investigations the irradiated material was dissolved for analysis in a solution of sulphide, sulphite and sulphate carriers. However, some later experiments lead to the conclusion that neutral ^{35}S atoms (S^0) are probably also present in the crystals (Milham 1952). Sulphur atoms in macro amounts and under normal conditions are combined into S_8 rings and therefore the reactions of S^0 are unknown and new analytical methods had to be developed for the separation of this fraction.

Since isotopic exchange between sulphides, sulphites and sulphates does not occur, these forms can be separated by various methods.

Most of the published results were obtained with simplified procedures. The sulphide fraction was separated as cadmium sulphide and all other possible forms of ^{35}S remaining in the filtrate were oxidized to sulphate. The sulphide fraction was then also oxidized to sulphate. So, after precipitation with barium, both fractions were in the same chemical form, ensuring identical self-absorption corrections for the low-energy beta rays of $^{35}S(E_{max} = 0.167$ MeV).

All three carriers can be separated as follows – at first barium sulphite and sulphate are precipitated in an alkaline medium and cadmium or copper sulphide precipitated in the supernatant. (Centrifugation is preferable to filtration since poorly filterable precipitates are formed in alkaline media.) The barium precipitate is then treated with acid and the sulphur dioxide released is absorbed in an alkaline solution of hydrogen peroxide. All fractions are then converted to barium sulphate precipitates (Milham 1952, Milham et al. 1965, Kasrai and Maddock 1970). Alternatively, each of these carriers can be precipitated separately, leaving all others in the solution; sulphate with benzidine, sulphite with aluminium ions and sulphide with cadmium ions (Meyer and Adloff 1967) for ^{35}S analyses.

The S^0 fraction was first determined from the difference of activity in the sulphide fraction under two conditions of separation. The irradiated target was dissolved in an outgassed dilute solution of sodium hydroxide containing either all three carriers or in the absence of carriers. In the latter case the activity in the sulphide fraction is lower than in the former case. The method is based on the assumption that the oxidation of S^0 activity to sulphite and sulphate is very fast in the absence of added carriers, while carrier-free sulphide is much more stable in de-aerated solutions. It is assumed that S^0 exchanges isotopically with sulphide in the presence of sulphide carrier. This assumption is based on the fact that isotopic exchange between elementary sulphur (S_8) and sulphide is very fast. There is evidence that some of the S^0 is not measured with this method.

Kasrai and Maddock (1970) proposed the dissolution of the target in a solution containing an excess of cyanide ions. S^0 can react with cyanide with formation of CNS^- anion, so that the activity can be separated into four fractions. Again sulphite and sulphate are precipitated first as barium salts, then cadmium sulphide is separated and thiocyanate remains in the solution. Meyer and Adloff used the reaction of the S^0 with an excess of sulphite carrier giving thiosulphate; this can be separated as nickel-triethylenediamine thiosulphate (Ames and Willard 1951).

All the above methods are convenient for analyses of material with low as well as high specific activities, as the amount of dissolved target does not influence the separation. More elegant methods like paper electrophoresis (Yoshihara et al. 1964) or column ion exchange (Abdel-Rassoul et al. 1969, 1970) have also been used. Unfortunately, these methods are inconvenient for materials of low activity and the investigators therefore had to irradiate their targets with high neutron doses. (See also Kasrai and Maddock 1970.)

For the separation of ^{35}S in the S^0 state Chiotan et al. (1964) have proposed using direct extraction with trichloroethylene before dissolution of the irradiated materials; however, the method is not suitable for quantitative analysis.

2.2. HANDLING OF SAMPLES

The distribution of ^{35}S can be influenced by many factors; most of them also change the type and number of radiation-induced defects in the crystals.

Unfortunately, the influence of these factors has only gradually been established and in many investigations it has not been taken into consideration. Moreover, the experimental conditions have not been sufficiently described in many papers for estimates of their effects to be made. To minimize these effects, the following principles should be observed.

The irradiated samples must be protected from light during storage, dissolution and processing. The storage temperature should be low, to decelerate the thermal annealing processes. Samples stored at temperatures below 0° C must be protected from moisture, otherwise, the composition of the surface layers of the crystals may be changed, and this can be serious when small crystals are used.

3. Factors Influencing the Distribution of the ^{35}S

The following subsections review the factors influencing the distribution of ^{35}S among the different valency states. The information derived from thermal annealing of samples is very pertinent to this matter. One must bear in mind that at present it is practically impossible to isolate individual effects and that each observed distribution of ^{35}S is a result of several superimposed effects.

3.1. EFFECT OF IRRADIATION TEMPERATURE

Most irradiations were carried out at the ambient temperature of the reactor, which is usually in the range of 40–60° C. In fact, this temperature concerns only the surface of the container. At high fluxes (10^{12}–10^{13} n·cm^{-2}sec^{-1}) the temperature of the samples is higher due to the absorption of fast neutrons and gamma radiation.

The effect of irradiation temperature was observed by Cifka and Kliment (1966) in ammonium chloride. These experiments were carried out in daylight. It was studied in more detail by Bračokova (1968) with polycrystalline chlorides, protected from light. The results presented in fig. 17.1 show that in samples irradiated at an average temperature of 80° C, the preliminary

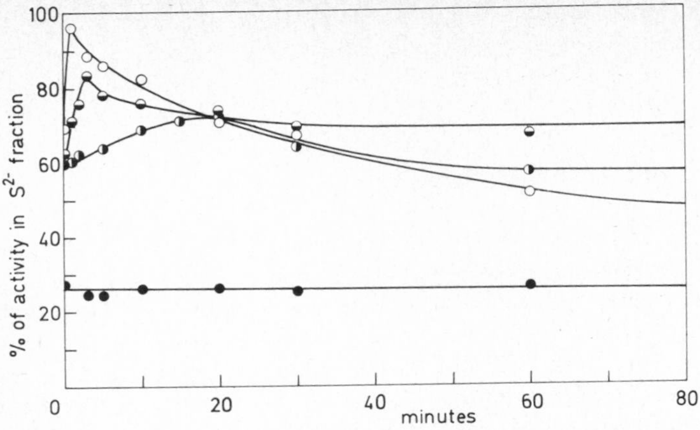

Fig. 17.1. Thermal annealing (at 150° C) of potassium chloride samples irradiated at various temperatures (Bračoková 1968). Polycrystalline material was irradiated at dry-ice temperature (0), at 20° C (◒) and at 80° C (●). Monocrystals were irradiated at dry-ice temperature. (◐).

increases in the sulphide fraction on thermal annealing is absent, and that the distribution of ^{35}S is constant for three hours. This means that thermal annealing takes place during the neutron irradiation and labile ^{35}S species are already transformed into more stable ones. Alternatively the labile radiation induced defects (colour centres) in the crystals could be annealed and therefore are not available to produce effects in the early stages of heating of the sample. The remaining more stable defects probably need a temperature higher than 150° C for reaction with the ^{35}S.

This observation explains the differences in annealing behaviour of target materials irradiated at dry-ice temperature (Cífka and Bračoková 1966, Bračoková and Cífka 1970) and at elevated temperature (Maddock and Mirsky 1965, Chiotan et al. 1968a, b).

3.2. EFFECT OF STORAGE TEMPERATURE

Because of the effects described in subsect. 3.1 the temperature of storage of irradiated samples is important. Samples irradiated at dry-ice temperature are more sensitive to thermal annealing than those which are partially annealed during neutron irradiation. Therefore, they must be stored at dry-ice temperature. The changes in ^{35}S distribution in polycrystalline potassium chloride irradiated at dry-ice temperature and stored at room temperature in darkness are shown in fig. 17.2 (Bračoková 1968). In barium chloride irradiated at dry-ice temperature, the ^{35}S distribution changes even at 10° C (Cífka and Bračoková 1970).

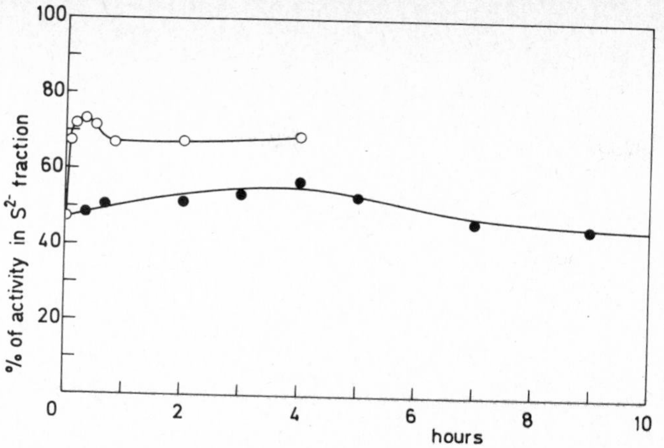

Fig. 17.2. Effect of storing conditions and of light on the distribution of ^{35}S (Bračoková 1968). Polycrystalline potassium chloride was irradiated at dry-ice temperature and stored at room temperature in darkness (●) and exposed to daylight (0).

3.3. EFFECT OF LIGHT

Maddock and Mirsky (1965) observed an increase of ^{35}S in the sulphide fraction, on exposing neutron-irradiated crystals of sodium and rubidium chloride to daylight, while the results for potassium and caesium chloride were not so conclusive. They used targets irradiated at 40° C. During the exposure to light the crystals were bleached, which means that some colour centres produced by the ionizing radiation disappeared. The bleached crystals showed no increase in the sulphide fraction in the early stages of thermal annealing. The difference can be explained by the absence of reducing centres in the bleached samples.

Bračoková (1968) found an increase in the sulphide fraction for polycrystalline potassium chloride irradiated at dry-ice temperature (fig. 17.2) when the crystals were exposed to daylight at room temperature. Kronrád and Kacena (1966) described similar changes in the ^{35}S distribution for silver chloride caused by light and by heating.

3.4. EFFECT OF NEUTRON AND IONIZING RADIATION DOSE

It has been found by Maddock and Mirsky (1965) that samples of sodium and potassium chloride receiving a lower gamma dose during irradiation contain a higher percentage of ^{35}S in the sulphide fraction.

Cifka and Bračoková (1966) studied the effect of absorbed radiation dose in more detail for polycrystalline potassium chloride at dry-ice temperature. The curve which was later completed by Bračoková (1968) is presented in fig. 17.3;

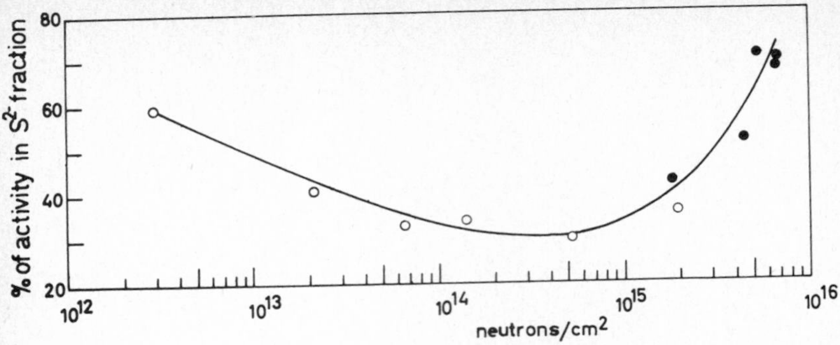

Fig. 17.3. Variation of the distribution of ^{35}S with increasing total thermal neutron dose Cifka and Bračoková 1966 (0), and Bračoková 1968 (●)).

the initial percentage of ^{35}S in the sulphide fraction is plotted against the thermal neutron dose, which was most readily determined, because the dosimetry of gamma radiation in an operating reactor is not easy; fortunately, the ratio of thermal to fast neutrons was constant for the given irradiation position, as well as the ratio of thermal neutrons to gamma radiation. Very roughly, with a dose of $1 \times 10^{15} n_{th} \cdot cm^{-2}$ the samples received a dose of gamma radiation of about 1 Mrad.

Increasing dose leads to an increasing concentration of radiation defects, which can react with ^{35}S during thermal annealing. Hence different annealing curves were obtained for these samples – the extent of the initial reduction increased with increasing dose (fig. 17.4).

In all the above-mentioned experiments only the sulphide fraction was separated from the other forms of ^{35}S. Recent results of Rajman (1971) show that the rest of the ^{35}S activity is in the S^0 fraction, when the analytical method

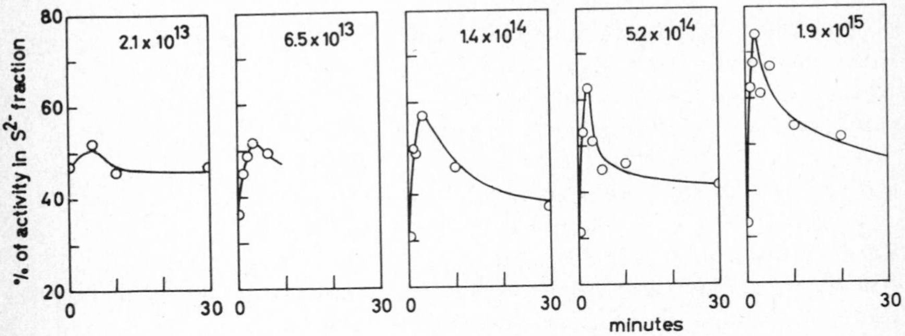

Fig. 17.4. Thermal annealing curves (at 150° C) of polycrystalline potassium chloride irradiated at dry-ice temperature. The thermal neutron doses are presented in n cm^{-2} (Cifka and Bračoková 1966).

ith cyanide is used; no activity was found in the sulphite and sulphate recipitate. In these experiments sodium chloride (polycrystalline samples as ell as monocrystals) was irradiated at dry-ice temperature; the thermal eutron dose was within the range 2–4 × $10^{14} n_{th} \cdot cm^{-2}$.

It is rather difficult to draw conclusions from the other published data ecause the experiments differ in too many other conditions. Nevertheless, the ata, arranged according to increasing total dose of thermal neutrons are eviewed in table 17.1. In general, the percentage of ^{35}S in the sulphide fraction ecreases with increasing dose of concominant ionizing radiation.

Heavily irradiated samples differ in annealing behaviour from those eceiving lower doses. Generally higher temperatures and/or longer heating eriods must be used to produce any change in the ^{35}S distribution. Usually, an ncrease of ^{35}S in the sulphide fraction was observed; sometimes followed by a ecrease. For bleached crystals of potassium, rubidium and caesium chloride a ecrease in sulphide fraction was observed, while bleached sodium chloride amples showed no change during the heating [Maddock and Mirsky (1965); eating at 180° C]. Owens and Hobbs (1969) observed an increase in sulphide action and a decrease in S^0 fraction in potassium chloride annealed at 215° C.

.5. EFFECT OF IMPURITIES

he presence of ^{35}S activity in the non-sulphide fractions was originally xplained as a result of the reaction of ^{35}S recoils with oxygen in the target naterial. It seemed to be confirmed by a difference in the ^{35}S distribution in ormal potassium chloride crystals and outgassed crystals (Koski 1949a, b arlson and Koski 1955, Yoshihara et al. 1964). The reaction with traces of vater was considered to be another possibility. However, other investigators ound a high proportion of ^{35}S in the sulphide fraction even for crystals that are ot degassed and a predominant role for oxygen in the oxidation of ^{35}S has een eliminated.

A possible reaction of ^{35}S with oxygen present in the crystals might occur uring post-irradiation heating of the crystals. In fact, heating in oxygen gave n increase in the ^{35}S sulphide fraction compared with heating in vacuo, in nost experiments (see ch. 18). However, Cífka and Kliment (1966) found even nore pronounced oxidation processes in ammonium chloride twice sublimed in acuo compared with the untreated material.

It has clearly been established that above about 150° C, oxygen can enter lkali chloride crystals as superoxide O_2^- ions, displacing chloride as chlorine. his may affect annealing in air at elevated temperatures and also, possibly, the ehaviour of samples of alkali chloride that have been heated in air (Kanzig nd Cohen 1959).

Concerning the possible action of water, Cífka and Bračoková (1966) did ot observe any effect of pre-heating, in the temperature range of 20–700° C,

Table 17.1

Target	Type of crystals	Total neutron dose (n-cm^{-2}), temperature (°C) gamma-dose					
		1.3×10^{14}	2.1×10^{14}	6×10^{14}	$0.4–8\times10^{14}$	1.2×10^{15}	1.5×10^{15}
		–	40	40	–	–	−70, +15
		0.1	9.6	25	–	23	1.5
		A	A	A	B	C	D
NaCl	poly-	–	85	70	45+31[a]	–	–
	mono-	60	–	26–70	–	–	–
KCl	poly-	–	–	6–12	38+40[a]	14+60[a]	–
	mono-	41–45	–	20–30	–	–	–
RbCl	poly-	–	–	–	35+50[a]	–	–
	mono-	–	–	40–60	–	–	–
CsCl	poly-	–	–	26–60	25+40[a]	–	–
NH_4Cl	poly-	–	–	–	–	–	70–80
AgCl	poly-	–	–	–	–	–	–
	mono-	–	–	–	–	–	–
	pptd-	–	–	–	–	–	–

A: Maddock and Mirsky (1965). B: Milham et al. (1965). C: Owens and Hobbs (1969). D: Cifka and Kliment (1966). E: Kronrád and Kacena (1966). F: Yoshihara et al. (1964). G: Chiotan et al.

on the initial distribution of ^{35}S in polycrystalline potassium chloride irradiated at dry-ice temperature.

On the other hand, the effect of water on the annealing behaviour of ^{35}S cannot be excluded. Chiotan et al. (1968a) observed different behaviour for polycrystalline sodium chloride, dried before irradiation at 130° C and 210° C; in the first case, at lower annealing temperatures, reduction of ^{35}S was followed by oxidation, while in the second case only the reduction processes were observed. Cífka and Kliment (1966) found for ammonium chloride crystals dried at 120° C a reduction phase in the annealing curves, while for the material sublimed in vacuo this reduction period was absent; if this effect was due to the water content, then in this case water must act in a different way in the sodium chloride target. (For an alternative explanation see Sect. 3.6).

Alkali chlorides are hydrolysed by water vapour to a perceptible extent (Rolfe 1958) and the hydroxide ions introduced in this way can have a considerable effect on the distribution of the ^{35}S. The doping of potassium chloride

Initial percentage of ^{35}S in sulphide fraction

(Mrad) and reference

2.7×10^{15}	1.2×10^{16}	1.6×10^{16}	6.5×10^{16}	3.6×10^{17}	5.8×10^{17}	1.9×10^{18}
-70	50–80	40	50–80	130	–	–
2.5	1	50	7.5	2000	1000	–
E	F	D	F	G	D	H
–	–	–	45	12	–	18
–	–	–	–	–	–	–
–	–	–	0	25	–	35
–	–	–	–	–	–	–
–	–	–	0	–	–	0.5
–	–	–	–	–	–	–
–	–	–	0	–	–	–
–	11	50–60	–	–	1–2	–
30	–	–	–	–	–	–
10–15	–	–	–	–	–	–
40–50	–	–	–	–	–	–

(1968a, b). H: Abdel-Rassoul et al. (1970).
[a]Percentage of ^{35}S in S^0 fraction.

with potassium hydroxide did not lead to a progressive increase in ^{35}S in the non-sulphide fraction (Bračoková and Cifka 1970), but in the range of concentrations studied (0.002–3 mole % of potassium hydroxide) a decrease of ^{35}S in the sulphide fraction was observed, followed by an increase back to the original value, followed finally by another decrease. These changes were explained as being due to the indirect effect of the OH^- impurity, which results in an increased production of F centres by the radiation (Schulman and Compton 1962) (see also ch. 18).

The influence of other chemical impurities cannot be easily evaluated, because the purification is usually accompanied by a change in the crystal state of the target material. Kronrád and Kacena (1966) observed nearly no change in the initial distribution of ^{35}S with increasing numbers of crystallizations of silver chloride monocrystals.

Target material doped with divalent cations or enriched in alkali metal can be considered as chemically impure material. However, it is preferable to analyze such data in terms of the lattice defects and therefore the results are reviewed in the following section.

3.6. EFFECT OF POINT DEFECTS AND DISLOCATIONS

The possibility was mentioned very early that the distribution of ^{35}S is influenced by the centres produced in the crystals by the action of ionizing radiation. At first, it was supposed, that free chlorine created by the gamma-radiation could be responsible for oxidation of ^{35}S (Levey et al. 1948). Then Caillat and Süe (1950b) observed an effect of F centres on the valence state of ^{32}P in sodium chloride monocrystals.

A systematic study of the influence of electron containing F centres on the valence state of ^{35}S was carried out by Maddock and Mirsky (1965). They prepared crystals of sodium chloride with an excess of metallic sodium and irradiated them with neutrons. They found more than 90% of total ^{35}S activity in the sulphide fraction, as compared with 25% for untreated control crystals.

In other experiments they irradiated potassium chloride crystals with increasing amounts of calcium or cadmium chloride. In such crystals, each divalent cation must be balanced by the absence of one potassium ion, to keep the crystal electrically neutral. So, the targets contained an increasing number of cationic vacancies, which are good traps for electrons. It is known, that such crystals contain more F centres when exposed to ionizing radiation than those of perfect crystals (Schulman and Compton 1962). In fact, Maddock and Mirsky (1965) found an increasing percentage of ^{35}S in the sulphide fraction with increasing concentration of divalent cation; this is shown in fig. 17.5.

Free electrons can also be trapped on dislocations. Crushing the crystals is one of the methods for obtaining material with a high density of dislocations

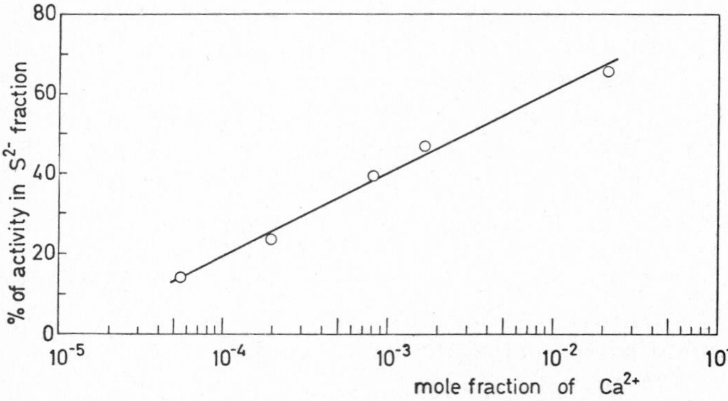

Fig. 17.5. Initial distribution of ^{35}S in calcium-doped potassium chloride crystals as a function of the mole fraction of calcium (Maddock and Mirsky 1965).

(and vacancies). A higher percentage of ^{35}S in the sulphide fraction was found in crushed crystals, compared with the untreated monocrystals and this can be considered as a further evidence for the effect of electron donor centres on the ^{35}S recoils (Maddock and Mirsky 1965).

The presence of defects in the lattice influence also the annealing behaviour of the ^{35}S. Bračoková (1968) obtained different annealing curves for polycrystalline potassium chloride and for monocrystals, both irradiated under identical conditions (fig. 17.1). She also observed a small but distinct difference in the second stage of the annealing curve for the ^{35}S in the sulphide fraction and in the plateau value for polycrystalline potassium chloride dried before neutron irradiation at 180° C for 4 and 24 h – in the latter case the oxidation processes were less pronounced.

Simple melting and cooling of polycrystalline potassium chloride, crystallized from water, before neutron irradiation also results in a change in the shape of the annealing curve; in the melted material the first fast reduction period was absent (Bračoková and Cifka 1970), as compared with the polycrystalline target (figs. 17.1 and 17.3). The observed difference can be again explained by various lattice imperfections. Still more apparent is the effect in silver chloride (Kronrád and Kacena 1966), as can be seen in fig. 17.6.

An attempt to study the effect of anion vacancies has also been made. Cifka and Bračoková (1970) irradiated barium chloride containing increasing amounts of potassium chloride. The study remained incomplete owing to the high content of monovalent cations in the 'pure' barium chloride. In the target material with the highest content of potassium 80% of the ^{35}S was initially present in the sulphide fraction but the decrease of the activity in this fraction on

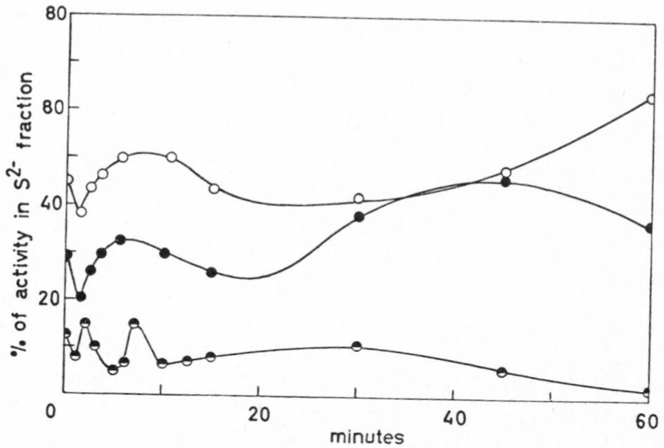

Fig. 17.6. Thermal annealing (at 100° C) of silver chloride irradiated at dry-ice temperature. Precipitated material (0), melted material (●), monocrystal (◐) (Kronrad and Kacena 1966).

thermal annealing was very rapid: in 20 min heating at 150° C the content fell below 10%.

3.7. EFFECT OF IRRADIATION WITH IONIZING RADIATION

The exposure of alkali metal chlorides to ionizing radiation results in the formation of colour centres. These can be classified, in principle, into centres containing trapped electrons and centres containing trapped holes. The centres differ according to the number, type and steric arrangement of vacancies. The formation, as well as destruction of individual types of centres depends markedly on the temperature.

Maddock and Mirsky observed an increase of activity in the sulphide fraction, in samples irradiated with 2 MeV electrons prior to neutron irradiation, compared with samples without such treatment. Bračoková (1968) studied the initial distribution and annealing behaviour of ^{35}S in polycrystalline potassium chloride which was exposed to cobalt-60 gamma rays prior to neutron irradiation (doses from 0.1 to 3 Mrad). She observed that the initial distribution was not substantially altered. On the other hand, the reduction processes during the thermal annealing at 150° C proceeded to a lesser extent and the subsequent oxidation processes were slower. There was no difference in samples pre-irradiated at dry-ice temperature and at room temperature.

Bračoková (1968) also carried out numerous experiments on the gamma-irradiation of neutron irradiated samples. The most interesting results are presented in fig. 17.7. The gamma irradiation caused a decrease of the ^{35}S in the sulphide fraction, but on subsequent thermal annealing of the samples, the same percentage of ^{35}S in sulphide fraction was reached for all samples including the untreated one. Hence, during the gamma irradiation ^{35}S recoils were oxidized but the electron excess defects which existed in the vicinity of the ^{35}S were not affected and their electrons could be released in the usual way during the heating of the crystals.

3.8. EFFECT OF CHEMICAL COMPOSITION OF THE TARGET

The influence of chemical composition of the target on the behaviour of ^{35}S, is sometimes called the cation effect. It is better to consider not only the chemical composition but also the structure of the crystal. In this sense, the effect of cation is indirect.

Chlorides of sodium, potassium and rubidium crystallize with a face-centered cubic lattice while caesium and ammonium chloride have a body-centered cubic lattice. However, the distribution of ^{35}S is more influenced by the individual composition and state of the given crystal, including chemical impurities and lattice imperfections, rather than by the general type of lattice. This is easily seen from the table 17.1.

The behaviour of ^{35}S in silver chloride during the annealing is more complex en in the alkali chlorides, although silver chloride has the same crystal ructure as sodium chloride. However, it differs substantially from all above-entioned chlorides, especially in plasticity, and in its defect structure. In silver loride at the room temperature Frenkel defects predominate, while in the kali chlorides one finds Schottky defects.

Owens and Hobbs (1969) prepared mixed crystals of potassium loride–potassium bromide, and studied the initial distribution and, to some tent, annealing behaviour of the ^{35}S. The progressive replacement of chlorine bromine results in a decrease in concentration of F centres produced by nizing radiation (Arends et al. 1965), so the effect of the anion can be again terpreted as indirect. In principle, increasing content of bromide, equivalent decreasing concentration of F centres, decreased the activity in the S^0 action. The majority of the rest of the ^{35}S was presumably in higher oxidation ates, and only a part was in the sulphide fraction. Also the annealing haviour was influenced by the composition of the mixture.

4. The Effect of Analytical Procedure

ıe main types of analytical methods used in the study of ^{35}S have been viewed in section 2. Because the various investigators used different alytical procedures, the comparability of results is poor.

There are some possible sources of discrepancies. According to present ıowledge the following processes can occur during the dissolution and ıalysis of the sample:

) When the irradiated target is dissolved in a mixture of sulphide and sulphate rriers and then sulphide is precipitated. The precipitate contains S^{2-} and obably the S^- and S^0 states of ^{35}S owing to rapid exchange.

) When the irradiated target is dissolved in a mixture of sulphide, sulphite and lphate carrier. Then it is possible that a part of the ^{35}S in the S^0 state will act with the sulphite with formation of thiosulphate (Milham et al. 1965). So, e percentage of ^{35}S in the sulphide fraction will be less than in the previous se. The subsequent method for distinguishing the sulphate and sulphite state n seriously affect the results. The usual method was to precipitate the barium lts and liberate the sulphur dioxide. In such a case, the ^{35}S originally present S^0 state remains in the sulphate fraction.

When the irradiated sample is dissolved in a solution containing an excess of lphite (Meyer and Adloff 1967) or an excess of cyanide ions (Kasrai and addock 1970), it is supposed, that the ^{35}S in the S^0 state forms either iosulphate or thiocyanate, but there is incomplete information concerning the mpleteness of these reactions although the $NC^{35}S^-$ yield is constant above ⁄I cyanide. In the absence of carriers and also in the absence of oxygen, it is supposed that S^0 state is immediately oxidized to sulphite and/or sulphate. Yoshihara et al. (1964) mentioned the possibility of action of V-type centres at the moment of dissolution on the ^{35}S recoils in the absence of carriers. It is known that these centres partially form elemental chlorine after dissolution of irradiated crystals in water (Hackskaylo et al. 1953, Burns and Williams 1955). Hence, in the absence of carriers, ^{35}S in sulphide form can also be transformed into higher oxidation states. Solution in cyanide should suppress such oxidation. It can therefore be concluded, that the three above-mentioned methods for the analysis of ^{35}S in S^0 state cannot be expected to give identical results. However, a detailed comparison of them has not been carried out.

5. Model Experiments

Several investigators tried to study the behaviour of ^{35}S in the crystal lattice without the initial rather undefined disordering due to the nuclear reaction. Kronrád and Kacena (1967) doped silver chloride with sulphide-^{35}S or sulphate-^{35}S. They found that at low concentrations (10^{-6}g of ^{35}S per gram of silver chloride) of doping substance, the distribution of ^{35}S between the sulphide and nonsulphide fractions is independent of the original valence of the ^{35}S used for doping. Also the annealing behaviour was very similar for both materials and even similar to neutron-irradiated materials.

Argo and Maddock (1969) succeeded in preparing alkali halide crystals doped with sulphide-^{35}S and hydrosulphide $^{35}SH^-$, and they used these materials for the evaluation of changes caused by ionizing radiation.

A new possibility arises in the implantation of labelled ions. Freeman et al. (1967) bombarded the alkali chlorides with S^+ ions labelled with ^{35}S and having an energy of 40 keV. They analyzed the materials using the cyanide method and obtained a distribution of ^{35}S between the S^{2-}, S^0, SO_3^{2-} fractions similar to that found for neutron-irradiated materials (Kasrai et al. 1971).

6. Processes Occurring During the Neutron Irradiation and Subsequent Treatment of some Inorganic Chlorides

The following description of the most likely processes in neutron-irradiated inorganic chlorides has been gathered from published data and discussed by numerous investigators who studied the reactions of the ^{35}S.

The kinetic energy of ^{35}S recoils after the nuclear reaction Cl(n,p)S is about 15 keV (Gilbert et al. 1944, Koski 1949a, b, Croatto and Maddock 1949, Milham 1952). The range of ^{35}S in inorganic chloride lattices is not certain, because it depends on the charge of the ^{35}S recoils. Neglecting the charge effects, the range of the ^{35}S atom would be approximately 20Å at maximum, for potassium chloride in which the masses of cations and anions are nearly the same and comparable with that of ^{35}S. For other chlorides the range would not be substantially greater. The behaviour of K^+ or Cl^- ions having energies up to 130 eV has been calculated recently for the potassium chloride lattice (Torrens and Chadderton 1967) and this model can be applied to ^{35}S recoils in the last stages of recoil. In any case, the thermalized ^{35}S recoils are surrounded by a highly damaged lattice containing cationic and anionic vacancies as well as interstitial atoms or ions.

The charge of the ^{35}S recoils changes during the thermalization. As seen in chs. 2 and 3 the escape of the proton from the compound nucleus leads to electronic excitation and at first the ^{35}S recoils probably have an appreciable positive charge. Most of this charge is quickly neutralized by electrons from surrounding ions, because the energy of the recoiling species is not sufficiently high for self-ionizing collisions. The final charge of the individual ^{35}S recoil depends on the position of the recoil in the lattice. Maddock and Mirsky (1965) have supposed that ^{35}S recoils in anionic sites can have one or two negative charges, while S^0 state can occupy cationic, anionic or interstitial sites and S^+ state will be restricted to cationic and interstitial sites. In addition, the formation of covalent bonds between the ^{35}S and surrounding chloride ions or atoms (due to trapped holes) must be taken into account for all states of sulphur with the exception of S^{2-} and possibly S^- (even the bond between potassium and chloride ions in the crystal has partially covalent character).

The thermal neutron bombardment is usually accompanied by gamma radiation (including the capture gamma photons from the (n,γ) reaction on chlorine) and sometimes by fast neutron irradiation. Fast neutrons produce vacancies and interstitials. Gamma radiation produces mainly free electrons and holes. Electrons and holes can travel through the crystal lattice until they are recombined or trapped at vacancies or impurities.

The surroundings of ^{35}S recoils are rich in defects and therefore both electrons and holes may be trapped there. As the neutron irradiation goes on, the ^{35}S recoils are slowly oxidized from the S^{2-} and/or S^- state to some higher oxidation state. On the other hand, the lattice imperfections like vacancies and interstitials should not be affected, if the neutron irradiation is carried out at low temperatures.

Thermal treatment of the irradiated samples results both in the release of trapped electrons and the annealing of the lattice disorder in the surroundings of the ^{35}S. The released electrons can convert the ^{35}S recoils into sulphide ions, or perhaps S^- ion. (This is demonstrated in fig. 17.7.) During the rearrangement of the disordered zones containing the ^{35}S recoils some ^{35}S atoms or ions move into new positions. If this movement terminates in cationic or interstitial sites, such transfer must be associated with a change in the charge state. The S^+ state is considered to be the precursor for the sulphate fraction,

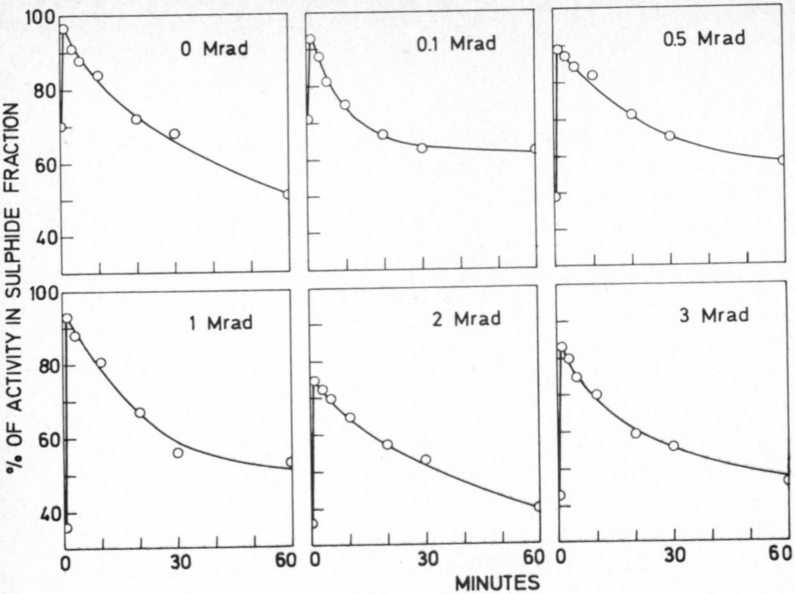

Fig. 17.7. The change in initial distribution and thermal annealing (at 150° C) behaviour in polycrystalline potassium chloride as a function of gamma radiation dose. Sample was irradiated with neutrons at dry-ice temperature, and then with cobalt-^{60}Co γ-rays again at dry-ice temperature (Bračoková 1968).

according to Maddock and Mirsky (1965). So, the oxidation of ^{35}S is observed. Bond formation with Cl^- or Cl^0 at trapped holes may also occur.

The combination of ionizing radiation and high temperature during the neutron irradiation therefore results in a higher percentage of ^{35}S in the non-sulphide fraction and because the labile defects have been annealed, the post-irradiation treatment must be carried out at a comparatively higher temperature to observe any effect.

In the case of neutron irradiation of samples pre-treated with ionizing radiation, the ^{35}S recoil takes place in a lattice with a different distribution and concentration of defects. However, the dissipation of recoil energy results in formation of a disordered zone in which the previously formed lattice defects are destroyed. Therefore the initial distribution of ^{35}S is nearly the same in samples irradiated at a dry-ice temperature as for untreated samples if the dose given is less or only about equal to the dose received during the neutron irradiation. On the other hand, the surroundings of the disordered zone contain more electronic defects compared with the untreated samples and therefore the supply of electrons and/or holes is greater and can affect the behaviour of ^{35}S during subsequent thermal annealing.

It is evident that the changes in valence states of the ^{35}S recoils depend most of all on the presence of free electrons or holes in the crystal. In their absence, the annealing changes are small, as it can be seen in fig. 17.4.

The number of free electrons and holes in targets irradiated under the commonly used conditions is much higher (by several orders of magnitude) than the number of chemical impurities in the crystal, and also their mobility in the crystal is much higher. Hence, the effect of chemical impurities (including oxygen) on the valence state of ^{35}S is most probably indirect.

7. Further Developments

Further studies of the $^{35}S^0$ fraction, using $S_2O_3^=$ scavenging, suggest that the S^0 estimated as NC^*S^- may not be derived from a single crystal precursor (Giulianelli 1969, Giulianelli and Willard 1974).

Ianovici and Taube (1975) have shown that the ^{35}S species that can be volatilised from the irradiated NaCl at high temperatures is not all derived from the $^{35}S^0$ content of the irradiated salt. The first material volatilising finds its origin in the sulphite–sulphate fraction, suggesting the formation of sulphur-chlorine compounds.

Further work on the role of small concentrations of OH^- in the irradiated salt has shown that the doped salt is especially sensitive to post n-irradiation radiolytic oxidation of the $^{35}(S^0+S^=)$ to sulphite and sulphate. The concentration of OH^- used was so low that an indirect (electronic) mechanism seems inevitable. At high γ doses some conversion of sulphate to sulphite precursor takes place (Baptista and Marqués 1974). An investigation of cyanide, sulphide and hydrosulphide doped alkali chlorides provides still stronger evidence that the sulphite and sulphate precursors arise from reaction of the ^{35}S with V type centres produced by the ionising radiation. In the cyanide doped salts both an in situ NC*S and a $^*S^0$ fraction could be distinguished showing that the *S does not diffuse far once the recoil associated events are over (Kasrai et al. 1976).

8. Concluding Remarks

An understanding of the processes influencing the chemical state of ^{35}S after nucleogenesis in inorganic chlorides necessitates first of all the use of more sophisticated chemical separation and identification procedures. At present, there are several analytical possibilities of differentiating between the S^{2-}, S^0, and S^{n+} states. A more detailed comparison would be desirable. Subsequently most of the experiments described in section 3 should be repeated, but one should take great care to define precisely the defect state of the target crystals, the irradiation conditions, and the handling of the irradiated targets, to avoid superpositioning of several effects.

The complete analysis is more laborious and time consuming, but there is no other way to obtain more accurate and quantitative conclusions concerning the behaviour of ^{35}S in crystalline chlorides.

CHAPTER 18

^{32}P IN CRYSTALLINE POLAR CHLORIDES

G.W.A. NEWTON

Department of Chemistry, University of Manchester, Manchester, UK

Chemical Effects of Nuclear Transformations in Inorganic Systems
Edited by G. Harbottle and A.G. Maddock

Contents

Introduction

^{32}P atoms can be produced in a chloride matrix by reactions such as $^{35}Cl(n,\alpha)^{32}P$ or by more complex reactions, e.g. $^{35}Cl(p,3pn)^{32}P$ using high-energy protons (Murin et al. 1961c). The reaction $^{32}S(n,p)^{32}P$ is used in solids containing sulphur. A large fraction of the work reported has been concerned with the (n,α) and (n,p) reactions; the recoil energies are about 30, 100 and 20 000 keV for the (n,p), (n,α) and (p,3pn) reactions respectively.

The chemical form of the ^{32}P atoms produced in these reactions has been the subject of many papers, but there is little agreement about either the initial valence distribution or how this is affected by pre- and post-irradiation treatment of the crystal, or by its purity and perfection.

1. Initial Valence Distribution

Unlike the atoms produced in Szilard–Chalmers reactions, the ^{32}P atoms present in a sulphate matrix, for example, are non-isotopic and their low concentration (about 10^{10} atoms g^{-1}), might lead to the valence distribution being sensitive to radiation conditions and even trace quantities of impurities.

The first non-isotopic reactions to be studied were those involving isotopes of sulphur by Andersen (1936) at Aarhus in Denmark and Kamen (1941) at the Radiation Laboratory, University of California, Berkeley. The earliest investigations of the chemical behaviour of ^{32}P (produced in NaCl) were reported by Aten (1942, 1947). This work was extended to LiCl and KCl by Caillat and Süe (1950a, b) who were the first to consider the possible importance of colour centres. They observed reductive thermal annealing of ^{32}P at 350° C and ascribed this to the discharge of electrons from F centres, the F centre being defined as an electron trapped at an anion vacancy in the lattice.

Much work has been carried out since these early experiments, and this has indicated strongly that intrinsic and extrinsic defects, impurity and radiation produced in the crystal are extremely important. However, there have been no experiments performed which definitively identify a specific defect being involved in a particular recoil reaction in the solid state. The experimental difficulties arise for three main reasons:

(i) there is no physical technique which is sensitive enough to fingerprint the ^{32}P recoil species in the solid.

(ii) the reactions taking place at the dissolution step (when the solid is analysed) are largely unknown and uninvestigated. Such experiments that have been performed (Milham et al. 1965, Baptista 1968, Meyer 1970a, b, Maddock and Mahmood 1973, 1976) indicate that variation of the solution conditions can have a profound effect on the measured valence distribution.

(iii) There are severe limitations in the analytical procedures used, and there is a need for the development of new methods. In the phosphorus system these usually involve analysis for phosphate, phosphite, hypophosphite, and occasionally phosphine. These are labelled P(V), P(III), P(I) and (P−III) respectively, and are identified in aqueous solutions in the presence of the appropriate carriers. In most discussions, and in this work, these labels refer to the unidentified precursors of these species in the solid and not to the measured solution stable species.

In view of the difficulties it is not surprising that there is a lack of agreement between different workers in the field, although results for a given sample irradiated and analysed under identical conditions are reproducible. The many results on the initial valence distribution are collected together in table 18.1, and some trends are apparent.

It would seem that the best agreement is obtained for material crystallised from solution, particularly for NaCl. In general, there is an increase in the lower valence precursors for single crystal samples, and for samples irradiated in a low gamma background (Carlson and Koski 1955), Butterworth 1964, Baptista et al. 1970). These results can be explained by assuming that, in the absence of other effects, the ^{32}P precursor is mainly of the P(I) type; this idea was first suggested by Butterworth (1964) to account for the high P(I) yields obtained in T(d,n) irradiations; it was also assumed by Bogdanov et al. (1969a) for proton irradiations. It was assumed further that any change in the P(I) concentration was due to extrinsic or intrinsic defects present in the crystal, or produced by radiation, behaving as electron traps or donors.

In most systems the initial valence distribution has little significance because it is just a measure of the extent of some annealing process at the particular temperature and radiation conditions of the experiment. In the pure KCl system there is no evidence of any such annealing until temperatures above 140° C (Andersen et al. 1971b, c, Baptista et al. 1968). However, doping with oxygen or removal of F centres by photobleaching results in the observation of an oxidative annealing process at temperatures less than 100° C (Andersen and Baptista 1971b, c). These differences in the photo and thermal annealing behaviour of ^{32}P in neutron irradiated KCl, as well as the observation of $^{32}PH_3$ by Bogdanov and Murin (1967a) have given weight to the idea of at least two precursors for the P(I) fraction. It is essential that more detailed experiments are carried out in which the purity of the sample and the radiation and

Table 18.1
The initial valence distribution of radiophosphorus in chloride matrices

Material	Condition of sample	Irradiation conditions	P(V)	P(III)	P(I)	Ref.
LiCl [a]	melted in air	Reactor –	49	50		Caillat and Süe (1950a)
LiCl [a]	–	Reactor $1.6 \times 10^{15} n_{th}\ cm^{-2}$	39	61		Butterworth (1964)
$LiCl \cdot H_2O$ [a]	–	Reactor $1.6 \times 10^{15} n_{th}\ cm^{-2}$	95	5		Butterworth (1964)
NaCl [a]	–	Reactor	–	66		Aten (1947)
NaCl [a]	–	Reactor 48 h	46	60		Caillat and Süe (1950b)
NaCl [a]	–	Reactor 15 d	41	52		Caillat and Süe (1950b)
NaCl [a]	–	Reactor	–	60		Lindner (1958)
NaCl [a]	vacuum	Reactor $1.6 \times 10^{15} n_{th}\ cm^{-2}$	45	55		Butterworth (1964)
NaCl [b]	–	Reactor	45	55		Caillat and Süe (1950a, b)
NaCl [b]	Optically decolourised	Protons(660MeV) $1.3 \times 10^{12} p\ cm^{-2}$	10	20	70	Bogdanov and Olevskii (1969a)
KCl [a]	melted in air at 900° C		48	51		Caillat and Süe (1950a)
KCl [a]	–	Reactor $1.4 \times 10^{15} n_{th}\ cm^{-2}$	45	32	23	Cifka (1964)
KCl [a]	–	Reactor $1.6 \times 10^{15} n_{th}\ cm^{-2}$	39	61		Butterworth (1964)
KCl [a]	–	Reactor $3.6 \times 10^{15} n_{th}\ cm^{-2}$	68	32		Butterworth (1964)
KCl [a]	–	Reactor $10^{18} n_{th}\ cm^{-2}$	99.2	0.8		Carlson and Koski (1955)
KCl [a]	–	Reactor $10^{18} n_{th}\ cm^{-2}$	95.3	2.8		Carlson and Koski (1955)
KCl [a]	–	Reactor –		50		Lindner (1958)
KCl [a]	irrad. in darkness	Reactor –	65	15	20	Cifka (1965)
KCl [a]	irrad. in daylight	Reactor –	60	–	–	Cifka (1965)
KCl [a]	irrad. at –78° C	Reactor $2 \times 10^{15} n_{th}\ cm^{-2}$	65	12	23	Cifka and Bračoková (1966)
KCl [a]	–	T(d,n) low gamma	12	88 P(–3)		Butterworth (1964)
KCl [a]		T(d,n) $5 \times 10^{10} n_f\ cm^{-2}$	18	43 39 P(–3)		Carlson and Koski (1955)

Table 18.1

The initial valence distribution of radiophosphorus in chloride matrices

Material	Condition of sample	Irradiation conditions	P(V)	P(III)	P(I)	Ref.
KCl [a]		(d,n) $10^{14}n_{th}\ cm^{-2}$	36	32	32	Carlson and Koski (1955)
KCl [a]	0.01 mole % Ca^{2+}	Reactor $1.4 \times 10^{15}n_{th}\ cm^{-2}$	21	36	43	Cifka (1964)
KCl [a]	0.06 mole % Ca^{2+}	Reactor $1.4 \times 10^{15}n_{th}\ cm^{-2}$	13	41	46	Cifka (1964)
KCl [a]	10^{-3} mole% KOH irrad. at −78° C	Reactor $5 \times 10^{15}n_{th}\ cm^{-2}$	60	25	12	Bračoková and Cifka (1970)
KCl [a]	0.10 mole% KOH irrad. at −78° C	Reactor $5 \times 10^{15}n_{th}\ cm^{-2}$	32	22	46	Bračoková and Cifka (1970)
KCl [a]	–	Reactor ($^{32}PH_3$)	13	12	43	Murin et al. (1961c)
KCl [b]	Harshaw single crystal	Reactor $3.6 \times 10^{15}n_{th}\ cm^{-2}$	88	12		Butterworth (1964)
KCl [b]	Kyropoulos single crystal	Reactor $4.1 \times 10^{14}n_{th}\ cm^{-2}$	53	47		Butterworth (1964)
KCl [b]	Stockbarger single crystal	Reactor $3.6 \times 10^{15}n_{th}\ cm^{-2}$	23	17	60	Baptista et al. (1968)
KCl [b]	–	Reactor $1.6 \times 10^{16}n_{th}\ cm^{-2}$	30	–	–	Andersen and Baptista (1971b)
KCl [b]	–	Protons(660 MeV) $5.7 \times 10^{11}p\ cm^{-2}$	–	–	32	Bogdanov et al. (1969b)
KCl [b]	optically decolourised	Protons(660 MeV) $6.5 \times 10^{11}p\ cm^{-2}$	30	17	54	Bogdanov et al. (1969a)
KCl [b]	–	Protons(660 MeV) $6.5 \times 10^{11}p\ cm^{-2}$	54	22	24	Bogdanov et al. (1969a)
KCl [b]	optically decolourised	Protons(660 MeV) $1.3 \times 10^{12}p\ cm^{-2}$	27	28	55 (9% PH_3)	Bogdanov and Olevskii (1969a)
KCl [b]	–	Protons(660 MeV) $2 \times 10^{12}p\ cm^{-2}$	21	19	51	Bogdanov and Murin (1967a)
KCl [b]	–	Protons(660 MeV) $2 \times 10^{12}p\ cm^{-2}$	34	10	56	Bogdanov et al. (1969a)
KCl [b]	–	Protons(660 MeV) $3.4 \times 10^{12}p\ cm^{-2}$	–	–	60	Bogdanov et al. (1969b)
KCl [b]	irrad. at 333 K	T(d,n) $1.8 \times 10^{12}n_f\ cm^{-2}$	21	–	–	Baptista et al. (1970)
KCl [b]	irrad. at 77 K	T(d,n) $1.8 \times 10^{12}n_f\ cm^{-2}$	12	–	–	Baptista et al. (1970)

KCl^{b}	irrad. at 77 K	Reactor $3.0 \times 10^{14} n_{th}\ cm^{-2}$	25	75		Baptista et al. (1970)
KCl^{b}	irrad. at 77 K	Reactor $3.6 \times 10^{15} n_{th}\ cm^{-2}$	42	58		Baptista et al. (1970)
KCl^{b}	7.5 ppm OH^{-}	Reactor $3.6 \times 10^{15} n_{th}\ cm^{-2}$	77	9	14	Baptista et al. (1968)
KCl^{b}	< 10 ppm O_2	Reactor $1.6 \times 10^{16} n_{th}\ cm^{-2}$	25	–	–	Andersen and Baptista (1971b)
KCl^{b}	0.5 mole % K_2S	Protons(660 MeV) $4.7 \times 10^{11} p\ cm^{-2}$	61	20	20	Bogdanov and Olevskii (1969b)
KCl^{b}	0.05 mole % Ca^{2+}	Protons(660 MeV) $6.5 \times 10^{11} p\ cm^{-2}$	31	23	47	Bogdanov et al. (1969a)
KCl^{b}	0.05 mole % Ca^{2+}	Protons(660 MeV) $2 \times 10^{12} p\ cm^{-2}$	28	13	59	Bogdanov et al. (1969a)
KCl^{b}	0.10 mole % Ca^{2+}	Protons(660 MeV) $2 \times 10^{12} p\ cm^{-2}$	40	17	40	Bogdanov and Murin (1967a)
$BaCl_2{}^{a}$	–	Reactor $2 \times 10^{15} n_{th}\ cm^{-2}$	10	20	70	Butterworth (1964)
$BaCl_2{}^{a}$	0.077 mole % M^{+}	Reactor $2 \times 10^{15} n_{th}\ cm^{-2}$	5	10	85	Cífka and Bračoková (1970)
$BaCl_2{}^{a}$	0.03 mole % M^{+}	Reactor $2 \times 10^{15} n_{th}\ cm^{-2}$	9	20	71	Cífka and Bračoková (1970)

[a] Polycrystalline material. [b] Single crystal material.

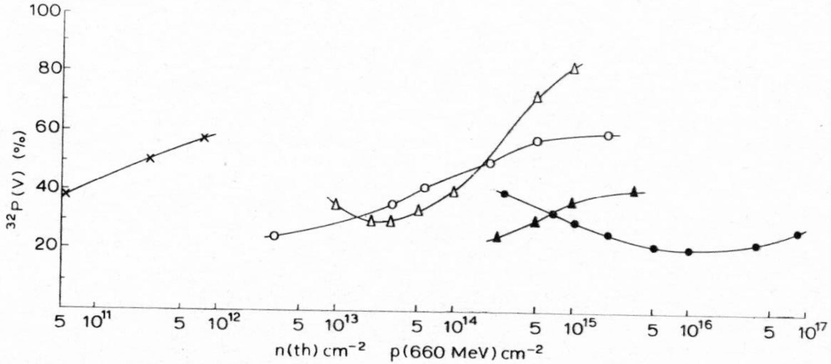

Fig. 18.1. Variation of the initial valence distribution of ^{32}P in KCl as a function of time of irradiation. (X) Bogdanov et al. 1969a, (O) Cífka and Bračoková 1966, (△) Butterworth 1964, (•), (▲), Baptista et al. 1970.

dissolution conditions are more carefully controlled before a better understanding of the initial valence distribution of ^{32}P can be obtained. Nevertheless, some indication of the role of impurities and radiation conditions have been obtained, and these will now be discussed.

Recent investigations using irradiated $AlCl_3$ and $(CH_3)_4N.Cl$ and non-aqueous methods of analysis indicate that the crystal precursors are all low oxidation state species of ^{32}P and that aqueous solution involves them in oxidation reactions. (Maddock and Mahmood, 1976.)

2. Effect of Impurities

As in the case of the initial valence distribution there is little agreement about the effects of impurities. There is evidence by Baptista et al. (1968) that the presence of 3 ppm OH^- increases the concentration of $^{32}P(V)$ in the crystal. On the other hand, very much larger concentrations of OH^- seemed to have no effect (Bračoková and Cifka 1970). The difference is possibly because there is a plateau concentration which was already exceeded with the lowest concentrations used in the latter experiment; 3 ppm is already a large excess over the number of ^{32}P atoms present.

Impurities other than OH^- seem to be important in determining the ^{32}P valence distribution as shown by Cifka (1964), Baptista et al. (1968) and Andersen and Baptista (1971b). Evidence has been given by Andersen and Baptista that electron excess centres can be introduced into oxygen doped KCl. These not only lower the thermal stability of the intrinsic radiolytic defects, but probably compete with ^{32}P recoils for V centres or their decomposition products (see also Kanzig and Cohen 1959).

3. Effects of Ionising Radiation

From the foregoing it can be seen that a study of the effects of ionising radiation is best carried out on pure samples in order to minimise the variables involved; this has not always been the case.

Changes of valence distribution with time of irradiation emphasises the importance of radiation-induced defects; these are shown in fig. 18.1 for KCl. There is some agreement between the two sets of data on polycrystalline samples, but these are different from the results on single crystal material. The fact that the slopes of all these curves change with irradiation dose, and that a different slope is obtained when irradiation is carried out at 77 K indicates that two or more competing processes are involved. The results for polycrystalline samples could be explained by the production of an oxidising radiolysis product of an impurity in the samples. The results on the pure single crystals can be explained in terms of six competing reactions (Baptista et al. 1970):

^{35}Cl	+ neutrons	$\rightarrow {}^{32}P_H \rightarrow {}^{32}P_T$	1
	gamma	$\rightarrow$ S,L,X	2
$^{32}P_H$	+ S	$\rightarrow {}^{32}P_{(ox)}$	3
$^{32}P_H$	+ L	$\rightarrow {}^{32}P_{(ox)}$	4
$^{32}P_T$	+ S	$\rightarrow {}^{32}P_{(ox)}$	5
S	+ X	$\rightarrow$ inactive species	6

where $^{32}P_H$ = low valence ^{32}P recoil species before it has lost all its kinetic energy (hot reaction); $^{32}P_T$ = low valence ^{32}P recoil species in thermal equilibrium with the KCl lattice; S = short-lived hole species; L = long-lived hole species; X = long-lived hole trapping species.

The time scales implied are those consistent with the idea that S,L and X could be produced by pre-gamma irradiation, but S species would decay away before the ^{32}P recoils were produced in-pile. These reactions can only be interpreted by considering the interrelation between pre- and post-gamma irradiations and reactor irradiations at ambient temperature and 77 K. The distinction between pre- and post-gamma should be emphasised. In the former case a decrease in the yield of ^{32}P(V) with increasing gamma dose is observed; whereas in the latter case the reverse is true. These facts have been observed by several workers (Carlson and Koski 1955, Butterworth 1964, Bogdanov et al. 1969a, Baptista et al. 1970) in KCl and other systems (Baumgärtner and Maddock 1968), but they are by no means universal; it would seem that ^{32}P in LiCl is insensitive to post-gamma irradiation for example (Butterworth 1964). The species present in the crystal are probably different for the two cases, because the irradiation conditions and the measured valence distribution for LiCl and KCl are different as shown in table 18.1. It would be of value to have a quantitative estimate of the species shown in the six reactions above, but so far this has not been possible because of the lack of information concerning their identity and their rates of formation.

Qualitative arguments suggest that pre-gamma irradiation increases the concentration of L and X prior to reactor or post-gamma experiments. When comparing liquid nitrogen and ambient temperature irradiations, the rates of trapping of electrons and holes will be different because this depends on temperature as well as dose rate among others (Schulman and Compton 1962). The observation that post-gamma irradiation causes oxidation at room temperature but not at 77 K (Baumgärtner and Maddock 1968, Boyd and Larsen 1968b, Baptista et al. 1970) could imply that an energy of activation was involved in the reaction between $^{32}P_T$ and radiolysis products (S), or that S is not produced at 77 K. The former seems the most likely. It is considered that S species rather than L are involved because post-gamma irradiations are very much more sensitive than pre-gamma irradiations. (One can only speculate as to the nature of S species; they could be free holes or short-lived hole species of the type observed in pulse radiolysis experiments [Ueta 1967, Baptista 1968].) There is no reaction between $^{32}P_T$ and S,L or X species at 77 K; therefore in reactor irradiations at 77 K any observed reactions must be between $^{32}P_H$ and radiolysis products, that is 'hot' reactions. Experimental evidence is available that both transient and stable radiolysis products are involved in reactions with $^{32}P_H$ during reactor radiations at 77 K (Baptista et al. 1970), but the former are more effective.

Further work by Baptista and Marqués (1974) shows that the OH^- content of the crystals has a profound effect on post γ irradiation oxidation of ^{32}P species. These effects are suppressed at low temperatures.

Bogdanov and Olevskii (1973) have found that pre γ irradiation of KCl enhances the yield of ^{32}P in low oxidation states. But the OH^- content of their material is not stated.

4. Photo and Thermal Annealing

Figure 18.2 summarises the position concerning thermal annealing and photo annealing in pure and doped KCl systems.

In the pure system, annealing commences at about 140° C, which coincides with the onset of cation vacancy transport; the activation energy for this stage measured by Baptista et al. (1968) also agrees with the concept that cation

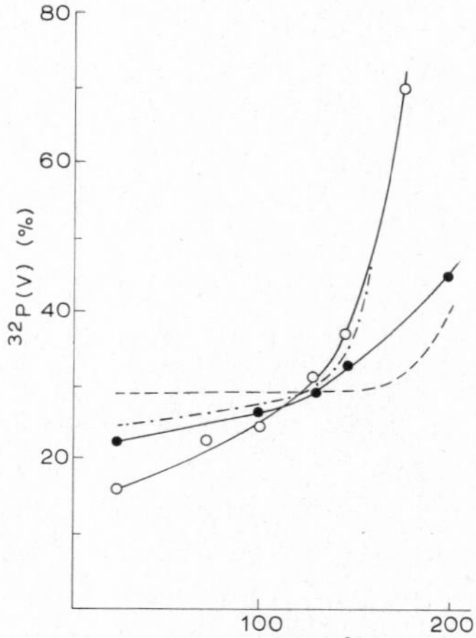

Fig. 18.2. Isochronal thermal annealing (4h) of ^{32}P in pure and oxygen doped KCl single crystals. O P(V) – pure KCl after bleaching of F centres, • P(V) – doped KCl after F centre bleaching, - - - -P(V) – pure KCl before bleaching,-·-·-· P(V) – doped KCl before bleaching.

diffusion is the rate controlling step. This is understandable, because any oxidative reaction in the crystal must involve a cation vacancy (equivalent to a single negative charge in a 1:1 ionic crystal) to maintain electrical neutrality.

Competitive reactions have been observed in some crystals, and these have been described as reductive and oxidative annealing stages. However, because of the complex nature of those crystals in terms of purity and crystal perfection it is difficult to unravel the different competitive processes involved.

In the earlier part of the curve shown in fig. 18.2, that is at temperatures below 140° C, there is evidence of a different stage appearing in the annealing process, which is brought about either by doping or bleaching. This is argued as evidence for the existence of at least two P(I) species in the irradiated crystals (Andersen and Baptista 1971b, c). It has previously been shown by Bogdanov and Bondarevskii (1967) that the species obtained using a P(I) carrier only can be subdivided into a P(−III) (phosphine) and a P(I) fraction, that is there are at least two types of P(I) species in the irradiated crystals.

Photo-annealing of crystals is not well understood; initially, Cifka (1965) suggested that the observed photoreduction was a direct reaction between P^{+5} and an electron in the crystal. It is unlikely that P^{+5} exists as such in the crystal. Bogdanov et al. (1967) have given a detailed account of both photoreduction and photo-oxidation of radio-phosphorus, both of which occur in their KCl crystals. They imply that major changes take place in the structure of the phosphorus species during the bleaching of F and V centres. More recent studies by Andersen and Baptista (1971b, c) imply that this drastic reorientation is unlikely because they observe an easy, direct and reversible conversion of a P(I) species to a P(V) species in the crystal below 140° C.

In conclusion it can be said that much has been achieved in understanding the role of various factors which control the valence distribution of radiophosphorus in various systems; much more remains to be done. Three main areas which could be investigated are:

(a) the controlled addition of defects to well-defined systems,
(b) analytical procedures for identifying forms of radio-phosphorus other than P(I), P(III) and P(V),
(c) the dissolution step; investigation of the reactions which take place between the precursor and the solvent.

CHAPTER 19

CHEMICAL REACTIONS OF RECOIL ATOMS GENERATED BY NUCLEAR FISSION

U. WAGNER-ZAHN

Physik-Department E15, Technische Universität München,
Garching bei München, F.R. Germany

Chemical Effects of Nuclear Transformations in Inorganic Systems
Edited by G. Harbottle and A.G. Maddock

Contents

1. Introduction

Energetic recoil atoms, such as ^3H and ^{14}C from $^6Li(n,\alpha)$, $^3He(n,p)$, and ^{14}N (n,p) reactions impinging on appropriate compounds have been widely used to produce radioactively labelled substances (Rowland et al. 1961b, Urch and Wolfgang 1961b, Wolf 1961, Stöcklin and Vogelbruch 1968). Synthesis of radioactively labelled molecules may also result from chemical reactions of fission fragments with suitable organic or inorganic 'catcher' materials. Chemical reactions of the primary fission products, which are formed with energies of about 100 MeV, are usually accompanied by reactions of the recoil atoms, which originate in the β^- decay of the primary products and possess recoil energies of the order of magnitude of chemical bond energies. This complicates the interpretation of the mechanism of complex formation in fission synthesis considerably.

The behaviour of both iodine and transition metal atoms recoiling into various catcher media following fission has been studied extensively. In the recoil chemistry of fission iodine organic catcher systems are used preferentially. In the present chapter these reactions will only be treated very briefly. The recoil synthesis of fission product labelled metal organic compounds, on the other hand, will be surveyed more extensively and the possible mechanisms leading to molecule formation will be discussed.

2. Recoil Chemistry of Fission Iodine Atoms

Walton et al. (Walton 1957, Walton and Croall 1955, Hall and Walton 1959, 1961) studied the radiolytic decomposition caused by fission products striking crystalline nitrates and iodates, as well as the reactions of iodine recoil atoms growing in by β^- decay of their precursor fission products. Ormond and Rowland (1961) observed the formation of iodobenzene after neutron bombardment of benzene in quartz tubes which were coated with U_3O_8. More than 50% of the secondary fission ^{131}I was found to be bound as iodobenzene, which indicates a high reactivity of iodine recoil atoms formed by β^- decay in benzene. Recoil reactions of iodine formed by the β^- decay of ^{132}Te ($T_{1/2} = 74$ h), without a preceding fission process, have indeed been observed in solutions of $^{132}TeCl_4$ in benzene (Bacher and Adloff 1962). The experiments of Denschlag et al. (1963) Denschlag and Gordus 1967) show that in CH_4 the primary fission iodine preferentially forms organic iodocompounds, whereas no iodination by primary fission iodine was observed in benzene. It was therefore possible to deduce values for cumulative fission yields from isotopic ratios of the iodine isolated as iodobenzene. Amiel and Paiss (1965, Paiss and Amiel 1964) have studied the substitution of primary fission iodine in methyl iodide, CH_3I, in which a high efficiency could be expected for the displacement of the stable iodine by fission iodine, since the replacing and the replaced atoms have nearly the same mass. Atoms created by β^- decay cannot displace the iodine in CH_3I by direct collisions, but can react with radicals produced by radiolytic damage. This radiation-induced isotopic exchange may mask the selectivity of the hot displacement reactions of the energetic primary fission products. Baumgärtner (1965) studied the reactions of fission iodine recoiling into halobenzenes and alpha- and betabromonaphthalene. The high yield of 85% of iodobenzene formation when the halobenzene, C_6H_5X, is used as collector, or 'catcher' shows that the substitution of the halogen atom by the fission iodine by far exceeds hydrogen substitution. In the bromonaphthalenes the substitution of the bromine by the fission iodine was found to take place with conservation of isomeric form. More recently Blachot and Carras (Blachot and Vargas 1967, Blachot and Carras 1969) have isolated and studied a number of fission iodine isotopes by using tetraphenyltin, $(C_6H_5)_4Sn$, as catcher for the fission recoil atoms. The newly formed phenylcompounds of iodine, tellurium and antimony can be separated from the fission products and even from the carrier compound by chromatography or fractional sublimation.

3. Recoil Chemistry of Fission Molybdenum Atoms

3.1. METAL HEXACARBONYLS AS CATCHERS

In the thermal fission of ^{235}U, ^{99}Mo is formed with a cumulative yield of 6.06% (Coryell and Sugarman 1951, Reed and Turkevich 1953, Terell et al. 1953, Zicha et al. 1967, Wiles and Coryell 1954). If it recoils into chromium hexacarbonyl, $Cr(CO)_6$, up to 60% of the ^{99}Mo fission nuclei can replace the central chromium atom of the carbonyl and form radioactively labelled $^*Mo(CO)_6$, which is isostructural with the catcher compound. Baumgärtner and Reichold (1961d) bombarded finely powdered mixtures of $Cr(CO)_6$ with 0.028 mol% to 40 mol% U_3O_8 for about 30 min in a thermal neutron flux of 10^{13} n/cm$^2\cdot$s. After 24 h the volatile fission $Mo(CO)_6$ was separated, together with the Cr-$(CO)_6$ carrier, from all of the other fission products and recoil fragments by vacuum sublimation at 90° C through a column of glass beads. Besides the γ-rays of ^{51}Cr, the γ-spectrum of the sublimate showed only lines due to ^{99}Mo

and its daughter product ^{99m}Tc, since the other short-lived fission molybdenum isotopes had decayed before the separation. Further resublimations and repeated purifications by chromatography do not change the amount of ^{99}Mo per mg $Cr(CO)_6$ carrier. The identity of the ^{99}Mo compound formed by recoil synthesis with $Mo(CO)_6$ could be proved by its analogous elution behaviour to a mixture of labelled $^{99}Mo(CO)_6$ and $Cr(CO)_6$ on an aluminium oxide column. The fact that no fission radioisotopes other than ^{99}Mo and its Tc daughter could be detected in the sublimate shows the high selectivity of the incorporation reaction.

The mechanism of the reaction

$$^{235}U\ (n,f)\ {}^{*}Mo \xrightarrow{Cr(CO)_6} {}^{*}Mo(CO)_6$$

is not well understood. Since the primary fission yield of ^{99}Mo in the thermal fission of ^{235}U is only 2% of the total (Pappas 1955, Coryell 1952) and since, on the other hand, depending on experimental conditions, about 60% of the fission molybdenum nuclei can be isolated as hexacarbonyl, secondary molybdenum must participate in the complex formation.

Zicha and Zahn (1972, Zicha 1966) have undertaken a number of experiments aimed at a better understanding of the mechanism of the fission synthesis of $Mo(CO)_6$. No γ-rays due to ^{99}Nb were observed in the sublimate of the samples, which were separated quickly enough after the end of the irradiation for the 2.4 min isotope ^{99}Nb to be still present in measurable amounts. This shows that no volatile niobium carbonyl compounds are formed in fission synthesis and that fission Mo present in the sublimate must have reached there as a volatile Mo-compound. In order to study the recoil chemical behaviour of primary fission Mo separately from that of the Mo originating in the β^- decay of Nb precursors the time interval between the end of the bombardment and the separation was varied from 6 s to 26 min and the Mo activities per unit mass of $Cr(CO)_6$ were compared. The ratio of the ^{99}Mo activity in the sample separated 6 s after the end of the irradiation to the activity in the sample which was separated after the decay of ^{99}Nb ($T_{1/2} = 2.4$ min) was 0.5. This proves that secondary fission Mo forms $Mo(CO)_6$ in a $Cr(CO)_6$ matrix, but it does not give information about the incorporation of primary fission Mo atoms, since ^{99}Mo is also formed, in about 50% of the decay events, from an isomer, ^{99m}Nb, with a half-life of 10 s.

Information on both the incorporation probability p of primary fission Mo and the incorporation probability s of secondary fission Mo could, however, be obtained (Zicha and Zahn 1972, Zicha 1966) from measurements of the ratios

$$k = (pY_M + sY_N)/(Y_M + Y_N),$$

for the masses 99 and 101, where Y_M and Y_N are the primary and secondary fission yields, respectively. The experimental values $^{99}k = 0.495$ and $^{101}k = 0.500$ turned out to be equal within the error of about 3% although the ratios of the primary and secondary fission yields are widely different for ^{99}Mo and ^{101}Mo (Wahl et al. 1962). Therefore the experimentally found equality of ^{99}k and ^{101}k can only be understood if $p = s = k$ to better than about 10%.

This result shows that the incorporation of fission Mo into $Cr(CO)_6$ occurs with the same probability for primary and secondary Mo atoms. It is independent of the recoil energy, which for the primary fission Mo is larger by a factor of about 10^7 than for Mo from β^- decay of Nb precursors. The incorporation probability depends somewhat on the mixing ratio of U_3O_8 and $Cr(CO)_6$ since, when the uranium content increases, fission recoil atoms are stopped in the uranium oxide (Baumgärtner and Reichold 1961d).

Considering these findings one can exclude impact substitution as a mechanism of hexacarbonyl formation. Baumgärtner and Schön (1963, 1964) suggest β^- decay of the Ru precursors as the cause of the fission synthesis of $^{103}Ru(C_5H_5)_2$ in a similar system. A similar hypothesis could not explain why $Mo(CO)_6$ formation occurs with the same probability for primary and secondary fission Mo atoms. It seems more reasonable to suggest a purely chemical mechanism, similar to some extent to the transfer annealing of ^{51}Cr in CrO_4^{-} observed by Collins et al. (1965). Recent experiments (Zahn 1972, unpublished) showing that about 5% of carrier-free ^{99}Mo can be incorporated in $Cr(CO)_6$ by simple heat treatment, tend to support the idea of a chemical, recoil energy independent reaction mechanism for the formation of $Mo(CO)_6$. The high yield of $^{99}Mo(CO)_6$ in the fission synthesis may be due to the perfect mixing achieved by one of the components being injected into the other by the fission recoil. A few eV of excitation energy, available both after β^- decay and the stopping of a fast primary fission product may play an important role in stimulating this reaction process. Moreover radiation decomposition, caused by the slowing down of the fission product itself, may provide additional partners which may recombine with fission Mo atoms.

It is more difficult to understand, in terms of a conventional chemical reaction mechanism, why the $^{99}Mo(CO)_6$ yield is completely independent of annealing at elevated temperatures (Baumgärtner and Reichold 1961a, d), whereas in neutron bombarded $Cr(CO)_6$ as well as in $Mo(CO)_6$ a strong increase of metal retention with the annealing temperature was observed (Zahn et al. 1969). It might, however, be possible that a solid state exchange mechanism leading to the incorporation of fission Mo in $Cr(CO)_6$ occurs immediately after the nuclear process and hence has reached its equilibrium state before separation and annealing procedures can be performed.

The incorporation of fission Mo into $Cr(CO)_6$ has found applications in nuclear chemistry. By means of the selective and extremely fast separation of Mo from all the other fission products in an automatic on-line sublimation apparatus (Meier-Komor 1968) a number of short-lived fission Mo species could be isolated and identified for the first time. The half-lives of ^{103}Mo through ^{107}Mo and the disintegration behaviour of these isotopes and of their Tc and Ru decay products were extensively studied (Zicha 1966, Kienle et al. 1962a, b, 1963a, b, Meier-Komor 1968, Maul 1969, Zicha 1969). New isotopes were identified on the basis of genetic relations established by measurements of the intensities of characteristic γ-lines and of β-decay curves. Moreover the half-lives of the Mo precursors ^{103}Nb through ^{106}Nb were determined or, at least upper limits could be given (Maul 1969). With this information cumulative fission yields for Mo isotopes could be calculated from the β^- decay curves of the Tc and Ru daughter isotopes. The relative fission yields for $^{102}Mo : ^{103}Mo : ^{104}Mo : ^{105}Mo$ were found to be $(1.48 \pm 0.40) : 1 : (0.49 \pm 0.10 : (0.27 \pm 0.05)$ (Zicha et al. 1967). This indicates that the yield of ^{102}Mo, presumably the fission product complementary to the double magic ^{132}Sn nucleus, is enhanced over the yield predicted by the theory of Wahl et al. (1962).

3.2. DIBENZENECHROMIUM AS CATCHER

Fission Mo atoms impinging on dibenzenechromium, $Cr(C_6H_6)_2$, react (Baumgärtner 1965) with formation of the isostructural, radioactively labelled compound $Mo(C_6H_6)_2$, which can be isolated from the fission products by sublimation. Moreover a certain fraction of the Mo atoms which undergo β^- decay has been found in the parent molecular structure as unsublimable $^{99m}Tc(C_6H_6)_2^+$ cation. The formation of $^{99m}Tc(C_6H_6)_2^+$ had originally been observed after the β^- decay of ^{99}Mo in neutron activated $Mo(C_6H_6)_2$ (Baumgärtner 1965). A successful conventional synthesis of dibenzenetechnetium compounds has been stimulated by this observation (Palm et al. 1962a). Experiments with dibenzenechromium as catcher for fission recoil atoms are complicated by the high oxidation sensitivity of $Cr(C_6H_6)_2$ and $Mo(C_6H_6)_2$, which requires complete exclusion of oxygen during irradiation and sublimation.

4. Recoil Chemistry of Fission Ruthenium Atoms

4.1. DICYCLOPENTADIENYL COMPOUNDS AS CATCHERS

The first observation of a fission synthesis of metal organic compounds was the formation of $^{103}Ru(C_5H_5)_2$ in a neutron bombarded mixture of U_3O_8 and ferrocene, $Fe(C_5H_5)_2$, reported by Baumgärtner and Reichold (1961b). At the same time hydrogen atoms in the cyclopentadiene ligands are substituted by fission iodine yielding iodoferrocene, $C_5H_5FeC_5H_4^{131}I$. Apart from the iodine incorporation, the ruthenium incorporation is highly selective, since of all the fission products only Ru and I form volatile compounds with the ferrocene catcher.

The separation of the radioactive $Ru(C_5H_5)_2$ from the fission products together with the iodoferrocene and the ferrocene carrier is accomplished by sublimation or by extraction with ether. For a mixture containing 1.3 mol% U_3O_8 per mol $Fe(C_5H_5)_2$ the yield of $^{103}Ru(C_5H_5)_2$ is about 60%. Due to self absorption of the fission recoil atoms in the uranium oxide it decreases slowly as the U_3O_8 content rises, until, for 33 mol% U_3O_8 the yield of $^{103}Ru(C_5H_5)_2$ and $^{105}Ru(C_5H_5)_2$ has dropped to about 26%.

Considering that the primary fission yield of ^{103}Ru is only $< 1\%$ of the chain yield of mass 103 (Wahl et al. 1962), the high yield of labelled $^{103}Ru(C_5H_5)_2$ shows that the essentially secondary ^{103}Ru formed by the β^- decay of the ^{103}Mo and ^{103}Tc precursors has reacted to form ruthenocene. As in the case of the formation of fission $Mo(CO)_6$, the reaction mechanism is not well understood. Baumgärtner (1965) and Baumgärtner and Schön (1964) have performed a number of experiments to study the mechanism of molecule formation in fission synthesis. They used mainly ^{105}Ru as an indicator, which has a primary fission yield of only 1% (Wahl et al. 1962). Their results and conclusions will now be summarized.

Conceivably the substitution of the central iron atom by a fission ruthenium atom in ferrocene might proceed via the formation of intermediate complexes with precursor atoms in the place of the iron atom. These intermediate complexes, formed by collisions of the fission products with the iron atoms, would then become stable ruthenocene molecules after the β^-decay of Mo or Tc precursors. However, neutron bombarded mixtures of 33 mol% U_3O_8 and $Fe(C_5H_5)_2$ heated to 230° C at a time when only about 5% of the ^{105}Ru is formed, yield as much $^{105}Ru(C_5H_5)_2$ as mixtures which were not heated. This means, that the intermediate complexes would have to be stable in the molten state, which is most unlikely. Therefore such intermediate complexes do not seem to exist. Baumgärtner and Schön (1964) also observed no temperature dependence of the $^{105}Ru(C_5H_5)_2$ yield in samples which were stored at 4.2 K, 77 K, and 200 K immediately after the end of the irradiation. From this they conclude, that thermal reactions between the fission Ru atoms and the surrounding molecules or molecular fragments would not occur. This does not seem entirely conclusive, since such reactions might have taken place at a high rate at the time of separation, which was done by dissolution in chilled solvents. Such recombinations can proceed even at low temperatures as has been shown by studies of the metal retentions after (n,γ) reactions in cyclopentadienylmanganesetricarbonyl, $C_5H_5Mn(CO)_3$. In this system 80% of the $C_5H_5{}^{56}Mn(CO)_3$ which is capable of recombination has reformed at −50° C (Zahn 1967b). The $^{105}Ru(C_5H_5)_2$ yield furthermore turns out to be constant over a γ-dose range from 5×10^4 to 10^7 r. This indicates that reactions induced by radicals, which were produced by the reactor γ-radiation, do not play an important part in complex formation.

All these observations make Baumgärtner and Schön assume that, for the cyclopentadienyl system, complex formation with secondary fission products is a result of β^- decay and the concomitant electronic excitation. The primary ruthenium fission yields are too small for any conclusion as to the reactions of primary Ru fission products to be drawn from these experiments. One should, however, note that the fission $Mo(CO)_6$ formation in $Cr(CO)_6$ proceeds with the same probability for primary fission products and for Mo formed in β^- decay (Zicha and Zahn 1972).

The use of different homologous compounds as catchers (Baumgärtner and Schön 1964) shows that chemical bond considerations have to be taken into account in the interpretation of the reaction probabilities. For mixtures containing 33 mol% U_3O_8 the $^{105}Ru(C_5H_5)_2$ yield in a ferrocene catcher is 25%, in $Ru(C_5H_5)_2$ it is 10% and at 8% it is lowest in $Os(C_5H_5)_2$, which has the highest bond stability. With Ru nearly complete energy exchange between the original central atom and the recoil atom would be possible. Therefore the yield should be highest in a $Ru(C_5H_5)_2$ matrix, if direct impact substitution were the reaction mechanism. However, the data show that complex formation takes place in an energy range where chemical bond stability seems to be of importance. On the other hand this behaviour does not rule out the possibility that the formation of radioactively labelled $Ru(C_5H_5)_2$ in a metal cyclopentadienyl matrix is due to a solid state exchange mechanism, similar to the one proposed earlier for the formation of fission $Mo(CO)_6$ in a $Cr(CO)_6$ catcher. This concept finds further support in experiments by Baumgärtner and Reichold (1961a), who observed the formation of radioactively labelled ruthenocene in neutron bombarded mixtures of ruthenium compounds and ferrocene. Although Baumgärtner and Reichold attribute the ruthenocene formation to the reaction of the more energetic recoil atoms from resonance neutron capture, one could as well assume a solid state exchange reaction which at room temperature has already reached its equilibrium state.

The experiments of Zahn and Harbottle (1966), who have studied the reactivity of ^{103}Ru from thermal fission of ^{235}U in liquid mono- and dicyclopentadiene and of $^{106}RuCl_3$ in C_5H_6, show that the important contribution to complex formation comes from conventional chemical reactions. Up to 80% $^{103}Ru(C_5H_5)_2$ formation was observed in C_5H_6 but only 2.5% in $C_{10}H_{12}$. The high yield of the fission synthesis in the monomer catcher can only be explained by thermal reactions, since the monomer as well as the dimer cyclopentadiene should present almost the same cross section to a hot recoil atom. The $^{106}Ru(C_5H_5)_2$ yield of 90% in a mixture of $^{106}RuCl_3$ and C_5H_6 can, of course, only be brought about by thermal reactions.

4.2. CHROMIUM HEXACARBONYL AS CATCHER

Rutheniumpentacarbonyl, $Ru(CO)_5$, is known as an unstable, highly volatile compound, which can be isolated at low temperature (Hieber and Heusinger

1957). Baumgärtner and Reichold (1961b, d) had not been successful in detecting fission $Ru(CO)_5$ in the sublimate from a neutron bombarded mixture of U_3O_8 and $Cr(CO)_6$. Cooling the condensate to −50° C Zicha (1966) could identify ^{103}Ru and ^{105}Ru in the sublimate of a sample which before the separation had been stored long enough to allow the ^{103}Mo and ^{105}Mo nuclei to decay completely. Under these conditions ^{103}Ru and ^{105}Ru cannot appear in the sublimate as decay products of their Mo precursors. They can, however, be accounted for by assuming the formation of sublimable $Ru(CO)_5$.

4.3. METAL ACETYLACETONATES AS CATCHERS

In neutron bombarded mixtures of UO_2 and cobalt acetylacetonate, $Co(C_5H_7O_2)_3$, Meinhold and Reichold (1968, 1969) have observed the formation of $^{103}Ru(C_5H_7O_2)_3$. The labelled compound was identified by its r_f value in column chromatography. The observed value coincides with that of carrier amounts of $Ru(C_5H_7O_2)_3$. The probability of fission iodine incorporation into the acetylacetonate ligands by hydrogen substitution is 44 to 49% (Meinhold and Reichold 1971). Labelling cannot be explained as a consequence of hot reactions, but occurs in a similar manner to conventional chemical reactions.

Contrary to the carbonyl and cyclopentadienyl systems, the initial $^{103}Ru(C_5H_7O_2)_3$ yield in the $Co(C_5H_7O_2)_3$ matrix can be enhanced by thermal treatment. A plateau value of 63% is reached after annealing for 5 h at 150° C. The isothermal annealing behaviour is similar to that observed after (n,γ) reaction in $Co(C_5H_7O_2)_3$ (Meinhold and Reichold 1970, Shankar et al. 1961c, b, 1965a). The formation of $^{103}Ru(C_5H_7O_2)_3$ can also be observed in other target materials like aluminium, chromium, iron, and rhodium acetonylacetonate. Whereas the $^{103}Ru(C_5H_7O_2)_3$ yield in the $Ru(C_5H_7O_2)_3$ catcher is almost the same as in the $Co(C_5H_7O_2)_3$ catcher, it is much lower in the other three compounds. The increase of the $^{103}Ru(C_5H_7O_2)_3$ yield with the annealing temperature is more pronounced in the cobalt and rhodium acetylacetonate matrices. A collision substitution mechanism for the reaction can be excluded, since the $^{103}Ru(C_5H_7O_2)_3$ yield is independent of the mass of the metal atom in the acetylacetonate catcher. Therefore the fission synthesis of ruthenium acetylacetonate seems to be a solid state exchange reaction rather than a hot reaction. Such solid state exchange reactions have so far only been known to involve isotopic atoms like ^{60}Co in $Co(C_5H_7O_2)_3$ (Nath et al. 1966) and ^{51}Cr in CrO_4^{--} (Collins et al. 1965). However Meinhold and Reichold could show (1970) that 23% of the ^{103}Ru atoms deposited on the surface of $Co(C_5H_7O_2)_3$ crystals have formed $^{103}Ru(C_5H_7O_2)_3$ after 12 h at 150° C and that the reaction velocity increases with increasing temperature.

5. Recoil Chemistry of Various Fission Atoms

A number of fission recoil atoms, which have not been dealt with in the preceding section are known to form complexes when recoiling into appropriate catcher media. Although these reactions may not have been studied extensively, they will be briefly summarized.

Meinhold and Reichold have observed the incorporation of fission ^{105}Rh into cobalt acetylacetonate, $Co(C_5H_7O_2)_3$, (1968, 1969, 1970). The newly formed $^{105}Rh(C_5H_7O_2)_3$ was identified by column chromatography. In the same catcher the fairly unstable $^{99}MoO_2(C_5H_7O_2)_2$ is also formed.

An appreciable formation of $^{97}Zr(C_5H_7O_2)_4$ was observed in, and only in, the tetravalent $Th(C_5H_7O_2)_4$ as catcher. This indicates that the fission product only interacts with the constituents of a single catcher molecule. Zvara et al. (1963) have produced $^{97}ZrCl_4$ with a yield of 75% by blowing Cl_2 gas and $ZrCl_4$ carrier through an ampoule which contains an uranium foil and is located in a Po–Be source. In a N_2 stream no ^{97}Zr is transported. In N_2 containing $ZrCl_4$, however, $^{97}ZrCl_4$ formation also takes place. This indicates that an exchange mechanism is responsible. Since the independent yield of ^{97}Zr is about half of its cumulative yield (Wahl 1958), $^{97}ZrCl_4$ is formed from primary and secondary fission atoms and it is not possible to distinguish between the contribution of hot and thermal reactions in the formation of $^{97}ZrCl_4$.

By chromatography of benzene solutions Blachot and Vargas (1967) isolated fractions containing $(C_6H_5)_2^{132,\,133,\,134}$Te, and $(C_6H_5)_3^{130,131}$Sb from a neutron bombarded mixture of UO_2 and tetraphenyltin, $(C_6H_5)_4Sn$. The volatile compounds $(C_6H_5)_2Te$ and $(C_6H_5)_3Sb$ can be separated from the fission products and even from the catcher compound much faster by fractional sublimation in a long tube along which a temperature gradient exists. Using this separation method the decay of ^{134}Te could be studied in detail (Blachot and Carraz 1969).

6. Conclusions

Although several extensive studies have been concerned with fission synthesis in solids, the mechanism leading to molecule formation is still rather ill-understood. The concept most compatible with the present data, however, seems to be that of a solid state exchange reaction, which is independent of the recoil energy of the primary or secondary fission products over a wide range reaching from a few eV to about 100 MeV. The implantation of the fragments in the host lattice, as well as low energy excitation of the implanted atoms at the time when they come to rest or undergo β^- decay may facilitate fission synthesis by conventional solid state exchange reactions.

Aside from the theoretical interest in recoil chemistry, however, fission synthesis has found practical application as a fast and highly selective separation method in nuclear chemical studies.

Addendum (February 1977)

No further work concerning the synthesis of molecules by reaction of energetic recoil atoms from fission with appropriate catcher matrices has come to the author's knowledge. However, fission synthesis has found an application in fast separation of fission products by selective production of volatile compounds in varying carrier gases (Bachmann et al. 1973, Hoffmann et al. 1973).

CHAPTER 20

HIGH-ENERGY AND SPALLATION PROCESSES

I. DEMA
Institute of Atomic Physics, Bucharest, Romania

Chemical Effects of Nuclear Transformations in Inorganic Systems
Edited by G. Harbottle and A.G. Maddock

Contents

1. Introduction

This chapter deals with chemical consequences of (inelastic) nuclear interactions between nuclei and nucleons with energies exceeding 100 MeV. Some interactions of heavy ions with nuclei are also considered. For details on nuclear reactions at high energies see Miller and Hudis (1959). We shall here discuss only the chemical aspects of the recoil atoms produced in the bombardment of inorganic solid targets, especially when high-energy protons were used.

In the study of the hot-atom chemistry of radioactive atoms occurring in high-energy bombardments, alkali halides are of special interest. This is due to the fact that the reaction products are often non-isotopic with the target and also because there is already a substantial body of data related to chemical and physical properties of irradiated alkali halide crystals. Alkali halides when bombarded with high-energy particles give rise to radioisotopes of elements like P, Sb, As, S, Se, Te, Br and I, quite suitable for use in hot atom studies. In addition, by spallation processes several different recoil atoms may be obtained in the same crystal lattice (e.g. in CsCl irradiated with protons, P, Sb, S, Te and I occur). Moreover, high-energy bombardments allow us to study the behavior of the same recoil atom in isotopic and nonisotopic crystal lattices (e.g. radioiodine from KIO_3 and from cesium salts). In many cases the behavior of the same recoil atoms obtained in the same target by both proton and neutron irradiation may be compared (e.g. the radioiodine occurring in CsCl, the radiophosphorus and radiosulphur in KCl).

The bombardment of alkali halides with high-energy protons causes displacement of atoms from their places in the lattice (Pearlstein 1953, Kobayashi 1956, 1957, Smoluchowsky 1956) and also excites the electronic structure. At high energies, besides primary proton effects, secondary nucleons coming from inelastic collisions and having energies higher than 10 MeV may give rise to strong local ionizations which also have to be taken into account (Smoluchowsky 1956). However, the temperature of NaCl crystals during 350 MeV proton irradiation was not above room temperature (Kobayashi 1956, 1957). As a result of the primary and secondary actions, color centers of F and V type are produced.

The initial chemical state of the newly formed atoms is of course not known.

They have energies in the range of several MeV (Borisova et al. 1959) and will in general be strongly ionized and will interact to produce lattice defects (electrons and holes) during the slowing down process. The first indications are that spallation products tend to appear in lower oxidation states (Rudstam 1956).

In section 2, hot-atom chemistry in alkali halides like NaCl, KCl and CsCl will be discussed and also several results obtained using CsBr, alkali halide mixed crystals and some oxygenated compounds ($CsNO_3$, Cs_2CO_3 and so on) will be mentioned. The hot atoms involved are radio-isotopes of P, Sb, Te and I.

In section 3 we will include proton-induced reactions in potassium iodate (with a short discussion of telluric acid) and in some perrhenates and hexachloro-rhenates (IV).

In section 4 the chemical behavior of radiophosphorus, radiosulphur, radioiodine, radiorhenium and radiomanganese atoms formed in various targets under high-energy bombardments will be treated.

Several conclusions pointing the direction for further studies will be presented at the end of the paper.

2. Alkali Halides

2.1. PHOSPHORUS

The first experiments concerning the chemical state of radiophosphorus atoms resulting from spallation reactions have been carried out by Murin et al. (1961a, b). They irradiated pure KCl monocrystals and also KCl crystals doped with $CaCl_2$ and K_2S, as well as NaCl crystals. Before irradiation, the crystals were annealed several hours at 450–770° C. For the chemical separation of the three radiophosphorus fractions (hypophosphite, phosphite and phosphate) classical methods with inactive carriers were used; in some experiments, besides the three fractions, a phosphine fraction, PH_3, was also measured.

In KCl crystals annealed before irradiation phosphate yields of about 10–20% were obtained, while in untreated crystals the yields rose to 50–60%. By heating irradiated crystals at 200–400° C the phosphate and phosphite fractions increased at the expense of hypophosphite. The large fraction of higher valence states in non-annealed crystals may be accounted for by the presence, in KCl, of defects related to its history, defects which compete with the phosphorus atoms for electrons. The chemical change undergone by radiophosphorus in irradiated crystals may be regarded as being due to its reaction with radiolysis products and disorder centers produced during irradiation.

The appearance of PH_3 in KCl–$CaCl_2$ crystals was observed (Bogdanov and Murin 1965). Phosphine yields varied from 10 to 22% depending on the amount of added carrier and pH. This dependence is explained by isotopic exchange reactions between phosphine and one of the phosphoric acids and by solvent reactions.

It is of interest to point out that the amount of radiophosphorus appearing as oxyanions depends also on the pH of the solution in which the crystals are dissolved (Bogdanov and Murin 1967a, b). By illuminating irradiated, pure and $CaCl_2$-doped crystals with light in the F band, a reduction of radiophosphorus takes place, enhancing P(I) and phosphine yields. By this treatment the optical density of the F and V bands decreases. This leads to the assumption that recoil atoms are able to capture photoelectrons. Further, the authors suppose that after recoil, P atoms become stabilized in one of no less than four different oxidation states. By F-bleaching radioactivity passes directly from the oxyanions to the phosphine fraction. A certain amount of recoil phosphorus in the crystal exhibits tetra-coordination with sp^3 bonds. Ionic phosphorus complexes like PCl_4^+, $KPCl_3^+$ and $K_2PCl_2^+$, corresponding to the sp^3 state, give on dissolution PO_4^{3-}, HPO_3^{2-} and $H_2PO_2^-$ ions, respectively. When exposed to light, these radicals capture an electron and the phosphorus becomes tri-coordinated (p^3); the p^3 bonds react by hydrolysis in a manner similar to those in K_3P.

It was found that oxidation of phosphorus recoils occurs only in the initial stages of proton irradiation (Bogdanov et al. 1969a, c). The most probable mechanism responsible for oxidation is the trapping of a positive hole.

In deformed NaCl and KCl crystals the influence of the crystal dislocation concentration on the chemical fate of radiophosphorus recoils was studied (Bogdanov and Olevskii 1969c), the effect appears only at high values of relative compression and leads to an increase of the P(I) yield. When a bivalent anion impurity (K_2S) is introduced into a KCl lattice an increased oxidation of recoils is found in comparison with pure crystals (Bogdanov and Olevskii 1969d).

2.2. ANTIMONY

Radioantimony was obtained by irradiation of some cesium salts with 600 MeV protons in the external beam of the synchrocyclotron. After irradiation and addition of suitable carriers, the tri- and pentavalent states were separated (Dema and Zaitseva 1966d). It was observed that for CsCl and CsBr a very high percentage of recoil atoms appear in the lower oxidation state. By heating irradiated crystals or by exposing them to light, oxidation of radioantimony recoil species takes place; this is attributed to interaction of the recoil atom with some electron deficient centers.

In oxygenated compounds (cesium nitrate and sulphate) the trivalent antimony yield is about 50%; on heating oxidation-reduction phenomena occur in which radiolytic fragments evidently take part.

2.3. TELLURIUM

After dissolving crystals of proton-irradiated CsCl and CsBr in 4–8 M HCl most of the radiotellurium recoils appear in the tetravalent state (97.3 and 97.7%, respectively) (Dema and Zaitseva 1971). By heating irradiated crystals a transfer of radioactivity from the tetra- to the hexavalent state is found. Thermal annealing curves for 100, 200, 340 and 425° C show in the early stages of heating some weak reduction reactions which disappear after further heating, the predominant process, after annealing one hour, being oxidation. The oxidation of radiotellurium fragments seems to be due to interaction with centers in the crystal which may be identified with V-type centers (chlorine atoms). The radiotellurium atoms could easily encounter positive holes which are known to become mobile on heating.

Irradiation of oxygen containing compounds like $CsNO_3$ and Cs_2SO_4 gives rise to phenomena similar to those described for antimony.

2.4. IODINE

When cesium salts are irradiated with 660 MeV protons radioiodine is obtained in higher yield than antimony and tellurium. The nuclear reactions from which iodine results are the following: ${}^{133}_{55}Cs(p,3\,pxn)\ {}^{120-130}_{53}I$ (the isotope ^{123}I obtained by Cs irradiation may also arise from ^{123}Xe decay). Hot-atom experiments were carried out on CsCl, CsBr and also on some oxygenated salts like Cs_2CO_3, $CsNO_3$, Cs_2SO_4 and Cs_2CrO_4. At room temperature for cesium salts containing oxygen in their structure, the yield of the higher valency states (IO_4^-, IO_3^- and possibly IO_2^-) falls between 40 and 60%, except for Cs_2CrO_4 (about 20%). If irradiated samples are exposed to daylight, changes occur in valency state distributions, pointing to oxidation-reduction processes. The radioiodine distribution is also found to depend on the concentration of inactive carriers in the dissolution medium and on the extent of annealing after irradiation.

These phenomena may be attributed to reactions of the radioiodine atoms with radiolytic fragments, which may be found in bombarded crystals in much larger concentrations than the recoil atoms. Their stability and nature changes with the crystalline matrix and temperature: thus, annealing reactions reflect chemical and structural differences found in the close vicinity of the recoil atoms (Dema and Zaitseva 1969a).

In CsCl and CsBr crystals irradiated in a flux of $10^8 p/cm^2 \cdot sec$, radioiodine is found in oxidized states in yields of about 10%; this yield rises with increasing proton flux. At the same time, with flux increase, radiolytic product concentrations become higher, and free halogen occurring during irradiation may oxidize iodine recoil fragments into higher valency states (Dema and Zaitseva 1969b). The formation of species with I–Cl bonds which by hydrolysis would lead to oxidized states is quite possible. Owing to their very high recoil energy (several MeV), immediately after the nuclear event, recoil atoms exhibit a large positive charge which is gradually neutralized by electron capture, when the fragments slow down. It is possible that such fragments will be stabilized in interstitial sites as neutral and/or negatively charged iodine. It does not seem that trapped species having larger positive charge than I^+ are very probable. If the fragments are localized in positive vacancy sites, the presence of species like I^0 and/or I^+ may be postulated. Dissolving leads to lower valency forms as shown experimentally. There is some experimental evidence for the formation of neutral iodine in irradiated crystals (Dema and Zaitseva 1966a).

The yield of oxidized forms increases with increasing proton flux as a consequence of irradiation defects. Free chlorine (0.0014 mg Cl_2 for 1 g CsCl at 10^{14} p/cm^2) appears only at higher proton fluxes, in agreement with the proposed mechanism (Dema and Zaitseva 1969b). The electron affinity of chlorine is higher than iodine so that free chlorine atoms or V-type centers tend to accept electrons from negatively charged iodine. Thus a positive hole associated with a vacancy will be the most probable oxidizing agent for the recoil atoms located in vacancy sites. The 'protective effect' of inactive carriers has been demonstrated for CsCl (Dema and Zaitseva 1966a) as well as for other alkali halides (Harbottle 1965).

Heating of irradiated CsCl or CsBr crystals leads to the decrease of oxy-form yields. Isothermal annealing curves have shapes similar to those usually observed in hot-atom studies. The proton dose does influence the degree to which radioiodine fragments anneal at a given temperature and in a given time. Low doses lead to high percentages of lower valency states and little annealing on temperature increase (Dema and Zaitseva 1965, 1966d).

Attempts to calculate the activation energy following the Vand–Primak treatment (Costea et al. 1963) led to values distributed between 0.7 and 0.9 eV (Dema and Zaitseva 1965). The activation energy was not found to change with increasing defect concentration in samples irradiated with various integrated proton doses. However, a gradual increase of the half width of annealing activation energy spectra with log of proton dose was observed (Dema and Zaitseva 1966a, b).

In CsCl–NaCl, CsCl–KCl and CsCl–RbCl mixed crystals, the yield of radioiodine oxidized states is higher than in pure CsCl. Several systems show an increased oxidized yield when dilution increases (Dema and Zaitseva 1967). For the CsCl–CsBr system the percent yield of oxidized states decreases when CsBr is added in small amounts (0.1–0.5 moles per mole): it exhibits a minimum at the dilution of 1:2, after which it rises again at dilution of 1:5 and

1:10 moles of CsBr per mole CsCl respectively (Dema and Zaitseva 1968b). The oxidized yield is again found to change on heating, light exposure and dissolving in the presence of inactive carriers. These results were interpreted by considering data found in the literature, which show that in a mixed crystal the concentration of defects responsible for radioiodine oxidation is much higher than in either pure component (Dema and Zaitseva 1967, 1968b). It seems very interesting that radiosulphur produced in KCl–KBr mixed crystals shows similar behavior (Owens and Hobbs 1969).

Crystal impurities present in the CsCl lattice also affect the final chemical state of radioiodine. Thus by irradiating four categories of CsCl samples, among which one was of high purity, phenomena analogous (Dema and Zaitseva 1969a) to those discovered by Butterworth and Campbell (1963) for radiophosphorus occurring in ordinary KCl crystals and monocrystals, may be observed.

In CsCl, irradiation with 14.6 MeV neutrons produces radioiodine (^{130}I) with more than 90% in a reduced state. With heating or exposure to light, oxidation–reduction processes are not found to appear (Dema et al. 1968, Dema and Zaitseva 1969a), the system resembling CsCl irradiated with low proton fluxes (10^8p/cm^2·sec). When CsCl samples were γ-irradiated before 14.6 MeV neutron irradiation, the fraction of reduced states diminished in comparison with non-treated crystals, and on heating reduction reactions like those occurring in CsCl irradiated with high proton doses took place.

These phenomena suggest that radioiodine recoil atoms tend to stabilize in the CsCl lattice in the lowest possible oxidation state, the subsequent annealing reactions being a consequence of the influence of defects produced during irradiation. The same conclusion may also be drawn from the behavior of NaCl crystals doped with Na ^{131}I and then exposed to γ-radiation (Dema 1971).

2.5. COMPARISON

If the behaviors of Sb, Te and I recoil atoms arising from proton irradiation of CsCl lattices are compared, we see that by thermal annealing and exposure to light, oxidation of radioantimony and radiotellurium takes place, while radioiodine formed using small proton doses does not change its chemical state. This difference in behavior may be explained by taking into account that the electron affinity of iodine is higher than that of Sb and Te, so that the tendency to form negative ions is much lower for the two latter elements. Therefore one would expect Sb and Te recoil atoms to exhibit a stronger tendency for oxidation than radioiodine nuclei, in agreement with experimental data.

3. Oxyanion Systems

3.1. POTASSIUM IODATE

Here a study was made in which the behavior of recoil atoms in an iodine-containing lattice irradiated with 660 MeV protons could be compared with the behavior of the same recoil atoms in the non-isotopic CsCl crystalline lattice mentioned earlier.

Experiments showed that for proton-irradiated KIO_3 the retention increased with increasing irradiation dose. Heating of irradiated crystals also led to an increase in retention, the corresponding curves exhibiting a conventional shape (Dema and Zaitseva 1966b, c, 1968a). A kinetic analysis of these curves revealed first-order processes with activation energy (Vand–Primak) varying around 0.7 eV for a frequency factor $B = 10^5$ sec^{-1}.

The increase of retention on heating was thought to be due to the interaction of radiolytic oxygen-containing species either with iodine neutral atoms (experimental evidence was given for their presence) or with fragments of IO_2^- and IO^- types, thus leading to iodate. A significant result of these experiments is that the influence of recoil energy on the final fate of the radioiodine is negligible: potassium iodate irradiated with protons exhibits a retention of about the same value (69.7%) as for neutron-irradiated samples.

Oxidation processes were also observed in thermal annealing of telluric acid crystals irradiated with protons (Dema and Zaitseva 1972). In this case radiotellurium behaves somewhat like that formed in the nonisotopic lattices of cesium salts.

3.2. RHENIUM COMPOUNDS

The experiments on the behavior of radiorhenium recoils produced in $NaReO_4$, NH_4ReO_4, $KReO_4$ and K_2ReCl_6 by the reaction (p,pxn) show that the recoil atoms are usually stabilized as the parent molecule, except for ammonium salts (Ianovici and Zaitseva 1969a). In these compounds recoil fragment reactions with radiolytic products take place and result in radicals whose interaction with water leads again to the parent molecule. Similarity between the behavior of radiorhenium produced by proton irradiation and that resulting from neutron capture is also found (Ianovici and Zaitseva 1969b).

4. Miscellaneous

There are two reports in the literature dealing with the chemical behavior of radiophosphorus resulting from bombardment of metallic aluminum with nitrogen ions multiply ionized (Chackett and Chackett 1954) and of radiosulphur formed in the bombardment of some chlorine-containing compounds with 300 MeV alpha particles and 550 and 680 MeV protons (Herber 1961).

Phosphorus-32 may be obtained by irradiation of ^{27}Al with $^{14}N^{6+}$ ions accelerated up to energies of about 100 MeV with a relatively large yield. By dissolving irradiated Al foils in the presence of air, about $\frac{2}{3}$ of the recovered activity is carried by gaseous phosphine while the remainder becomes stable as hypophosphite ion. The absence of ^{32}P as phosphate shows that phosphorus atoms as expected are found in the lattice in lower oxidation states and also that not all atoms are in the phosphide state, because if they were, the whole activity ought to be obtained as phosphine.

Sulfur-35 (87.1 d) was obtained by the reactions $^{37}Cl(\alpha, pn)$, $^{35}Cl(\alpha,3pn)$ and $^{37}Cl(p,2pn)$, and ^{38}S(172 min) resulted from the reaction $^{37}Cl(\alpha,3p)$. In NH_4Cl solid targets a considerable fraction of the radiosulphur was found in a non-sulphate state. In the case of $NaClO_4$, almost the whole activity appears in the sulphate state. The results lead to the conclusion that the amount of energy transferred to a sulphur atom in the nuclear reaction does not have a decisive influence on the final chemical state of radiosulphur.

It is also worthwhile to mention the work of Zaitseva and Chou (1963), historically the first study of the chemical state of radioiodine formed in cesium salts, iodates and tellurates irradiated with 660 MeV protons. The yields of higher oxidation states were shown to depend on the target molecule structure.

The chemical state of radiorhenium formed in several inorganic iridium compounds was studied by irradiation with 660 MeV protons (Ianovici and Zaitseva 1970a, b, c). In sodium and ammonium hexachloroiridates, a considerable amount of activity is found in the heptavalent state: the yield is higher for hydrated salts than for anhydrous ones.

Finally let us mention the work of Veljkovic and Harbottle (1962), who performed several $KMnO_4$ irradiations using the 3 GeV proton flux available in the Brookhaven Cosmotron. Radiomanganese was obtained formed by the nuclear reaction $^{55}Mn(p,p3n)^{52}Mn$.

5. Conclusions

Considering the studies carried out up to the present at high energies, it appears that the large amount of energy transferred to the nascent atom immediately after the nuclear event does not exclusively determine the final chemical state of the recoil products. The original charge does not affect the final chemical state inasmuch as several electron exchanges take place between the recoil atom and matrix atoms during the slowing-down process.

It is of interest to point out the possibility of introducing recoil atoms having different chemical properties into the same crystalline matrix, by high-energy nuclear reactions. For instance, in the CsCl lattice phosphorus, antimony, sulphur, tellurium and iodine can all be produced by high-energy bombardment; a comparative study of these atoms in CsCl monocrystals might clear up many phenomena taking place when energetic atoms stabilize in irradiated crystalline alkali halide lattices.

Also interesting, but perhaps less important is the possibility of comparing the behavior of the same recoil atom in isotopic and nonisotopic lattices.

Naturally the investigations are difficult not only because of the rarity of high-energy accelerators but also because of the elaborate methods which must be used to remove the activity of the undesired radio elements which always occur in spallation reactions.

Addendum (December 1976)

Some recent experiments carried out at Dubna by Zaitseva et al. (1973, 1974, Zaitseva and Ianovici 1971, 1972) are devoted to the chemical behavior of zirconium and yttrium recoil atoms formed by irradiation with high-energy protons of dispersed metal powders (Zr and Nb) in beta-diketones and also in several fluorinated deta-diketonates of Y and Zr.

For the bombardment they have used suspensions of Zr and Nb finely dispersed in acetylacetone, trifluor-acetylacetone and hexafluoro-acetylacetone and also chelates of Zr and Y with trifluoro-acetylacetone (HTFA), trifluoro-acetyl-iso-butytyl-methane (HTIBM) trifluoro-acetyl-iso-valeryl-methane (HTIVM) and hexafluoro-acetyl-acetone (HHFA).

After chemical analysis of the irradiated suspensions, the radiozirconium and radioyttrium atoms found in the organic material as chelate molecules have the largest yield in fluorinated beta-diketones; the yield of ligand-deficient states in different beta-diketones remained practically constant, while the percentage of inorganic forms changed with changes in the yield of chelate molecules. In the irradiated fluorine-chelates of the type $Zr(TIVM)_4$, $Zr(TIBM)_4$ and $Y(TFA)_3$ having an asymmetric structure, the ^{86}Zr atoms tend to stabilize as chelate molecules with a large yield (40–50%); the value for other fluorinated chelates varies from 10 to 30%. Higher yields (40–50%) for ^{86}Zr and ^{87m}Y ligand-deficient forms were observed in the case of proton irradiated beta-diketonates.

All these experiments show the role of the chemical environment in which the energetic radioactive atoms are released and also the role of subsequent chemical reactions which influence the final chemical state of the recoil products.

Acknowledgement

The author wishes to thank Dr. T. Costea from the Institute of Atomic Physics, Bucharest, for helpful discussions and a critical reading of the manuscript.

CHAPTER 21

STUDIES OF THE NATURALLY RADIOACTIVE ELEMENTS

A.G. MADDOCK

University Chemical Laboratory, Cambridge, UK

Chemical Effects of Nuclear Transformations in Inorganic Systems
Edited by G. Harbottle and A.G. Maddock

Contents

Introduction

Not unexpectedly studies of both the physical and chemical effects of nuclear transformations began with the naturally radioactive elements. Use was made of the mechanical recoil following alpha decay, as well as the charge state of the recoiling atom, in separating carrier-free preparations of the decay products, especially those of the radon isotopes. Such direct effects of the recoil played a part in the discovery of the transuranic elements and have been used in the characterisation of all the most recently discovered actinide and trans-actinide elements.

1. Recoil Following Radioactive Decay

The recoil following alpha decay amounts to from 70 to 150 KeV. The first observation of such recoil was due to Wood (1914). More recently the ranges of the recoiling atoms following alpha decay have been measured. Earlier data and references are tabulated in papers by Baulch and Duncan (1957) and Marx (1966). Wiechmann and Biersack (1967) measured the ranges of ^{224}Ra recoil atoms in various gases and Cano (1968) measured those of ^{206}Pb. Ambrosi and Wolfson (1969) have measured the energy loss of recoiling ^{237}Np on passing through plastic films.

1.1. EFFECTS OF RECOIL ON APPARENT VOLATILITY

One of the more remarkable effects of the large recoil following alpha decay was the anomalous volatility observed with many alpha active materials of high specific activity. First observed by Lawson (1918) these effects are particularly noticeable with polonium sources whose properties suggested an improbably high volatility for the polonium compound, usually the element, concerned (Brooks 1955, Weigel 1959). It is observed that much greater amounts of the alpha active material appear in the gas phase than the true volatility of the compound would warrant. It is now clear that this effect arises in part because the substantial recoil can detach aggregates of atoms or molecules of the alpha active compounds from the surface of the source and disperse it in the gaseous phase as a radiocolloidal 'smoke' and, perhaps, also because the sputtering action of the alpha particles leads to a similar effect (Riehl and Sizmann 1964, Diethorn 1965, Cano and Dressel 1965).

1.2. THE CHARGE STATE OF ALPHA DECAY PRODUCTS

In the absence of any electronic excitation the product would carry a charge of −2. The relatively slow and presumably adiabatic character of alpha-particle emission (see ch. 2) should give much less 'shake off' effects than in beta decay but some loss of electrons could still be anticipated.

Experimental studies have been conducted since the earliest period of the study of radioactivity. Amongst the difficulties complicating such studies may be mentioned the strongly ionising events that occur if the alpha decay yields an internally converting excited state of the daughter nucleus. The data prior to 1964 have been analysed in detail by Wexler (1965a, b).

Mund et al. (1931) found a most probable charge of +2 for the ^{218}Po from the decay of radon, but the method used did not give weight to uncharged daughter species. Szucs and Delfosse (1965) found that uncharged daughter atoms were most frequently formed in the same system, but that positively charged species up to +9 could be detected, with the highest probability of +1. Cano and Dressel (1965) studied the recoil atoms from alpha decay of ^{210}Po and ^{241}Am present as thin deposits on a solid surface and found a most probable positive charge of +1 with similarly decreasing probabilities of more highly charged species.

Gunter et al. (1966 and Gunter 1968) also explored the charge distribution on the ^{226}Th atoms recoiling from a thin ^{230}U deposit and compared it with the distribution when the same ^{226}Th is produced in an internally converting excited state. The pure alpha decay gave a fairly sharply peaked probability distribution rising from 0.20 for zero charge to a maximum of about 0.5 for +1 and 0.27 at +2 and then falling off rapidly so that more than 99% of the events produced atoms of charge less than +6. When internal conversion takes place the distribution is flatter and the most probable charge is as high as +10. These authors note that for pure alpha decay the charge distribution will generally be different for gaseous and plate mounted sources. Further in the latter case the distribution depends on the nature of the substrate and is quite sensitive to diffusion of the source material into the substrate. Thus it depends on the age of the source. Similar conclusions have been reached by Perrin and De Wieclawik (1966, 1969, also De Wieclawik 1968, 1969a, b). It has also been pointed out that because the daughter atom can be regarded first as a negatively charged ion the distribution may change with the electron affinity of the daughter species even if all other aspects of the decay process are the same. Meyer et al. (1972) have investigated the charge state of ^{208}Pb and ^{208}Tl from ^{212}Po and ^{212}Bi decay with similar conclusions. They also showed that only a very small (< 0.1%) number of events lead to negatively charged species.

1.3. SEPARATIONS DIRECTLY DEPENDENT ON THE MECHANICAL RECOIL

Wood's discovery was soon applied to the separation of ThC″ (^{208}Tl) from the thoron decay products (Wertenstein 1914) and other separations of this kind were used from time to time. Even the much smaller recoil following β decay (0 to ~ 17 eV for these species) can be utilised for this purpose. Mundschenk has recently collected the recoiling products on a moving belt to explore their decay characteristics. Both alpha (^{208}Tl) and beta (^{211}Po) recoil products were investigated (Mundschenk 1970, 1971). Generally the daughter species recoils from a solid phase through a gas but some studies have been made of recoil from a solid into a liquid. These were important in the study of the effects of injection of colloidal thorium oxide into animal systems. Measurements have been made on the proportion of ^{228}Ra and ^{224}Ra passing into solution, as well as that of ^{233}Pa produced by the beta decay of ^{233}Th from the (n,γ) reaction (Beydon and Gratot 1968, Heyder and Kaul 1971).

Because charged products are often formed the alpha decay products are better collected in gases on charged plates. The behaviours of the actinon and thoron decay products have been reported by Schwartz et al. (1966). In the presence of methane they found evidence for the formation of volatile compounds by the lead isotopes.

Briand and Chevalier have shown how the successive recoils in an alpha decay chain can be used to separate the different decay products (1970).

1.4. RECOIL OF FISSION FRAGMENTS

The very large recoil energy of the fission fragments was quickly used in the isolation of fission products (Joliot-Curie 1939). Several more recent studies have used the recoil to transfer the fission products to another phase, for example from U_3O_8 particles to surrounding gelatin (Bruyn 1955, Wolfgang 1956, Naki and Yajima 1958). Ranges of the fission products have been estimated in the same way (Henry and Herczeg 1958, Henry et al. 1961).

Perhaps one of the most important experiments of this kind depended on the failure of an activity to suffer a substantial recoil. McMillan (1939) was led to the identification of the beta active ^{239}U, and subsequently to the discovery of neptunium, by the failure of this activity to recoil from a uranium oxide target.

The heavily ionising, high LET, fission products can produce radiolytic decomposition of the medium in which they dissipate their energy. Such radiolysis has been studied by Walton and his collaborators (Walton and Croall 1955, Hall 1958a, b, Hall and Walton 1958, 1959, 1961, Bertocci et al. 1961).

1.5. RECOIL SEPARATIONS WITH THE POST-CURIUM ELEMENTS

Practically all of these elements have first been made by heavy ion reactions and the product has been collected after recoil from the target substance. An account of a typical arrangement is given by Harvey (1962) (see also Silva 1972). Since the half-lives are commonly short, collection on a moving belt facilitates detection.

The selectivity of such processes has been enhanced by Zvara who has arranged for the product atom or atoms to recoil into a reactant gas stream so that some fractionation of products according to volatility, or rather adsorbability, becomes possible (e.g. Zvara et al. 1964, 1969). Using a chlorinating gas stream and comparing the movement of the radioactive product in the presence and absence of alkali halide adsorbents, Zvara has reported chemical evidence for the identity of the post-actinide products (Zvara et al. 1970). Exploratory experiments have also been reported on the movement of ^{224}Ra decay products in a stream containing methyl radicals from the thermal cracking of lead tetramethyl. More than 90% of the ^{212}Bi was carried away in the gas stream with a lower, but still substantial, proportion of ^{212}Pb and ^{208}Tl. These experiments are intended to develop procedures useful for the characterisation of synthetic super-heavy elements (Hoffmann et al. 1971a, b).

1.6. MÖSSBAUER SOURCES FED BY ALPHA DECAY

The discovery that americium alloys and compounds could be used to feed the 59 keV Mössbauer excited level of ^{237}Np provides interesting experimental evidence of the time scale of the events following alpha decay in solids (Stone and Pillinger 1964, Dunlap et al. 1968). Although the line widths obtained were at best about fifteen times the theoretical value (i.e. the value determined by the lifetime of the excited state) the fact that sharp absorption lines were observed shows that the neptunium atom has settled down in a metastable well defined lattice situation, not necessarily a normal lattice site, before the decay of the Mössbauer level (half-life 63 ns). Further any serious local heating has been dissipated. Kaplan (1966) has pointed out that these observations are in keeping with theoretical estimates of the time scale for the dissipation of the recoil energy in a solid matrix (see ch. 3).

2. Geochemical Consequences of Alpha Recoil

There are numerous examples of anomalous isotopic abundances in natural materials. For instance, the ^{234}U/^{238}U ratio has been observed to be very high in some natural waters (e.g. Cherdyntsev 1955, Syromyathikov 1965,

Sakanoue et al. 1968). Rowland et al. (1961a) drew attention to the possibility that some of these anomalies might arise from recoil effects. Thus the recoil of the UX_1 (^{234}Th) might be expected to enhance the amount of this species, and its beta decay products UX_2 (^{234}Pa) and UZ (^{234m}Pa), in an aqueous phase in contact with a uraniferous mineral. Under some conditions these products might be carried away to decay to ^{234}U elsewhere, or, if they adsorb back on the uranium mineral, because both thorium and protactinium readily adsorb at most natural pH values, the surface of the mineral will be enhanced in ^{234}U and subsequent solution will yield an abnormal $^{234}U/^{238}U$ ratio. Variations of this kind of cycle of events have been discussed by Kigoshi (1971) and Sakanoue and Komura (1971).

Rowland et al. made some measurements on the Pb II/Pb IV ratio for the ^{210}Pb in some uranium minerals. Uraninite and pitchblende gave a ratio of about 140 but in carnotite it was only 1.5 (1961a).

The role of alpha recoil was also discussed by Zaborenko and Babeshkin (1961) who pointed out its importance in relation to the loss of gaseous decay products from rocks and minerals.

3. Emanating Power

As indicated in section 2 the large recoil after alpha decay can contribute an abnormal term to diffusion processes. This phenomenon has been studied extensively in investigations of the emanating power of solids – the proportion of a radon isotope, produced from a radium parent incorporated in the solid, that escapes into a gaseous phase. It is not proposed to enlarge on this subject. Reviews by Zaborenko (1973) and Bussière (1969) should be consulted.

4. The Chemical Effects of Alpha Decay

It is surprising that very little attention has been paid to the chemical effects of alpha decay. Haïssinsky and Cottin (1948) investigated the state of UX_1 (^{234}Th) grown from uranyl benzoyl acetonate and dibenzoylmethane complexes, both in solution and in the solids; uncomplexed UX_1 was separated by adsorption on barium carbonate. They found retentions of 20–50% for the solutions and 80–90% for the solids. The remarkably high value for the solutions suggests that either the analytical technique is unreliable or that reformation of UX_1 complex takes place with traces of free ligand. Similar experiments by Govaerts and Jordan (1950) on solutions in pyridine of the Schiffs base complex uranyl salicyladehyde o-phenylenediimine showed that 90% of the UX_1 would deposit on the cathode and 10% on the anode on electrolysis.

Aten et al. (1960) have investigated the state of ^{208}Tl (ThC″) formed by alpha decay of ^{212}Bi (ThC) in aqueous solutions. As indicated in section 1.2 alpha decay is not a very strongly oxidising process and one could imagine the Bi^{3+} would yield the isoelectronic Tl^{+}. In fact about 30% of $^{208}Tl^{3+}$ was found. However, an internally converting 40 keV level of ^{208}Tl is formed in a substantial proportion of ^{212}Bi decays, so that the observed $^{208}Tl^{3+}$ is perhaps lower than would be anticipated.

Weber et al. (1972) have studied the effect of a cathodic potential on the release into solution of the same ^{208}Tl species from a ^{212}Bi source deposited on a platinum disc. A high proportion of the ^{208}Tl could be brought into solution even from rather thick ^{212}Bi sources. It is difficult to analyse the different factors involved in such a process.

Most recently Gal et al. (1970) have investigated the state of ^{237}Np formed by alpha decay of ^{241}Am by Mössbauer emission spectroscopy. In solid oxide hosts, both Am_2O_3 and AmO_2, the neptunium appeared in the +4 and +5 states. In the solid AmF_3, $AmCl_3$, AmOCl and $Am_2(C_2O_4)_3$ it appeared as Np^{3+} as was also the case in precipitated Am^{4+} hydroxide. In frozen aqueous solutions the neptunium appeared as Np^{3+} independently of the oxidation state of the americium compound solute. Substantially the same results were obtained for measurements at 77 and 4.2 K.

5. Other Nuclear Reactions Involving These Elements

In view of the relative amounts of published work the effects of beta decay will be treated separately.

5.1. THALLIUM

Sarrach and Vormum (1961) investigated the effects of thermal neutron capture in $R_2TlOOC{\cdot}CH_3$ (R = alkyl group). A large proportion of the ^{204}Tl could be separated as thallous acetate, but measurements of the retention were complicated by radiolytically initiated exchange reactions. Thallous compounds are generally susceptible to radiolysis and Frediani and Lo Moro (1969) decomposed much of their target material ($TlClO_4$ and $TlNO_3$) in high dose pile irradiations. Wheeler et al. (1968) found several products with Ph_2TlCl but their identification was uncertain. Thallous compounds are particularly well suited to a search for oxidising effects following neutron capture. An early attempt to detect such oxidation was unsuccessful, possibly because of annealing. It was observed that thallous carbonate was also sensitive to radiolysis (Maddock and de Maine 1956a). Successful results were obtained by Butterworth and Harbottle (1966). They irradiated solid thallous salts at −80° C and found 15% $^{204}Tl^{3+}$ in $TlNO_3$; 21% in $TlClO_4$; ~ 2% in

Tl_2SO_4, and 0.9% in TlCl. The proportion of $^{204}Tl^{3+}$ was not sensitive to the pile dose, but suffered thermal annealing, returning to $^{204}Tl^{+}$ at modest temperatures, below 100° C for $TlNO_3$ and at 150° C for $TlClO_4$. Subsequently measurements were reported for the ^{202}Tl produced by the (n,2n) reaction in the same salts: 12% $^{202}Tl^{3+}$ in $TlNO_3$; 18% in $TlClO_4$ and about 1% in Tl_2SO_4 and TlCl (Ackerhalt et al. 1972).

Relevant to such studies is the investigation of the effect of ionising irradiation on thallium doped potassium chloride. The thallium was labelled with ^{204}Tl to facilitate the analysis. As the irradiation proceeds the Tl^{+} is converted to Tl^{3+} and F centres but the production of Tl^{3+} slows down and reaches a limiting proportion at very high doses (~600 Mrads). This limiting proportion, about 40% at low concentrations of thallium dopant, decreases as the thallium concentration increases. The disappearance of Tl^{+}, measured optically in situ, is more rapid than the similarly measured appearance of Tl^{3+}. On thermal annealing of the irradiated system the F centre absorption disappears much more rapidly than the Tl^{3+} absorption. Optical bleaching of the F centres reduces the Tl^{3+} absorption rapidly but without apparent corresponding increase in Tl^{+}. The oxidation–reduction balance in the system is difficult to establish, as indeed in most radiolysed solids, because of surface and dislocation species, impurities and formation of aggregate centres (Butterworth and Harbottle 1968).

5.2. BISMUTH

An enrichment of the ^{210}Bi (RaE) produced by neutron irradiation of a bismuth compound was reported by Maurer and Ramm (1942). More details about the same system were given by Popplewell (1963) and Wheeler et al. (1968). The latter also examined $BiPh_3Cl_2$ and found a very low retention. About half the ^{210}Bi could be separated in inorganic form. Murin and Nefedov (1955) report an even higher yield of inorganic product (80%).

5.3. ASTATINE

Recoil chemistry provides interesting preparative possibilities with astatine, which can be produced by orbital electron capture in ^{211}Rn. The radon can be dissolved in a hydrocarbon and the products examined by gas or thin layer chromatograph. Several RAt compounds have been characterised in this way. Yields of 30–50% of organic products containing At are obtained. Thus 44% of PhAt could be found with ^{211}Rn in benzene (Nefedov et al. 1970, Kuzin et al. 1972).

5.4. URANIUM

Most of the earlier papers were directed towards the preparation of strong sources of ^{239}U and its decay product ^{239}Np. Irvine obtained a very modest enrichment of 10, irradiating ammonium uranyl acetate and precipitating the ^{239}U on a basic salt, presumably as U^{4+} (1939). Much higher enrichments were found for β diketonate and Schiffs base complexes (Breslow and Hamaker 1942, Starke 1942a, b, Götte 1946, 1948, Melander 1947). But the yields were generally poor. The (n,2n) product has been separated in a similar way by Melander and Slatis (1948) using uranyl salicylaldehyde o-phenylenediimine. An enrichment of more than 500 and a yield of 15–25% could be obtained by adsorbing the product on charcoal from a solution of the irradiated complex in pyridine.

More recent studies have concerned changes in oxidation state of the uranium. In neutron irradiated U_3O_8 the U IV and U VI fraction have the same specific activity. In uranyl sulphate solution most of the ^{239}U appears as uranyl ion but in uranous sulphate solution more than 50% appears in the uranyl state (Saito and Sekine 1958). A similar result is found for solid anhydrous uranous sulphate and the proportion of ^{239}U VI, initially ~ 60%, is decreased by thermal annealing. A still larger proportion (~ 77%) of ^{237}U VI is obtained after the (n,2n) reaction. For the hydrated salt the (n,2n) results are similar but the (n,γ) process gives rather less ^{239}U VI. In uranyl salts about 98% of the product was always in the hexavalent state (Aten and Kapteyn 1968b). Heitz and Ruffenach (1971 and Ruffenach 1971) have also found oxidation following the (n,γ) process (see Table 21.1). In the uranyl acetate system there were difficulties with exchange processes.

Table 21.1

Compound	UO_2SO_4	$U(SO_4)_2$	$UO_2(OOC{\cdot}CH_3)_2$	$U(OOC{\cdot}CH_3)_4$
^{239}U IV	2	40	30– 2	90
^{239}U VI	98	60	70–98	10

5.5. AMERICIUM

Enrichment of ^{242}Am by neutron irradiation of ^{241}Am adsorbed on a zeolite has recently been reported. A yield of 20–70% with an enrichment of greater than 10 has been obtained. The americium was adsorbed onto the zeolite from 0.01 M nitric acid and then dried and calcined before neutron irradiation (Shaviev et al. 1973, Campbell 1973).

6. Effects of Beta Decay

The most profitable and most extensive studies on the naturally radioactive elements have concerned the effects of β decay.

6.1. LEAD ALKYLS

One of the oldest studies of the chemical effects of nuclear changes concerned the decay of ^{210}Pb (RaD) in lead tetramethyl and the appearance of ^{210}Bi (RaE) and ^{210}Po (RaF) in organic combination (Mortensen and Leighton 1934). The system has proved a difficult one and several features remain incompletely explained. Early measurements in both gaseous and liquid phases, showed appreciable yields of volatile bismuth and polonium products (Edwards and Coryell 1948, Edwards et al. 1953). Attempts to associate the yield of volatile ^{210}Bi compounds with cage effects or collisional deactivation of $Bi(CH_3)_3$ formed from $Bi(CH_3)_4^+$ were incompatible with the pressure dependance of the yield. At this time the decay scheme of the ^{210}Pb was imperfectly known and it was not appreciated that about 80% of the decays yield a highly internally converting excited state of the daughter bismuth. Baulch and Duncan (1958a, b, Baulch et al. 1961) attempted a more detailed analysis of the consequences of the decay processes. However, the possibility of hot or thermal exchange reactions of the daughter product was not very seriously considered. Kay and Rowland (1958 and Kay 1960) showed that the ^{212}Pb (ThB) formed by the alpha decay of thoron, via ^{216}Po (ThA), taking place in methane gave about 20% of the activity as a volatile lead product. They suggested that more attention be paid to the possibility of related processes in the $^{210}Pb(CH_3)_4$ system.

More recently Duncan and Thomas (1967) have shown that, after correction for exchange processes, at low enough pressure virtually all the ^{210}Bi is present in an involatile form. They conclude that the ground state ^{210}Bi formed as $^{210}Bi(CH_3)_4^+$ cannot yield appreciable quantities of $^{210}Bi(CH_3)_3$. However, the actual form of the bismuth was not explored. Carlson and White (1968) have examined the X irradiation products of $Pb(CH_3)_4$ mass spectrometrically. Such irradiation produces deep orbital vacancies in the lead atoms. The most complex species found in appreciable yield ($> 0.1\%$) only contains one carbon atom still bound to the lead so that practically no alkyl species will survive internal conversion in the gas phase.

6.2. LEAD ARYLS

Nefedov and collaborators (Nefedov and Andreev 1957, Nefedov and Bel'dy 1957, Nefedov et al. 1959, 1969, 1961a, b, Nefedov and Grachev 1960) have made an extensive study of beta decay in the lead aryls in solid and solution phases. Experiments have been made with ^{210}Pb, ^{214}Pb and ^{212}Pb (RaD, RaB and ThB). The differences in the yields of inorganic bismuth isotopes in $^*PbPh_3Cl$ were related to the proportion of daughter bismuth atoms suffering internal conversion (see table 21.2). $^{210}PbPh_4$ gave substantial yields of $^{210}BiPh_3$. Much lower yields of the organic bismuth compounds were found if decay took place in solution.

Table 21.2

Solid compounds	Internal conversion	Inorganic *Bi	*$BiPh_3Cl_2$
$^{212}PbPh_3Cl$	30 %	17 %	50 %
$^{210}PbPh_3Cl$	80 %	43 %	19 %
$^{214}PbPh_3Cl$	15 %	11 %	

The earlier studies used classical carrier techniques for separation but Nefedov and Grachev (1960) introduced carrier-free chromatographic procedures, $^{210}PbPh_3Cl$ yielded $^{210}BiPh_2Cl$, $^{210}BiPh_3Cl_2$ and $^{210}BiPh_3$ as well as inorganic bismuth. When $^{210}PbPh_2X_2$ ($X = NO_3$ or Cl) was used the yields of organic ^{210}Bi were lower.

The Adloffs (1964) have shown that particular care is necessary in the use of a carrier in these systems, because in the absence of a carrier a substantial proportion of $BiPh_3^{2+}$ is found which is not seen if $BiPh_3$ carrier is present.

7. Beta Decay Synthesis of Organic Polonium Compounds

Murin and Nefedov and collaborators (1960) have extended the procedures described in section 6 to the characterisation of a large number of organic polonium compounds after preparation by beta decay of ^{210}Bi (RaE) in aryl derivatives. Chromatographic separations have been used and the related tellurium compounds used as a guide to the identification of the polonium compounds. The ^{210}Bi has been used in compounds of the general formulae $BiAr_3$ and $BiAr_3X_2$ (X = F, Cl, Br). The polonium appears as $PoAr_2$, $PoAr_3X$, $PoAr_2X_2$ and in inorganic form (Nefedov et al. 1963a, b, 1964c, 1965a, b, c, d, e, 1966a, b, 1972). With $BiPh_5$ there was some evidence for $PoPh_4$, although the major product was $PoPh_3^+$.

Nefedov suggests that the distribution of the polonium amongst the various product species reflects the stability of the cationic polonium analogue of the parent bismuth compound. This view is supported by the results for various aryl groups and different X. For example, see table 21.3 for results on bismuth tri-tolyls and table 21.4 for results on the Ar_3BiX_2 compounds.

Table 21.3

Compound	Ar_3Po^+ (as % of organic Po)	Ar_2Po
$^{210}Bi(o\text{-}C_6H_4{\cdot}CH_3)_3$	67	33
$^{210}Bi(m\text{-}C_6H_4{\cdot}CH_3)_3$	39	61
$^{210}Bi(p\text{-}C_6H_4{\cdot}CH_3)_3$	68	32

Table 21.4

Compound	$Bi(o\text{-}C_6H_4CH_3)_3F_2$	$Bi(o\text{-}C_6H_4CH_3)_3Cl_2$	$Bi(o\text{-}C_6H_4CH_3)_3Br_2$
$\%Ar_3Po^+$ found	5	33	73
Compound	$Bi(m\text{-}C_6H_4CH_3)_3F_2$	$Bi(m\text{-}C_6H_4CH_3)_3Cl_2$	$Bi(m\text{-}C_6H_4CH_3)_3Br_2$
$\%Ar_3Po^+$ found	6	29	87
Compound	$Bi(p\text{-}C_6H_4CH_3)_3F_2$	$Bi(p\text{-}C_6H_4CH_3)_3Cl_2$	$Bi(p\text{-}C_6H_4CH_3)_3Br_2$
$\%Ar_3Po^+$ found	5	44	93

8. Other Syntheses Using Beta Decay

Evidence for various complexes of protactinium and neptunium has been obtained by beta decay of the parent thorium and uranium complexes. The sublimable phthalocyanines have been obtained in reasonable yields (Lux and Ammentorp-Schmidt 1965, Lux et al. 1970, Endo and Sakanou 1972). But ^{233}Pa formed in thorium acetylacetonate does not appear to be volatile (Kawazu and Sakanou 1974).

8.1. DISTRIBUTION OF OXIDATION STATES FOLLOWING BETA DECAY

The ^{233}Pa IV/^{233}Pa V ratio for ^{233}Pa grown in the thorium tetrahalides changes with the pH of the solution used to conduct the analysis: it increases with the pH. It changes in a complicated way with the dose received by the thorium halide during neutron irradiation and appears to reflect differences in the radiolytic behaviour of the thorium halides (Carlier and Genet 1972 and Carlier 1971).

An investigation of the state of ^{239}U and daughter ^{239}Np in neutron irradiated $K_2UO_2Cl_4 \cdot 2H_2O$, $UO_2(NO_3)_2 \cdot 6H_2O$ and in a uranyl solution has produced the unexpected result that most of the ^{239}Np appears in the Np IV and NpO_2^+ states (Peretrukhin et al. 1967a, b). They suggest coulombic explosion of the NpO_2^{3+} but extensive charging only accompanies a small proportion of beta decay events and Np VII compounds are known.

CHAPTER 22

TRITIUM RECOIL REACTIONS IN INORGANIC SOLID COMPOUNDS

Constance MANTESCU* and Tudor COSTEA

Institute of Atomic Physics, Bucharest, Romania

*Deceased.

Chemical Effects of Nuclear Transformations in Inorganic Systems
Edited by G. Harbottle and A.G. Maddock

Contents

1. Introduction

The behavior of recoil tritium produced in inorganic solid compounds is an esoteric chapter of the already esoteric field of Hot Atom Chemistry. A very small group of investigators have devoted their attention to this problem. Rowland and his collaborators were the first to study the reactivity and the stabilization forms of recoil tritium in neutron-irradiated LiCl by dissolving the irradiated compound in various solvents (Hoff and Rowland 1958) or by irradiating lithium salts in the presence of organic molecules (Lee and Rowland 1963). Măntescu, Costea (1962, 1963) and coworkers studied the release of tritium after irradiation of a wide range of lithium compounds; data were reported on the post-irradiation thermal release from lithium-doped nickel oxide, LiCl, lithium halides, carbonate, chromate, nitrate, etc. (Costea and Măntescu 1966) and from metallic lithium (Măntescu et al. 1971), while for LiCl the kinetics of thermal release were established (Măntescu and Costea 1966). Lindner et al. have studied recoil reactions of tritium in phosphates, phosphites and hypophosphites (Lindner et al. 1965). As an extension of the work of Lindner et al. (1965), Van Urk (1970) investigated the role of the bond character of the hydrogen atom in a variety of solid inorganic compounds in determining the final tritium distribution (Van Urk 1970).

It is the aim of this chapter to summarize the results of these studies.

2. Experimental Techniques

With a few exceptions, in which the $^{3}He(n,p)^{3}H$ nuclear reaction was used (Van Urk 1970), the $^{6}Li(n,\alpha)^{3}H$ nuclear transformation was the reaction used to produce recoil tritium.

A variety of techniques was used in order to determine the distribution and chemical state of recoil tritium in the irradiated compound: these included dissolution in inorganic or organic compounds, associated or not with specific chemical reactions (Hoff and Rowland 1958, Costea and Măntescu 1966, Măntescu and Genunche 1963, Măntescu et al. 1976, Lindner et al. 1965, Van Urk 1970), thermal release followed by fractional condensation at low temperatures (Măntescu and Costea 1962, 1963, 1966, Costea and Măntescu

1966, Măntescu et al. unpublished), radio-gas chromatography (Lindner et al. 1965, Van Urk 1970) and ion exchange (Van Urk 1970). The radioactivity was determined with gas counters or liquid scintillation spectrometers.

3. Chemical State of Recoil Tritium

Tritium formed by nuclear reactions in different solid lithium compounds presents us, at least hypothetically, with the possibility that it may exist in atomic, cationic, anionic, molecular or combined forms. With the exception of the hydrides (tritides) the ordinary combinations of tritium are volatile and easily identified and analyzed.

Different groups of investigators have utilized widely differing irradiation conditions (neutron flux and gamma dose rate) and unfortunately the various methods of analysis adopted by them have also made it difficult to compare the results.

Van Urk (1970), in a very detailed thesis, has considered that tritium formed by nuclear reaction could be recovered in the molecular form HT, as tritide T^-, and as labile tritium T^+ (this form being able to exchange rapidly with a hydrogen of the water molecule leading to TOH). The separation of the initially molecular form and that resulting from the hydrolysis of the tritide was achieved by dissolution in D_2O, the resulting DT being separated from HT by gas chromatography.

The Bucharest group applied the technique of thermal release of the different volatile forms of tritium followed by fractional condensation and dissolution of the solid residue as well as a series of specific chemical reactions for each species considered. In this way the following species were determined: TH, TX (in halides, X = F, Cl, Br, I), TOH (in oxysalts, in compounds containing water of crystallization or as an impurity), LiT and T. The last two species were also identified by specific chemical reactions (Măntescu et al. 1976).

3.1. LITHIUM METAL

In metallic lithium a large fraction of the tritium has been found to be stabilized as tritide, although this fraction is not very reproducible (Măntescu et al. 1971). Van Urk (1970) has reported a considerable fraction of labile tritium, while the present authors (Măntescu et al. unpublished) have found atomic tritium through the agency of some specific labelling reaction. The results are given in table 22.1.

It is interesting to note the appearance of tritiated organic molecules following the dispersion of irradiated metallic lithium in different organic compounds as evidence of the existence of atomic tritium. The T atoms arise from a reactive zone in the metallic lithium ca 0.3–1.0 μ thick (Măntescu et al. 1976).

Table 22.1
Distribution of tritium in metallic lithium

Target	% HT	% T^-	% T^+	% T	Ref.
Li powder	3.4–6.0	55.8–82.0	12.4–39.7	–	b
Li shot	4.4–6.5	52.6–81.6	14.0–41.7	–	b
Li wire	2.3–13.2	32.4–47.5	42.7–65.3	–	b
Li platelet	35.1[a]	26.5	–	40.4	c

[a] This fraction was called molecular tritium; the analysis to differentiate between HT and T_2 was not done; the content of hydrogen as impurity in the high-vacuum distilled metallic lithium was also not determined.
[b] Van Urk (1970)
[c] Măntescu et al. unpublished.

3.2. HYDRIDES

The determination of the tritium distribution in the different groups of hydrides was especially interesting because the nature of the hydrogen–metal bond could vary in the series from an ionic character to a covalent one. The results obtained by Van Urk (1970) in the case of simple or complex lithium hydrides as well as the alkaline and alkaline-earth hydrides irradiated in the presence of ^3He are given in table 22.2.

Table 22.2
Distribution of tritium in hydrides

Target	% HT(DT)	% T^-	% T^+
LiH	5 ± 2	88 ± 3	7 ± 2
LiD	2 ± 2	88 ± 2	10 ± 2
$LiBH_4$	30 ± 4	57 ± 4	13 ± 4
$LiBD_4$	34 ± 3	54 ± 1	12 ± 4
$LiAlH_4$	47 ± 1	45 ± 2	8 ± 3
$LiAlD_4$	41 ± 2	51 ± 1	8 ± 2
NaH	19 ± 1	44 ± 1	37 ± 1
$MgH_{1.6}$	9 ± 2	66 ± 5	25 ± 5
$CaH_{1.8}$	45 ± 3	29 ± 1	26 ± 2
$BaH_{1.9}$	13 ± 2	17 ± 1	70 ± 2
$MgNiH_{3.8}$	29 ± 1	35 ± 1	36 ± 1

Van Urk suggested that in simple hydrides tritium prefers the tritide to the molecular form. In the complex lithium hydrides the converse is true. The Bucharest group have shown that by heating the irradiated $LiAlH_4$ to 900° C, the whole activity may be recovered as HT (Costea and Măntescu 1966), thermal decomposition being responsible for this. The greater ratio of labile tritium in the case of alkaline and alkaline-earth hydrides has been attributed to the presence of hydroxides in the target as impurities.

3.3. HALIDES

The behavior of recoil tritium in the neutron irradiated halides is interesting from many points of view. In the halides substantial fractions of volatile compounds TX (X = F, Cl, Br, I) appear beside the molecular tritium and the tritide. In compounds containing water of crystallization (or water as an impurity) TOH has been identified. It has also been interesting to attempt to find other possible states of tritium, as for instance atomic tritium. In table 22.3 are given the results obtained by Costea and Mantescu (1966) by thermal release techniques and some additional chemical reactions for TX and T (Mantescu et al. 1971). Van Urk's results on the same type of compounds are also given in table 22.3.

Due to the difference in analytical methods used in the case of halides by the two groups it seems to be extremely difficult to compare the results obtained independently at Bucharest and Amsterdam. Måntescu et al. found tritium halide, atomic tritium and tritide only in LiF and LiCl, and great quantities of molecular tritium, whereas Van Urk did not identify the first two species, finding tritide in all the halides studied, small quantities of molecular tritium and great quantities of labile tritium (much more than the sum of TX + TOH). The effect of preheating (presumably the removing of the water of crystallization as impurity) leads on the one hand to an increase in the percentage of atomic tritium (Măntescu et al. 1976) and on the other hand to a decrease in the percentage of labile tritium as well as of tritide (Van Urk 1970). The increasing percentage of TX from fluorine to iodine found after heating the irradiated samples showed that hydrogen (tritium) halides could be formed both during the irradiation and after, and the yields could perhaps be connected with the bond dissociation energy of H(T) X (X = F, Cl, Br, I).

The results for sodium halides irradiated in the presence of helium obtained by Van Urk were similar to those of the same author for lithium halides. A small quantity of T_2 appears in these cases, the author explaining this phenomenon by reference to the filling of the crystal vacancies produced by the radiation damage.

3.4. OXIDES AND HYDROXIDES

In Li_2O the bulk of the activity (~93%) is found as labile tritium, the tritide being small (~6%) while in anhydrous and hydrated LiOH the whole activity has been found as labile tritium (>99.4%), and Van Urk concluded that the whole activity in metallic lithium has to be present as tritide, the labile tritium

Table 22.3
Distribution of tritium in lithium halides

Target	Treatment	Percent yield % HT	% T^-	% T^+	% TX	% T	% TOH	Ref.
LiF	Without preheating (0.5% H_2O)	–	–	–	–	6.9	–	[a]
	300° C preheating for 24 h	–	–	–	4.0	1.8	–	[a]
	After thermal release up to 1000° C	61.7	29.2	–	5.7	–	3.5	[b]
		2	21	77	–	–	–	[c]
	Dried at 80–400° C	2	10–50	90–50	–	–	–	[c]
LiCl	Without preheating (water of crystallization)	–	–	–	–	9.4	–	[a]
	300° C preheating for 24 h	–	–	–	7.9	0.4	–	[a]
	After thermal release up to 1000° C	59.2	27.9	–	11.4	–	1.4	[b]
		2	26	72	–	–	–	[c]
	Dried at 150–400° C	2	41–49	57–49	–	–	–	[c]
LiBr	300° C preheating for 24 h	–	–	–	20.4	1.4	–	[a]
	After thermal release up to 700° C	67.7	0.6	–	29.9	–	1.8	[b]
		4	10	86	–	–	–	[c]
	Dried at 200–400° C	2	50–62	48–36	–	–	–	[c]
LiI	Without preheating	–	–	–	19.9	1.1	–	[a]
	After thermal release up to 700° C	65.9	0.0	–	28.2	–	5.9	[b]
	Dried at 400° C	6	45	49	–	–	–	[c]
	1–3 molecules of cryst. water	10–7	2–0.1	88–92	–	–	–	[c]

[a] Măntescu et al. unpublished.
[b] Costea and Măntescu (1966)
[c] Van Urk (1970).

resulting only from the action of impurities in the metal (oxide, hydroxide). This data seems in contradiction with the observation of a large percentage of atomic tritium by another type of experiment (Măntescu et al. unpublished).

3.5. OXY-SALTS

Some lithium oxy-salts have been studied. Generally large yields of TOH were obtained, and sometimes TH resulted. These data are given in table 22.4.

Table 22.4
Distribution of tritium in lithium oxy-salts

Target	% TH	% T	% T^+	% TOH	Ref.
Li_2CO_3	19.1	7.2	–	73.8	a
Li_2CrO_4	83.3	1.9	–	14.9	a
$Li_2SO_4 \cdot H_2O$	~2.5	–	~97.5	–	b
$LiClO_4 \cdot H_2O$	~1.1	–	~98.9	–	b

[a] Costea and Măntescu (1966).
[b] Van Urk (1970).

Unlike the anhydrous chromate, hydrated chromate releases only TOH (Costea and Măntescu 1966). The same is true of the nitrate, the presence of water in the sample leading to the formation of TOH.

A recent investigation (Kudo and Tanaka 1975) gives fresh data on the distribution of products in irradiated lithium oxide.

3.6. PHOSPHORUS COMPOUNDS

Some interesting experiments were initiated by Lindner et al. (1965) on lithium salts containing orthophosphoric, phosphorus and hypophosphorus ions, following the forming of the P–T bond. The results given by Van Urk (1970) are shown in table 22.5.

Table 22.5
Distribution of tritium in phosphorus compounds

Target	% HT(DT)	% T^+	% P–T
Li_3PO_4	1 ± 1	96 ± 1	3 ± 1
LiH_2PO_4	3 ± 0.1	97 ± 0.1	~0.3
$Li_2(HPO_3)$	17 ± 2	61 ± 5	22 ± 3
$LiH(HPO_3)$	3 ± 2	59 ± 4	38 ± 4
$LiD(HPO_3)$	3 ± 1	66 ± 1	31 ± 1
$LiH(DPO_3)$	12 ± 4	57 ± 4	31 ± 7
$Li(H_2PO_2)$	35 ± 1	22 ± 41	43 ± 3

The results indicate that the percentage of T found in P–T bonds, increases in the following sequence of the anions; phosphoric, phosphorous and hypophosphorous, while the labile tritium decreases. In some cases the percentage of molecular tritium is considerable.

4. Kinetic Analysis of the Thermal Release of Tritium

Except for lithium tritide all the tritium compounds encountered are volatile and ought to be released by heating the solid irradiated target.

It is well established (Costea and Măntescu 1966) that a great quantity of volatile tritium is released when crystallochemical reactions such as the thermal annealing of radiation damaged crystals (halides, chromate, lithium-doped nickel oxide) or thermal decomposition (carbonate, chromate, lithium aluminium hydride) take place. The same phenomenon was observed when some modification in the state of aggregation occurred: melting (lithium nitrate); volatilization (halides). In some cases two or more intensive releases of tritium associated with the processes mentioned above were observed.

The problem of the kinetics of the release of tritium associated with annealing of irradiated crystals is similar to that resulting from studies of the release of gases from uranium fission. A more detailed study was done for neutron irradiated LiCl by analyzing the kinetic release curves for both uncondensable and condensable fractions (Măntescu and Costea 1966).

It was difficult in both cases to interpret the kinetics by classical diffusion governed by concentration gradients because the diffusion coefficient cannot be considered constant during the experiment. At the same time the phenomena of 'bursts' as well as of pseudo-plateaux depending on the temperature appeared. It proved to be more interesting to employ two treatments used with success in the annealing of damaged solids; in both cases the diffusion was governed by the jump frequency of the diffusive entities and by the occurrence of processes with distributed activation energies.

Subsequent studies of the diffusion of tritium from neutron irradiated LiF confirmed both the 'bursts' and the pseudo-plateaux (Cohen and Diethorn 1965), the classical diffusion theory being applied only at low temperatures.

It is worthwhile emphasizing that the behavior of recoil tritium has to be correlated with other data concerning the action of ionizing radiation on the solid lithium targets obtained by different methods of analysis, for example by ESR, NMR, X-rays, etc. The literature contains a series of results especially for LiF. Using this data a model of the behavior of recoil tritium was set up describing its interaction with the defects created in the radiation-damaged compound (Măntescu et al. 1976).

CHAPTER 23

CHEMICAL EFFECTS FOLLOWING ION IMPLANTATION

T. ANDERSEN

Institute of Physics, University of Aarhus,
DK-8000 Aarhus C, Denmark

Chemical Effects of Nuclear Transformations in Inorganic Systems
Edited by G. Harbottle and A.G. Maddock

Contents

1. Ion Implantation versus Nuclear Recoil Technique

The application of the ion implantation technique (Carter and Colligon 1968, Nelson 1968) to the study of the chemical effects associated with nuclear transformations in solids is based upon the theory (Wolfgang 1965a) that no significant difference should exist between the chemical reactivity of a recoil atom produced by nuclear means with a recoil energy of 50 keV and the corresponding ion electrically accelerated to the same energy. Since the chemical fate of a recoiling atom will be determined in the last part of its track, when the energy is well below 100 eV, it is reasonable to assume that ion implanted atoms can simulate the chemical behaviour of recoil atoms produced by the (n,γ) process with recoil energies of a few hundred eV.

The energy of (n,γ) generated recoils will be lost by nuclear collisions of the hard sphere or shielded Coulomb type. The slowing down of an accelerated ion involves energy loss by electronic and nuclear stopping. The energy of the ion will determine the ratio between these two modes of energy loss. The dominant loss is by nuclear collision in the final part of the track.

While similar to the recoil method in many respects the ion implantation technique in principle offers two advantages: elimination of the unwanted effects of the extraneous radiation invariably associated with the nuclear generation of recoil atoms, and control of the energy and charge of the recoil atom.

Ion implantation at moderate energies produces a much larger local concentration of foreign atoms than is obtained by nuclear reaction. The lattice still suffers damage by the ion implantation, but the particular disadvantage of the nuclear reaction, the creation of a large number of radiolytic defects in the lattice by the concurrent dose of ionizing radiation, is avoided.

If the ion implantation technique is combined with sectioning techniques on single crystals, ion implantation allows differential measurements, revealing changes in chemical behaviour as a function of range, whereas only integral measurements are possible with the recoil technique. Studies (Andersen and Ebbesen 1971) of (1) the relationship between the range distribution of the implanted radioactive ions and their chemical valence state determined after dissolution of the implanted crystals, (2) diffusion of the implanted ions during storage and various annealing procedures, (3) the influence of the crystal orientation during implantation, show that it is possible to gain more insight into some aspects of the chemical behaviour of radioactive recoil atoms in solids using the ion implantation technique, since these problems cannot be studied by the nuclear recoil technique.

2. Ion Implantation

Isotope separators and heavy ion accelerators have been used for implantations since Croatto and Giacomello in 1954, working with the Uppsala separator, first tried to simulate the chemical effects of the $^{14}N(n,p)^{14}C$ process by irradiating organic compounds with accelerated beams of $^{14}C^+$ and $^{14}CO^+$ ions. Since then implantation of $^{51}Cr^+$ in K_2CrO_4 (Andersen and Sørensen 1965, 1966a), $^{32}P^+$ and $^{35}S^+$ in alkali halides (Andersen and Ebbesen 1971, Andersen and Sørensen 1966b, Freeman et al. 1967, Kasrai et al. 1971, Andersen and Baptista 1971a), tritium in organic compounds (Ascoli and Cacace 1965, Ascoli et al. 1967, Paulus 1966, 1967, Paulus and Adloff 1965) $^{57}Co^+$ in trans- and cis-cobalt $[Co\,en_2\,Cl_2]NO_3$ (Andersen et al. 1968, Wolf and Fritsch 1969), and $^{64}Cu^+$ in α and β-copperphthalocyanine have been reported (Andersen et al. 1968).

The targets were either single crystals or thin foils prepared by vacuum evaporation or by electrophoretic techniques. Energetic beams ranging from a few to 450 keV have been used. The induced radioactivity depends on the isotope used and the problem investigated, but usually $10^{-4} - 1\ \mu Ci$ has been used. With targets suffering great radiation damage, which is the case for several organic compounds, the radiation conditions chosen have to be a compromise between the need to introduce into the target a sufficient activity for the subsequent analysis and the necessity to keep the radiation damage to the target surface at the lowest level. Often the beam is made to sweep the target surface to eliminate the effects of a concentrated beam. For single crystals such as NaCl (Andersen and Ebbesen 1971, Andersen and Sørensen 1966b, Freeman et al. 1967, Kasrai et al. 1971) the crystals are usually aligned so the beam enters parallel to a major plane or another preferred orientation.

The irradiation time can range from a few seconds to $\sim$ 1 h. The temperature is usually room temperature, but low temperature experiments have been performed (Andersen and Sørensen 1966b). For longer irradiations with beams containing non-radioactive molecular ions of the same mass number as the implanted radioactive ion the temperature may rise to $\sim 100°$ C during implantation.

Chemical analysis and kinetic investigations are performed in the same manner as in recoil studies of related systems produced by nuclear reactions.

3. Ion Implantation Studies of Organic Systems

Croatto and Giacomello's (1954) initial studies concerned $^{14}C^+$ ion irradiation of benzoic acid, stearic acid, cholesterol, and vitamin B and resulted in an incorporation of some of the ^{14}C in the target compound in each case. Following the first isotope separator experiments, other workers from the same group have since irradiated benzoic acid (Aliprandi et al. 1956) and cholesterol (Aliprandi and Cacace 1956) with $^{14}C^+$, $^{14}CO^+$, and $^{14}CO_2^+$ ions. Yields of 2.8% and 1.4% of labelled benzoic acid and cholesterol, respectively, resulted from the carbon-14 ion irradiation. Labelling of organic molecules by ion implantation has also been reported for benzene (Lemmon et al. 1956) using C-14 and for a number of organic substances using tritium ions (Wolfgang et al. 1956).

Later work by Lemmon et al. (1961), Mullen (1961), Tz-Hong Lin (1969) and Ascoli et al. (1967, and Cacace 1965) has been oriented more towards the fundamental problems associated with hot-atom chemistry in the solid phase. For benzene irradiated at $-196°$ C (Lemmon et al. 1961, Mullen 1961) it was found that the yield of ^{14}C-labelled benzene, toluene, and cycloheptatriene decreased with decreasing energy and increased with decreasing energy density (eV) per target molecule. It was also seen that the toluene–^{14}C produced had the same ring and methyl distribution of activity as that formed from ^{14}C nuclear recoil labelling in benzene–2 methylpyrazine solution.

The severe degradation (Lemmon et al. 1961, Mullen 1961) of the labelled products formed by the $^{14}C^+$ irradiation of benzene severely limits the possibility of drawing firm conclusions with regard to the mechanism of formation of the primary products. However, Tz-Hong Lin (1969) has obtained evidence for a mechanism of product formation involving intermediate carbenes and radicals resulting from the C–H and C=C bond insertion of hot $C(^3P)$, $C(^1D)$, and $C(^1S)$. The hot and thermal $C(^1S)$ species reacted similarly, whereas the reactions of hot $C(^3P)$ and $C(^1D)$ are different from those of the thermal counterparts.

A comparison (Ascoli and Cacace 1965, Ascoli et al. 1967) of the chemical effects of 40 keV tritium ions implanted in solid organic targets, such as sodium m-iodo-benzoate, with recoil tritons from the $^6Li(n,\alpha)^3H$ process indicates that the radiochemical yield and the intramolecular tritium distribution is the same for these methods, in agreement with the hypothesis that the initial energy has no influence on the product distribution, when it is well above the threshold value for product formation. The results obtained with 40 keV T^+, T_2^+ and T_3^+ ions (Ascoli et al. 1967) were the same, a result consistent with the view that high-energy polyatomic species dissociate into the constituent atoms in their initial collisions with the target molecules before their energy is sufficiently lowered to make the formation of stable products possible.

The study of solid organic systems will probably benefit in the coming years from the recent interest (Carter and Colligon 1968, Nelson 1968) in physical studies of ion implantation in single crystals of organic materials.

4. Ion Implantation Studies of Inorganic Systems

4.1. THE K_2CrO_4 SYSTEM

Preliminary $^{51}Cr^+$ implantation studies (Andersen and Sørensen 1965) were initiated to gain more information about the recoil fragments formed upon neutron irradiation of oxyanion systems. The nature of these fragments had played a predominant role in the understanding of the chemical fate of the radioactive atoms during and after the nuclear event. The alkali metal chromates were the most studied systems (Andersen 1968) and the same analytical technique was used in all investigations. This technique separated the radioactive products into two groups, chromic and chromate ions. The radioactive chromic fraction was supposed to result from the reduction of $^{51}CrO^{4+}$ fragments by water, whereas the hydrolysis products of $^{51}CrO_3$, $^{51}CrO_2^{2+}$ together with $^{51}CrO_4^{=}$ accounted for the chromate fraction assuming Libby's hypothesis (1940) regarding the formation of fragments in neutron irradiated oxyanion systems. This hypothesis predicts rupture of O^{2-} fragments from the CrO_4^{2-} molecule leaving the valence state of the chromium atom in the residual molecule unchanged in the irradiated crystal.

Chromate containing systems offered a good opportunity (Andersen and Sørensen 1966a) to test the validity of Libby's hypothesis and by choosing the K_2CrO_4–K_2BeF_4 mixed crystal system it was possible to test the validity of the assumption (Maddock and Vargas 1961a, b) that reformation of the original molecule occurs during thermal annealing in this system, too.

Since chromous ions exhibit strong reducing properties, detection of radioactive chromous ions upon solution of an irradiated chromate-containing system would disprove Libby's theory. If $^{51}Cr^+$ ions introduced into very dilute mixed K_2CrO_4–K_2BeF_4 crystals undergo thermal annealing in a manner similar to the ^{51}Cr atoms generated in this system by thermal neutrons, it is unlikely that reformation of the original molecule disrupted by neutron capture can be occurring during thermal annealing, as previously assumed.

The combination of a new analytical technique (Andersen and Olesen 1965, Gütlich and Harbottle 1966), ion implantation and mixed crystals settled (Andersen and Sørensen 1966a) the chromous problem and disproved Libby's hypothesis.

Figure 23.1 shows a comparison between the chemical distribution of ^{51}Cr atoms generated by thermal neutrons and ^{51}Cr implanted as 70 keV $^{51}Cr^+$ ions in the mixed K_2CrO_4–K_2BeF_4 crystals.

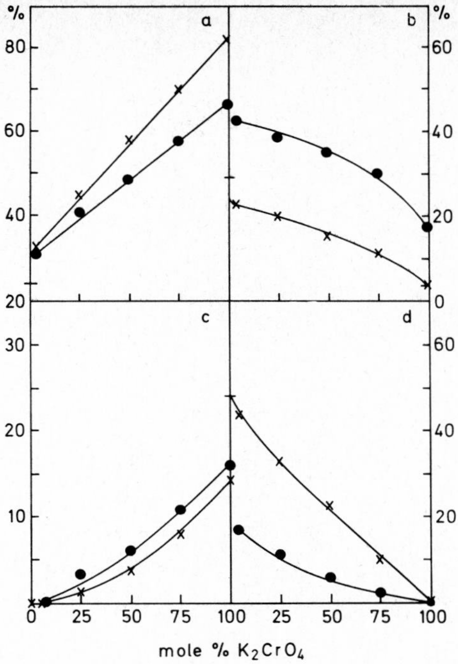

Fig. 23.1. Chemical distribution of ^{51}Cr in mixed crystals of potassium fluoberyllate after irradiation with 70 keV $^{51}Cr^{+}$ ions (+) or thermal neutrons (o). (a) $^{51}CrO_4^{2-}$, (b) $^{51}Cr^{2+} + {}^{51}Cr^{3+}$, (c) dimeric and polymeric Cr(III), (d) insoluble Cr compound.

If the energy of the accelerated $^{51}Cr^{+}$ ion was reduced to a few keV the $^{51}CrO_4^{=}$ yield was ~ 65% for pure K_2CrO_4 (Andersen and Sørensen 1965) in good agreement with the (n,γ) data. The energy dependence observed for ion implantation in K_2CrO_4 does not represent a real energy dependence, but reflects the initiation of annealing reactions, due to the larger energy deposition, for more energetic ions.

The data (Andersen and Sørensen 1966a) in fig. 23.1 proved the value of the implantation method for studies of thermal annealing reactions in ionic crystals. The implantation data disproved the assumption that real reformation of the disrupted molecules from the parent fragments took place during thermal annealing and gave support to the hypothesis that the observed changes in the valence of the radioactive atoms reflect electronic rearrangements in the crystals, since ~ 30% $^{51}CrO_4^{=}$ forming fragments were detected by both methods of introducing ^{51}Cr atoms into 5 mole % K_2CrO_4–K_2BeF_4 mixed crystals.

The very much higher local concentration of ^{51}Cr atoms produced by the implantation than by nuclear reaction accounts for the observed higher content of more complex molecules or insoluble compounds in several systems (Andersen and Sørensen 1966a), e.g. the data in fig. 23.1 d.

4.2. THE NaCl SYSTEM

Investigations of implanted $^{32}P^{+}$ and $^{35}S^{+}$ in NaCl (Andersen and Sørensen 1966a, Freeman et al. 1967) have shown that the same species are formed as seen in neutron activated material. Table 23.1 gives the result for $^{35}S^{+}$ (Freeman et al. 1967). The most noticeable difference is the much larger amounts of the sulphate precursor produced in the neutron irradiations. The greater part of the sulphur introduced by the ion implantation penetrates less than 5 μm: nonetheless the presence of air or oxygen after implantation has no appreciable effect on the ^{35}S distribution. The species yielding sulphite and sulphate upon solution are not increased by aerial oxidation and provide still more convincing proof that the sulphite and sulphate do not pre-exist in the crystals.

Table 23.1

Mode of production	Sulphur fraction (%)			
	I	II	III	IV
	Measured as			
	S^{2-}	CNS^{-}	SO_3^{2-}	SO_4^{2-}
1. ^{35}S introduced by (n,p) reaction	46.1	9.6	21.5	22.8
2. ^{35}S introduced by ion implantation	46.4	26.4	17.7	9.5

Another, and still more pronounced, difference lies in the annealing behaviour of the $S^{=}$ and S^{0} (thiocyanate) fractions (Kasrai et al. 1971). These differences in annealing behaviour have been utilized in the identification of the annealing mechanisms as well as of the entities present in the solid phase. It is assumed that $S^{=}$ and S are located on anionic sites whereas the sulphite and sulphate fractions may be formed by the hydrolysis of S–Cl complexes and are assumed to arise from ^{35}S at cationic or interstitial sites.

$^{32}P^{+}$ ions implanted in NaCl exhibit (Andersen and Ebbesen 1971) the same annealing reactions with nearly the same onset temperature for the dominant $^{32}P(I) \rightarrow {}^{32}P(V)$ annealing reaction as seen for neutron activated crystals. Bleaching the F-centres before thermal annealing proved the existence of two P(I) precursors, which can both undergo the $P(I) \rightarrow P(V)$ reaction during annealing, but at markedly different temperatures. This observation was also made on neutron activated samples (Andersen and Baptista 1971b).

Table 23.2

Sample	Mode of ^{35}S-production and treatment	% ^{35}S			
		$S^=$	CNS^-	$SO_3^=$	$SO_4^=$
A	by (n.p) reaction	50.8	10.7	27.7	10.7
A′	A sample, annealed 1 h,205°C	65.0	1.9	10.3	22.7
B	by ion implantation	68.6	21.3	7.7	2.3
B′	B sample, annealed 1 h,205° C	22.3	57.0	15.9	4.8

Figure 23.2 shows the range distribution for different implantation angles. The crystals are aligned with the ⟨100⟩ plane parallel to the beam for the 0° curve. The three other curves were obtained rotating the crystals 30, 45, and 60°, respectively around the ⟨001⟩ with respect to the ⟨100⟩ plane before implantation.

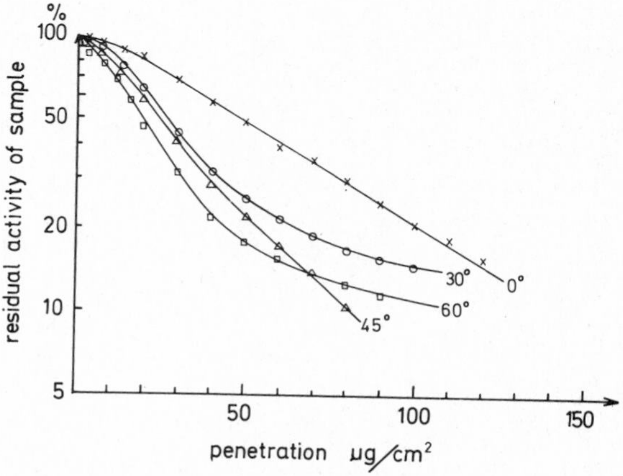

Fig. 23.2 Range distribution for 40 keV $^{32}P^+$ ions implanted into NaCl single crystals. The angles given represent the angle between the <100> and the incident beam.

The marked difference in the range curves (fig. 23.2) for the most deeply penetrating 25% of the ^{32}P atoms is also reflected in the chemical behaviour of these atoms upon dissolution. Table 23.3 shows the results.

The precursor of the P(V) fraction is enriched, relatively, in the most deeply penetrating part of the atoms, particularly in the 30° and 60° experiments. The crystals were kept at room temperature for the same period of time before chemical analysis, and the implantations were performed in daylight. It is important only to compare crystals kept for the same period of time after implantation, since diffusion occurs even in crystals kept at −196° C. Figure 23.3 shows the range distribution for crystals kept at −196° C as a function of

Table 23.3

		Implantation angle			
	%	0°	30°	45°	60°
Most deeply penetrating 25% ^{32}P atoms	P (V)	46.8	63.3	46.7	70.2
	P (III)	10.1	5.8	8.2	8.1
	P (I)	43.1	30.9	45.1	21.7
Total crystal 100% ^{32}P atoms	P (V)	32.5	–	–	42.8
	P (III)	6.6	–	–	9.0
	P (I)	60.9	–	–	48.2

storage time, but no changes in the total chemical distribution was seen at −196° C.

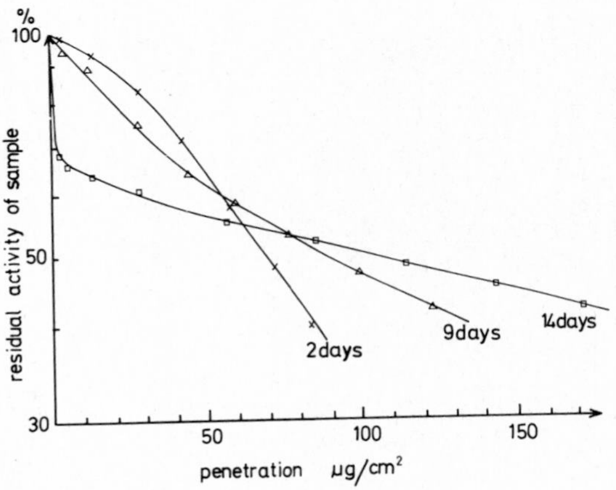

Fig. 23.3. Range distribution for 40 keV $^{32}P^{+}$ ions after storage at room temperature.

The deeply penetrating tails (fig. 23.1, table 23.3) are enriched in respect of the P(V) precursor. An analysis of the layers near the surface exposed to implantation and of the part of ^{32}P atoms penetrating the crystal to the rear side gave the results shown in table 23.4.

This correspondence between the most deeply penetrating 40% of the ^{32}P atoms and the high content of P(V) presursor was also seen during storage experiments (fig. 23.4).

The high content of the $^{32}P(V)$ precursor in the most deeply penetrating tails and the close connection between changes in range distribution and the $^{32}P(V)$ yield indicate that the ^{32}P species able to diffuse in the crystal are detected as

Table 23.4

Remaining % of ^{32}P atoms	% P(V)	% P(III)	% P(I)
100	32.0	9.8	58.2
98	31.5	10.0	58.5
88	42.0	10.5	47.5
61	48.5	12.5	39.0
1.5	73.2	1.3	25.5

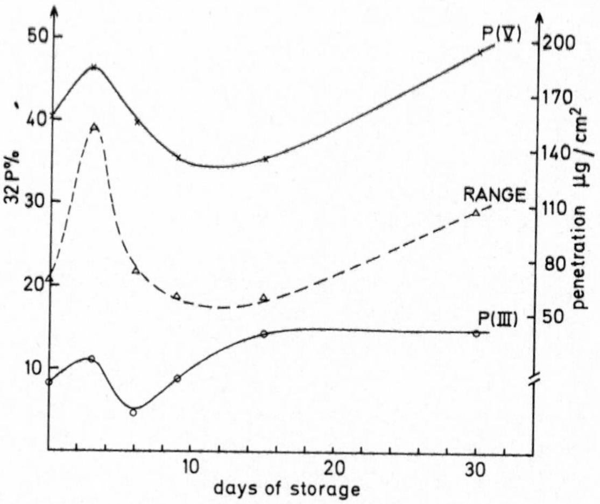

Fig. 23.4. Chemical distribution for the most deeply penetrating 40% of the implanted $^{32}P^{+}$ ions as a function of storage time at room temperature. The thickness of the layer peeled off before chemical analysis is shown (dotted line).

$^{32}P(V)$ after dissolution, if they come to rest in a region with nearly no radiation damage. It seems unreasonable that the mobile $^{32}P(V)$ precursor should be a P–Cl compound. It is more reasonable to account for the data by assuming interstitial diffusion of the $^{32}P(V)$ precursor. It is not possible from the present data to identify the $^{32}P(V)$ precursor, but its behaviour would coincide with the behaviour of singly or doubly charged phosphorus ions. The presence of P^{+} or P^{++} would be possible in an alkali halide lattice using Lidiard's criterion (1957).

The data in fig. 23.4 and the fact that the yield of the $^{32}P(III)$ fraction is 10–12% in the tails of the range distribution, although these tails are generated by diffusion, may suggest that the precursor for the $^{32}P(III)$ fraction may be associated in some manner with the $^{32}P(V)$ precursor. The $^{32}P(III)$ fraction could be generated during dissolution as the result of a reaction between the

$^{32}P(V)$ precursor and a charge carrier released from a nearby defect. Table 23.4 indicates that reactions occurring during dissolution have an influence on the chemical distribution obtained. 32% $^{32}P(V)$ was determined when the total activity was analysed. Stripping of the surface layers containing 12% of the implanted atoms, and also the most damaged regions, leads to an increase in the $^{32}P(V)$ fraction. The 42% $^{32}P(V)$ present in the remaining 88% of the implanted ^{32}P atoms would be equal to 37% for the total crystal, assuming that the removed layers only contained non-$^{32}P(V)$ generating species. Since the $^{32}P(V)$ yield can be determined with an accuracy of $\pm 1\%$ these data show the role of the dissolution process. Charge carriers present in the more damaged regions are able to react with the $^{32}P(V)$ precursor during dissolution and reduce the $^{32}P(V)$ yield.

This influence of dissolution is an unwanted effect and may seriously limit the possibilities for further research in the field of chemical effects following ion implantation unless non-destructive solid-state analytical techniques are developed.

4.3. OTHER INORGANIC SYSTEMS

Implantations of $^{57}Co^{+}$ in cis and trans- [Co $en_2Cl_2]NO_3$ (Andersen et al. 1968, Wolf and Fritsch 1969) and of $^{64}Cu^{+}$ in α- and β-Cu-phthalocyanines (Andersen et al. 1968) have shown a marked influence of the stereo-chemical configuration and of the crystal structure. Labelling of these molecules by ion implantation takes place by a mechanism equivalent to the mechanism for labelling by nuclear recoil technique. For 20 keV $^{57}Co^{+}$ ions implanted in the trans-compound 33% of the radioactivity is detected in the trans-form and only 2% in the cis-form. For 30 keV Cu^{+} implantation 8.5% is detected in the β-form and only 1.8% in the α-form of Cu-phthalocyanines.

5. Recent Developments (1972-1976)

The lack of non-destructive analytical techniques to determine the chemical fate of the radioactive ions implanted in solids in small quantities has limited the application of the ion implantation method to simulate chemical effects of nuclear transformations in solids. In the period 1972–76 the technique has not been used to study simple inorganic systems, but only systems for which the dissolution processes play a minor role, such as organic (Cailleret 1973, Cailleret et al. 1971, 1972, 1973, 1974, 1975, Akcay et al. 1973) or metal–organic (Jenkins and Wiles 1972, Yoshihara et al. 1974) systems.

The Strasbourg group (Cailleret 1973, Cailleret et al. 1971, 1972, 1973, 1974, 1975, Akcay et al. 1973) has studied in great detail the chemical fate of radioactive I^{+} and I_2^{+} ions implanted into solid butane in the energy range between 4 eV and 3000 eV. The total organic yields and the relative yields of the radioactive iodides formed by both types of ions compare well for energies above 75 eV and also reflect the properties observed from neutron activation of butyl-iodides. At energies below 75 eV marked discrepancies between the two implanted systems are seen. It is very likely that the mechanism in the organic-halide formation can be attributed to a two-step reaction. The first step involves the formation of ions, radicals or excited species, through the transfer of kinetic energy or charge transfer reactions, followed by a second step in which these species react with halogen.

The metal-organic systems (Jenkins and Wiles 1972, Yoshihara et al. 1974) have been studied in order to be able to prepare radioactive-labelled metal-organic molecules or ions. By implanting In^+ ions into copper- or metal-free phthalocyanines, Yoshihara et al. (1974) were able to form indium-phthalocyanine complexes, which were compatible with the complexes formed by the β process

$$^{115}\mathrm{Cd\ Pc} \longrightarrow {}^{115m}\mathrm{InPc}^+.$$

6. Conclusion

The implantation technique has been a valuable tool for studying certain aspects of chemical effects of nuclear transformations in solids. Further progress in applying this technique, particularly to simple inorganic systems, is closely associated with the development of non-destructive analytical methods for determining the chemical identity of the implanted ions.

CHAPTER 24

IN SITU STUDIES

J.P. ADLOFF

Laboratoire de Chimie Nucléaire, Centre de Recherches Nucléaires
Strasbourg, France

Chemical Effects of Nuclear Transformations in Inorganic Systems
Edited by G. Harbottle and A.G. Maddock

Contents

Introduction

The radiochemical analysis of irradiated targets or labelled compounds inevitably starts with the dissolution of the sample. By this act the nature and the distribution of recoil species can be significantly altered. Furthermore a long time elapses between the occurrence of the nuclear process and the analysis of the formerly "hot" species. These limitations are avoided by the use of "in situ" investigations which rely on various physical techniques. Advantage is taken somehow or other of the radiations emitted by the radioactive recoil atoms. In situ methods provide direct information from within the solid itself of the valency, bonding and surroundings of nucleogenic atoms, shortly after their formation, i.e. after a period which is often of the order of the lifetimes of excited nuclear levels. Some of these methods provide dynamic data on the properties of the hot species as a function of the time elapsed since the nuclear decay.

Mössbauer spectroscopy is by far the most used in situ technique in hot-atom chemistry. The measurement of the angular correlation of the successive radiations emitted by a nucleus is another way to measure the hyperfine fields at the location of the recoil atoms. A few other promising methods are still scarcely explored.

1. Mössbauer Spectroscopy

The chemical applications of Mössbauer spectroscopy have been the subject of several treatises and numerous reviews (see e.g. Goldanskii and Herber 1968). Mössbauer studies of the chemical effects of nuclear transformations have been reviewed by Adloff and Friedt (1972), Bondarevskii et al. (1971) and Maddock (1972). In the usual Mössbauer experiments the substance to be investigated is used as the absorber and the source is chosen to give the narrowest possible single line emission. Reciprocally, in *Mössbauer emission spectroscopy* ("source" experiments) the spectra obtained with a single line absorber reveal on a 1 to 100 ns time scale the chemical effects, or "after effects", of the nuclear process (radioactive decay, nuclear reaction, Coulomb excitation) which populates the Mössbauer level in the source. The oxidation state of the emitting atom and the symmetry of the new environment of the nascent atom can be inferred from the emission spectra by way of the usual parameters: the isomer shift, determined by the electronic density at the nucleus (E0 interaction) and the quadrupole (E2) and magnetic (M1) splittings resulting from the electric field gradient and the magnetic field acting on the nucleus.

The Mössbauer effect has been observed on about 70 isotopes of some 40 elements, but the application of Mössbauer spectroscopy to hot-atom chemistry has hitherto been restricted to a few systems, more particularly the two most convenient Mössbauer source nuclides ^{57}Co and ^{119m}Sn.

In most practical cases, the Mössbauer level is populated by radioactive decay or by a long-lived isomeric transition. The after-effects of the nuclear process may be investigated in stoichiometric compounds of the source atoms [^{57}Co-labeled cobalt compounds or ^{119m}Sn-labeled tin compounds] or in any material doped with the radioactive parent atoms. In the former case, the chemical state of the radionuclide is identical with that of the isotopic stable carrier atoms; however the preparative method may influence the chemical state of the daughter atoms. In doped materials considerable doubt may exist as to the chemical state of the parent since charge compensation is necessary if the lattice has to accommodate an aliovalent dopant. Post-preparation treatments of the source such as annealing, firing, irradiation can alter the type and the distribution of the defects.

Mössbauer parent atoms with long enough half-lives can be generated directly in the source material containing the target atoms. This procedure has been applied to long-lived isomers of the Mössbauer nuclei produced by neutron capture, e.g. ^{118}Sn(n,γ)^{119m}Sn in tin compounds. The irradiated material serves as Mössbauer source and can also be submitted to a radiochemical analysis. The spectra show the effect of both the activation process and the decay to the Mössbauer level. Such experiments require a high activation cross section, a large recoil free fraction and a sufficiently narrow line for the resolution of the hyperfine parameters.

Alternately the Mössbauer level can be populated from a nuclear reaction or by coulombic excitation in "on line" experiments; this is necessary if a suitable radioactive parent atom is lacking. Typical examples are ^{39}K(n,γ)^{40}K, ^{56}Fe(d,p)^{57}Fe, ^{56}Fe(n,γ)^{57}Fe, ^{157}Gd(n,γ)^{158}Gd, Coulomb excitation of ^{57}Fe with 3 MeV α particles, of Hf with 6 MeV α, of Ni with 20 MeV oxygen ions. These processes involve high recoil energies and the "hot" source atoms are likely to be displaced in the lattice. In the Coulomb recoil implantation procedure the excited nuclei are propelled into a backing material or implanted through vacuum into a selected matrix. States with lifetimes longer than a few ns will decay chiefly in the implanted matrix. By this method the Mössbauer atoms are collected in an environment free of direct heating and radiation damage due to the incident particle beam. The oxidation state, the symmetry of the surroundings and the binding mode of the recoil atom at the location where it comes to rest are inferred from the Mössbauer emission spectrum. A relevant example is the partial oxidation of ^{57}Fe after thermal neutron capture in ^{56}Fe in a hydrated ferrous sulphate target (Berger et al. 1967).

Mössbauer spectroscopy has given rise to the possibility of detecting metastable states formed in radioactive decay and of studying the relaxation processes of these transient species. However in the few instances where the Mössbauer emission spectra of ^{57}Fe have been traced as a function of the time elapsed after ^{57}Co EC (in cobalt oxide, sulphate, chloride and ^{57}Co doped ferrous ammonium sulphate) (Triftshauser and Craig 1967, Triftshauser and Schroerr 1969) there was no evidence of any change in the chemical form of the iron between 5 ns and 300 ns after the decay. In one case however [$CoSO_4 \cdot 7H_2O$] a time-dependent recoilless fraction has been observed at the Fe^{3+} sites but not at the Fe^{2+} sites. The relaxation time of f is estimated to be between 10 and 100 ns (Hoy and Wintersteiner 1972). The Auger charging cascade, the electronic redistribution and the charge relaxation are achieved in a time shorter than the 10^{-7}s mean life of the Mössbauer level. The short duration of the immediate consequences of nuclear transformations in solids, as compared to the lifetimes of Mössbauer levels, has been supported from observations on other systems. The Mössbauer effect is readily observed even if the initial recoil energy of the excited nuclei amounts to hundreds of keV, such as after α decay (Stone and Pillinger 1964). Within high accuracy the isomeric shift, quadrupole splitting, and the Debye–Waller factor are found to be the same following the $^{56}Fe(d,p)$ activation of the Mössbauer level in Fe_2O_3 and the absorption spectrum of iron oxide (Christiansen et al. 1973). Thus the majority of the recoil atoms come to rest and are strongly bonded in the lattice before the nuclear de-excitation. Time-dependent perturbation has also not been apparent in angular correlation measurements on the 136 keV, 8.9 ns level of ^{57}Fe in an iron foil; clearly recoil and radiation damage effects are completely recovered about 10 ns after the nuclear reaction (Jones 1970a).

The recoil free fraction f is strongly temperature-dependent. Local heating of the lattice due to the energy brought by the recoil atom should lower the f value. In metals the heat dissipation is so fast that no f lowering is observed, even for the high recoil energies associated with thermal neutron capture or particle induced nuclear reactions. However in more insulating solids, like oxides, smaller f values have been observed for the more exoergic nuclear decays (f values in α-decay populated ^{237}Np sources are lower compared to β decay, Stone and Pillinger 1964) which has been attributed to a local heating of the lattice (Mullen 1965). On the other hand, a delayed coincidence Mössbauer spectroscopy experiment has shown that the f value can increase as a function of the time elapsed after a radioactive decay. The increase is due to the relaxation of a localized vibration mode of the Mössbauer nucleus excited by the nuclear process (Hoy and Wintersteiner 1972). No direct support of the hot-zone model nor evidence for the melting of the lattice around the recoil atoms has yet emerged from Mössbauer spectroscopic data (Jones 1970a). A few typical results will be now presented.

1.1. ELECTRON CAPTURE DECAY OF ^{57}Co TO ^{57}Fe

^{57}Co EC decays to the 136 keV level of ^{57}Fe which in turn deexcites to the 14.4 keV, 140 ns Mössbauer level. The highest recoil energy imparted to the excited iron nucleus is 3.4 eV, i.e. substantially below the displacement threshold energy. The electron capture triggers an Auger cascade, leading to an estimated maximum charge of +8, when starting from Co^{3+} (Pollak 1962). As has been indicated, these charge states do not survive long enough to be detected in the Mössbauer spectrum. Aliovalent iron states (i.e. charge states differing from that of the parent cobalt atom) are stabilized in a time shorter than 10^{-7}s and survive for an appreciably longer time.

Emission spectra of many cobalt compounds have been reported, including hydrated and anhydrous ionic salts (e.g. Ingalls et al. 1965, 1966, Friedt and Adloff 1967, 1968, 1969a, b, Cavanagh 1969), complexes and chelates (e.g. Friedt and Asch 1969, Fenger et al. 1970a, b, Nath et al. 1970, Saito et al. 1971, Friedt et al. 1971) and oxides (e.g. Wertheim 1961, Ok and Mullen 1968, Kundig et al. 1969, and Blomquist et al. 1971). The spectra of complex ligand ^{57}Co compounds (i.e. with polyatomic ligands) are best interpreted by reactions of the nascent iron with radical fragments originating from the local autoradiolysis of the environment of the decay site by the low energy-Auger electrons. Oxidizing fragments such as OH, SO_4^- stabilize Fe^{3+} (fig. 24.1, lower part), whereas organic radicals are mostly reducing towards Fe^{3+} so that in Co(III) molecular complexes the Fe(II) form is favoured (fig. 24.2, upper part). This view is supported by the increasing yield of Fe^{3+} with the number of H_2O ligands in the nearest coordination shells due to the higher density of OH radicals, the absence of after-effects in compounds of Co(III) with strong radiolytically resistant ligands (phenyl derivatives), and in Co(II) molecular complexes for which a further reduction of Fe^{2+} appears hardly possible (fig. 24.3). Furthermore the spectra of these compounds are analogous to the absorption spectra of the corresponding γ- or electron-irradiated iron complexes (Friedt et al. 1970) (fig. 24.2). For example, Fe^{3+} has been observed in all hydrated Co^{2+} salts; similarly, Fe^{2+} is oxidized to Fe^{3+} in radiolyzed hydrated ferrous salts (Wertheim and Buchanan 1969). Internal and external irradiation provide the same qualitative transformation, the yields depend on the radiation dose delivered to the system.

Other mechanisms have been proposed. One of the models is an extension to the solid phase of the coulombic fragmentation process (Nath et al. 1970). The latter has been primarily established in the gas phase, but appears much less credible in a condensed system in which the intramolecular charge transfer will

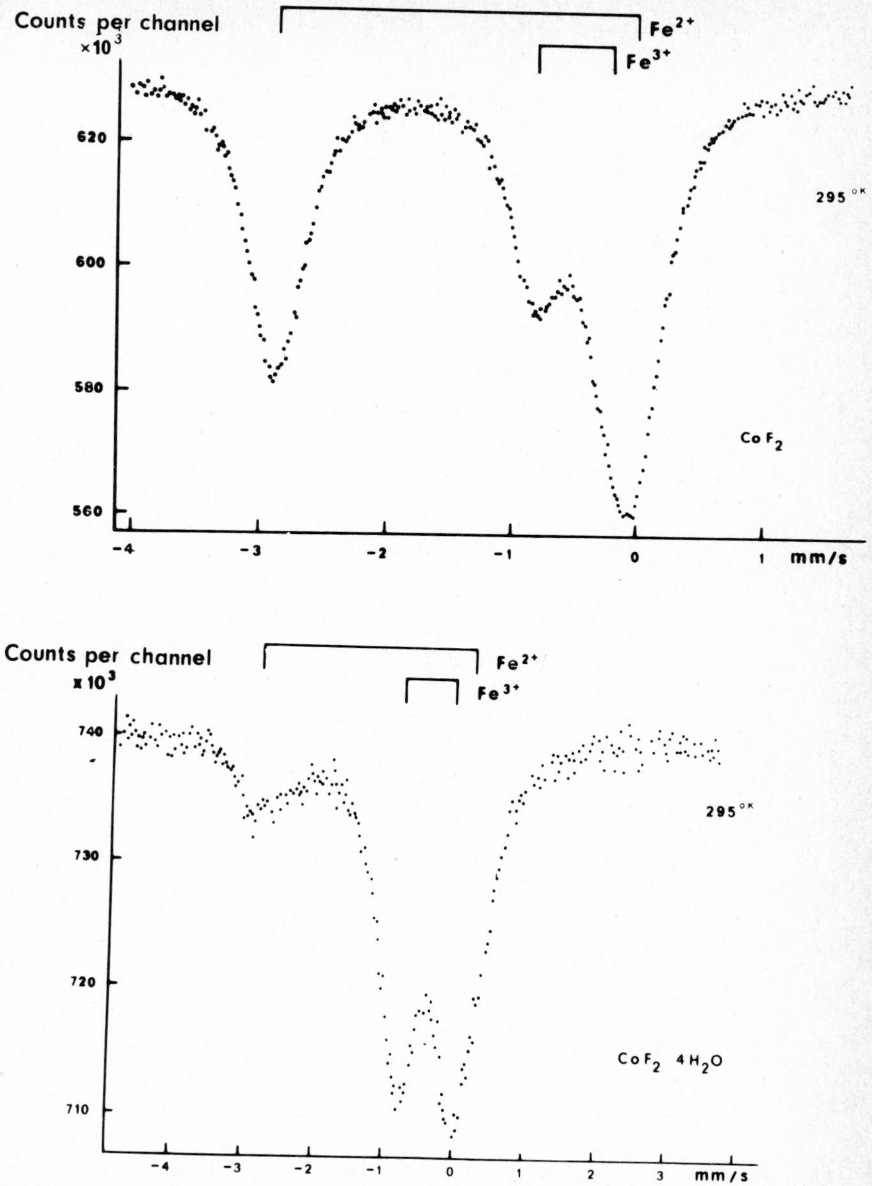

Fig. 24.1. Mössbauer emission spectra of $^{57}CoF_2$ (upper figure) and $^{57}CoF_2$, $4H_2O$. The Fe^{2+}/Fe^{3+} ratio deduced from the areas of the peaks is 1.27 in the anhydrous fluoride and 0.34 in the tetrahydrate. The spectra are taken against a single line potassium ferrocyanide absorber. The increased yield of Fe^{3+} in the hydrated fluoride is due to the oxidizing properties of $OH^{\cdot}$ radicals.

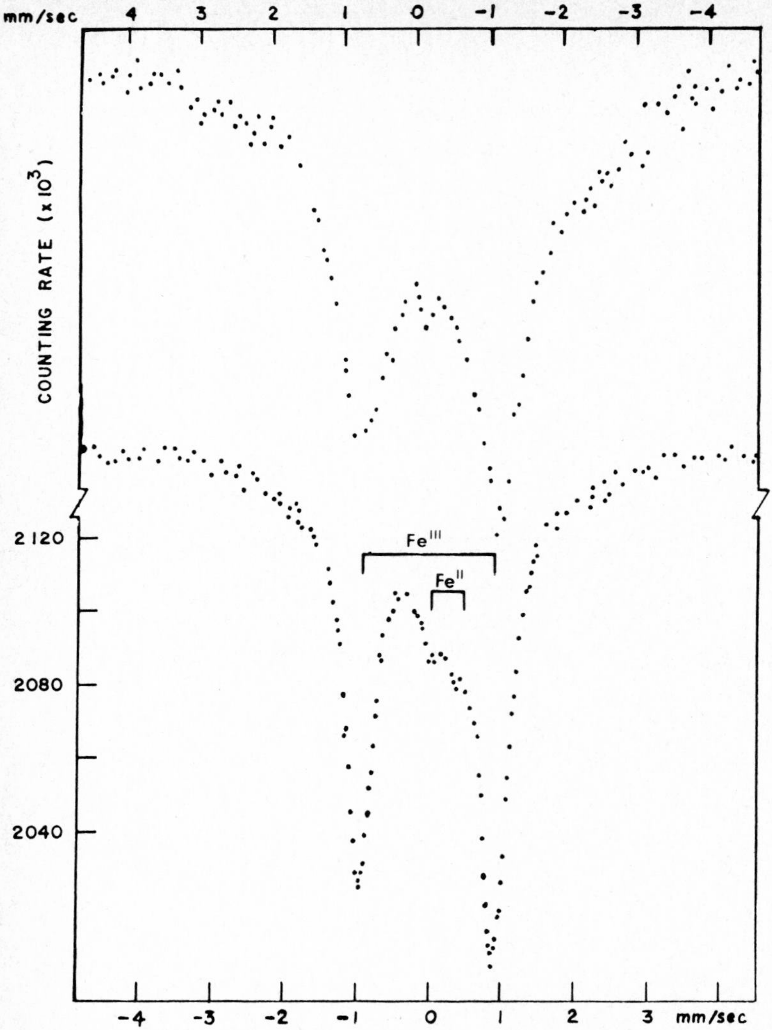

Fig. 24.2. Emission spectrum of ^{57}Co(III) trisdipyridyl perchlorate (upper figure, Nath et al. 1970) and absorption spectrum of electron-irradiated Fe(III) trisdipyridyl perchlorate with a dose of 1800 Mrads.

be perturbed by the surrounding molecules or ions. An alternative explanation is based on the observed pressure induced reduction of Fe^{3+} into Fe^{2+} (Hazony and Herber 1969). The $^{57}Fe^{3+}$ ions replacing $^{57}Co^{3+}$ ions after EC have a slightly larger ionic radius (0.64 Å as compound to 0.63 Å for Co^{3+}) and might experience an internal pressure which could stabilize the reduced species.

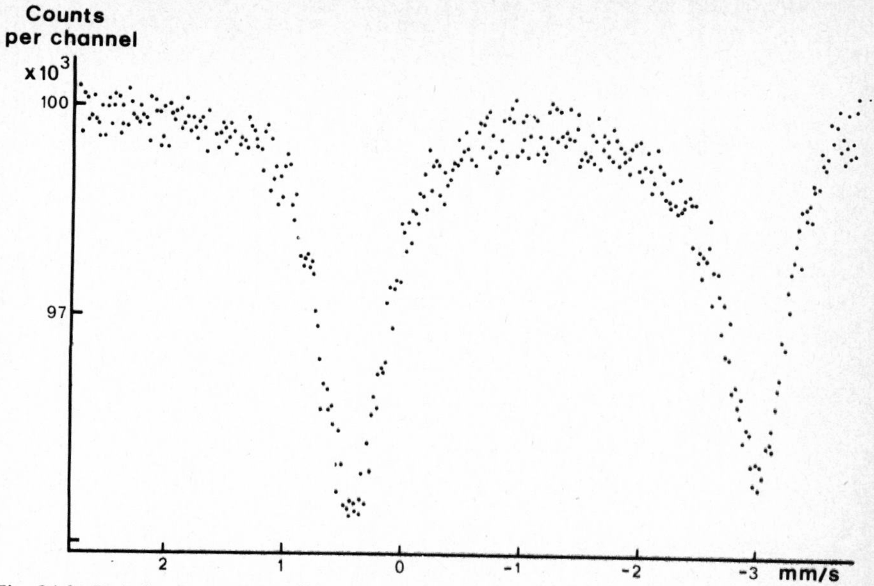

Fig. 24.3. Emission spectrum of $^{57}CoCl_2$, 6 pyridine at 80 K (Friedt et al. 1971). The iron is stabilized as Fe II. This spectrum is similar to the absorption spectrum of $FeCl_2$, 6 pyridine.

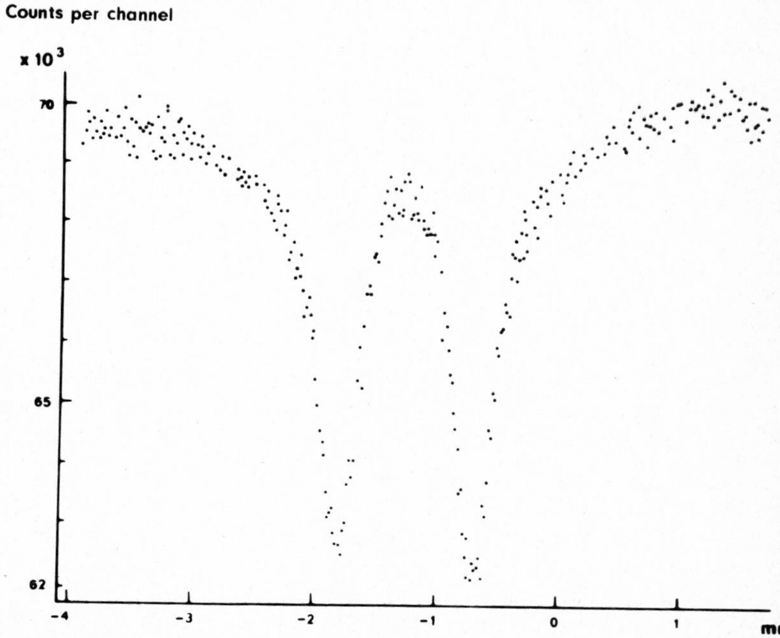

Fig. 24.4. Emission spectrum of $^{57}CoCl_2$. The iron is formed solely as Fe^{2+} (Friedt et al. 1970).

However this mechanism is not consistent with the observation of after-effects in tin compounds (see below) and in $^{57}Co^{3+}$ doped ferric complexes, since in both cases this size effect obviously cannot be invoked.

Simple ligand compounds, such as oxides and fluorides, show a marked radiation stability and the autoirradiation can merely create defects and minor radiation damage around the decay sites. The results reported for ^{57}Co-cobalt oxides and ^{57}Co-doped metal oxides (NiO, FeO,. . .) by various authors are in poor agreement. These compounds are known to be defect solids and the discrepancies arise from the difficulty of obtaining well-defined stoichiometric oxides with cobalt atoms in a single valence state. According to their mode of preparation, the Fe^{3+} yield in CoO, NiO, etc. sources ranges from 0% to 100%, depending on the nature and density of charge compensating defects (cation defects and excess oxygen ions) introduced in the solid.

Data on labelled or doped ionic salts are less confused. In these samples, the iron is found either in the charge state of the cobalt ($CoCl_2$, fig. 24.4) or in an oxidation state higher by one unit ($^{57}CoF_2$, fig. 24.1, upper part) (Cavanagh 1969, Friedt 1970a, b, Friedt and Danon 1970). Occasionally the formation of Fe^{+} has been suggested (e.g. Chappert et al. 1969). The role of lattice parameters and of thermodynamic properties of the solid now become predominant. In ^{57}Co-doped rutile structure metal (II) fluorides it is apparently the misfit of the Fe^{2+} ions ($r = 0.76$ Å) in the host lattice which favours the stabilization of the smaller Fe^{3+} (0.64 Å) (Wertheim et al. 1969, Wertheim and Buchanan 1969). Fe^{3+} would survive preferentially in cation lattice sites smaller than those of Fe^{2+} in FeF_2. This trend has been established more quantitatively by considering the lattice energy of the host matrix. In a series of isostructural compounds, the Fe^{3+} yield increases with the local electrostatic potential at a cation site (Cruset and Friedt 1971a, b).

When neutral ^{57}Co atoms are isolated in a xenon matrix, ^{57}Fe is found in two states: Fe° ($3d^6 4b^2$) and Fe^{+} ($3d^7$) (Micklitz and Barrett 1972).

1.2. ISOMERIC TRANSITION OF ^{119m}Sn AND THE $^{118}Sn(n,\gamma)^{119m}Sn$ REACTION

A highly converted ($\alpha > 6000$) IT populates the 23.8 keV 19 ns Mössbauer level of ^{119}Sn. Thus the primary consequences of the decay are analogous to those of ^{57}Co EC. No aliovalent species have been observed after IT in tin compounds with simple ligands SnO_2, SnO, SnI_4, $SnCl_2$ and $BaSnO_3$. In complex ligand compounds, as expected from ^{57}Co EC data, the oxidized form Sn^{4+} is observed in $^{119m}SnSO_4$ (Llabador and Friedt 1971) (fig. 24.5) and hydrated stannous salts, while the reducing properties of organic radicals favour the formation of Sn^{2+} in complex oxalates of Sn(IV) (Sano and Kanno 1969).

A few source experiments have been reported on neutron-irradiated tin compounds. In the oxide SnO_2, the recoil atoms come to rest in substitutional

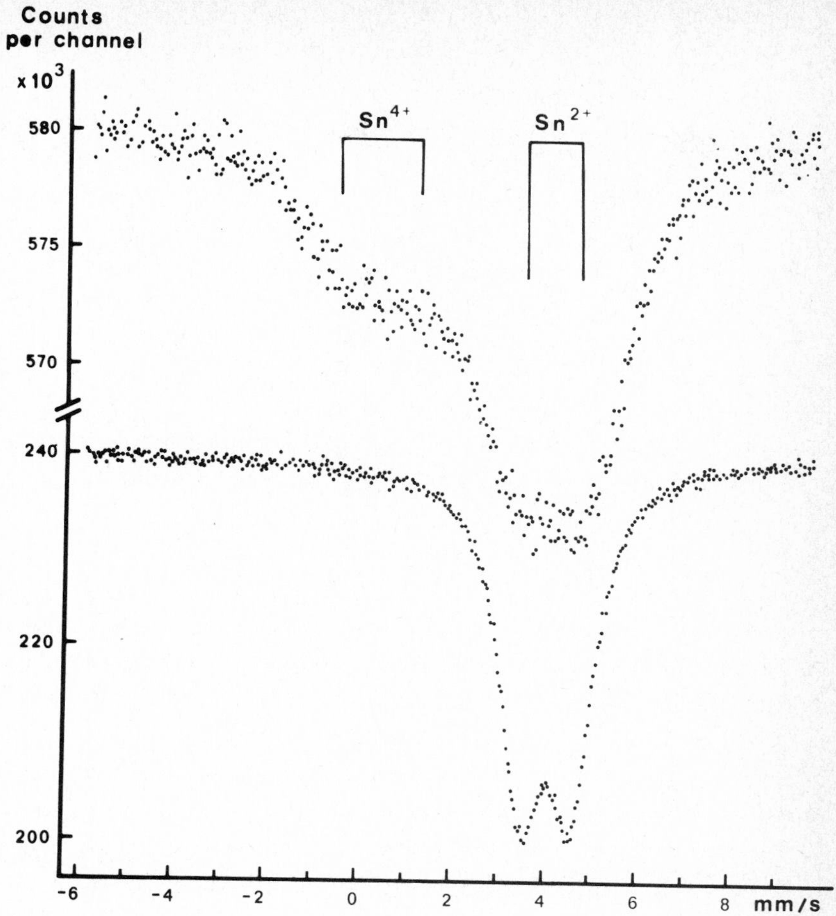

Fig. 24.5. Emission spectrum at 100 K of $^{119m}SnSO_4$ (upper figure) against a $BaSnO_3$ absorber and absorption spectrum of the same compound with a $Ba^{119m}SnO_3$ source. The emission spectrum exhibits besides the Sn^{2+} peaks a doublet characteristic of Sn^{4+} (Llabador and Friedt 1971).

lattice sites and in the neighbourhood of crystal defects as is apparent from the decrease of the *f* factor (as compared to the *f* value in the source compound as absorber) (Yoshida and Herber 1969). Thermal annealing removes the defects and increases *f*. A small proportion (~ 5%) of SnO_2 appears in neutron-irradiated SnO. Reduced Sn^{2+} recoil fragments are observed in irradiated magnesium stannate (Hannaford et al. 1965, Hannaford and Wignall 1969) (fig. 24.6) but it is noteworthy they are not detected by the radiochemical

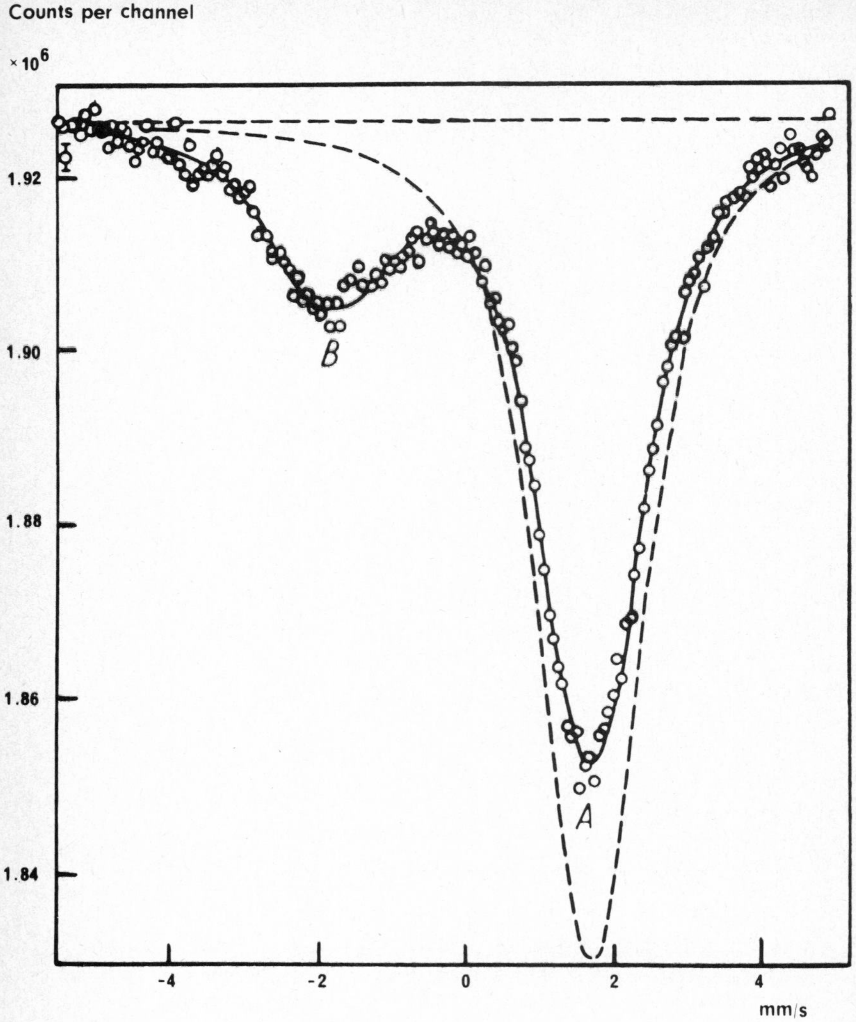

Fig. 24.6. Emission spectrum at 77 K of neutron irradiated magnesium stannate relative to a Mg_2 Sn absorber at 293 K. Line B is characteristic of divalent tin charge compensated by an oxygen vacancy, line A corresponds to Sn^{4+} in Mg_2SnO_4. The recoil defect is annealed at 600° C (Hannaford et al. 1965).

analysis (probably because of the interfering Sn^{II}-Sn^{IV} exchange reaction) (Andersen and Østergaard 1968). The annealing behaviour of irradiated Mg_2SnO_4 remains controversial.

1.3. NOBLE GASES

In situ investigations have been performed on the effects of the ^{131}I and ^{129}I β decay to the isobaric xenon nuclides in solid iodine compounds. Both isotopes populate Mössbauer levels. The molecular perturbation due to pure β decay is very mild and the source experiments give evidence that xenon compounds are produced with a high efficiency. Formation of $XeCl_4$, XeO_3, $XeCl_2$, $XeBr_2$ could be inferred from the emission spectra of several iodo-compounds and details of their structure deduced from the hyperfine parameters (Perlow and Perlow 1964a, 1965, 1968, Perlow and Yoshida 1968). Source experiments with ^{83}Rb and ^{83}Br (which both populate the ^{83}Kr Mössbauer level via a highly converted IT) labelled compounds have been less conclusive with respect to the synthesis of krypton compounds. Most sources, including alkali bromides and bromates (Pasternak and Sonnino 1967), rubidium halides (Krasnoperov et al. 1969, Gütlich et al. 1970) and the krypton-hydroquinone clathrate exhibit broad emission lines, attributed to a local distortion of the crystalline field as a consequence of the IT of ^{83m}Kr. In all these matrices Kr appears in the atomic form. On the other hand the high f value and the larger linewidth of $^{83}SeO_2$ sources (^{83}Se decays to ^{83}Kr via ^{83}Br) have been attributed to the formation of chemical bonds between the Kr and O atoms. Similar bonding has been suggested for the Kr hydroquinone clathrate (Hazony and Herber 1971).

1.4. MISCELLANEOUS

A few source experiments on other systems have been reported. After-effects have been observed in rare earth sources. When ^{161}Tb β-decays in a CeO_2 matrix, the daughter atoms are found as Dy^{3+} and Dy^{4+} (Khurgin et al. 1970). No after-effects are noted in the $^{151}Gd^{3+}$ EC decay to ^{151}Eu in Er_2O_3 and $ErF_3,2H_2O$ but in $Er_2(C_2O_4)_3$, $10H_2O$ part of the daughter atoms are reduced to Eu^{2+} by the oxalate radicals (Glentworth et al. 1971).

Quite interesting are the results of the source experiments on ^{129m}Te and ^{129}Te (β decaying to ^{129}I) labelled ammonium hexachlorotellurate $(NH_4)_2TeCl_6$. The spectra and f values are identical with both parent atoms, and are characteristic of the "normal" $^{129}ICl_6^-$ ion. Thus the highly converted IT of ^{129m}Te is irrelevant for the fate of ^{129}I and there is no evidence for a Coulombic explosion of the molecules in the solid state although it is actually observed in aqueous solutions of ^{129m}Te tagged compounds (Jones and Warren 1970a).

The chemical consequences of the β^- decay of ^{193}Os to ^{193}Ir have been studied in source experiments. In OsO_4, $K_2OsO_4,2H_2O$ and osmocene the isoelectronic iridium compounds are formed. However in the Os(IV) hexahalides several charge states of the daughter atoms have been observed [Ir(IV), Ir(III) and possibly Ir(II)]. Sources prepared by neutron irradiation of osmium compounds give the same pattern as that observed after the radioactive decay (Rother et al. 1969).

2. Perturbed Angular Correlations

The principle and the chemical applications of PAC have been reviewed by De Benedetti et al. (1966) and Vargas (1972), the latter with special attention to hot-atom chemistry. The angular correlation between two successively emitted radiations (which in the present context will always be two photons) by the same nucleus (i.e. a sequence $A \xrightarrow{\text{decay}} B \xrightarrow{\gamma} C \xrightarrow{\gamma} D$) is described by the function

$$W(\theta) = \sum_{k \text{ even}} A_k \, P_k(\cos\theta) = 1 + b_2 \cos^2\theta + b_4 \cos^4\theta,$$

where the A_k are numerical parameters depending on the properties of the nuclear levels and of the emitted radiations and $P_k(\cos\theta)$ are Legendre polynomials. The number of terms is usually limited to $k = 2$ and 4. The correlation is easily obtained from the number of photons γ_1 and γ_2 emitted in coincidence as a function of the angle of the direction of the two γ rays and the anisotropy is defined as

$$A = W(180°)/W(90°) - 1 = b_2 + b_4$$

In the case of M1 and E2 interactions on the intermediate state C, the correlation is perturbed (because of the precession of the intermediate spin around the field directions) and

$$W(\theta,t) = \sum_{k \text{ even}} A_k \, G_k(t) \, P_k\,(\cos\theta),$$

where $G_k(t)$ is the time-dependent perturbation function which includes all information on the type of the perturbation. The $G_k(t)$ can be determined and compared to theoretical expressions which have been established for the various types of interactions. The perturbation diminishes the anisotropy of the correlation, but of much more interest is the measurement of the perturbation as a function of the time elapsed since the emission of the first photon [differential perturbed angular correlations (DPAC) experiments].

In hot-atom chemistry, the interest of PAC depends on the possible direct measurement of the effects of the radioactive decay which populates the excited levels, and of the effects of the nuclear reactions by which the parent atoms are produced. Thus PAC and Mössbauer spectroscopy are obviously complementary techniques. Until now the application of PAC to hot-atom chemistry has been limited to the γ cascades of ^{188}Os and ^{181}Ta. In the first case, the sequence is

$$^{187}Re(n,\gamma)^{188}Re \xrightarrow[16.9h]{\beta^-} {}^{188}Os^* \xrightarrow{\gamma\gamma} {}^{188}Os$$

Sato et al. (1966) have shown that the anisotropy from an irradiated potassium perrhenate sample changes on annealing the irradiated sample. Thus despite radiochemical analysis never detecting any Szilard–Chalmers effect in this compound, a substantial fraction of the recoil atoms is displaced in the lattice. The integral perturbation factors G_2 and G_4 have been determined by Bădică et al. (1970) in K_2ReCl_6, $NaReO_4$ and NH_4ReO_4. A strong attenuation of the anisotropy in the irradiated samples is observed. The changes of G_2 and G_4 with the irradiation conditions, the dissolution, recrystallisation and annealing of the samples have been measured. A correlation between the attenuation factors and the retention R measured by standard radiochemical procedures, has become apparent:

	G_2	G_4	R(%)
$NaReO_4$	0.423	0.381	100
K_2ReCl_6	0.731	0.520	60–70

The attenuation factors increase on annealing, indicating the restoration of the initial structure. The character of the perturbation is electric and/or magnetic, changing in the course of the annealing.

A better understanding of the perturbation is gained from DPAC measurements. Such experiments have been performed on the $\gamma\gamma$ cascade of $^{181*}Ta$:

$$^{180}Hf(n,\gamma)^{181}Hf \xrightarrow[42.5d]{\beta} {}^{181}Ta^* \xrightarrow{\gamma\gamma} {}^{181}Ta$$

in several complexes of hafnium. The source is prepared either by neutron irradiation of hafnium compounds, or by chemical synthesis from radiohafnium.

Table 24.1

DPAC analysis of irradiated and synthesized hafnium acetylacetonate. The two asymmetry parameters η_1 and η_2 refer to two different sites with the corresponding abundances P_1 and P_2 (Béraud et al. 1969)

	Synthetic source	Irradiated source
G_2(integral)	0.225 ± 0.005	0.220 ± 0.007
ω_0(Mrad/s)	710 ± 60	640 ± 80
V_{zz}(V/cm²)	$(12.3 \pm 1.5) \times 10^{17}$	$(11.2 \pm 1.8) \times 10^{17}$
η_1 (P_1)	0.1 (0.4)	0.45 (0.3)
η_2 (P_2)	1 (0.6)	1 (0.7)

PAC experiments can be completed by the radiochemical analysis of the recoil species.

An example of DPAC in irradiated and "synthesized" hafnium acetylacetonate is shown on fig. 24.7 (Béraud et al. 1969). The differential attenuation function for a pure E2 interaction is a function of the quadrupole coupling constant; from the decomposition of the experimental curve one extracts the quadrupole frequency $\omega_0 = 6\omega Q$, the electric field gradient V_{zz} and the asymmetry parameter, η, of the different recoil species as well as their relative proportions. The modification of the environment of the excited Ta nuclei formed by (n,γ;β^-) or β^- is apparent from the results in table 24.1.

Detailed radiochemical and DPAC measurements on the complex of hafnium with n-benzoylphenylhydroxylamine have shown a close agreement between the results of the two techniques (Vargas et al. 1969, Vulliet 1970). Three hafnium species are found by chromatography of the targets, and the characteristic quadrupole frequencies of each fraction have been computed from the $G_k(t)$ curve. A correlation is apparent between the value of the frequency and the chemical behavior of the recoil fragment; for instance the more labile fragment has the smaller quadrupole coupling constant.

Similar studies, including thermal annealing have been performed on the complex of Hf with tropolone (Tissier 1970).

New fluorocomplexes of tantalum formed by the beta decay of ^{181}Hf in HfO_2, $(NH_4)_3HfF_7$, K_3HfF_7 and K_2HfF_6 have been detected by DPAC measurements. A comparison between Mössbauer parameters on ^{176}Hf and DPAC results on K_2HfF_6 indicate that no exceptional charge states are produced in the β^- decay of ^{181}Hf (Berthier et al. 1971).

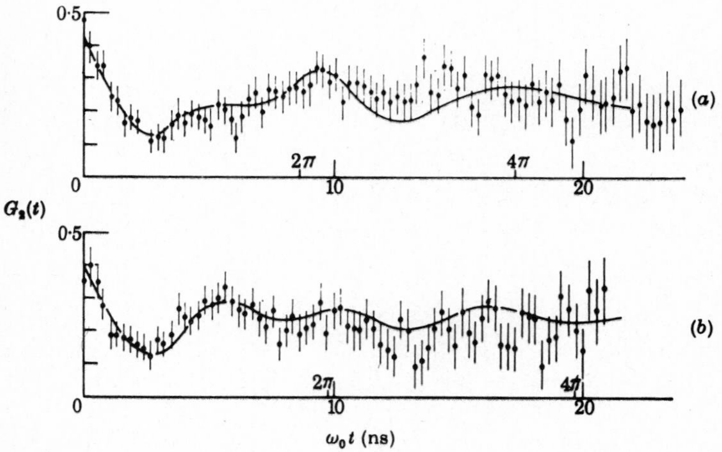

Fig. 24.7. Measured $G_2(t)$ attenuation function for synthetic (a) and irradiated hafnium acetylacetonate (b) (Béraud 1969).

3. Miscellaneous

3.1. HALF-LIFE CHANGES

The alteration of the decay probability λ of a radionuclide with the chemical bonding of the atom has been observed on several occasions (De Benedetti et al. 1966, Vargas 1972). Thus the chemical state of recoil species can be inferred from half-life measurements. For EC decay, λ depends on the K or L electron density at the nucleus which may be altered directly by chemical bonding, or indirectly by the changes in screening parameters. Low Z radioisotopes with a few days half-life are the best candidates for the detection of the variation of λ. In neutron-irradiated copper phthalocyanine, the λ value of recoil ^{64}Cu ($T = 12.8$ h) is smaller than that of the chemically bound radiocopper, the relative change being $\Delta\lambda/\lambda = (-8.8 \pm 1.3) \times 10^{-4}$ (Auric and Vargas 1971).

Alteration of IT decay rates could in principle be used for the same purpose. In that case, highly converted low-energy transitions (i.e. conversion occurring in the outer shells) are the most appropriate. In all cases thorough differential measurements must be performed since the expected effect is of the order of 10^{-3} to 10^{-4} (Perlman and Emery 1969). More subtle information on the chemical bonding of the radioactive atoms may be gained from the measurement of the intensities of electron conversion lines which are proportional to the electron density of the individual shells and subshells. As a typical example, the 5s/4s line intensity ratio is 0.107 ± 0.003 in white tin, but 0.073 ± 0.004 in SnO_2 for the 23.87 keV M1 transition of ^{119m}Sn.

These measurements are not easily made nor do they represent universal methods, but they may provide, in specific cases, an interesting in situ investigation of the chemical environment of hot-atoms.

3.2. SHAKE-OFF INDUCED LUMINESCENCE

A few experiments have been described on the shake-off monopole excitation following the β decay in the solid phase. The spectra of the photons accompanying the sudden change of the nuclear charge in the β decay of ^{85}Kr in solid krypton (Micklitz 1968) or trapped in solid rare gases (Micklitz et al. 1969, Luchner and Micklitz 1970) give information on the influence of the surroundings on the levels of the excited daughter ion $^{85}Rb^+$, and on the displacement energies of this ion.

3.3. MATRIX ISOLATION

Matrix isolation in conjunction with in situ determination appears as an interesting technique. Unstable entities may be produced by a nuclear reaction or decay and trapped at low temperature in an inert solid matrix. Mössbauer and PAC measurements can be made at low enough temperatures to limit the molecular motions and to prevent any reaction with an activation energy greater than a few hundredths of an eV.

3.4. NUCLEAR RESONANCE FLUORESCENCE (NRF)

The basic difficulty for observing the resonant absorption or scattering of γ rays emitted by radioactive nuclei is the shifting of the emission and absorption lines due to energy loss of the photons by recoil. In some cases the Döppler shift of the photon energy due to the impulse given to the excited nucleus by the preceding decay may be sufficient to overcome the energy loss and the resonance fluorescence of the photon can be observed. NRF experiments can be achieved in the solid phase if the recoil time of the radioactive atom is longer than the lifetime of the excited level. This restricts the experiment to τ values in the picosecond range (Metzger 1959).

The application of NRF to hot-atom chemistry has been discussed by Adloff (1971). The slowing down process of atoms with KE below 300 eV, which can hardly be examined by any other method, may be inferred from the shape of the γ lines emitted by the recoiling nuclei. The intensity of the resonantly scattered photons depends on the direction of emission of the recoil atoms in a monocrystal, and also on the nature and concentration of defects. NRF experiments appear to be an interesting in situ technique in hot-atom chemistry.

Addendum (December 1976)

During the 1971–1976 period, Mössbauer emission spectroscopy still remained overwhelmingly the principal in situ technique in hot-atom chemistry. Most of the literature dealt with the effects of ^{57}Co EC decay in a great variety of compounds. New experiments have started on the EC decays of ^{119m}Te and ^{119}Sb feeding the Mössbauer level of ^{119}Sn, in addition to the isomeric transition of ^{119m}Sn. Those trends are reflected in the additional (not exhaustive) literature which also includes the very few entries on PAC applications to hot-atom chemistry.

The observation of a change in the decay rate of recoil atoms is still restricted to the single case of ^{64}Cu (Auric and Vargas 1972). However another group of authors, as a matter of fact with a different counting device, did not find such an effect within experimental errors (Johnson et al. 1974).

A change in the K_α/K_β intensity ratios in various ^{51}Cr-labelled compounds has been observed and a possible application to hot-atom chemistry has been suggested (Tamaki et al. 1975).

The optical emission spectrum of rubidium atoms formed in the β decay of ^{85}Kr in solid krypton is a good example of in situ studies on matrix-isolated radioactive atoms (Micklitz and Luchner 1974). The same holds for the optical excitation of Sc after ^{45}Ca β decay in solid xenon (Gerth et al. 1972). Here again Mössbauer emission spectroscopy of ^{57}Co isolated in a krypton matrix deserves a special mention (Micklitz and Barrett 1972).

The nuclear resonance fluorescence method has been shown to give information about the slowing down in solids of hot atoms with energies below 15 eV (Langhoff 1971). Dynamic data on recoil species could in principle also be inferred from Doppler shift attenuation and related methods (Hauser et al. 1974).

Addendum to the Literature (up to October 1976)

REVIEWS

Adloff (1975), Friedt and Danon (1972), Nath (1975), Wertheim (1971) and Zahn et al. (1973).

APPLICATION OF MÖSSBAUER SPECTROSCOPY TO THE STUDY OF CHEMICAL EFFECTS OF NUCLEAR TRANSFORMATIONS

1. EC decay of ^{57}*Co*

Co in a halide matrix: KF, NaCl, NaF, LiF, MgF_2
Maddock et al. (1976)

Co in solid xenon
Micklitz and Barrett (1972)

Co in glasses
Kulikov et al. (1972b)

Co on cationites
Ablesimov et al. (1975)

Co on cationic resins
Kumer et al. (1972)

Co spinels
Spencer and Schroeer (1974)

Co implanted in iron compounds
Fleisch et al. (1975)

Co in magnetite
Ito (1974)

Co doping Al, Cr, Mn, Fe and Co. acetylacetonates
Ramshesh et al. (1972a)

Co doping Mn(acac)$_3$, Fe(acac)$_3$, Co(acac)$_3$
Sano and Iwagami (1971)

Co doping Fe(acac)$_3$
Sano et al. (1971)

Co doping $FeCl_3$
Cruset (1974)

Co doping Cd_3 [Fe(CN)$_6$]$_2$, 12 H_2O
Fenger and Frees (1974)

Co doping [Fe(CN)$_6$]K_4 and [Fe(CN)$_6$]K_3
Prusakov et al. (1974)

Co doping various complexes
Siekierska et al. (1972)

Tetravalent Co and Fe compounds
Perfiliev et al. (1974)

Frozen organic solutions of $CoCl_2$
Afanasov et al. 1975a)

$CoCl_2$, 2H_2O and $CoCl_2$, 6 H_2O
Friedt et al. (1973)

$Co(NO_3)_2$ and $CoSO_4$
Bondarevskii et al. (1973)

Ba_2CoO_4
Kulikov et al. (1972a)

Co_3 $(PO_4)_2$, 8H_2O and Co_3 $(PO_4)_2$, 4H_2O
Friedt and Llabador (1972)

Hexacyanocobaltate (III) complexes: Mn, Fe, Co, KNi, Cu, Zn and Cd[Co(CN)$_6$]
Fenger et al. (1973)

Cd_3[Co(CN)$_6$]$_2$, 12H_2O
Fenger and Olsen (1974)

Hexacyanocobaltate (III) complexes of Ni, Cu, Co, Fe^{2+} and Fe^{3+}
Prusakov et al. (1973)

[Co(NH$_3$)$_6$] $_2$(C_2O_4)$_3$, 4H_2O and [Co(en)$_3$]$_2$ (C_2O_4)$_3$, 9H_2O
Sano and Ohnuma (1974a)

$[Co(NH_3)_6]$ $[Fe(CN)_6]$, $[Co(NH_3)_6]$ $[Cr(C_2O_4)_3]$, $3H_2O$, $[Co(NH_3)_6]$ $[Fe(C_2O_4)_3]$, $3H_2O$
Sano and Ohnuma (1974b)

$[Co(NH_3)_6]$ $(NO_3)_3$, $[Co(en)_3]$ $(NO_3)_3$, $K_3[Co(CN)_6]$, $K_3[Co(NO_2)_6]$ etc.
Sano and Ohnuma (1975)

Co^{11} salicylate and glycolate
Ladriére and Apers (1974)

Co diarsines
Cruset and Friedt (1972)

Co^{11} acetylacetonates
Afanasov et al. (1975b)

$M^ICo^{II}(acac)_3$ (M = Li, Na, K, Rb, Cs)
Ho Hsang Wei et al. (1974)

$Co(CH_3COCXCO\ CH_3)_3$ X = Br, H, NO_2, SCN
Afanasov et al. (1975c)

$[Co(bipy)_3]$ $(ClO_4)_3$, $[^{57}Co(bipy)_3]$ $[Co(CN)_6]$, $2H_2O$, $K_3[Co(CN)_6]$ etc.
Siekierska et al. (1972)

Co dipyridyl complexes
Afanasov et al. (1975d)

$CoCl_2$, npy and $[Co(bipy)_3]$ $(ClO_4)_3$
Sanchez et al. (1973)

Co phthalocyanines
Srivastava et al. (1974)

Co^{II} phthalocyanines
Thompson et al. (1972)

Co porphines
Prusakov et al. (1972)

$[Co(1-10\ phen)_3]$ $(ClO_4)_3$, $2H_2O$
Srivastava and Nath (1974)

$[Co(phen)_3]$ $(ClO_4)_2$, $2H_2O$
König et al. (1972)

$[Co(phen)_3]$ $(ClO_4)_3$, $2H_2O$, $[Co(phen)_2\ (en)]$ Cl_3, H_2O, $[Co(phen)\ (en)_2]$ $(NO_3)_3$, $[Co(en)_3]$ Cl_3
Misroch et al. (1976)

$[Co(phen)_3]$ $(ClO_4)_3$, $2H_2O$ and $[Co(bipy)_3]$ $(ClO_4)_3$, $3H_2O$
Srivastava and Nath (1976)

$[Co(phen)_2]$ $(NCS)_2$, $[Co(bipy)_2]$ $(NCS)_2$ (cis)
Ensling et al. (1976)

[Co(2 methylphen)$_3$] $(ClO_4)_2$, $2H_2O$
Fleisch and Gütlich (1976)

Biomolecules
Cardin et al. (1974)
Marchant et al. (1972)
Munck and Champion (1974a)

^{57}Co EC decay and radiation effects
Baggio-Saitovich et al. (1972)

Time differential spectroscopy in $[Co(phen)_3]$ $(ClO_4)_3$
Grimm et al. (1975)

^{57}Co in solid state exchange reactions
Calusaru et al. (1973)
Ramshesh et al. (1972b)

2. Isomeric transition of ^{119m}Sn

^{119m}Sn on ion exchangers
Ablesimov and Bondarevskii (1973b, 1975)

^{119m}Sn in frozen solutions
Bondarevskii and Tarasov (1972)
Murin et al. (1972)

$SnCl_2$ and $Sn_3(PO_4)_2$
Friedt and Llabador (1972)

$Sn(ClO_4)_2$, $3H_2O$, $SnBu_4$, $SnBu_2Cl_2$ and $SnBu_2SO_4$
Llabador and Friedt (1973)

Sn (benzyl)$_3$X
Mahieu and Llabador (1974)

$SnBr_4$, 2py, $SnBr_4$, bipy, $SnCl_2$, bipy
Sanchez and Friedt (1973)

^{118}Sn(n,γ) and ^{119m}Sn IT in tin chalcogenides
Seregin and Savin (1973)

^{118}Sn (n,γ) and ^{119m}Sn IT in $SnSO_4$
Friedt and Vogl (1974)

3. ^{119m}Te and (or) ^{119}Sb EC decays

SnTe
Ambe and Ambe (1976b)

Sb, Sb_2Te_3 and Sb_2S_3
Ambe et al. (1972b)

Sb_2Te_3
Ambe and Ambe (1973a)

Sb and Te metals, oxides and chalcogenides
Ambe et al. (1974)

H_6TeO_6
Ambe and Ambe (1974)

H_6TeO_6 and TeO_2
Ambe et al. (1973)

$Sb(OH)C_2O_4$
Ambe and Ambe (1975b)

SbI_3 and TeI_4
Ambe and Ambe (1974b)

Sb_2O_3, $Sb_2(SO_4)_3$, SbF_3, $SbCl_3$, $SbBr_3$ and SbI_3
Llabador (1974)

4. ^{129m}Te β decay

^{129m}Te in alkali halides
Seregin and Savin (1972)

$(NH_4)_2TeX_6$ and $(NH_4)_2$ TeX_4, X = Cl, Br, I
Johnstone et al. (1972)

TeO_3
Warren and Jones (1971)

^{128}Te(n,γ) and ^{129m}Te β decay in H_6TeO_6, $(H_2TeO_4)_n$, α-TeO_3, TeO_2, H_3TeO_3 and Te
Warren et al. (1971)

5. Miscellaneous decays

Isomeric transition of ^{83m}Kr
Kolk (1975)

^{121m}Sn β decay
Ambe and Ambe (1975a)

^{125}I EC
Bochkarev (1972)

^{151}Gd EC and ^{151}Sm β
Glentworth et al. (1973)

^{151}Sm β
Probst et al. (1972)

6. On-line experiments

^{56}Fe (n,γ)
Fenger (1972)
Jeandey and Peretto (1975c)

^{56}Fe (d,p)
Christiansen et al. (1973)

^{177}Hf (n,γ)
Jeandey and Peretto (1975b)

PERTURBED ANGULAR CORRELATIONS

^{180}Hf (n,γ)^{181}Hf
Abbe and Marques-Netto (1975)
Boussaha et al. (1976)

^{139}La (n,γ) ^{140}La and ^{140}Ba β^-
Vasudev and Jones (1972)

^{133}Ba EC
Glass and Klee (1976)

CHAPTER 25

CHEMICAL EFFECTS OF NUCLEAR TRANSFORMATIONS IN MIXED CRYSTALS

Horst MÜLLER

Chemisches Laboratorium der Universität Freiburg, Freiburg, F.R. Germany

Chemical Effects of Nuclear Transformations in Inorganic Systems
Edited by G. Harbottle and A.G. Maddock

Contents

1. Introduction

Investigations of the chemical effects of nuclear transformations in solids generally suffer from the restriction that diluted solutions containing inert or reactive substances cannot be used. This method was of great value in the case of liquids and gases. In the solid state doping with foreign atoms or atoms of a different valence state may change the distribution of the different recoil species, as it produces electronic or atomic defects. New species, however, are expected to be formed in solid-state investigations only if homogeneous mixed crystals or double salts are used. If one restricts one's consideration to coordination compounds as useful objects for recoil chemistry investigations, three types of mixed crystals may be differentiated:

(1) mixed crystals from components with different central atoms, e.g.: $KMnO_4$–$KClO_4$ (Rieder et al. 1950, Owens and Lecington 1973, Lecington and Owens 1971), K_2CrO_4–K_2SO_4 (Andersen and Sørensen 1966a, Andersen 1963a, Maddock and Vargas 1961a, b, Green et al. 1953, Harbottle 1954, Sherif et al. 1974), $(NH_4)_2CrO_4$–$(NH_4)_2SO_4$ (Maddock and Vargas 1961b), KH_2PO_4–KH_2AsO_4 (Claridge and Maddock 1963b), Me′(acac)$_3$–Me″(acac)$_3$ (Me′, Me″ = Co, Cr, Al, Rh, Mn) (Shankar et al. 1960, 1961d, 1965a, 1966, Venkateswarlu et al. 1962, Machado et al. 1965), K_2IrCl_6–K_2PtCl_6 Van Ooij 1971), K_2IrCl_6–$(NH_4)_2PtCl_6$ (Van Ooij 1971), $Na_2IrCl_6 \cdot 6H_2O$–$Na_2PtCl_6 \cdot 6H_2O$ (Van Ooij 1971), $NaBrO_3$–$NaClO_3$ (Hobbs and Owens 1966, Owens and Boyd 1976), $K_4Os(CN)_6$–$K_4Fe(CN)_6$ and related systems (Kay and Diefallah 1966, Diefallah and Kay 1973, Diefallah 1968), K_3Feox_3–K_3Alox_3 (Dharmawardena and Maddock 1970a), cobalt–chromium ammine complexes (Lazzarini and Fantola-Lazzarini 1971a, Fantola-Lazzarini and Lazzarini 1973a), Me′Pc–Me″Pc and Me′Pc–H_2Pc (Me′,Me″ = Co, Cu, Zn; Pc = phthalocyanine) (Kujirai and Ikeda 1973, 1974, Ikeda and Kujirai 1975, Kudo 1972).

(2) mixed crystals from components with different ligand atoms, e.g.: K_2ReBr_6–K_2ReCl_6 (Bell et al. 1969a, b, 1972, Rössler et al. 1972, Müller 1963a, 1968b, Müller and Martin 1969, 1976, Müller et al. 1977), K_2IrBr_6–K_2IrCl_6 (Van Ooij 1971).

(3) mixed crystals from components with both different central atoms and ligands, e.g.: K_2ReBr_6–K_2SnCl_6 (Müller 1962a, 1965a, c, 1968a, 1969b, 1976,

Müller and Cramer 1970), K_2ReCl_6–K_2SnBr_6 (Müller 1965a), K_2ReBr_6–K_2OsCl_6 (Müller 1967b, c), K_2OsBr_6–K_2SnCl_6 (Müller 1966a, b), K_2CrO_4–K_2BeF_4 (Andersen and Sørensen 1966a, Andersen 1963a, de Maine et al. 1957, Maddock and de Maine 1956b, Maddock and Vargas 1961a, b, Green et al. 1953, Harbottle 1954), $KMnO_4$–KBF_4 (Lecington and Owens 1971), K_2IrBr_6–K_2PtCl_6 (Van Ooij 1971), $Na_2IrCl_6 \cdot 6H_2O$–$Na_2PtBr_6 \cdot 6H_2O$ (Van Ooij 1971).

2. Mixed Crystals from Components with Different Central Atoms and Identical Ligands

The discussion of chemical effects of nuclear transformations in mixed crystals of type (1) may be restricted to some examples that have been published recently.

The chemical effects of the $^{55}Mn(n,\gamma)^{56}Mn$ nuclear reaction in $KMnO_4$–$KClO_4$ mixed crystals have been compared with the chemical fate of $^{54}MnCl_2$ dopant in such mixed crystals after irradiation with 10 Mrad (Owens and Lecington 1973, Lecington and Owens 1971). The yields of $^*Mn^{2+}$, $^*MnO_2$, and $^*MnO_4^-$ agree quantitatively in both experiments, $^*Mn^{2+}$ decreasing and $^*MnO_4^-$ increasing with increasing $KMnO_4$ content of the mixed crystals. The conclusions are: the ^{56}Mn recoil atom loses all its oxygen ligands. The ultimate chemical fate of the recoil depends mainly on the condition of the immediate environment of the atom at rest in the crystal.

In $NaBrO_3$–$NaClO_3$ mixed crystals the $^{82g}BrO_3^-$ retention is a polynomial function of third order of the mixed crystal composition for each thermal treatment (Owens and Boyd 1976). These results can be interpreted in terms of a model requiring reaction of the recoil atom with radiolysis fragments trapped at host anion sites. Self-radiolysis resulted from capture radiation, recoil and decay radiation, the target being shielded from external reactor γ-radiation and fast neutrons.

Recently experiments have been performed with mixed crystals of different metal phthalocyanines, Me′Pc–Me′′Pc, and with mixed crystals of a metal phthalocyanine with metal-free phthalocyanine, MePc–H_2Pc, with the possibility of investigating both the α and β modifications.

The results for the systems α- and β-CuPc–CoPc are shown in table 25.1. While for the α crystals the retention changes linearly with the composition it changes logarithmically for the β crystals. The retention of ^{60}Co is always higher than that of ^{64}Cu (Kujirai and Ikeda 1973).

In α- and β-ZnPc–H_2Pc the chemical fate of ^{65}Zn and ^{69m}Zn has been investigated (Ikeda and Kujirai 1975). The retention increases with increasing H_2Pc, the difference between ^{65}Zn and ^{69m}Zn remaining constant for α and β crystals, respectively. The retentions for the pure and the infinitely diluted

Table 25.1
Retention (in %) in CuPc–CoPc mixed crystals

	CuPc		CoPc		Ret = f[CuPc]
	pure	infinitely diluted	pure	infinitely diluted	
α-CuPc–CoPc	4.5	3.0	3.0	8.7	Ret = $a+b$[CuPc]
β-CuPc–CoPc	13.0	4.0	3.9	22	log Ret=$a'+b'$[CuPc]

substances, respectively, are given in table 25.2. The higher retention of the ^{69m}Zn is attributed to the lower recoil energy.

Similar results have been obtained for CoPc–H_2Pc mixed crystals (Kujirai and Ikeda 1974). For both α and β crystals the initial retention increases with increasing H_2Pc content, but the degree of increment is larger for the β mixed crystals. In annealing experiments the annealing rate for β crystals is much faster than for α crystals. At least for mixed crystals up to 40 mol % CoPc the pseudo-plateau of the annealing is higher for the β crystals. Generally it can be said that β mixed crystals are more susceptible to annealing. Obviously the reason for this behaviour is the difference in the type of conduction and the influence of adsorbed oxygen.

Very interesting are experiments with the CuPc–H_2Pc system (Kudo 1972b). For the α crystals the retention of 5.5% increases with H_2Pc content up to 45% at 20 mol % H_2Pc and then remains constant. The data are identical for mixed crystals and for mixed-disk samples pressed under a pressure of 10^7 Pa. Even higher retention values are found for β mixtures. α-CoPc irradiated in contact with α-H_2Pc shows a higher initial retention of 17 to 21% compared with 5.5%. Addition of H_2Pc to both α- and β-CuPc sensitizes the target to thermal annealing. The addition of the electron acceptor chloranil results in an increase of the retention for both α- and β-CuPc, the addition of the electron donor tetramethyl-p-phenylene-diamine has no effect.

The experiments with phthalocyanine systems give strong support to an electronically controlled annealing mechanism.

Table 25.2
Retention (in %) in ZnPc–H_2Pc mixed crystals

Modification	Nuclide	Pure substance	Infinitely diluted with H_2Pc
α	^{65}Zn	7.5	28
α	^{69m}Zn	15.5	36
β	^{65}Zn	14	60
β	^{69m}Zn	23	48

3. Mixed Crystals from Components with or without Different Central Atoms but with Different Ligands

The most instructive research work has been done with mixed crystals of the types (2) and (3). In these systems recoil atoms may react to give species that cannot be formed in the pure substances which implies that the recoil atoms are labelled by inactive atoms, in a manner dependent on special features of the recoil event. A liquid organic system, $CH_3{}^{127}I–C_3H_7{}^{129}I$ and $CH_3{}^{129}I–C_3H_7{}^{127}I$, respectively, resembling mixed crystals of type (3), has been investigated (Iyer and Martin 1961, Willard 1965).

3.1. MIXED CRYSTALS AS A TOOL FOR INVESTIGATING RECOIL REACTION MECHANISMS

Mixed crystals of the type $K_2ReBr_6–K_2SnCl_6$ have been introduced into recoil chemistry research with the intention of discriminating between different models leading to retention (Müller 1962a, 1965a, b, 1970). It has been proved that such homogeneous mixed crystals may be prepared and handled – even under reactor conditions – without ligand exchange. Re(IV) with its d^3 configurations is, indeed, expected to be kinetically inert (Müller 1963a, 1965a, c). Rhenium recoil atoms in $K_2ReBr_6–K_2SnCl_6$ mixed crystals are expected to produce different species dependent upon the retention controlling mechanism. Five such mechanisms must be discussed (Müller 1962a, 1965b, c, 1967a, b, 1970).

(1) *Primary retention* results from recoil events with an energy not sufficient for displacement of the radioactive rhenium atom. The displacement energy necessary for an ejection of the central atom out of the cage of its six ligands may be estimated following physical calculations for metals or ionic substances at 25 eV (Müller 1967a, b). $*ReBr_6^{2-}$ should be the result of such primary retention for mixed crystals of any composition. Primary retention here signifies in a slightly incorrect manner the sum of the "true" primary retention (with no intermediate bond breaking) together with the retention due to direct recombination (i.e. reformation of the original species from exactly the same atoms in exactly the same geometrical arrangement as before the nuclear event).

(2) The *billiard-ball* mechanism (Libby 1947) predicts displacement of a ordinary inactive central atom out of the cage formed by its ligands by a recoiling atom in such a way that the cage remains intact. $*ReBr_6^{2-}$ and $*ReCl_6^{2-}$ are to be expected as the only reaction products from displacement of a Re atom out of a bromide cage or displacement of a Sn atom out of a chloride cage. The relation of $*ReBr_6^{2-}/*ReCl_6^{2-}$ should depend on the mixed crystal composition, and the $*ReBr_6^{2-}$ proportion is expected to be approximately proportional to the K_2ReBr_6 content of the mixed crystals.

(3) The *ligand-loss model* (Libby 1940) predicts that in the first step the recoiling atom loses only part of its ligands or attached atoms. After it has come to rest the recoiling atom fills up its ligand sphere with atoms from the new surroundings. In this case mixed labelled forms such as $*ReBrCl_5^{2-}$, $*ReBr_2Cl_4^{2-}$ etc. can be expected to be formed from $*ReBr^{n+}$, $*ReBr_2^{m+}$ etc. intermediates. The bromine content of the mixed crystals used and the probability of failure of bond breaking during the recoil event will influence the distribution of the mixed forms.

(4) The *hot-spot model* (Yankwich 1956, Harbottle and Sutin 1958) predicts a kind of melt in which the atoms are highly disordered as a result of the energy release by the recoil atom. For an event of 300 eV energy a reaction zone (the "hot zone") of 1000 atoms is heated for 10^{-11} s to a temperature of at least 1000 K (Seitz and Koehler 1956). This seems to be sufficient to produce large disorder – that means an entire mixing or even a real melt – at least in the central parts of the hot zone. The interpretation of the Harbottle–Sutin hot-spot model as given here seems justified because of its relationship to the displacement spike concept of Brinkman (1954). This model predicted an even larger disorder. It should be mentioned, however, that Harbottle and Sutin themselves were concerned primarily with thermal or even thermodynamic effects instead of disorder phenomena. For the mixed crystals under consideration all mixed ions $*ReBr_nCl_{6-n}^{2-}$ $(0 \leqslant n \leqslant 6)$ are expected with a distribution that should be controlled mainly by statistics, i.e. the proportion of Br and Cl present determine the distribution.

A recently proposed model of "Super-Hot Zones" including 10^6 displaced atoms (Gardner et al. 1970a, b) has been shown to be unfounded because of erroneous measurements (Rössler and Otterbach 1971a).

(5) The *disorder model* (Fehlordnungs-Modell) is based upon the calculations of Vineyard et al. (Gibson et al. 1960, Vineyard 1961) about radiation damage in crystalline solids, especially metals. Calculations on the coordination compounds K_2ReCl_6 and K_2ReBr_6 have been performed recently (Robinson et al. 1974, Rössler and Robinson 1974). It predicts – at least for events with energies smaller than some keV – a reaction zone ten times smaller than the hot-spot model and negligible temperature effects because of very rapid loss of energy by focusson and crowdion transport (Silsbee 1957). Disorder is small, only a few displacements and replacements arise while the order of the lattice and even the original positions of the atoms are essentially preserved. For K_2ReBr_6–K_2SnCl_6 mixed crystals mixed ions $*ReBr_nCl_{6-n}^{2-}$ are expected as in the two preceding models but the distribution should be different.

A schematic representation of these models is shown in fig. 25.1 (Müller 1970).

Mixed hexabromochlororhenates(IV) can be separated by electrophoresis or by a thin layer method (Müller 1962a, 1963a, 1965a, 1969a). It has been

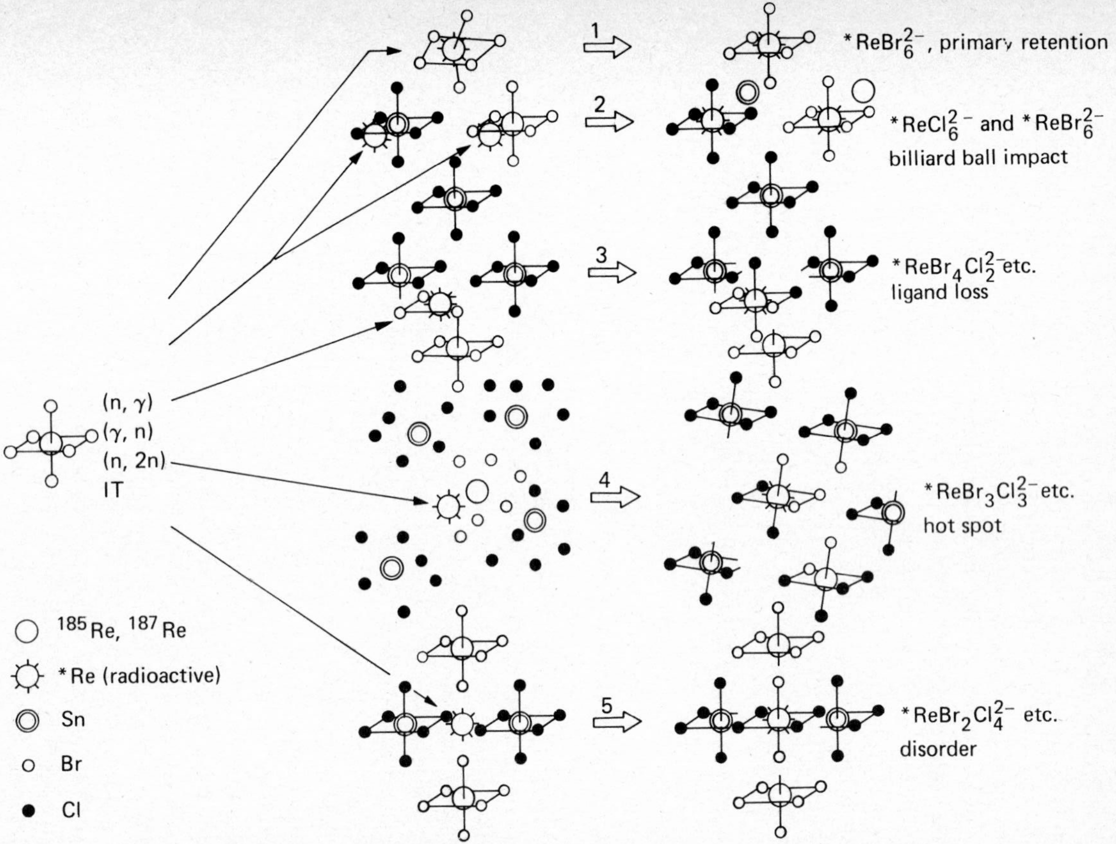

Fig. 25.1. Models relating to the retention in K_2ReBr_6–K_2SnCl_6 mixed crystals. For the ligand-loss model it has been assumed that the surrounding lattice remains largely undisturbed (Müller

checked that during the irradiations and the analytical separations no observable ligand exchange occurs. It has been further confirmed that the observed effects happen in the solid and not during or following dissolution (Müller 1965a). In fig. 25.2 a thin layer separation of the mixed hexabromochlororhenates (and the perrhenate) is shown. The amount of the perrhenate yield may change in mixed crystals compared with the value for the pure Szilard–Chalmers substances; no explanation for this can be derived as yet from the properties of the mixed crystals.

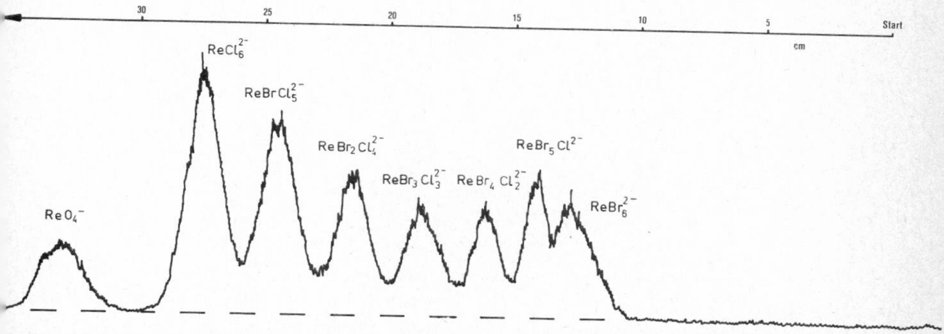

Fig. 25.2. Separation of the radioactive ions $ReBr_nCl^{2}_{6-n}$ and ReO_4^- using thin layer cellulose foils and 3.2 m H_2SO_4 (Müller 1970).

3.2. CENTRAL ATOM RECOIL

In fig. 25.3 the results are shown for the nuclear process $^{187}Re(\gamma,n)^{186}Re$ in K_2ReBr_6–K_2SnCl_6 mixed crystals of different composition (Müller 1969b). Irradiations have been made with 50 MeV γ quanta. The recoil energy of the Re atoms may be estimated at 10 keV, the recoil range should be approximately 100 Å. The distribution of the different hexabromochlororhenates is nearly statistical with a slight preponderance of the chloride containing forms. The primary retention is zero because the species $*ReBr_6^{2-}$ vanishes for mixed crystals with low K_2ReBr_6 concentrations. The formation of mixed forms excludes the billiard-ball model. The ligand-loss model must be excluded because $*ReBrCl_5^{2-}$ vanishes for low K_2ReBr_6 content. This implies that no intermediate $*ReBr^{n+}$ species with one bromide ligand left, that could fill up its ligand sphere, is formed. The same argument is valid for all other bromide containing species. Hot-spot and disorder models are not precluded by the experiments and apparently the hot-spot model seems to be in good agreement with the experiments. The billiard-ball retention is probably zero although the amount of $*ReBr_6^{2-}$ + $*ReCl_6^{2-}$ does not fall below 13%.

The results for the nuclear reaction $^{185}Re(n,\gamma)^{186}Re$ in K_2ReBr_6–K_2SnCl_6 mixed crystals are presented in fig. 25.4 (Müller 1965a). The recoil energy

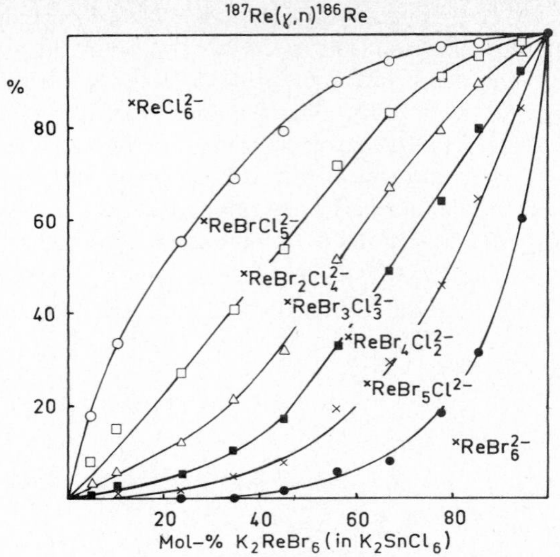

Fig. 25.3. Proportions of the different hexabromochlororhenates-(^{186}Re) resulting from the nuclear reaction ^{187}Re(γ,n)^{186}Re as a function of the mixed crystal composition. The sum of the hexahalogenorhenates has been normalized to 100%; the ReO_4^- yield amounts from 9 to 14% dependent on the mixed crystal composition.

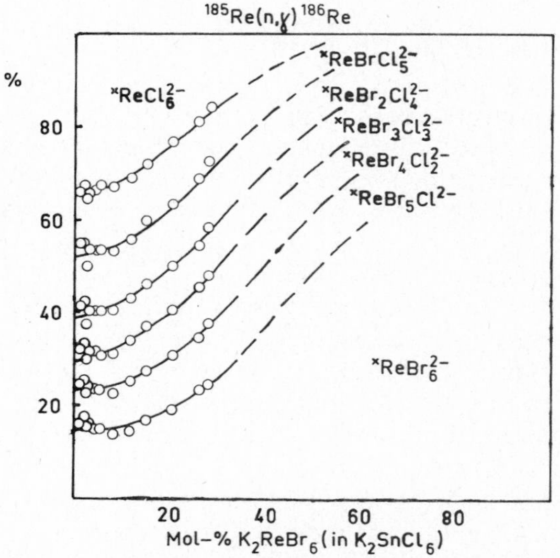

Fig. 25.4. Proportions of the different hexabromochlororhenates (^{186}Re) resulting from the nuclear reaction ^{185}Re(n,γ)^{186}Re as a function of the mixed crystal composition. The sum of the hexahalogenorhenates has been normalized to 100%; the ReO_4^- yield amounts from 9 to 11% dependent on the mixed crystal composition.

amounts to ca. 100 eV, the recoil range is expected to be of the order of one or two lattice spacings. The distribution of the mixed hexabromochlororhenates is quite different from the preceding experiment with 10 keV recoil atoms. Even for the most dilute mixed crystals a distinct enrichment of the bromide containing species is found. Neglecting details one third of the ligands of the recoil atoms are bromide and the rest chloride even for the most dilute mixed crystals. Such an enrichment can only be explained if it is assumed that recoil range and reaction zone radius are both very small. A schematic representation of this behaviour is shown in fig. 25.5. From the lattice constants of the mixed crystals, the number of atoms in the unit cell, and the enrichment, a reaction zone can be calculated with a content of ca. 25 atoms and a radius of ca. 5 A. The recoil range is of the same order. These results are contradictory to the hot-spot model but in good accordance with the disorder model. 12% $*ReBr_6^{2-}$ is regarded as primary retention. Computer calculations are in accord with these considerations (Robinson et al. 1974, Rössler and Robinson 1974). From fig. 25.5 it can be seen, too, that for large recoil ranges, however, no discrimination can be made between a hot-spot model (large reaction zone) and disorder model (small reaction zone) as in both cases a statistical distribution of the reaction species is expected.

Similar results have been obtained for (n,γ)-Re recoils in K_2ReBr_6–K_2OsCl_6 (Müller 1967b), (n,γ)-Os recoils in K_2OsBr_6–K_2SnCl_6 (Müller 1966a) and (n,γ)-Ir recoils in K_2IrBr_6–K_2PtCl_6 (Van Ooij 1971) mixed crystals. In

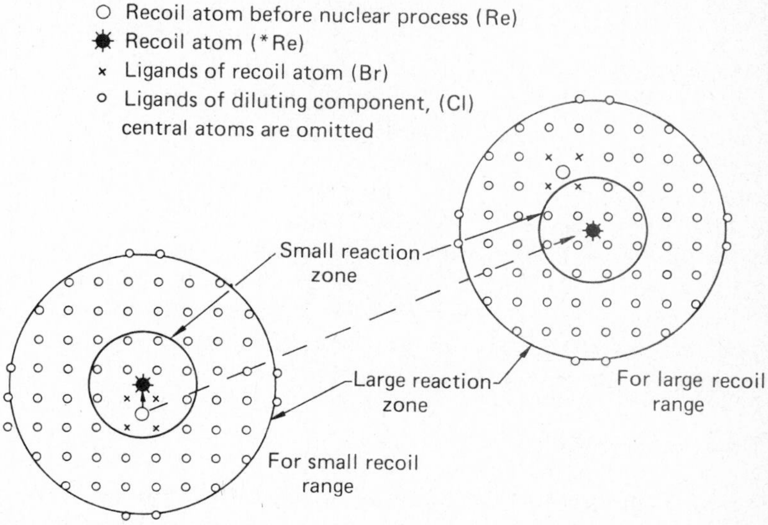

Fig. 25.5. Schematic representation of the content of the reaction zone in diluted K_2ReBr_6–K_2SnCl_6 mixed crystals for the four possible combinations small or large recoil range with small or large reaction zones, respectively. K and Sn are omitted for better clarity (Müller 1968a).

K_2ReCl_6–K_2SnBr_6 mixed crystals (Müller 1965a) an enrichment of the chloride containing forms was observed; this proves that in any case the initially coordinated ligands are enriched. The results are summarized in table 25.3. The experiments only supply us with an upper limit for the billiard-ball retention; it must be assumed, however, that this kind of retention is zero because the mean recoil range is smaller than the distance of ca. 7 Å between two neighbouring central atoms in the lattices.

Recently the perrhenate annealing has been investigated in neutron activated K_2ReBr_6–K_2SnCl_6 mixed crystals (Müller 1976). Approximately one third of annealable ^{186}Re interstitials anneal to $ReBr_6^{2-}$ mainly in the temperature range 40–60° C, two thirds form the other six species $ReBr_nCl_{6-n}^{2-}$ $(O \leqslant n \leqslant 5)$ without any significant preference and at no higher temperature jump. These results may be interpreted that one third of the recoil ^{186}Re-atoms are still situated inside the original ligand cage and that the pertinent annealing reaction is controlled mainly by an electronic process.

The distribution of the different mixed bromochloro complexes is expected to be distinctly influenced by changing the distance between the complex anions in the irradiated substance. Using $Na_2IrCl_6 \cdot 6H_2O$–$Na_2PtBr_6 \cdot 6H_2O$(1:10) mixed crystals, Van Ooij (1971) reported the formation of only 5% $^*IrCl_6^{2-}$. No mixed forms have been observed in this system because of the greater distance between the $IrCl_6^{2-}$ and $PtBr_6^{2-}$ ions.

K_2ReF_6–K_2SnCl_6 mixed crystals should be of interest for the investigation of the influence of the ligand on the recoil behaviour of the central atom. One would expect a larger recoil range shown by a less pronounced enrichment of the originally coordinated ligands in the recoil species. Efforts to prepare K_2ReF_6–K_2SnCl_6 mixed crystals were unsuccessful, but resulted in the preparation of a new compound $K_3Cl[ReF_6]$ crystallizing in an anti-perovskite structure.

The nuclear processes $^{185}Re(n,\gamma)^{186}Re$ and $^{187}Re(\gamma,n)^{186}Re$ both produced mixed $^*ReCl_nF_{6-n}^{2-}$ ions in very similar amounts and corresponding to a statistical distribution (Müller and Abberger 1973). The reason for this at first sight unexpected behaviour is the very thorough mixture of Cl and F already in the double salt. Figure 25.6 presents the ionophoretic separation of the mixed forms, table 25.4 the distribution of the different species. Mixed chlorofluororhenates have been unknown until now.

3.3. LIGAND ATOM RECOIL

The mixed crystal systems under discussion have the advantage that besides the recoil behaviour of the central atoms, the recoil behaviour of the ligands may be examined, too. Therefore they represent one of the rare cases in recoil chemistry where two elements may be investigated in the same compound. Experiments with bromine and chlorine recoils in mixed crystals are, indeed, very instructive.

Table 25.3
Results for central atom recoil reactions in mixed crystals of the type K_2ReBr_6–K_2SnCl_6

Nuclear process	System	Recoil energy (eV)	Recoil range (Å)	Reaction zone (Å^3);atoms	ReO_4^- yield for low concentration (%)	Enrichment of ligand (%)	Primary retention (%)	Billiard-ball retention (%)
$^{185}Re(n,\gamma)^{186}Re$	K_2ReBr_6–K_2SnCl_6	100		740;27	9	34	⩽ 12	⩽ 33
$^{185}Re(n,\gamma)^{186}Re$	K_2ReBr_6–K_2OsCl_6	100		440;17	35	53	⩽ 14	⩽ 4
$^{190}Os(n,\gamma)^{191}Os$	K_2OsBr_6–K_2SnCl_6	100	ca.5, experimental	730;26	–	34	⩽ 12	⩽ 25 probably 0
$^{185}Re(n,\gamma)^{186}Re$	K_2ReCl_6–K_2SnBr_6	100		620;19	8	48	⩽ 18	⩽ 21
$^{191}Ir(n,\gamma)^{192}Ir$	K_2IrBr_6–K_2PtCl_6	100		690;27	–	ca. 34	⩽ 5	⩽ 31
$^{191}Ir(n,\gamma)^{192}Ir$	$Na_2IrCl_6 \cdot 6H_2O$–$Na_2PtBr_6 \cdot 6H_2O$	100		–	–	100	⩽ 5	0
$^{187}Re(\gamma,n)^{186}Re$	K_2ReBr_6–K_2SnCl_6	10000	100, estimated	?	10	0	0	⩽ 13, probably 0

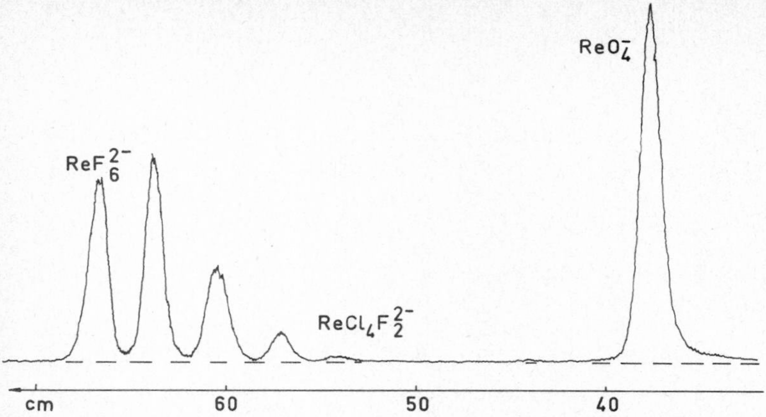

Fig. 25.6. Ionophoretic separation of the radioactive ions $^{186}ReCl_nF_{6-n}^{2-}$ and $^{186}ReO_4^-$ resulting from the nuclear reactions $^{185}Re(n,\gamma)^{186}Re$ and $^{187}Re(\gamma,n)^{186}Re$ in $K_2ReF_6 \cdot KCl$.

Table 25.4
Distribution of rhenium recoil species in $K_2ReF_6 \cdot KCl$ following nuclear processes (in %)

	$^{185}Re(n,\gamma)^{186}Re$	$^{187}Re(\gamma,n)^{186}Re$	Statistical distribution for $\sum ReCl_nF_{6-n}^{2-} = 50\%$
ReO_4^-	49.6	48.7	} 50
?	0.7	1.7	
$ReCl_6^{2-}$	0	–	0
$ReCl_5F^{2-}$	0.3	–	0
$ReCl_4F_2^{2-}$	0.8	–	0.2
$ReCl_3F_3^{2-}$	3.0	2.3	1.8
$ReCl_2F_4^{2-}$	10.4	10.8	8.3
$ReClF_5^{2-}$	18.0	16.2	19.8
ReF_6^{2-}	17.0	20.6	19.8

The chemical effects of bromine recoils with low energy have been investigated in labelled (unirradiated) mixed crystals $K_2Re^{80m}BrBr_5$–K_2SnCl_6 (Müller and Cramer 1970). ^{80m}Br decays into ^{80g}Br which is radioactive, too, so the effects of the nuclear transformation can be followed. The not unexpected result was that the decay does not change the chemical form of the coordinated bromide ligands; the retention in this case is 100%. An upper limit to the recoil energy originating from inner coulombic repulsion as a consequence of the vacancy cascade that follows the decay is of the order of 10 eV which is in fact too small to produce any displacement in ionic solids. These results therefore are in very good accordance with physical models.

The results of the chemical effects of the nuclear process $^{81}Br(n,\gamma)$ $^{82m}Br \rightarrow {}^{82}Br$ in K_2ReBr_6–K_2ReCl_6 mixed crystals (Müller and Martin 1969, Bell et al. 1972, Rössler et al. 1972) are shown in fig. 25.7. Besides a yield of 12% free bromide the main products of the recoil reaction are $Re^{82}BrCl_5^{2-}$ and $Re^{82}BrBr_5^{2-}$ while the other mixed forms are absent. This result is contradictory to the hot-spot model because for this model the distribution of the mixed forms should be the same for bromine and for rhenium recoil. The only explanation for the results is a solid-state reaction following the billiard-ball model:

$$^{82}Br + ReX_6^{2-} \rightarrow Re^{82}BrX_5^{2-} + X \ (X = Cl,Br)$$

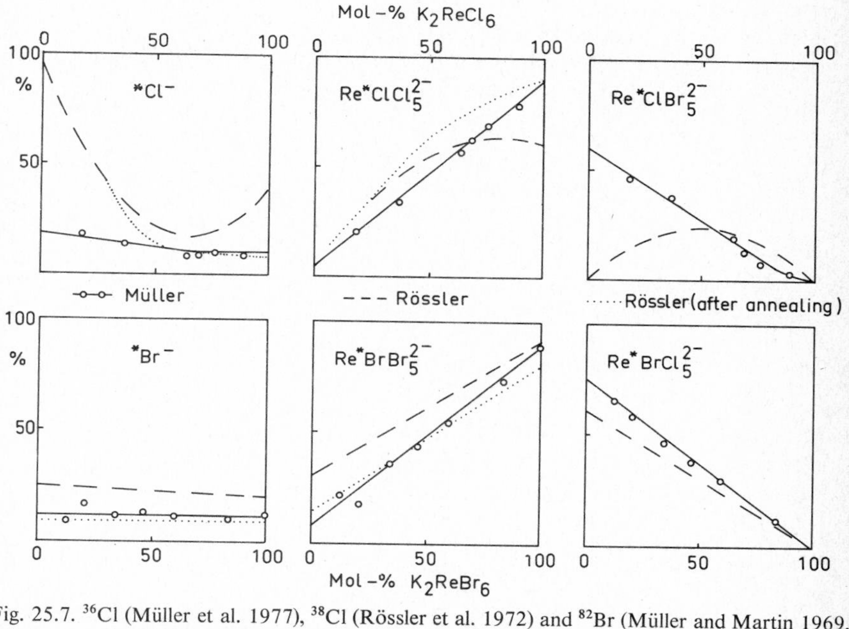

Fig. 25.7. ^{36}Cl (Müller et al. 1977), ^{38}Cl (Rössler et al. 1972) and ^{82}Br (Müller and Martin 1969, Rössler et al. 1972) labelled recoil products resulting from thermal neutron capture in K_2ReBr_6–K_2ReCl_6 mixed crystals as a function of the composition.

If the bromine recoil hits a $ReBr_6^{2-}$ ion, $Re^{82}BrBr_5^{2-}$ results, from $ReCl_6^{2-}$ $Re^{82}BrCl_5^{2-}$ will be formed. The probability of the formation of these two ions is roughly proportional to the molar concentration of the parent complex ion in the mixed crystals: the recoil bromine does not discriminate between the two halide ligands. It is assumed that the process runs directly as a "hot" reaction or via an intermediate ReX_5 – ^{82}Br in a zero order reaction. For mixed crystals with vanishing content of K_2ReBr_6, ca. 8% $Re^{82}BrBr_5^{2-}$ is found which represents the primary retention (in the sense explained at the beginning). Rössler (Rössler et al. 1972) obtained 20–25% free $^{82}Br^-$ and 14% primary retention.

The reactions of ^{38}Cl recoils in K_2ReBr_6–K_2ReCl_6 mixed crystals seem to follow much more complicated paths. This must be concluded from the nonlinear variation of the products $*Cl^-$, $Re*ClCl_5^{2-}$ and $ReBr_5*Cl^{2-}$ with mixed crystal composition (see fig. 25.7) following the expressions:

$$Y(^{38}Cl^-) = 95 - 237\gamma + 181\gamma^2,$$
$$Y(Re^{38}ClCl_5^{2-}) = 5 + 56\gamma + 87\gamma(1\text{-}\gamma),$$
$$Y(ReBr_5{}^{38}Cl^{2-}) = 94\gamma(1\text{-}\gamma),$$

where γ = mol fraction of K_2ReCl_6 and Y the yield in percent of the respective species. The most astonishing fact is that for zero K_2ReCl_6 content the extrapolated yield of $^{38}Cl^-$ amounts to 95% ! Bell (Bell et al. 1972) and Rössler (Rössler et al. 1972) have suggested the following steps:

(1) Formation of secondary ligand vacancies

$$^{38}Cl + ReCl_6^{2-} \rightarrow ReCl_5\square^- + Cl + {}^{38}Cl^-.$$

(2) Ligand exchange with formation of tertiary ligand vacancies

$$ReCl_5\square^- + ReBr_6^{2-} \rightarrow ReCl_5Br^{2-} + ReBr_5\square^-.$$

(3) Product formation by

(a) Substitution reaction (first order) as a direct replacement or as a rapid vacancy–interstitial recombination of entities lying within a reaction cage

(b) Multistep process via exchange of ligand vacancies (second order) as a vacancy–interstitial combination preceded by a ligand–vacancy exchange

$$ReCl_5\square^- + {}^{38}Cl^- \rightarrow ReCl_5{}^{38}Cl^{2-}. \qquad \text{(a) and (b)}$$
$$ReBr_5\square^- + {}^{38}Cl^- \rightarrow ReBr_5{}^{38}Cl^{2-} \qquad \text{(b)}$$

The primary retention [≡direct correlated recombination] (zero order) amounts to 5%, a direct "hot" reaction between $ReBr_6^{2-}$ and ^{38}Cl must be excluded. These differences between ^{82}Br and ^{38}Cl recoil are mainly attributed to the differences in recoil energy and energy transfer: With ^{82}Br recoils 0.66 (or more) secondary ligand vacancies are produced per nuclear event, 40% of them undergo a ligand–vacancy exchange, and the product formation is mainly by first order substitution. With ^{38}Cl the data are 2.37 (or more) secondary ligand vacancies, 76% ligand–vacancy exchange, and the product formation is mainly by the multistep second order process.

Computer calculations have been performed for infinitely diluted mixed crystals K_2ReBr_6–K_2ReCl_6 (Robinson et al. 1974, Rössler and Robinson 1974). The median vector ranges of Cl recoils in K_2ReBr_6 and of Br recoils in K_2ReCl_6 are greater than those in the respective parent substances. This reflects the fact that the energy transfer Cl/Br amounts at most to 85% in a single collision, so that the number of collisions is enlarged compared with the pure substances and hence the recoils travel farther. The comparison of the experiments with the calculations for the primary retention and the yield of the free halide is shown in the tables 25.5 and 25.6. Generally the agreement is good with the exception of $^{36}Cl^-$ yield in K_2ReBr_6, the reason for this discrepancy is not yet known. It must be mentioned, however, that the recombination radius of 8 Å (that means the maximum distance within which a vacancy and a recoil atom recombine) does not follow from the calculations but is deduced from a comparison of the experimental data with the recoil range distributions, and that a displacement threshold of 5 eV is assumed.

Table 25.5
Primary retention of halide recoil in K_2ReCl_6 and K_2ReBr_6

Crystal	Recoil atom	Primary retention (in %)	
		calculated [a]	experimental
K_2ReCl_6	Cl	4	5 Rössler et al. (1972)
K_2ReBr_6	Br	12	8 Müller and Martin (1969), 14 Rössler et al. (1972)

[a] Recombination radius 8 Å

Table 25.6
Yield of free halide for the systems *Br/K_2ReCl_6 and *Cl/K_2ReBr_6

Crystal	Recoil atom	Yield of free halide (in %)	
		calculated [a]	experimental
K_2ReCl_6	Br	14	25 Rössler et al. (1972) 12 Müller and Martin (1969)
K_2ReBr_6	Cl	14	95(!) Rössler et al. (1972)

[a] Recombination radius 8 Å.

Recently the chemical effects of the $^{35}Cl(n,\gamma)^{36}Cl$ nuclear reaction in K_2ReBr_6–K_2ReCl_6 mixed crystals have been investigated (Müller et al. 1977). Due to the long half-life of ^{36}Cl(3×10^5 y) irradiations with 6×10^{18} neutrons/cm^2 were necessary. The results are contradictory to Rössler's experiments with ^{38}Cl (Rössler et al. 1972) and resemble the experiments with ^{82}Br: free $^{36}Cl^-$ amounts to between 10 and 19% dependent of the mixed crystal composition, $ReBr_5{}^{36}Cl^{2-}$ decreases linearly and $ReCl_5{}^{36}Cl^{2-}$ increases linearly with increasing K_2ReCl_6 content of the mixed crystals. The more intimately mixed species $Re^{36}ClCl_{n-1}Br_{6-n}^{2-}$ $(1 < n < 6)$ amount to 16% or less. The reason for the discrepency with the ^{36}Cl and ^{38}Cl recoil experiments is not yet known. The results are shown in fig. 25.7 together with the ^{82}Br experiments.

For the $^{81}Br(n,\gamma)^{82}Br$ nuclear process in K_2ReBr_6-K_2SnCl_6 mixed crystals, in principle results are to be expected similar to the K_2ReBr_6–K_2ReCl_6 experiments with the ion $Sn^{82}BrCl_5^{2-}$ instead of $Re^{82}BrCl_5^{2-}$. As $Sn^{82}BrCl_5^{2-}$ will hydrolyse during the dissolution of the irradiated crystals the recoil bromine will appear as free bromide-(82), a result which was in fact observed (Müller, unpublished). ^{82}Br is mainly formed by the way $^{81}Br(n,\gamma)^{82m}Br \xrightarrow{6.1\ min} {}^{82}Br$. From the preceding experiments on the $^{80m}Br \longrightarrow {}^{80g}Br$ isomeric transition one can conclude that this process has no consequences for the metal–bromine bond, therefore the observed variations in the case of ^{82}Br are really due to the (n,γ) recoil energy which may be estimated at 250 eV.

Bromine recoils, with a larger recoil energy than 250 eV, can be produced by the nuclear reaction $^{79}Br(\gamma,2n)^{77}Br$. In K_2ReBr_6–K_2ReCl_6 mixed crystals the yield of $Re^{77}BrBr_5^{2-}$ increases, the yield of $Re^{77}BrCl_5^{2-}$ decreases significantly with increasing K_2ReBr_6 concentration, all the other mixed hexabromochlororhenates and the free bromide-(^{77}Br) are only slightly dependent on the mixed crystal composition (Müller and Martin 1976). For zero K_2ReBr_6 content the following could be obtained by extrapolation:

$^{77}Br^-$	16%
$Re^{77}BrBr_5^{2-}$	5%
$Re^{77}BrBr_4Cl^{2-}$	2%
$Re^{77}BrBr_3Cl_2^{2-}$	8%
$Re^{77}BrBr_2Cl_3^{2-}$	7%
$Re^{77}BrBrCl_4^{2-}$	18%
$Re^{77}BrCl_5^{2-}$	44%

The total bromine content of the recoil labelled species amounts to 31% at zero K_2ReBr_6 concentration clearly indicating that on average two bromide ligands (one of them being ^{77}Br, the other inactive) of the original six survive the nuclear process and are still bonded to their original Re central atom. Especially the appearance of the more intimately mixed species can only be explained by a new mechanism which has been named the "Inverse Cage Effect". Following this model Br recoils that are directed towards the central atom transfer their kinetic energy via the central atom to all or part of the other ligands which are driven in the surrounding lattice. The newly formed

Table 25.7

Results for ligand atom recoil reactions in mixed crystals of the type K_2ReBr_6–K_2ReCl_6

Nuclear reaction	E_R (eV)	System	Primary retention (%)	Free halide yield for pure substance (%)	Free halide yield for diluted substance (%)	Substitution (billiard ball reaction) (%)	Inverse cage effect (%)
$^{80m}Br \rightarrow {}^{80g}Br$	10	K_2ReBr_6–K_2SnCl_6	100	0	0	0	0
$^{81}Br(n,\gamma)^{82m}Br \rightarrow {}^{82}Br$	250	K_2ReBr_6–K_2SnCl_6	7	12	93	–	–
$^{81}Br(n,\gamma)^{82m}Br \rightarrow {}^{82}Br$	250	K_2ReBr_6–K_2ReCl_6	8 (Müller and Martin 1969) 14 (Rössler et al. 1972)	12	12	75	5
$^{37}Cl(n,\gamma)^{38}Cl$	400	K_2ReBr_6–K_2ReCl_6	5	16	95	[a]	5
$^{35}Cl(n,\gamma)^{36}Cl$	400	K_2ReBr_6–K_2ReCl_6	5	10	16	68	11
$^{79}Br(\gamma,2n)^{77}Br$	1000?	K_2ReBr_6–K_2ReCl_6	5	5	16	45–50	30–35

[a] Small, product formation mainly by multistep process via exchange of ligand vacancies.

vacancies are then filled by halide ions from the environment including the ligands just pushed away. The experimental results may be summarized as follows: Besides a very small yield of primary retention ($Re^{77}BrBr_5^{2-}$) one third of the recoil bromine atoms react following the inverse cage effect, two third are caught as interstitials ($^{77}Br^-$) or undergo simple displacements ($Re^{77}BrCl_5^{2-}$). The recoil energy of the ^{77}Br is estimated to be not larger than ca. 1 keV. This suggests that the nuclear reaction $^{79}Br(\gamma,2n)^{77}Br$ is far from a compound nucleus mechanism. The results of ligand recoil experiments are summarized in table 25.7. It will be very interesting to proceed to even greater energies.

CHAPTER 26

THE RETENTION AND ITS COMPONENTS

J.I. VARGAS and A.G. MADDOCK

University Chemical Laboratory, Cambridge, UK

Chemical Effects of Nuclear Transformations in Inorganic Systems
Edited by G. Harbottle and A.G. Maddock

Contents

1. Introduction

Throughout the preceding chapters a characteristic feature of the effects of nuclear changes in condensed media has been the appearance of some part of the product species in the same, or a closely related, state of chemical combination to the parent species.*

At least notionally the retention, or proportion of product species found in the parent state of combination, can be considered to arise from four contributions: the first, R_i, arises from nuclear events that fail to rupture the parent molecule; the second, R_{ii}, from events in which dissociation of the molecule is followed very quickly by reformation from largely the same constituent atoms; the third, R_{iii}, from reformation reactions involving hot product atoms; and the last, R_{iv}, from reformation processes involving thermalised product species. Thus $R_0 = R_i + R_{ii} + R_{iii} + R_{iv}$.

In principle, it ought to be possible to estimate each of these contributions, but this has not so far been achieved. For a process such as the (n,2n) reaction R_i and R_{ii} must be negligible. But the R_0 values measured with this reaction are not greatly different from those found after the much smaller recoil following the (n,γ) process (Saīto et al. 1967a, Ambe et al. 1968, Ambe and Saīto 1970, Ackerhalt and Harbottle, 1972, Ambe and Ambe, 1973c). Although the larger part of the experimental data refer to the thermal neutron capture process, this may well be one of the more difficult cases to analyse. The mechanical recoil is not invariably large compared with the energy needed to both rupture the molecule and separate the product atom sufficiently to avoid an appreciable R_{ii} contribution. In addition even when the capture gamma spectrum is well established the incidence and extent of internal conversion of low lying states of the product nucleus may not be.

It is generally assumed that the recoil necessary to render R_i and R_{ii} negligible is about 25 eV. But it must be recognised that this figure is based on theoretical, and some experimental, data for very different kinds of crystals. At present the only direct experimental evidence stems from the experiments of Yoshihara et al. (Yoshihara 1970, Yoshihara et al. 1970, Yoshihara and Kudo 1970) and these indicated a rather higher figure, perhaps 40–50 eV, was

*Except where specially indicated this chapter will confine itself to the (n,γ), (n,2n) and (γ,n) reactions.

necessary for the hydrated sodium salt of the indium ethylene diamine tetracetate anion. These data were obtained using the (γ,γ') excitation of the nuclear isomer of ^{115}In. It is unfortunate that there are so few elements for which this technique is possible (ch. 10) (Yoshihara and Mizusawa, 1972).

The above conclusion is also not incompatible with the important body of data obtained by Müller using solid solutions. This material has been discussed in ch. 25. It shows that the recoil atom becomes thermalized in a small zone that is not heavily fragmented. For the fairly modest recoil following many (n,γ) reactions the disturbed zone only extends over a few lattice units. These data are incompatible with the formation of a hot zone, and pyrolytic fragmentation, in the vicinity of the now static radioactive atom.

One might have hoped that at least for systems involving a very large recoil (e.g. n,2n process) where R_i and R_{ii} must be negligible that R_{iii} could be established, for instance by its temperature independent character. Few attempts seem to have been made (see Hillman et al. 1968, 1969, and ch. 12) and at present there are few data on solid systems that can be attributed undeniably to hot-atom effects. To a large extent this is because some of the annealing processes seem to take place even at very low temperatures, certainly as low as 195 K and possibly even at the temperature of liquid nitrogen. Nonetheless some participation of hot reactions seems probable in most systems.

At present the most direct evidence lies in the appearance in some systems of products that do not feature in the annealing reactions, for instance, the formation of products containing P–P bonds, such as diphosphite, in irradiated phosphates (ch. 9). Other examples may be the formation of metal–nitrogen bonds in some irradiated ammonium salts of metal halo-anions (Cabral 1965), $[^{60}Co(NH_3)_5NO_2]^{++}$ in $Co(NH_3)_6(NO_3)_3$ (Harbottle 1961) and part of the $^{59}Fe(CN)_6^{4-}$ found in irradiated $K^{56}Fe^{58}Fe(CN)_6$ (Fenger et al. (1970b).

The classical radiochemical methods of investigation used in most of these studies suffer two major limitations. As noted in several of the previous chapters the radioactive product may suffer various reactions during analysis, especially if this involves solution. Indeed the ligand deficient products that may well survive in the lattice cannot survive solution unchanged. Most other reactive products, for example, atomic forms, will also react upon solution. The other difficulty is that radiochemical analysis is usually a slow process and cannot compete with rapid changes occurring in the irradiated solid.

In situ physical methods are at first sight much more attractive, but, as pointed out elsewhere (Maddock 1972), they have serious limitations. Neither Mössbauer emission spectra nor perturbed angular correlation studies will reveal poorly defined sites for the radioactive product atoms. Useful data can only be obtained for those atoms reaching well-defined, reproducible sites of definite coordination and oxidation state. Further one cannot hope to find a simple correlation between the radiochemical and physical data because they can never refer to the effects of exactly the same nuclear transformation. For instances, even supposing ^{57}Fe were radioactive a radiochemical analysis of a ^{57}Co labelled compound would measure the effects of the orbital electron capture in the ^{57}Co plus the partly internally converted transitions of the excited states of the ^{57}Fe produced. The Mössbauer spectrum would indicate the state of the 14 keV level of the ^{57}Fe but would not include the effects of this last appreciably internally converted transition.

However important evidence on the time scale of events following nuclear transformations in solids has been obtained from these sources. The Mössbauer technique shows that the atoms may reach metastable but only slowly changing states within 100 ns of the nuclear change (chs. 21 and 24). Thus there is experimental justification for distinguishing the very fast processes leading to R_{ii} and R_{iii} from the thermal reactions responsible for R_{iv}. However at a sufficiently high temperature these thermal reactions should also proceed sufficiently rapidly that the time constant for the reaction is not long compared with the life-time of the ^{57m}Fe Mössbauer level. Such a situation should be distinguishable from an analysis of the line widths in spectra recorded for different ages of ^{57m}Fe. The first attempts to identify such rapid annealing processes were negative (Triftshäuser and Craig 1966, 1967) but more recently positive results have been reported for the Mössbauer fraction in $CoSO_4 \cdot 7H_2O$ (Hoy and Wintersteiner 1972), and in Co(1.10 Phenanthroline)$_3$ $(ClO_4)_2$ (Grimm et al. 1975). But there are at present uncertainties in the interpretation of such data (Kankeleit 1975).

Thus, except at very low temperatures, thermally activated annealing reactions may superimpose any hot reactions and the separation of R_{iii} and R_{iv} is far from simple. When the recoil energy is low then the identification of R_i and R_{ii} becomes difficult, although solid solution experiments may be helpful. It must be noted, however, that the very satisfactory theoretical interpretation of the metal halo-anion mixed crystal data (ch. 25) does not seem applicable to phosphate–arsenate mixed crystals (Claridge 1965) for instance.

2. Annealing Processes

The annealing reactions can be defined as changes in the state of the product species from a nuclear change in a solid compound taking place appreciably after the nuclear change – say more than 1 ns later. These processes must involve thermalised species; any electronically excited species involved must either be very long-lived or generated by the treatment initiating the annealing. They are not hot reactions as usually understood.

The purpose of the study of these reactions is to identify the nature and environment of the metastable radioactive and other fragments produced by the nuclear change and to understand the mechanisms by which the reactions take place.

These reactions were called annealing reactions because in the majority of cases they reincorporate the radioactive product in the form of the parent compound. However since this reaction may take place in several steps it is convenient to generalise the term to include all the reactions of the radioactive fragments.

2.1. THE SPECIES IN THE SOLID

As the previous chapters have shown there are very few systems for which the nature of the fragment containing the radioactive atom is rigorously established.

It was shown in chs. 2 and 3 that for many nuclear changes both mechanical recoil and deep ionisation events, such as arise from internal conversion, can lead to molecular rupture.

Bond rupture by recoil could occur homolytically or heterolytically. If the recoil is big enough all the ligands should be lost by the recoiling atom. In a molecular solid, in the absence of reaction with the normal lattice components, the recoiling atom might be expected to thermalize as a neutral atom or, at most, a singly charged cation. In a polar solid survival of cationic species is more likely.

In solids it is unlikely that deep ionisation will lead to the coulombic explosion that can occur in gases. What happens probably depends to a large extent on the radiolytic effect of the Auger electrons on the surrounding lattice. Fortunately the effects of deep ionisation can often be studied in the absence of important recoil effects. The information from Mössbauer emission spectra (ch. 24) has been especially useful. The work of Nath et al. (1968a, b, 1970) and others has shown that if the ligands delocalise the metal orbitals substantially, for instance phthalocyanine and o-phenanthroline complexes, a considerable proportion of the ^{57}Co orbital electron capture decay events lead to the formation of the ferrous analogue. If, however, the ligand is liable to radiolysis, such as for instance oxalato groups, more profound chemical changes ensue. In still simpler polar solids, with monatomic ions, the effects will depend on the nature and number of electron traps in the solid, including that represented by the oxidised species that suffered the deep ionisation. There is evidence that in some systems electronically excited states of the iron complex, formed by ^{57}Co decay, may survive at least for the life-time of the excited Mössbauer level of the ^{57}Fe (Ensling 1970, Ensling et al. 1976).

It is somewhat surprising that even for such an essentially oxidising process as deep ionisation and subsequent Auger cascades oxidised radioactive products are very seldom observed. It is only in the case of β^- decay that oxidation seems fairly general.

For the commonly studied (n,γ) process there is seldom any detailed information about internal conversion in the capture gamma cascade.

Certainly oxidation of the radioactive fragment is unusual. It has been found for *Re in chlororhenate IV salts (Herr 1952a, Schweitzer and Wilhelm 1956, Apers and Maddock 1960); for *Pt in Pt en_2Cl_2 (Haldar 1954); for *Tl in simple salts such as the nitrate (Butterworth and Harbottle 1966); for *U in uranous salts (Aten and Kapetyn 1968b, Heitz and Ruffenach 1971) and *Se in selenites (Al Siddique et al. 1972). Much smaller yields appear in several other systems.

In the great majority of cases, processes accompanied by appreciable mechanical recoil, including especially the (n,γ) process, lead to ligand stripping and appearance of the radioactive product in a lower oxidation state upon analysis. The completely stripped products are the most difficult to identify with certainty in the lattice. A classical radiochemical distinction between a metal atom and its lower cationic states is difficult to establish (see ch. 8, the difficulty of distinguishing Cr^{2+} and Cr^{3+}). Further there is the difficulty that for most complex ions after removal of one or two ligand groups the residual fragment is liable to solvolysis on solution so that appearance of an aquated cation after solution of the irradiated salt in water does not necessarily imply either metal atoms or ions in the solid. However the attachment of different ligand groups after solution in different solvents or solutions does comprise good evidence for ligand deficient species in the solid.

2.2. EXCITATION OF ANNEALING

As the name implies the annealing processes usually lead to reincorporation of the radioactive species in the target species. There are at present very few exceptions. Labelled selenate is formed on annealing irradiated selenites (Al-Siddique and Maddock 1972). In a very few cases some reattachment of ligands is possible, but the parent compound is not reached (see e.g. Hassan and Heitz 1975).

2.2.1. Radiation annealing

The first kind of annealing reaction to be discovered was the in-pile radiation annealing of NH_4SbF_6, converting *Sb III back to *Sb V (Williams 1948b). At the same time the possibility of thermal annealing was foreseen. There are much less data for radiation than for thermal annealing and several simple features are not firmly established. For instance only very crude information is available on the dependence of radiation annealing on the L.E.T. of the radiation (Harbottle 1954, Maddock et al. 1963). Similarly the effect of dose rate has not been properly investigated. (But v. Tanaka 1964a, Arnikar et al. 1973b.)

The process of radiation annealing in most cases involves a thermally activated step, since it is usually suppressed entirely at the temperature of liquid nitrogen (Cobble and Boyd 1952, Harbottle 1954, Baumgärtner and Maddock

1968). Thus reactor irradiations conducted at this temperature usually show distributions of radioactive products that are independent of the duration of irradiation and consequent concurrent dose of ionising radiation. However, not all materials behave in this way, the alkali selenates and phosphates are, for example, exceptions (Al Siddique and Maddock 1972, Claridge, unpublished). Even if the retention is independent of dose at the temperature concerned it may still be substantial. More data are needed on these points.

Taking account of both radioactive decay and nuclear reactions there are always two kinds of annealing study possible: (i) studies after an irradiation or a rapid decay process where the age of the fragments lies between certain maximum and minimum values, (ii) studies of annealing accompanying irradiation or radioactive decay where the annealing can affect nascent as well as aged fragments. The greater part of the published work relates to studies of type (i). There are usually considerable experimental difficulties in arranging type (ii) studies with reactor irradiations.

2.2.2. Thermal annealing

A clear example of thermal annealing was described soon after that of radiation annealing. Radioactive chromium separable from irradiated potassium chromate before heating as *Cr III species, was annealed back to $*CrO_4^=$ on heating to 150° C (Green and Maddock 1949). It soon became apparent that these two processes accompanied most irradiations (Green et al. 1953) and the retentions measured included unknown proportions due to such reactions.

By comparison with these two modes of annealing the data on the other processes are fragmentary.

2.2.3. Photo annealing

Annealing by light was first noticed by Herr (1952a) in potassium chlororhenate IV. The phosphates appear to be especially liable to photo annealing (Claridge and Maddock 1959, 1961). In these salts photo annealing converts *P I to *P III and *P III to hypophosphate and polyphosphates, reactions which only occur at high temperatures as purely thermally excited processes (ch. 9).

The temperature dependence of photo annealing has not been investigated systematically, but in at least one case it can proceed at the temperature of liquid nitrogen (Lin and Wiles 1970).

2.2.4. Pressure annealing

Annealing can also be brought about by compression, but care is necessary to use isotropic and isothermal compression so that the effect can be distinguished from enhanced thermal annealing following crushing (see below) (Andersen and Maddock 1963c, Andersen 1963b, Kacena and Maddock 1965).

2.2.5. *Ultrasonic and shock wave annealing*

Getoff (1963a, b) has shown that ^{35}S produced by the (n,γ) reaction in potassium sulphate anneals under the influence of ultrasonic excitation. Detonation of attached lead azide led to some annealing of irradiated potassium chromate. However the crystal was also pulverised by this treatment (Andersen and Maddock 1963c).

2.2.6. *Hydration, dehydration and phase change*

Isothermal hydration or dehydration usually lead to some change in retention, but there is no uniform pattern of results. For phosphates Lindner (1958) finds no marked effect for either treatment. With lithium and sodium permanganates Bolton and McCallum (1957) find both treatments lead to a reduction in retention. A similar effect is observed with $Na_2CrO_4 \cdot 4H_2O$ (Yeh et al. 1970). It hardly seems probable that $*MnO_4^-$ once formed will decompose on either treatment so that it must be concluded that this result reflects changes in the proportion of some reactive fragment entity that the analytical technique estimates with the permanganate. In $Co(dipy)_3(ClO_4)_3$ however, hydration leads to an increased retention (Shankar et al. 1965b). A similar result accompanies dehydration of irradiated $Na_2IrCl_6 \cdot 6H_2O$ (Cabral and Maddock 1967), but hydration of irradiated Na_2IrCl_6 has little effect.

It must be noted that the hydration and dehydration processes are not really comparable. Hydration involves the introduction of a new reactant, water, into the crystal with the possibility of different reactions of the radioactive fragments to those they would experience on direct dissolution of the irradiated anhydrous solid in water. Dehydration must be more a lattice change effect since no new reactant species is involved.

Besides these isothermal data there are several studies in which dehydration has accompanied thermal annealing. Bell and Herr (1965) have reported data for $Na_2IrCl_6 \cdot 6H_2O$ and $Ca(BrO_3)_2 \cdot 2H_2O$ has been studied in some detail (Maddock and Müller 1960, Müller 1961, Arnikar et al. 1970a, 1971a).

Annealing accompanying a phase change has been far less clearly established, but since a discontinuous annealing of purely radiolytic damage seems to take place during a phase change it seems likely also to occur after nuclear changes (Maddock and Mohanty 1963). Kudo has attributed unusual inflexions in the thermal annealing of neutron irradiated α copper phthalocyanine to conversion to the β form on heating (1972a, b, c). However it is also possible that these irregularities are similar to those reported by Dimotakis (Dimotakis and Kontis 1963, Dimotakis and Maddock 1964).

2.3. FACTORS INFLUENCING THE ANNEALING REACTIONS

A great deal of effort has been devoted to identifying the factors influencing the rate and extent of the annealing reactions. Although probably most of the more important effects have been identified, little progress has been made in the detailed interpretation of the data.

2.3.1. Influence of the environment

The crystal lattice has a considerable influence both on the initial retention and on the rate and extent of annealing reactions. A convenient method of exploring such effects is to study a series of isomorphous salts of a particular complex ion. The salts of the oxyanions have been especially popular for such studies. The earliest data related to alkali iodates (Cleary et al. 1952, see also Arnikar et al. 1970a, Ambe and Saīto 1970) and permanganates (McCallum and Maddock 1953). There are also data for chromates, bromates, chlorates and phosphates (Harbottle 1954, Saīto et al. 1967a, Ambe et al. 1968, Ambe and Saīto 1970, Ambe and Ambe 1973a, Lindner 1958) (see chs. 6–9). Other investigators have explored the effect of changing the anion in a series of cobalt III complexes, such as $[Co(NH_3)_5NO_2]X$ (Saīto et al. 1963b).

Except for the studies using small (α,n) and (γ,n) neutron sources most of these investigations suffer from insufficient control of irradiation conditions. Where simultaneous irradiations were made the relative differences are significant but there is no evidence whether the differences reflect differences in annealing concurrent with the irradiation or are due to differences in the contributions $R_{\rm i}$, $R_{\rm ii}$ and $R_{\rm iii}$, as defined above.

The data of Ambe, Saīto and Sano on the halates relate to irradiations at liquid nitrogen temperature and include comparisons of both (n,γ) and (n,2n) reactions. The data emphasize the difficulty of interpretation. For bromates the retentions for the ^{81}Br(n,γ)^{82}Br reaction, which includes also the isomeric transition in the ^{82m}Br formed in a large proportion of capture events, is always about the same as that for the ^{79}Br(n,2n)^{78}Br reaction. But with iodates the ^{127}I(n,2n)^{126}I retentions are always less than for the ^{127}I(n,γ)^{128}I reaction. No clear pattern of change in retention with change in cation species for series with different common anions emerged. The same studies included a great deal of data on comparisons of hydrated and anhydrous salts. Differences in retention were sometimes in one direction, sometimes the other. However lower retentions for hydrated species predominated (Saīto et al. 1967a, Ambe et al. 1968, Ambe and Saīto 1970).

There are far less data on the changes in annealing characteristics in such series of compounds and no one has yet reported whether the annealing isochronals show related fine structure.

But some studies with cobalt complexes indicated that the initial retentions, under the same conditions of neutron irradiation, do not seem simply related to the susceptibility of the irradiated complex to further thermal or radiation annealing (Dimotakis et al. 1967).

2.3.2. *Isotopic effects*

Even isotopic modification of the environment seems to lead to changes in retention, although probably not to changes in annealing characteristics. The effects are small and the reported data are conflicting. Lindner at first reported small differences between the initial retentions for hydrated and deuterated phosphates (1958) but later work found the differences were not statistically significant (1962b). More recently Ambe and Saīto (1970) have reached similar conclusions for monohydrated and deuterated sodium iodate. However Andersen and Maddock (1963a) report significant differences for hydrated zinc bromates. Simultaneous irradiations of hydrate and deuterate were made at room temperature (table 26.1). These results certainly indicate a small difference in retention and, possibly in annealing rate.

Table 26.1

	R_0	R(12 h at 56° C)	R(16 h at 56° C)
$Zn(BrO_3)_2 \cdot 6H_2O$	29.0 ± 0.3	55.5 ± 0.5	59.1 ± 0.5
$Zn(BrO_3)_2 \cdot 6D_2O$	27.7 ± 0.4	52.1 ± 0.5	53.4 ± 0.6

Differences of comparable magnitude were also found for changes in isotopic composition of the cation (table 26.2) and also for $^{64}Zn(BrO_3)_2 \cdot 6H_2O$ and $^{68}Zn(BrO_3)_2 \cdot 6H_2O$.

Table 26.2

	R_0	R(2 h at 210° C)	R(4 h at 210° C)
$^{24}MgCrO_4$	56.7 ± 0.3	74.0 ± 0.8	76.3 ± 0.3
$^{26}MgCrO_4$	52.7 ± 0.4	69.7 ± 0.2	71.9 ± 0.3

In each case the heavier cation leads to a lower retention. The differences in capture cross section of the different isotopes are too small to lead to appreciable differences in the ionising radiation dose received during neutron irradiation because of self irradiation by capture gamma radiation. Therefore radiation annealing during irradiation should be constant.

In these last instances there is no significant evidence for changes in annealing properties and the changes in retention might well reflect changes in R_{ii} and/or R_{iii}.

2.3.3. *Structural effects*

Both the initial retention and the sensitivity to thermal annealing are dependent on the geometry of the crystal lattice.

Cook (1960) found that the retentions of the α and β modifications of zinc phthalocyanine were very different. Yoshihara and Ebihara (1964, 1966) found similar differences for the corresponding forms of copper phthalocyanine. In each case the β form showed the higher retention and annealed more readily in the early stages of annealing. Similar results were obtained for (n,γ) and (γ,n) produced isotopes. In fact all four, α, β, γ and δ, modifications of copper phthalocyanine behave differently (Yang et al. 1970a, b).

It has also been shown that monoclinic and triclinic forms of potassium dichromate behave differently (Andersen and Maddock 1963b) (Table 26.3). Unfortunately these were ambient temperature irradiations and the precise values of R_0 are of little significance; but the triclinic form certainly anneals more rapidly than the monoclinic.

Table 26.3

$K_2Cr_2O_7$	R_0	R(16 h at 130° C)
monoclinic	82.4	87.0
triclinic	89.5	96.3

Together the above data show that the annealing reactions depend not only on the chemical nature of the target crystals but also on the geometry of the lattice and, possibly, the masses of the entities present.

2.4. INFLUENCE OF CRYSTAL DEFECTS

In the earlier work on this subject great difficulty was experienced in obtaining any reproducibility. The recognition of the annealing reactions indicated one source of irreproducibility: differing temperatures and radiation doses accompanying the neutron irradiation. However when the conditions of irradiation were standardised, for example by simultaneous irradiation of several samples, reproducibility of data from different preparations of a given target material was still poor. A search for the reasons for such irreproducibility revealed the effects of crystal defects.

2.4.1. *In pure materials*

It was found by Maddock and Vargas (1959, Maddock et al. 1963) that similar effects on the initial retention and acceleration of thermal annealing could be brought about both by irradiating the material with ionising radiation and by crushing the material before neutron irradiation. A more direct demonstration of the effect of crystal defects on annealing was obtained by interrupting a thermal annealing after a sufficient time for further annealing to be proceeding very slowly and to crush or irradiate with gamma radiation, using too small a dose for appreciable radiation annealing. On continuing heating at the same temperature as the initial annealing a further rapid annealing occurs as shown in fig. 26.1 (Andersen and Maddock 1962). A great deal of further work has confirmed that in most materials annealing is sensitive to the density and nature of the defects in the crystals. Thermal annealing isochronals show a great deal more detail if single crystals are used and there even appear to be qualitative differences in the annealing reactions (Andersen and Baptista 1971a, ch. 8). These experiments do not identify the kind of defect responsible, nor do they provide a distinction between the behaviour of point defects, defect aggregates and dislocations.

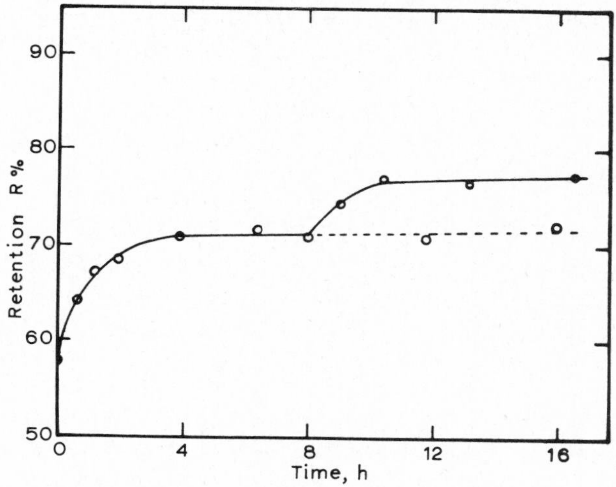

Fig. 26.1. Annealing isotherm. Full line —— untreated sample. Dashed line – – – annealing of sample interrupted at 8 h and sample either (i) crushed or (ii) irradiated with ionising radiation. (Andersen and Maddock 1962)

2.4.2. Doped materials

Since doping enables one to introduce controlled concentrations of some kinds of point defects it was soon used to identify one kind of defect enhancing thermal annealing. Introduction of polyvalent cations into K_2CrO_4 in amounts small enough to permit solid solutions of essentially the same crystal structure as the pure material (a slight change in lattice constants takes place) introduces cation vacancies. Such doped material anneals more rapidly than the pure crystals (Andersen and Maddock 1963d, Costea and Negoescu 1965, Costea et al. 1970). It is clear therefore that cation vacancies can enhance thermal annealing. The extensive data of Andersen and Baptista (1971a) and Ackerhalt et al. (1969, 1971) suggest that the defects are most important in affecting the low temperature annealing processes.

It is possible the higher temperature processes are really true exchange reactions while the low temperature annealing may involve reactions with defects and, perhaps, fragments in the terminal regions of the recoils. There are few data on the effect of defects on radiation annealing. Andersen and Maddock (1963d) found that the calcium doped potassium chromate annealed less readily than the pure crystals. Andersen and Baptista (1971a) find considerable differences between preannealed and ordinary single crystals of K_2CrO_4, again in the sense of radiation annealing being impeded by the defects.

Anion vacancies have also been investigated using similar phosphate doping of potassium chromate. The rate of annealing is again enhanced but the extent of annealing appears to be reduced. Presumably anion vacancies can also have a role in thermal annealing (Milenkovic and Maddock 1967, Costea et al. 1971a).

Unfortunately neither of the investigations of the role of vacancy point defects used the more refined analytical techniques which reveal the complexity of the chromate system.

There are no specific data on the role of dislocations but the large effect of crushing suggests they may be important.

The initial retention, and especially the annealing propensities, can also be changed by thermal treatments which modify the vacancy and, or, dislocation densities. Pre-annealing the crystals at a slowly reducing temperature before neutron irradiation usually makes subsequent thermal annealing more difficult. At the very least it changes the kinetic details of the reaction. On the other hand heating the crystals to a high temperature and freezing in the defects by quenching before irradiation makes them anneal more readily (Andersen and Maddock 1963d, Campbell and Jones 1968a). It seems very probable that the bulk defects in the irradiated crystals are important in the annealing processes because their concentration is generally much greater than that of the reacting radioactive fragments or atoms.

2.4.3. Generality of effect

Nearly all the data refer to the products of the (n,γ) or (n,2n) reactions and it is uncertain how far defects influence other systems. Andersen et al. (1967) and Jones and Warren (1968) find no effects of pre-annealing or pre-irradiating isomer labelled telluric acid on the equilibrium retention of a new generation of the daughter ground state tellurium species. This suggests that bulk defects have no role in the annealing in this material. However telluric acid is a very different type of solid from most of those studied in this connection: it can hardly be described as a hard polar insulator. Nor does the isomeric transition reaction always resemble the (n,γ) process.

It has been suggested that the annealing of ^{80}Br in ^{80m}Br labelled $NaBrO_3$ is insensitive to defects in the bulk of the crystals (Halpern and Dancewicz 1969). But Campbell and Jones (1968b) and later Jones (1970b) attribute the anomalous maximum in the annealing isotherm to such defects (fig. 26.2). This maximum is not repeated on a repeat annealing with a new generation of ground state fragments.

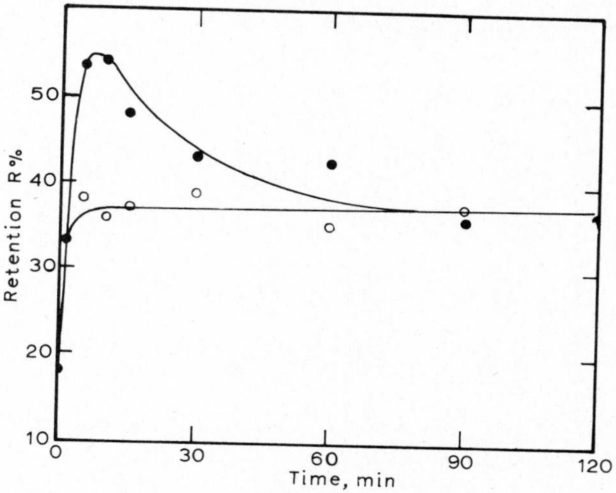

Fig. 26.2. Unusual annealing behaviour of ^{80}Br forming in labelled $K^{80m}BrO_3$ • First isotherm, o subsequent isotherms, all at 216° C. (Jones, 1970)

2.5. INFLUENCE OF THE AMBIENT ATMOSPHERE

The disentangling of the effects of the ambient atmosphere on thermal annealing has spread over a number of years. A good account of the earlier data has been given by Venkateswarlu (1969a). The first positive report of such effects was made by Nath et al. (1964a) who found that cobalt acetylacetonate and other complexes annealed more slowly in oxygen or air than in nitrogen or a vacuum. Several other gases and vapours were found to produce such changes.

A given gas or vapour does not always lead to the same direction of change of rate of annealing compared to that found in vacuo with different neutron irradiated compounds. Thus nitric oxide appears to enhance the thermal annealing in cobalt acetylacetonate (Thomas 1972) but impedes the annealing of salicylaldehyde triethylenetetramine Co III chloride (Rao and Nath 1966). Similar effects are observed with the metal phthalocyanines. With cobalt phthalocyanine oxygen inhibits and hydrogen enhances the annealing (Scanlon and Collins 1971), the effects being greater for the β than the α modifications. These effects are apparently due to adsorption of the gas or vapour on the annealing solid, but such adsorption has not been measured. However Odru and Vargas (1971) find that the effects increase as the crystal size decreases and the specific surface increases, for copper phthalocyanine.

Such effects may indeed account for the previously reported differences in the annealing of phthalocyanines with size of crystallites (Kudo 1971, Yoshihara and Yang 1969a). The annealing of all these phthalocyanines is very sensitive to the purity and crystal perfection of the metal complex (Sakanou and Endo 1970, Kudo 1972a, b, c).

The most extraordinary results of the ambient atmosphere are those reported for potassium chromate (ch. 8). The proportions of mono and polynuclear radioactive chromium III found on analysis seem to be affected by the presence or absence of gases during neutron irradiation. Air, oxygen, wet and dry nitrogen give a larger proportion of the monomeric product. The effect was found for single crystals neutron irradiated at the temperature of liquid nitrogen and is far less apparent with polycrystalline samples with more defects. For this material the presence of the defects apparently outweighs the above mentioned increasing effect of the ambient atmosphere with increasing specific surface of the material.

There are also differences between annealing in air and in vacuo; these also principally affect the changes in the proportions of mono and polynuclear chromium III species above about 120° C. Effects on the retention as radioactive chromate are small and only noticeable above 150° C. All these data were obtained for materials irradiated at −78° C.

In some systems, such as the enhanced annealing of $Co(NH_3)_6^{3+}$ salts in the presence on ammonia gas (Yasukawa and Saito 1965, Yasukawa 1966), one is inclined to suspect a direct chemical effect of the gas, implying penetration of the irradiated crystal. This might imply a much reduced rate of annealing in thoroughly degassed materials. On the other hand an electronic mechanism seems more reasonable to account for the effect of adsorbed ferric ions on the thermal annealing of Co III $(dipy)_3(ClO_4)_3 \cdot 3H_2O$ (Khorana and Wiles 1969).

3. Kinetic Aspects of the Annealing Processes

The classical method used by the chemist for the investigation of reaction mechanisms is to explore their kinetic characteristics and a lot of work has been devoted to the kinetics of annealing reactions. The well-known form of the annealing isotherms, established nearly a quarter of a century ago, immediately exclude the simpler patterns familiar in gas kinetics. Nor do they correspond to any of the rather more complex patterns found in solid-state thermal decompositions.

3.1. THE ORDER OF REACTION

The annealing isotherms clearly do not allow one to deduce the order of reaction from the progress of the reaction. In principle the order of reaction will indicate whether the annealing of given radioactive atom involves only the fragments it generated during and just after nucleogenesis or whether it can involve species connected with another nuclear event. Because only small concentrations of radioactive atoms are concerned (usually 10^{10}–10^{13} atoms/cm^3) one might perhaps expect a first-order process, only the entities from a given event and the normal lattice species being involved.

Although the point does not appear to have been systematically investigated published data suggest that the annealing isothermals for a given material are independent of both the initial and instantaneous concentration of radioactive atoms: thus the same isothermal is obtained after half the radioactive species has decayed as if measured immediately. Such behaviour suggests an essentially first-order process.

There are indeed a few annealing data that can be fitted adequately by simple first-order kinetics (Shankar et al. 1961a) or by a combination of two activation energies (Kudo and Yoshihara 1970).

3.2. KINETIC MODELS FOR ANNEALING

There is a large body of data available on the annealing kinetics of radiation damage, both in systems with chemical bonding and in simpler systems such as metals. Our present annealing behaviour has much in common with these data (Maddock 1975a). Basically there seem to be two models that reproduce most of the characteristics of the experimental data.

3.2.1. First-order processes with distributed energy of activation and/or frequency factor

A detailed treatment of these kinetics can be found in Maddock (1975a).

In this model it is supposed that the different reactive centres have either a spectrum of energies of activation or of frequency factors for first-order annealing processes. Considering first the case where the frequency factors are constant, let us suppose the spectrum of energies of activation can be represented by $f'(E) = dN_0/dE$ so that δN_0 is the initial number of unreacted centres with energies of activation for annealing lying between E and $E + \delta E$. Since the data are usually expressed in terms of the retention which is the fraction (or percentage) of reacted centres, $R = 1 - N/N_0$, it is convenient to normalise $f'(E)$ to obtain $F'(E)$, such that $\int_0^\infty F'(E)dE = 1$. Then $\theta = N/N_0 = \int_0^\infty F'(E) \exp[-\nu t \exp(-E/kT)]dE$, where k is Boltzmann's constant, T is the temperature, t is the time and ν the frequency factor. At fixed E, $n = n_0 \exp(-kt)$ and $k = \nu \exp(-E/RT)$

$$\frac{d\theta}{dt} = -\int_0^\infty \nu\, F'(E)\, \exp(-E/kT)\, \exp[-\nu\, t\, \exp(-E/kT)]\, dE$$

3.2.1.1. Isothermal data. The experimental data yield $\theta(t)$. Let us define a function $\varphi(E) = \exp[-\nu\, t \exp(-E/kT)]$. Its value at any given t and T gives the proportion of reactive centres surviving for chosen values of E. Thus φ plotted as a function of E for $\nu = 10^{10}\ s^{-1}$ and E expressed in units of kT has the form shown in fig. 26.3.

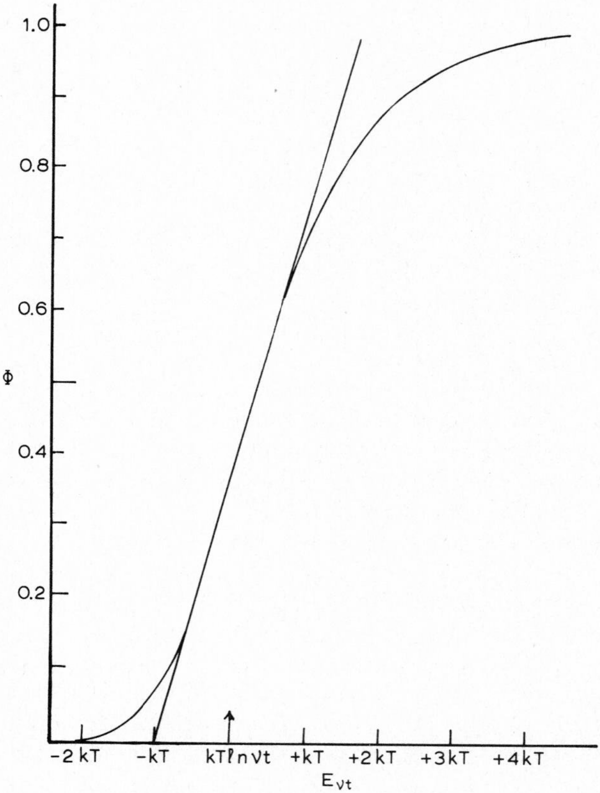

Fig. 26.3. The annealing function φ for a first-order process, at constant T and t, as a function of E. (Maddock 1975a)

It is assumed that $F'(E)$ does not change as annealing proceeds. The φ function shows a point of inflexion, $(\partial\varphi/\partial E)_t$ a maximum, and since $(\partial\varphi/\partial E)_t = (-t/kT)\,(\partial\varphi/\partial t)_E$ the entities with this value of E are annealing most quickly. Putting $(\partial^2\varphi/\partial E^2)_t = 0$ gives $E_0 = kT \ln \nu\, t$ where E_0 is the value of E at the point of inflexion, and $\varphi(E_0) = 1/e$.

One can now envisage the progress of the annealing as shown in fig. 26.4.

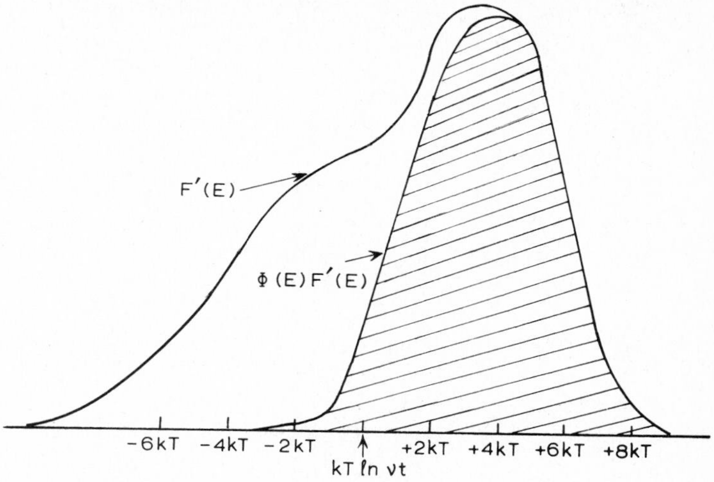

Fig. 26.4. Modification of $F'(E)$ as annealing proceeds – hatched area as yet unannealed. (Maddock 1975a)

The function $\varphi(E)F'(E)$ will define the fraction of the spectrum unannealed at t and T, the hatched portion being unannealed. As t increases at fixed T, the φ curve moves towards higher energies, its point of inflexion moving linearly with ln t. The function $\varphi(E)F'(E)$, determining the unannealed fraction $\int_0^\infty \varphi(E)F'(E)\mathrm{d}E$, moves in a related fashion.

The objective is to extract $F'(E)$ from $\theta(t)$. A simple approximation enabling one to do this was suggested by Vand (1943) and extended by Primak (1955, 1960). It is valid if $F'(E)$ extends over an interval that is large compared to kT and only changes slowly over intervals of two or three kT. In these circumstances one can approximate φ by a step (Heaviside) function located at $E_0 = kT \ln \nu t$ ($\varphi = 0$, $E < kT \ln \nu t$; $\varphi = 1$, $E > kT \ln \nu t$).

Then $\theta(t) \approx \int_0^\infty F'(E)\, H(E - E_0)\, \mathrm{d}E \approx \int_{E0}^\infty F'(E)\, \mathrm{d}E$

and $\dfrac{\mathrm{d}\theta}{\mathrm{d}t} \approx -F'(E_0)\, \dfrac{\mathrm{d}E_0}{\mathrm{d}t}$ but $E_0 = kT \ln \nu t$

$$\frac{\mathrm{d}E_0}{\mathrm{d}t} = \frac{kT}{t}, \quad \text{and} \quad -\frac{1}{kT}\left(\frac{\partial\theta}{\partial \ln t}\right)_T \approx F'(E_0).$$

Hence if θ is plotted against ln t one can obtain $F'(E_0)$. But to evaluate E_0 one must know ν. Now if isothermals have been measured at not too great a separation of T there will found pairs of points $(t_1 T_1)$ $(t_2 T_2)$ etc. for which $F'(E_0)$ is the same. Hence $(\ln \nu + \ln t_1)kT_1 = kT_2\, (\ln \nu + \ln t_2)$ or $\ln \nu = (T_2 \times$

$\ln t_1 - T_1 \ln t_2)/(T_1 - T_2)$. With several pairs of points a best value for ν can be obtained.

One also notes $\partial(\partial\theta/\partial \ln t)\, E_0/\partial T = kF'(E_0)$ and therefore the slopes of the $\theta - \ln t$ plots are proportional to T at corresponding points (E_0 constant).

It follows from the above relations that a set of isothermals plotted as $\theta - \ln t$ can be made to overlap on one curve by appropriate displacement along $\ln t$ axis.

Further for the particular case of $F'(E) = 1/\Delta E$, giving a block spectrum for $E_i < E < E_j$ and $\Delta E = E_j - E_i$ if the interval ΔE is several kT the $\theta - \ln t$ plot will be linear over an appreciable range of t.

More refined treatments have been developed along these lines (Maddock 1975a), but the limited accuracy of most annealing data seldom justify their use.

Thus a set of isothermal annealing data, in principle, enables one to extract $F'(E)$. Line "spectra" of energies of activation will only be reliably resolved when adjacent lines are separated by $\sim 2\,kT$ and closely spaced line spectra cannot be distinguished from continuously distributed spectra.

3.2.1.2. Isochronal data. An alternative approach is to measure the annealing after a constant time at different temperatures, giving $\theta(T)$. Such data can be treated along similar lines. The annealing function φ shows a point of inflexion and $(\partial\varphi/\partial T)_{E_0} = -(E/T)\,(\partial\varphi/\partial E)_T$ with $E_0[1 - \nu t \exp(-E_0/kT)] = 2\,kT$. A similar approximation gives

$$\left(\frac{\partial\theta}{\partial T}\right)_t \approx \frac{E_0}{T} F'(E_0) \quad \text{and} \quad F'(E_0) = -\frac{1}{E_0} \frac{\partial\theta}{\partial \ln T}$$

ν must be determined as before.

This method has some advantages in economy in the number of measurements and the structure of the $F'(E)$ is more apparent from the $\theta(T)$–$\ln t$ plots, but it is subject to the same limitations as the previous method. In either case a discrete energy of activation will appear as a distribution with a half width of about $2\,kT$ (i.e. 0.065 eV at 100° C).

3.2.1.3. Other methods. It is also possible to develop methods for treating data from linear tempering. In this method the sample is measured at various intervals during heating at a linear rate of increase of temperature. There are some experimental difficulties in ensuring such conditions especially if any exo or endothermic processes accompany annealing.

Step annealing consists in sequential isothermal annealing of a sample at a series of increasing temperatures. It is useful in providing a rather more accurate value for ν than the above method.

3.2.1.4. In the above process the frequency factor has been assumed constant. It is not difficult to develop a treatment to establish $F'(\nu)$ if E is assumed constant but this case seems physically less meaningful.

Unfortunately the case of both distributed E and ν cannot be resolved unless there are further constraints on acceptable distributions (Kimmel and Uhlmann 1969).

3.2.1.5. Other orders of reaction. Surprisingly extension to other orders of reaction leads to much the same results and $F'(E)$ can be extracted from isothermal data in the same way. However the resolution deteriorates and a single energy of activation will give an apparent distribution n times that obtained for a first-order process where n is the order of reaction.

3.2.2. Processes involving diffusion

Diffusion controlled kinetics, where the rate of reaction is controlled by the rate at which reactants encounter one another are well known (e.g. Smoluchowski 1916, 1918, review v. North 1966). But a model appropriate to the present problem needs to allow for a number of complicating factors: (i) Diffusion in a crystalline solid is not isotropic but takes place by jumps between a limited number of possible sites determined by the lattice type; (ii) Still further it may be modified by attractive interactions between the reacting entities due to electrostatic attraction, existence of a strain field, etc; (iii) The initial distribution of reacting entities is probably not homogeneous, pairs from the same initiating event are closer than those from different events; (iv) All pairs of reacting entities approaching within a certain distance may react or some proportion may diffuse away, this latter case corresponds to an energy of activation for recombination.

All these factors can be taken into account but in view of the limited precision of the data and increasing number of usually adjustable parameters involved it is doubtful how significant fits achieved in this way really are.

Treatments allowing for the quantised nature of crystalline diffusion are all purely numerical in character and will not be pursued further.

In a simple treatment an attractive interaction can be represented by a larger minimum separation before reaction occurs.

A simple treatment supposed all reactants in correlated pairs of separation b and with reaction inevitable for separation a and all other pairs much further apart than b. Fletcher and Brown (Brown et al. 1953, Fletcher and Brown 1953) showed that the change in retention due to recombining correlated entities

$$R = \frac{a}{b}\left(1 - \operatorname{erf}\frac{b-a}{2(Dt)^{1/2}}\right)$$

The part $(1 - a/b)$ that escapes correlated recombination could still recombine by a second-order process. It will be seen (i) At large t, $R \rightarrow a/b$ corresponding to the plateau value. (ii) Since D, the diffusion coefficient, always occurs as a product with t, and $d = D_0 \exp(-E/kT)$ all annealing curves due to this process can be superimposed by displacement with a $R - \ln t$ plot. (iii) This function gives a point of inflexion on an $R - t$ plot at low t and the slope $\to 0$ as $t \to 0$.

The constant plateau at a/b which is incompatible with experimental data on recoil annealing could be met by a distribution of b values which is physically reasonable. But the last two characteristics, which are not observed in the experimental data, are not avoided.

More elaborate models are developed by Waite (1957a, b, 1960) and recently by Peak and Corbett (1972). The expressions obtained usually contain two terms, one describing the correlated recombination at short times and the other the slower second-order process. At short times R will generally have to increase linearly with $t^{1/2}$.

3.2.3. *Other patterns*

Several other models have been explored (e.g. Harbottle and Sutin 1959). One rather simple one provides a physical basis for a distributed energy of activation. It proposes that the energy of activation is reduced because of some attractive interaction of the fragments. The latter could arise from electrostatic attraction or, for example, a strain field. It must be inversely dependent on the distance separating the reacting fragments, or on some power of this distance. If the separating distance is uniformly distributed between x_0 and x_∞ then

$$1 - R = \int_{x_0}^{x_\infty} \frac{1}{x_\infty - x_0} \exp\left[-\nu t \exp\left(-E/kT + \nu/x\right)\right] \mathrm{d}x.$$

This will yield plateaux at large t for any given T. It represents a special case of the model generalised in section 3.2.1.

3.2.4. *Structure in annealing isotherms*

There is some evidence that even simple annealing isotherms show some structure, that is to say $\mathrm{d}\theta/\mathrm{d}t$ does not decrease steadily with increasing t.

Dimotakis et al. (Dimotakis 1968, Dimotakis and Kontis 1963, Dimotakis and Stamouli 1964, 1967, Dimotakis and Papadopoulos 1970, 1972) have found several examples of such behaviour. One should distinguish two cases, (i) $\mathrm{d}\theta/\mathrm{d}t$ always $\geqslant 0$, (ii) $\mathrm{d}\theta/\mathrm{d}t$ changes sign, so that there is a maximum somewhere on $\theta(t)$. The latter case must either imply that labelled parent species formed during annealing may decompose again on further heating, which seems improbable; or that the analytical procedure is not selective enough to measure only the activity of the parent species (Dimotakis and Maddock 1964).

3.3. RADIATION ANNEALING

The radiation annealing process at first appeared more simple than the thermal annealing (Green et al. 1953). The number of annealable centres decreases exponentially with the dose of ionising radiation in the early stages of the process, suggesting that the probability per unit time of a centre annealing was proportional to the dose rate.

However neither the dependence of the process on dose rate, linear energy transfer, nor concentration of annealable centres have been very systematically investigated.

The observation that the radiation annealing involved a temperature dependent step (Cobble and Boyd 1952, Harbottle 1954) and that doses of ionising radiation too small to produce much annealing sensitise the irradiated material to further thermal annealing (Maddock et al. 1963), suggested that the ionising radiation might affect the distribution of energies of activation for the thermal annealing, an individual event reducing the energy for the affected centre. Indeed kinetic analysis has shown that defects introduced into the irradiated solid act in this way (e.g. Costea and Podeanu 1968, Costea et al. 1970, 1971a, b, Costea 1969). See fig. 26.5.

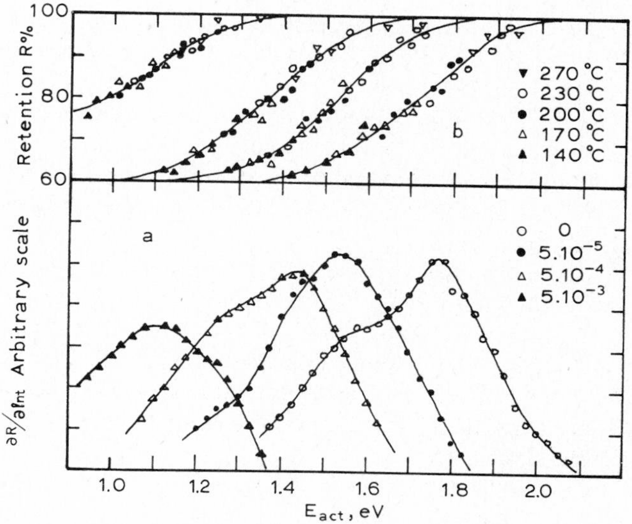

Fig. 26.5. (a) Distribution of energies of activation for thermal annealing of lanthanum doped potassium chromate with different mole fractions of lanthanum. (b) Superposed annealing data for lanthanum doped potassium chromate, R–ln *t* with displacement along ln *t* axis. Decreasing lanthanum concentrations from left to right. (Costea et al. 1970)

But there are further complexities. The exponential dependence on dose fails at higher doses and the process slows down eventually almost stopping (Baumgärtner and Maddock 1968). Presumably some other species produced by the ionising radiation in the crystals eventually interfere with the radiation annealing, possibly by competing with the radioactive centre for some kind of excitation (ionisation, radical production, etc.) introduced by the radiation. This view is substantiated by the observation that if the radiation annealed crystals are gently thermally annealed when the radiation annealing has practically stopped, using too low a temperature to give much thermal annealing, on recommencing radiation annealing it again proceeds rapidly and roughly exponentially with dose.

There are also some interesting details of the way in which the spectrum of activation energies is affected. As was shown in the previous section if one thermally anneals a sample of neutron irradiated material at temperature T, for time t, one anneals practically all radioactive centres up to an energy of activation $E = kT \ln \nu t$. Suppose the corresponding change in retention is produced by a dose of ionising radiation d. If a sample of neutron irradiated compound is given a dose of ionising radiation d followed by thermal annealing at T for time t a still further increase in retention accompanies this thermal annealing. Therefore the ionising radiation cannot simply progressively reduce the energies of activation for thermal annealing of those radioactive centres at the low-energy end of the spectrum (Maddock et al. 1963). Similar conclusions can be drawn from the qualitative observation that radiation annealing principally effects a process corresponding to the higher temperature, and therefore higher energy of activation, annealing process in the phosphates (Claridge and Maddock 1961a, b).

Finally the data are incompatible with all radioactive centres being equally likely to have their energy of activation reduced below some chosen value $E = kT \ln \nu t$, that is they will be thermally annealable in time t at temperature T. If this were the case the fraction of available centres annealing per unit dose would be independent of the retention, or number available to anneal. The experimental data for the alkali selenates does not agree with such a situation (Al Siddique and Maddock 1972).

There are few data on radiation annealing concurrent with the neutron irradiation but Costea (1961a) found that the retention increased rapidly at first and then slowed down to reach a plateau value, rather like the progress of an isothermal annealing, at a retention dependent on the dose rate. (See also Bulbulian and Maddock 1976.)

3.4. SUMMARISED CONCLUSIONS

Although it cannot be said to be rigorously proved it appears that the thermal annealing process is best described by a set of first-order processes with a spectrum of energies of activation. The superposition of annealing isothermals by displacements along the ln t axis required by this model is never as good as the accuracy of the data warrants at long annealing times. There are indications that some changes in frequency factor are also needed to account for the data (Siekierska, Maddock and Vargas, unpublished work).

The energy spectrum seems to be affected by the defects in the lattice of the irradiated material and, apparently, by the atmosphere in which annealing takes place. The radiation annealing seems to be effective by deforming and displacing the spectrum towards lower energies probably by the provision of some defect or radical in the vicinity of the radioactive centre. Other entities produced by the ionising radiation seem to interfere with this action. Possibly the radiation produced excitation, or ionisation, must be transferred across a region of perfect lattice to produce its effect near the radioactive centre. Other radiation produced defects could interfere with this transfer process.

The reactions probably only involve very small movements by the radioactive atom and its neighbours.

4. The Nature of the Irradiated Solid

Following reactions leading to substantial recoil, including all (n,2n) and (γ,n) reactions and possibly most (n,γ) processes, the affected molecule is ruptured and the recoiling atom is ejected into the surrounding lattice. It eventually comes to rest in a not very heavily damaged region, in doing so it may form new bonds by hot reactions with the normal lattice species, or by reaction with fragments it has generated from such species in the terminal damaged zone. These reactions although notionally different will probably not be distinguishable and will together constitute the hot reactions. In most systems some proportion of radioactive product atoms will fail to form new bonds at this stage and will be trapped in the lattice as monoatomic species. The identity of such species is difficult to establish and one can only surmise that more polar environments will favour ionic species and molecular crystals neutral atoms. The ionic species will be restricted to species of very few units of charge.

The (n,2n) and (n,γ) comparisons suggest that the environment and distribution of radioactive species is not radically different at lower recoil energies, where the average is still large compared to chemical bond energies, although the R_{i} and R_{ii} terms may be appreciable in these cases.

The precise pattern of degradation of the recoil energy, fragmentation of lattice species and reaction of recoiling atom can be expected to depend both on the kind and dimensions of the lattice, the masses of the atoms in the lattice and the chemical properties, especially bond energies, of the lattice components. The same factors determine the number and character of the metastable sites at which the radioactive atom can be trapped and the possibility of reaction with the surrounding species. Thus both the initial retention and annealing characteristics might be expected to be influenced by these factors. With certain simplifying assumptions the above model could be tested by Monte Carlo type computer calculations but realistic calculations would be very time consuming indeed.

Bunker and Van Volkenburgh (1970) have made simpler but interesting calculations for phosphates. They consider three cases (i) A free PO_4 unit with Morse potential interactions between the phosphorus and the oxygen atoms, an estimated P–O dissociation energy, parabolic O–O repulsive interactions which fall to zero at the equilibrium O–O distance for phosphate and spectroscopic values for the constant in the exponential of the Morse function as well as the equilibrium P–O distance and an estimated value for the parameter in the repulsive function. (ii) A "bound" phosphate case based on a hypothetical cubic Na_3PO_4 with further Morse terms tending to bind the oxygen to alternate corners of the cube and electrostatic interactions as required by the polarity of the lattice. (iii) An hydrated phosphate approximated by atoms of mass 18 located on the faces of the PO_4 tetrahedron with additional repulsive terms. The calculations were made for recoil energies corresponding to the principal capture γ rays from the ^{32}P; 6.79, 3.90 and 2.10 MeV.

As would be expected the chance of the recoiling phosphorus atom leaving its coordination sphere depends on the chance it moves in a direction avoiding too direct a collision with neighbouring atoms. Thus it decreases as the number of such atoms increases and the hydrate model leads to a higher chance than the bound phosphate case. Except for the lowest recoil energy almost all events lead to ejection of the phosphorus. P–O units are practically never ejected. Naturally the complexity and diversity of events grows rapidly with the complexity of the phosphate model chosen. In general a comparatively large amount of energy is dissipated within the unit cell so that for recoils of a few hundred electron volts the recoil atom only travels a very few lattice units and there is not a great deal of local fragmentation. The conclusions from this greatly simplified model are in agreement with the experimental data of Yoshihara et al. (1970) and of Müller (1970) referred to above. However such simple models do not allow for any bond reformation by hot reactions in other lattice cells.

A more favourable case for computer simulation is, perhaps, the behaviour of monatomic ligands following the (n,γ) reaction. As described in ch. 12, bromine isotopes produced from $ReBr_6^=$ in solid solutions of K_2ReBr_6 in K_2ReCl_6 appear only as $^*Br^-$, $ReBr_5^*$ Br and $ReCl_5{}^*Br^=$. Robinson et al. (1974) have made calculations for such crystals supposing that there is a critical energy, E_d, such that a recoiling bromine atom transferring E_d to another halogen atom in a collision will eject this atom from its lattice site. If its own energy is reduced below E_d a displacement reaction will occur, but if such reduction occurs without transfer of $\geqslant E_d$ the recoiling atom will end up as an interstitial. Rather good agreement with the experimental data is achieved supposing rapid recombination of ligand vacancies and interstitials over a volume containing about 50 halogen sites. The calculation suggests that the initial recoil energy has little effect on the average environment of the thermalised radioactive recoil atom. The results indicate that contributions R_{iii}, and for lower recoil energies R_{ii}, are very significant. The recoiling atom normally comes to rest in a region of comparatively small lattice damage as supported by Müller's investigations. A curious feature, perhaps consequent on the chemical character of this system, is that isomeric transition effects can be ignored. This would suggest the absence of any isotope effects (see below) in such systems.

4.1. IONISATION

Besides the bond breaking and reformation processes electronic changes are usually important. The nuclear process may be accompanied by an internally converted de-excitation of the product nuclei. This will most likely occur for a low-lying state and if the spin change is large may have a very appreciable half-life compared to the time of thermalisation of the recoil atom (for example 1 ns compared to a few ps). Thus such electronic effects will follow the dissipation of the recoil energy.

Even in the absence of such nuclear effects recoil bond rupture may occur heterolytically and the charged fragments may undergo charge transfer reactions. Such redox effects should be sensitive to the nature and number of lattice defects and to the presence of adsorbed electron trapping or releasing molecules on the surface of the crystallites. An effect of this kind might explain such observations as the influence of air on the distribution of radiochromium species after neutron irradiation at low temperatures and other low-temperature radiation effects.

However not all low temperature effects need be electronic (redox). Some recombination of the recoil atom with fragments it has generated or even with normal lattice species may occur with very low energy of activation. Vacancy–interstitial reactions may be possible at very low temperatures.

4.2. THE ROLE OF THE DEFECTS IN ANNEALING

The present evidence does not permit a rigorous identification of the role of the crystal defects in molecular terms.

4.2.1. Increased free space

Several kinds of point defect involve lattice vacancies and these and some dislocations increase the free space in the crystal. Since the annealing processes must involve some movement of atoms additional free space could facilitate such processes. Such free space will be most effective if present in the close vicinity of the radioactive atom. If the defects are introduced after the radioactive atoms, for example by crushing after neutron irradiation, the defects and consequent free space may tend to be formed near the lattice abnormality of the radioactive atom or fragment. But if neutron irradiation follows introduction of defects, as in the experiments with doped potassium chromate, there is no reason for such association and one would expect a much smaller effect if the mechanism involves simply easier movement around the radioactive atom (Maddock et al. 1963).

4.2.2. Participation in redox reactions

Both point defects and dislocations function as electron and hole traps and sources. One might expect such action to be effective over a number of lattice units so that a mechanism involving their participation in redox reactions during annealing need not require any close association of defects with the radioactive atom or fragment. Adsorbed gases might influence annealing in a similar fashion.

It can be objected that not all annealing processes appear to involve redox steps. For example it is not immediately obvious that annealing in say chromium III acetylacetonate or copper phthalocyanine involves any redox process. However existing evidence would not exclude the separable recoil atoms being present as Cr II, Cu I or even as neutral atoms. It may be significant that the annealing of potassium ferrocyanide is insensitive to oxygen and it is a Fe^{2+} precursor that anneals to ferrocyanide in the complex. It would be interesting to know if this system is also insensitive to crystal defects. Thus concealed redox steps may still occur. More experimental data would be useful on this point.

4.2.3. As centres for local excitation

Defects can act as centres for dissipation of exciton energy and for charge recombination. The local excitation of the lattice arising in this way might provide activation for the annealing recombination.

Annealing models based on electronic (Nath 1964, Nath et al. 1964a, b) or exciton (Shankar 1968, Venkateswarlu 1969a, b, Venkateswarlu and Kishore 1971) initiated steps have been elaborated in some detail, especially in relation to the effects in cobalt complexes. Although such steps seem quite plausible these models are far too tied to cobalt chemistry to account for what appears to be rather general behaviour.

5. Chemical Features of the Annealing Processes

Most systems show at least some thermal annealing, although it may be very slight with very complex target materials. Highly endothermic compounds may constitute exceptions but more data are need to substantiate this hypothesis (e.g. Hassan and Heitz 1975). Thermal annealing can often compete very favourably with thermal decomposition taking place in the bulk of the compound (Getoff and Maddock 1963, Costea and Negoescu 1966, Jach 1968, Jach and Chandra 1970) but annealing properties show no correlation with ease of thermal decomposition.

5.1. SEQUENTIAL OR ONE STEP ANNEALING

Besides the stripped atom or ion the irradiated solid normally contains radioactive atoms combined in ligand deficient species. These can be identified by their reaction products on solution in different solvents and solutions. The question arises how far do the annealing reactions proceed by sequential reattachment of ligands and how far is some other kind of process involved? The answer, confusingly, depends on the chemical nature of the target material.

In cobalt complexes of the Co III ammine type there is little evidence for sequential reattachment. For example although $^{60}Co(NH_3)_4(NO_2)_2^+$ and $^{60}Co(NH_3)_2(NO_2)_4^-$ as well as $^{60}Co(NH_3)_3(NO_2)_3$ are present in neutron irradiated $Co(NH_3)_3(NO_2)_3$, as well as $^{60}Co^{2+}$, or its precursor, the annealing seems to convert $^{60}Co^{2+}$ directly to the target compound. Several other Co III complexes behave in like fashion (Saito et al. 1960a, b, Harbottle 1961, Maddock et al. 1965).

On the other hand the halo and cyanometallate anion salts show clear evidence of sequential addition, so far as stability of the aquo intermediates in water permit investigation (Rauscher and Harbottle 1957, Cabral 1965, 1966a).

One might, perhaps, have expected that annealing involving reaction of an anionic ligand with a cationic radioactive species would proceed more easily than steps involving two anionic species. However there is no clear evidence that such is the case.

5.2. STEREOSPECIFICITY

A remarkable feature of the thermal annealing reactions is their stereospecificity. An irradiated *d* or *l* enantiomorph, for example *d*Co $en_3(NO_3)_3$ shows preferential annealing to yield the same enantiomorph (Zuber 1956). Similar preferences are shown by the *cis* and *trans* isomers of $Co(NH_3)_4XY$ species (Rauscher et al. 1961, Dimotakis and Maddock 1961).

This observation places severe limitations on the possible mechanisms of the annealing process that must be implicit in any acceptable model. The most reasonable suggestion at present is to attribute the preference to a lattice template effect.

5.3. COMPETITIVE ANNEALING

Some interesting conclusions can be drawn from experiments in which an element is present in two forms in a crystal, such as $Co(NH_3)_6Co(CN)_6$, and from compounds where two complex ions are present, such as $Co(NH_3)_6Fe(CN)_6$. The first experiments were made on these two compounds by Saito et al. (1964). The initial distributions of ^{60}Co were similar for the two compounds and included all members of the $Co(NH_3)_n(CN)_{6-n}$ series from $n = 0 - 6$, as well as $^{60}Co^{2+}$. $[Co(NH_3)_6]_2[CrO_4]_3 \cdot 5H_2O$ and a mixture of finely ground $Co(NH_3)_6Cl_3$ and K_2CrO_4 both gave small yields of $Cr(NH_3)_6^{3+}$ (Ikeda et al. 1970). The yield of $Cr(NH_3)_6^{3+}$ at first increased on annealing and then decreased (fig. 26.6) in both materials.

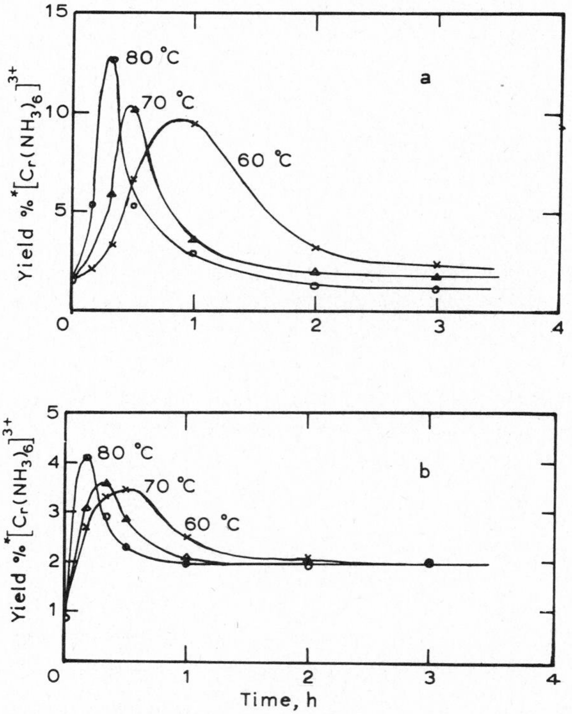

Fig. 26.6. (a) Growth of $[^*Cr(NH_3)_6]^{3+}$ on isothermal annealing of neutron irradiated $[Co(NH_3)_6]_2[CrO_4]_3$. (b) Growth of $[^*Cr(NH_3)_6]^{3+}$ on isothermal annealing of neutron irradiated mixture of $Co(NH_3)_6Cl_3$ and K_2CrO_4. (Ikeda et al. 1970.)

Similarly in neutron irradiated hydrated [Cr en_2Cl_2] [Co $glyc_2(NO_2)_2$] the ^{60}Co anneals into both cationic and anionic complexes (Lazzarini and Fantola-Lazzarini 1974). Correlated isochronals are obtained for the two processes, the two retentions being linearly related.

Clearly the recoiling atom can give the corresponding complex to another complex present in the target material and this process can occur by a thermal reaction as well as by processes accompanying nucleogenesis.

Several interesting results on such systems have been published by the Lazzarinis (1971a, b, 1972, 1974a, b, 1975). On annealing neutron irradiated [cis Co $en_2(NO_2)_2$] [Co EDTA]$\cdot 3H_2O$ the ^{60}Co anneals to reform both complex ions and the retentions in the two ions are linearly related, the slope of the line changing at the temperature at which dehydration occurs (fig. 26.7). The slopes are also changed by annealing in air rather than in vacuo. Similar effects occur in other complexes of this kind but are often more difficult to explore because

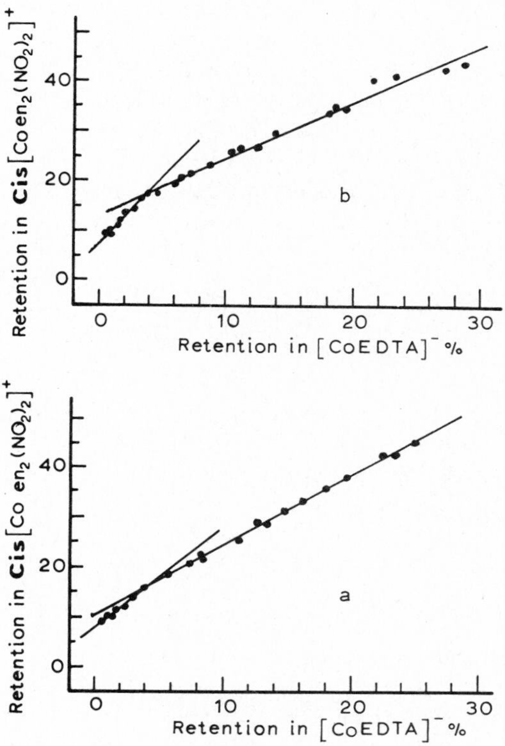

Fig. 26.7. Correlated thermal annealing in neutron irradiated cis [Co $en_2(NO_2)_2$] [Co EDTA]$3H_2O$. (a) annealing in vacuo (b) annealing in air. Dehydration occurs at intersection of lines (a) slopes 1.98 ± 0.13 and 1.41 ± 0.02, (b) slopes 2.61 ± 0.10 and 1.15 ± 0.04. (Lazzarini and Fantola-Lazzarini 1972.)

annealing to form one or other complex ion (in [cis Co $en_2C_2O_4$] [Co EDTA] the cation and in [cis Co $en_2(CN)_2$] [Co EDTA] the anion) predominates.

Both the $[Co\ EDTA]^-$ salts of cis and trans $[Co\ en_2(NO_2)_2]^+$ show correlated annealing of ^{60}Co into the cationic and anionic complexes. Slopes of the linear plots of the two retentions are different for the hydrates and anhydrous salts. The annealing is substantially different in air and in vacuo (Fantola-Lazzarini and Lazzarini 1972, 1973b).

These results, showing correlated isochronals, imply that the spectra of energies of activation for the two progresses are similar but the frequency factors differ.

In the solid solution $[Co_{0.2}Cr_{0.8}en_3]\ (NO_3)_3$ the ^{51}Cr and ^{60}Co show different structures in their isochronals for annealing. The two retentions are not linearily related and the annealing is not very different when conducted in air or in vacuo (Fantola-Lazzarini and Lazzarini 1973a).

Still further information can be obtained using isotopic changes in the target compounds. Using samples containing ^{58}Fe in only one or other ion it was shown that in $[Fe\ II\ dipy_3]_3[Fe\ III\ (CN)_6]$ the recoil ^{59}Fe anneals to form the cationic species but not the anionic one irrespective of the origin of the ^{59}Fe (Siekierska et al. 1973). In similar experiments with $KFe{\cdot}Fe(CN)_6{\cdot}H_2O$ (soluble prussian blue) the retention of the ^{59}Fe was only a few percent irrespective of its origin and practically no annealing occurs. $K_4Fe(CN)_6{\cdot}3H_2O$ shows a much higher retention and appreciable thermal annealing. This suggests that in the prussian blue the recoil ^{59}Fe has to compete with the cationic iron in the annealing reactions (Fenger et al. 1970b).

6. Solid State Exchange Reactions

The fact that the annealing reactions occur in environments that are not heavily damaged leads one to enquire how far exchange reactions may be possible between the recoiled atom and the normal lattice species: how far is annealing a function of the fragmentation of the lattice species due to recoil? The precedent of exchange data in solution made such processes seem highly improbable.

There are practical difficulties in introducing entities similar to the recoiled radioactive atom into the crystal and one can probably not hope to simulate the irradiated system in more than a crude fashion.

6.1. CHEMICALLY DOPED SYSTEMS

The first clear results were obtained by Kaucić and Vlatković (1963) who cocrystallised traces of labelled iodide with iodate at a pH avoiding exchange or other reactions. Such anionic systems are especially favourable: a similar process with cationic dopants will often introduce an hydrated species which is still further from simulating the irradiated systems. The iodide doped system underwent exchange on heating in a fashion resembling the thermal annealing reactions. Since that time a large number of other doped systems prepared generally by co-crystallisation have been shown to behave in the same way.

Related reactions can be brought about in non-isotopic systems, for example, $^{131}I^-$ in $CsClO_4$ yields $^{131}IO_3^-$ and $^{131}IO_4^-$ on heating (Khorana and Wiles 1969).

Not surprisingly chromic doped chromate, prepared by aqueous co-crystallisation, has been investigated in some detail. In this system there is not only the complication due to the probable hydration of the chromic species, but also formation of a not very well characterised chromic chromate. Apers et al. (1964) found an exchange process resembling, but not identical with, the annealing process. Mahieu et al. (1971) have extended this work and found a still closer relation between the processes. They have also provided important, evidence that it has the characteristics of an exchange reaction in that the chemical concentrations of Cr III and Cr VI remain constant throughout. The similarity of the exchange and annealing processes is especially marked in the high-temperature region. There are indeed data that show a ready exchange process between the otherwise distinguishable chromium atoms in "chromium chromate" on heating above 170° C (Rao and Maddock, unpublished).

A better approximation to the neutron irradiated material has been obtained in an ingenious experiment of Collins et al. (1972b). They prepared a batch of K_2CrO_4 crystal doped with ^{51}Cr III. In one portion of the crystals they produced more recoil ^{51}Cr by neutron irradiation. The other portion was given a dose of ionising radiation from ^{60}Co equal to that accompanying the neutron irradiation. The reactions of the dopant and recoil ^{51}Cr in the two samples was compared. Above 150° C the production of ^{51}Cr VI was practically the same. Both samples showed some fast processes below 0° C. But in the interval 0–150° C the recoil atoms anneal rather faster than the dopant exchanges. They suggest that these differences may be due to differences in environment especially arising from fragmentation effects due to the recoil atom.

A particularly clear demonstration of the close connection between the two processes was given by Andersen and Sørensen (1966a) and also Ackerhalt et al. (1969) who showed that after recrystallisation of neutron irradiated K_2CrO_4 the high temperature annealing or exchange process occurs in the same way as before such treatment. However the low temperature and particularly the monomer-dimer (polymer) changes in Cr III species are not observed in the recrystallised material. One can also conclude from these experiments that the high temperature process is probably not affected by the defects produced during irradiation, since recrystallisation should remove or at least, modify such defects.

It is important to observe that these exchange processes are not restricted to such simple complexes as the oxyanions but have been found with all sorts of complicated complexes. A convincing demonstration in such systems has been provided by Nath and Klein (1969). They doped Co III (dipy)$_3$ $(ClO_4)_3 3H_2O$ with ^{57}Co and showed that after annealing the Mössbauer emission spectrum corresponds to that of labelled ^{57}Co III (dipy)$_3$ $(ClO_4)_3 3H_2O$. In this case one would not anticipate any quantitative similarity between the exchange and annealing processes as the $^{57}Co^{2+}$ dopant was applied superficially. In other experiments of the same kind (Ramshesh et al. 1972a, b) it has been shown that $^{57}Co^{2+}$ dopant exchanges with Fe III and Mn III, but not Al III, Cr III or Co III acetylacetonates. Except for the anomaly of the last of these results ^{60}Co only anneals when Co III acetylacetonate is dissolved in the Fe III and Mn III complexes as matrices.

These exchange processes do not follow the normal kinetics for exchange, a combination of one or two first-order processes, but duplicate the unusual kinetics characteristics of the annealing reactions.

6.2. EFFECT OF CONDITIONS ON EXCHANGE IN DOPED SYSTEMS

These exchange processes show all the characteristics of the annealing processes. They can be enhanced by ionising irradiation although the purely radiolytic exchange process has not yet been studied. Such an effect was found for $^{60}Co^{2+}$/Co III (Acac)$_3$ (acetylacetonate) and $^{60}Co^{2+}$/Co III (dipy)$_3$ $(ClO_4)_3 3H_2O$ (Nath et al. 1966) and subsequently for many other cobalt complexes.

A similar enhancement occurs on crushing the doped crystals before heating (loc. cit.). It is also possible to stimulate the exchange photolytically in the $^{60}Co^{2+}$/Co (dipy)$_3$ $(ClO_4)_3 3H_2O$ system using absorption by the Co^{2+} ion (Nath and Khorana 1967). These effects all prove smaller in anhydrous Co(dipy)$_3$ $(ClO_4)_3$.

Finally the exchange processes are sensitive to the atmosphere in which the crystals are heated and to the presence of superficial (adsorbed) or incorporated hole or electron trapping species. Thus exchange in the $^{60}Co^{2+}$/Co (dipy)$_3$ $(ClO_4)_3 3H_2O$ system takes place more slowly in air than in vacuo (Nath et al. 1966). The presence of adsorbed Fe^{2+} slightly enhances and Fe^{3+} markedly impedes the exchange. As with the corresponding effect on annealing the sign of the effect of a given gas depends on the system concerned (Sen Gupta 1969). These data led Nath et al. (1966) to suggest that the annealing reactions are essentially related to exchange reactions. Such differences as are observed arise because chemical doping does not duplicate exactly the distribution and the environments of the radioactive atoms in the neutron irradiated solid. This view is now widely held. (Khorana 1968, Khorana and Nath 1969.)

The solid-state exchange processes not only show similar kinetics to the annealing but also duplicate its stereospecificity (Aalbers and LeMay 1974).

All the above experiments refer to exchange between a normal lattice component and an intruding component introduced by doping. So far very few experiments have been made on exchange between two normal lattice species. In $[Co(H_2O)_6][Co\ EDTA]_2 4H_2O$ very little exchange occurs unless the material is irradiated with ionising radiation before heating. It is important that more information on such processes be obtained (Lazzarini and Fantola–Lazzarini 1975).

The mixed valence compounds offer numerous opportunities for such studies. Unpublished work shows that thallium exchange and recoil annealing in $Tl^{1}Tl^{3}Cl_6$ are very similar processes (Fernandez, Duplatre and Maddock 1978).

Besides the studies in isotopic systems a good deal of work has been carried out on solid substitution reactions which show the same general characteristics.

Some attempts have been made to explore in a similar way the reactions of ligand deficient species that are present in the irradiated solids. Thus if the ligand deficient species formed in $Cr(Acac)_3$ is used as dopant a very rapid exchange process can be observed at modest temperatures (Găinar and Ponta 1971a, Găinar 1972).

6.3. ION IMPLANTED AND OTHER SYSTEMS

Amongst the other techniques one can use to simulate the neutron irradiated crystals is ion implantation. Unlike the chemical doping the ion probably comes to rest in a situation which will have suffered more local damage than in a low dose reactor irradiation. However if the implanted activity is kept very low and fairly short-lived species are used the difference need not be too great. There is no question of incorporation of solvated species and the charge state of the implanted species should approximate to that of the recoiled radioactive atom.

Some valuable results have been obtained in this way. When ^{51}Cr is implanted in K_2CrO_4 it undergoes thermal exchange like the annealing reactions. In K_2BeF_4 and K_2SO_4 the ^{51}Cr is present as Cr^{2+} but it exchanges in much the same way as it anneals after recoil in K_2CrO_4/K_2BeF_4 or K_2CrO_4/K_2SO_4 dilute mixed crystal. This suggests the possibility that the ^{51}Cr is also present as Cr^{2+} in irradiated K_2CrO_4 and the defect sensitive monomer–dimer/polymer changes on annealing may assume redox characteristics (Andersen and Sørensen 1966a).

Implantation of radiocobalt in *cis* and *trans* $[Co\ en_2Cl_2]NO_3$ gave material which showed partly stereospecific exchange on heating, but less marked than the neutron irradiated material. Stereospecificity decreased as the dose of cobalt increased (Andersen et al. 1968a, b, Wolf and Fritsch 1969). Ion implanted copper exchanged more readily in the β than the α phthalocyanine, as indicated by the annealing of the neutron irradiated material.

6.4. SOME THERMODYNAMIC CONSIDERATIONS

There are still some aspects of the solid exchange process that require further investigation. It is generally observed that the extent of exchange decreases as the chemical concentration of the dopant increases (e.g. Khorana and Nath 1967). But it has seldom been demonstrated that the dopant has been homogeneously incorporated in the host matrix. For most systems only a very small chemical concentration of dopant can be tolerated by the host lattice as a single phase system. Once a two phase system is reached the special features of this kind of process are lost and any exchange must resemble that between mixtures of the phases, if this can occur (Annoni and Lazzarini 1970, Lazzarini et al. 1970, 1971).

It is also curious to find that the radioactive atoms sometimes pass almost entirely into the chemical form of the host lattice. An exchange reaction proceeds until isotopic equilibration is reached. If the solid state exchange occurs with minimal movement of atoms one might expect that equilibration would occur only with the surrounding molecules, for instance if 6 normal molecules surround the radioactive dopant ion the limiting exchange would be $6/7 \equiv 85.7\%$. A higher value demands that isotopic equilibration occurs with a larger fraction of lattice occupants. If two or three coordination spheres are involved the maximum exchange will soon reach 100%.

The non-isotopic process such as the formation of chromate in ^{51}Cr III doped potassium nitrate, iodate and sulphate (Khorana and Wiles 1971) and the replacement cobalt and chromium in their oxinates by chromium and cobalt, respectively, involve other considerations. Unfortunately the relevant thermodynamic data, presumable choosing as standard states the dilute solid solutions, are not available. But one is inclined to doubt the thermodynamic feasibility of forming $CrO_4^=$ from Cr III in K_2SO_4 in the absence of ionising radiation. More data are needed on these questions.

7. Effects Due to the Nuclear Reaction

It has already been noted that the retention and annealing reactions are not highly sensitive to the recoil energy following the nuclear reaction as far as the (n,γ), (n,2n) and (γ,n) processes are concerned. There are however differences that are significant.

From a very early date it appeared interesting to compare the retentions of two radioactive isotopes of an element generated in one neutron irradiation in a compound and measured after the one radiochemical separation. To simplify the interpretation of the results it was desirable to avoid situations where substantial differences in epithermal resonance capture could occur. This implied a well thermalised neutron irradiation and avoidance of certain elements.

Unlike organic liquids indisputable effects were soon established (Jach and Harbottle 1958, Apers and Maddock 1960). In the bromates the differences were sometimes quite large and affected both the initial retention and the thermal annealing. In potassium hexachlororhenate IV there was a small difference in retention but no difference in annealing behaviour.

Another system which displayed appreciable differences is zinc phthalocyanine. Differences, both of retention and annealing, between ^{65}Zn and ^{69}Zn in zinc phthalocyanine were reported by Capron and Apers (1961). Yang and Yoshihara (Yoshihara and Yang 1968, Yang 1969) have extended these studies to measurements on the α and β forms and following (n,γ), (n,2n) and (γ,n) reactions as well as separate measurements of the ^{69m}Zn and ^{69g}Zn. They concluded that the initial retentions for the different isotopes depended on the magnitude of the recoil accompanying formation. The substantial retentions they report and their conditions of irradiation suggest that their values include some annealing. By contrast practically no differences are found for products of the same reactions in copper phthalocyanine (Sakanoue and Endo 1970).

Hillman and collaborators (1968, 1969) have also made comparisons of retentions for products of the (n,γ), (n,2n) and (γ,n) reactions in zirconium and hafnium phthalocyanines and metallocene dichlorides. The initial retentions although small vary over a considerable range. The values probably include some annealing component but post irradiation annealing was not studied.

From these and other observations one can conclude that (i) different (n,γ) products produced in a single neutron irradiation show different retentions and annealing behaviour, (ii) large differences in recoil energy do not necessarily produce much difference in these properties.

Hillman et al. observed that the biggest differences are found for those isotopes for which there is a large spin change during the nuclear reaction compared to those for which there is not. This observation suggests that the major cause of the differences may be the incidence of internally converting deexcitation processes occurring in the product nuclei after the thermalisation of the recoil energy. Such a process would probably modify the state of the recoil atom and its surroundings much more than the differences arising from different recoil energies. However smaller effects of recoil energy may also exist. Certainly at the low energy end one must reach a region where $R_i - R_{ii}$ increase rapidly. This will be relevant to the low recoil processes such as β decay and isomeric transition.

Important differences in annealing behaviour of different (n,γ) products from molybdenum in $Mo(CO)_6$ have been revealed by Groening and Harbottle (1970), fig. 26.8, suggesting quite different environments and distributions for the two kinds of recoil atom.

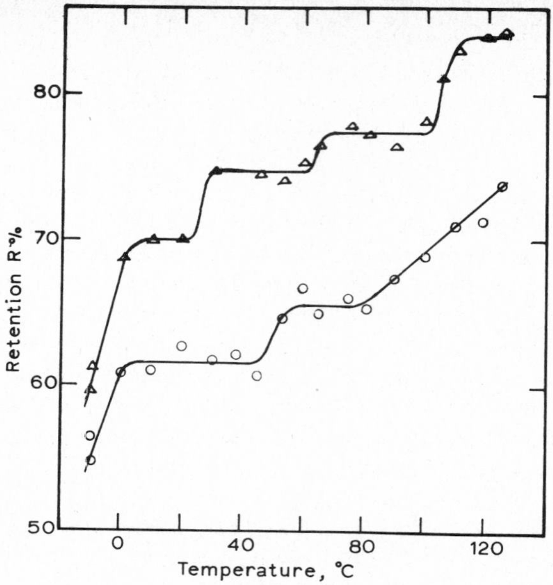

Fig. 26.8. Annealing isochronals for ^{99}Mo, △, and ^{101}Mo, O in neutron irradiated $Mo(CO)_6$. (Groening and Harbottle 1970.)

8. The Molecular Mechanism

The previous sections outline the principal characteristics of the annealing/exchange reactions and the retention processes. There is far less information at the level of molecular mechanisms.

Much attention has concentrated on the role of the defects and annealing atmosphere and the light they throw on the mechanism. Maddock et al. (1963) proposed that both purely electronic steps, determined by the electron and hole trap levels and the semi-conducting properties of the material, or an exciton stimulated process were possible explanations of the data.

The electronic model has been developed considerably by Nath and collaborators (Nath et al. 1964a, b, 1966, Shankar 1968). The filling or emptying of traps provides an obvious mechanism for a redox step in annealing. Even in the absence of such a step it might provide local excitation facilitating annealing/exchange by ligand transfer (Nath et al. 1968a). Electron–hole recombination would provide still more energy. A good example of the application of this approach is to be found in Costea et al. (1969, 1970, 1971a, b) treatment of the annealing in non-stoichiometric chromium oxides and chromates.

Venkateswarlu et al. have developed the alternative proposal that exciton discharge is involved (Venkateswarlu 1969a, b, Venkateswarlu and Kishore 1971). An advantage of this model is that it is equally reasonable where no redox step is involved but, as has been shown, it may be that most of these processes involve a redox step (Shankar, 1968).

Venkateswarlu et al. (Venkateswarlu 1969a, b, Venkateswarlu and Kishore 1971) whose work has mostly concerned the cobalt complexes have also proposed a more detailed model accounting for the oxygen atmosphere effect in terms of the formation of an intermediary which is an oxygen bridged dinuclear cobalt complex. It is certainly true that cobalt rather readily forms such complexes, but there is no certain evidence at present that the oxygen penetrates the solid as this model would imply. Further it leads to a peculiarly ad hoc explanation of phenomena which appear to be of wide spread incidence.

One of the noticeable omissions from the data, and a cogent reason why the electronic or exciton based mechanism is not yet resolved, is the absence of many radiochemical studies associated with physical measurements of the properties of the crystals. Some time ago Andersen et al. (Andersen and Olesen 1965, Andersen et al. 1966b) found a correlation between the glow curves, conductivity peaks and thermal annealing isochronals in potassium chromate and bromate. It is unfortunate that more data of this kind are not yet available. They strongly support the electronic model.

The phthalocyanines appear favorable compounds for such studies and preliminary data have been reported (Odru and Vargas 1971, Auric and Vargas 1973).

At present the electronic hypothesis seems more viable. It is easily extended to account for data such as the variable effect of hydration (Cabral and Maddock 1967) and the effect of impurity cations (Sen Gupta 1969).

Only two attempts have been made at a model for the bond making and breaking processes. Maddock and Collins (1968) have suggested that the limited free space in most hard solids will encourage formation of bi or poly nuclear complexes by the recoiling radioactive atom or fragment. This hypothesis finds support in the different partially annealed/exchanged products obtained with cobalt and other complexes (e.g. Stucky and Kiser 1969). The complex forms between the radioactive recoil atom or fragment and the normal lattice entities. The hypothesis demands potentially chelate or bridging ligand species. On appropriate excitation, and possibly after a redox step, ligand transfer occurs and annealing results. A rather similar specific application to potassium chromate has been proposed by Gütlich et al. (1971a, b).

There is one system of a slightly different character where a very satisfactory detailed explanation of both initial retentions and annealing seems possible. It concerns ligand recoil following the (n,γ) reaction in K_2ReCl_6, K_2ReBr_6 and there solid solutions. Substantially all the radioactive halogen appears as $^*X^-$, $Re^*X_6^=$ or $ReY_5{}^*X^=$. The initial retention data are shown in fig. 12.1, p. 251.

In the expressions for the yields of the different species shown $a =$ mole fraction of K_2ReCl_6 and $b = 1 - a =$ mole fraction of K_2ReBr_6. It is suggested that the zero order term arises from correlated recombination involving the original ligand deficient fragment. Terms involving a or b arise from hot substitution reactions and terms in ab involve ligand vacancy migration followed by combination. Thus $^*Cl + ReCl_6^{=} \rightarrow ReCl_5^{-} + Cl^{-} + {}^*Cl$, followed by $ReCl_5^{-} + ReX_6^{=} \rightarrow ReCl_5X^{=} + ReX_5^{=}$ several times and finally $^*Cl^{-} + ReX_5^{-} \rightarrow ReX_5^{*}Cl^{=}$.

The chlorine recoil is greater than the bromine and it is smaller in size, thus the bromine undergoes more correlated recombination.

The systems do not seem sensitive to radiation annealing, possibly because no obvious redox step is involved, but they do show thermal annealing to yield only the $Re^*X_6^{=}$ species. The annealing of the bromine leads to a similar increase in $Re^*Br_6^{=}$ at all compositions of crystals indicating correlated recombination with the initial ligand deficient entity. But the chloro compound only anneals when $\gamma \geqslant 0.25$ and the process is thought to involve a ligand vacancy transfer process. Exchange with halide doped pure complexes (either isotopically or not so) shows similar kinetic characteristics to the chloride annealing and is supposed to proceed through ligand deficient species produced in the lattice by thermal dissociation and ligand vacancy transfer (Bell et al. 1972, Rössler et al. 1972).

The mechanism proposed seems to fit the data well and further studies of this kind seem desirable. The role of ligand vacancies might be explored by fast neutron bombardment. Repetition of some measurements with low temperature irradiation seems especially desirable.

The greatest need at present seems to be more data on systems where parallel radiochemical and physical, for example luminescence or conductance, measurements are made and for more data testing models for the exchange-annealing reactions.

Data on exchange between normal lattice components is necessary to decide how far vacancy defects are essential to these processes. It is not clear whether the exchange between dopant and matrix is restricted to trace amounts of dopant.

It is particularly important to devise some way of determining how much movement of atoms is involved in these processes and to distinguish redox processes involving defects from free-space effects.

CHAPTER 27

CHEMICAL EFFECTS OF NUCLEAR TRANSFORMATIONS IN INORGANIC SYSTEMS: CONCLUSION

Garman HARBOTTLE

Department of Chemistry, Brookhaven National Laboratory, Upton, New York 11873, USA

Chemical Effects of Nuclear Transformations in Inorganic Systems
Edited by G. Harbottle and A.G. Maddock

The pervasive accomplishments of interdisciplinary research are a mark of present-day science: work in that field of research made up of the intersection of two or more other fields is an activity so taken for granted that we are coming more and more to regard it as the norm rather than the exception. It is easily forgotten that this was not always true. In many sciences, interactions that we see today are in fact still in a state of rapid growth. But the study of radioactivity, including nuclear physics and chemistry, is an exceptional case. From Becquerel's discoveries in 1896, through all the investigations of the newly-found radioisotopes of the natural decay series, there was always a strong collaboration between nuclear physicists and chemists, typified by the work of Madame Curie and by the chemical studies of Soddy, Fleck and von Hevesy leading to the concept of isotopes, and the generalization of Russell, Fajans and Soddy in 1913, which correctly identified α decay as a change in place in the Periodic Table two places lower, and β decay a change one place higher.

In a certain sense, this physical–chemical interaction is inherent in the materials investigated: the radioisotopes themselves, generally present in such minute quantities that their chemical properties are traceable only by physical measurements, e.g. of their radiations. Given all this, it is easy to understand how the extraordinary process of nuclear fission came to be discovered by two chemists, Hahn and Strassmann.

In this context, it also seems natural that Chadwick's discovery of the neutron should have been followed so quickly by Szilard and Chalmers' observation which forms the jumping-off point for this work. The enormous ramification of the research on chemical effects of nuclear transformations, one part of which, dealing with inorganic substances, is our present subject, probably stems in part from the tremendous range of possible systems for investigation – innumerable chemical molecules in all three phases containing a large number of radioactive, labelling atoms decaying by, or generated by, a great variety of nuclear reactions. But another motivation has been much more inductive and mechanistic: the opportunity for the chemist to investigate reactions of atoms in unusual states of charge, or possessing very high kinetic energies, corresponding to exceedingly great "temperatures", as was mentioned in ch.1, and discussed in several other chapters. Then, there is the possibility of forming labelled, unusual ions or molecules, whether they be of transitory existence like collision complexes, or more stable products. In some cases the hot-atom reaction is a convenient, short-cut method; in many cases it leads to labelled species having a valuable, high specific activity.

All these themes are, we trust, traced out in the preceding chapters. Where hot-atom chemistry has failed, it has probably done so because the models chosen were too simple – atoms are not, after all, billiard balls. One has, however, some hope that future, more realistic descriptions of hot-atom phenomena will follow in the train of two lines of research starting from opposite ends of the scale of molecular complexity. At the complex end better-designed, intensive experimentation utilizing in some cases new physical, in situ techniques, and leading to firm quantitative results such as rate constants and energies of activation is needed: here there is an opportunity for valuable interaction with gas and liquid-phase kinetics research, and with solid-state physics. At the simple end, there is already rapidly emerging a highly improved computer-based theory of elementary collision reactions. How far this can ultimately lead, we do not know. We have tried to present the reader with a picture of where it stands today.

BIBLIOGRAPHY

Aalbers, S. L. and LeMay, H. E., 1974, Inorg. Chem. **13**, 904.

Abbé, J. C. and Marques-Netto, A., 1975, J. Inorg. Nucl. Chem. **37**, 2239.

Abdel-Rassoul, A. A., Abdel-Aziz, A. and Aly, H. F., 1969, J. Inorg. Nucl. Chem. **31**, 3043.

Abdel-Rassoul, A. A., Abdel-Aziz, A. and Aly, H. F., 1970, Radiochim. Acta **14**, 113.

Abel, F. and Kalus, J., 1969, Phys. Stat. Solidi **32**, 619.

Ablesimov, N. E. and Bondarevskii, S. I., 1973a, High Energy Chem. **7**, 481.

Ablesimov, N. E. and Bondarevskii, S. I., 1973b, Soviet Radiochem. **15**, 472.

Ablesimov, N. E. and Bondarevskii, S. I., 1975, High Energy Chem. **9**, 147.

Ablesimov, N. E., Bondarevskii, S. I. and Kirin, I. S., 1975, High Energy Chem. **9**, 315.

Ache, H. J. and Wolf, A. P., 1965, Symp. Chem. Effects of Nucl. Transformations IAEA, Vienna **1**, 107.

Ache, H. J. and Wolf, A. P., 1966, Radiochim. Acta **6**, 132.

Ackerhalt, R. E., 1970, Thesis, State University of New York at Buffalo.

Ackerhalt, R. E. and Harbottle, G., 1972, Radiochim. Acta **17**, 126.

Ackerhalt, R. E., Collins, C. H. and Collins, K. E., 1969, Trans. Faraday Soc. **65**, 1927.

Ackerhalt, R. E., Collins, C. H. and Collins, K. E., 1970, Radiochim. Acta **14**, 49.

Ackerhalt, R. E., Collins, C. H. and Collins, K. E., 1971, Trans. Faraday Soc. **67**, 1459.

Ackerhalt, R. E., Ellerbe, B. and Harbottle, G., 1972, Radiochim. Acta **18**, 73.

Adams, J. T. and Porter, R. N., 1973, J. Chem. Phys. **59**, 4105.

Adamson, A. W. and Grunland, J. M., 1951, J. Amer. Chem. Soc. **73**, 5508.

Adloff, J. P., 1966, Radiochim. Acta **6**, 1.

Adloff, J. P., 1971, Radiochim. Acta **15**, 135.

Adloff, J. P., 1975, Hot Atom Chem. Status Rep., IAEA, Vienna p. 61.

Adloff, J. P. and Bacher, M., 1962, J. Chromat. **9**, 231.

Adloff, J. P. and Friedt, J. M., 1972, Panel on Applications of Mössbauer Spectroscopy, IAEA, Vienna p. 301.

Adloff, M. and Adloff, J. P., 1964, Compt. rend. **259**, 141.

Adloff, M. and Adloff, J. P., 1966, Bull. Soc. Chim. Fr. 3304.

Adloff-Bacher, M. and Adloff, J. P., 1964, J. Chromat. **13**, 497.

Afanasov, M. I., Nagy, S., Perfiliev, Yu. D., Vertes, A. and Babeshkin, A. M., 1975a, Radiochem. Radioanal. Lett. **23**, 181.

Afanasov, M. I., Perfiliev, Yu. D. and Babeshkin, A. M., 1975b, Vestn. Moscow Univ. Ser. II Khim. **16**, 749.

Afanasov, M. I., Perfiliev, Yu. D. and Babeshkin, A. M., 1975c, J. Radioanal, Chem. **27**, 125.

Afanasov, M. I., Perfiliev, Yu. D. and Babeshkin, A. M., 1975d, High Energy Chem. **9**, 250.

Akaboshi, M. and Kawai, K., 1971a, Biochim. Biophys. Acta **246**, 194.

Akaboshi, M. and Kawai, K., 1971b, Viva Origino **1**, 31.

Akaboshi, M., Kawai, K. and Waki, A., 1971, Biochim. Biophys. Acta **238**, 5.

Akcay, H., Cailleret, J., Abbé, J. C., Paulus, J. M. and Llabador, Y., 1973, J. Chim. Phys. **70**, 970.

Alfassi, Z. B. and Feldman, L., 1976, Int. J. Appl. Rad. Isotopes **27**, 125.
Alfassi, Z. B., Baer, M. and Amiel, S., 1971, J. Chem. Phys. **55**, 3094.
Aliprandi, B. and Cacace, F., 1956, Ann. Chim. (Rome) **48**, 1204.
Aliprandi, B., Cacace, F. and Giacomello, G., 1956, Ricerca Sci. **26**, 3029.
Allen, J. S., 1958, The Neutrino (Princeton Univ. Press, New Jersey) p. 20.
Al-Siddique, F. R. and Maddock, A. G., 1972, J. Inorg. Nucl. Chem. **34**, 3007.
Al-Siddique, F. R., Maddock, A. G. and Palma, T., 1972, J. Inorg. Nucl. Chem. **34**, 3015.
Altenburger-Siczek, A. and Light, J. C., 1974, J. Chem. Phys. **61**, 4373.
Amaldi, E., D'Agostino, O., Fermi, E., Pontecorvo, B., Rasetti, F. and Segré, E., 1935, Proc. Roy. Soc. (London) **A149**, 522.
Amano, R. and Sakanou, M., 1974, Radiochem. Radioanal. Lett. **19**, 197.
Ambe, F. and Ambe, S., 1973a, c, Radiochim. Acta **19**, 42.
Ambe, F. and Ambe, S., 1973b, Phys. Lett. **A43**, 399.
Ambe, F. and Ambe, S., 1974a, Radiochim. Acta **20**, 141.
Ambe, F. and Ambe, S., 1974b, Bull. Chem. Soc, Japan **47**, 2875.
Ambe, F. and Ambe, S., 1975b, Inorg. Nucl. Chem. Lett. **11**, 139.
Ambe, F. and Ambe, S., 1976, Chem. Phys. Lett. **39**, 294.
Ambe, F. and Saito, N., 1970, Radiochim. Acta **13**, 105.
Ambe, F., Sano, H. and Saito, N., 1968, Radiochim. Acta **9**, 116.
Ambe, F., Shoji, H., Ambe, S., Takeda, M. and Saito, N., 1972b, Chem. Phys. Lett. **14**, 522.
Ambe, F., Ambe, S., Shoji, H. and Saito, N., 1974, J. Chem. Phys. **60**, 3773.
Ambe, S. and Ambe, F., 1975a, J. Chem. Phys. **63**, 4077.
Ambe, S. and Saito, N., 1971, Radiochim. Acta **16**, 40.
Ambe, S., Ambe, F. and Saito, N., 1973, Radiochim. Acta **19**, 121.
Ambe, S., Tominaga, T. and Saito, N., 1971, Monatsh. Chem. **102**, 539.
Ambe, S., Tominaga, T. and Saito, N., 1972a, Radioisotopes (Tokyo) **21**, 543.
Ambrosi, D. A. and Wolfson, J. C., 1969, Nucl. Inst. Methods **74**, 251.
Ames, D. P. and Willard, J. E., 1951, J. Amer. Chem. Soc. **76**, 164.
Amiel, S. and Yellin, E., 1964, J. Inorg. Nucl. Chem. **26**, 2285.
Amiel, S. and Paiss, Y., 1965, Radiochim. Acta **4**, 157.
Andersen, E. B., 1936, Zeit. Physik. Chem. **B22**, 237.
Andersen, T., 1963a, Trans. Faraday Soc. **59**, 2625.
Andersen, T., 1963b, Nature (London) **200**, 1094.
Andersen, T., 1968, Thesis, University of Aarhus.
Andersen, T. and Baptista, J. L., 1971a, Trans. Faraday Soc. **67**, 1203.
Andersen, T. and Baptista, J. L., 1971b, Trans. Faraday Soc. **67**, 1213.
Andersen, T. and Baptista, J. L., 1971c, Phys. Stat. Solidi **44**, 29.
Andersen, T. and Ebbesen, A., 1971, Trans. Faraday Soc. **67**, 3540.
Andersen, T. and Knutsen, A. B., 1961, J. Inorg. Nucl. Chem. **23**, 191.
Andersen, T. and Maddock, A. G., 1962, Nature (Lond.) **194**, 371.
Andersen, T. and Maddock, A. G., 1963a, Radiochim. Acta **1**, 220.
Andersen, T. and Maddock, A. G., 1963b, Radiochim. Acta **2**, 93.
Andersen, T. and Maddock, A. G., 1963c, Trans. Faraday Soc. **59**, 1641.
Andersen, T. and Maddock, A. G., 1963d, Trans. Faraday Soc. **59**, 2362.
Andersen, T. and Olesen, K., 1965, Trans. Faraday Soc. **61**, 781.
Andersen, T. and Østergaard, P., 1968, Trans. Faraday Soc. **64**, 3014.
Andersen, T. and Sørensen, G., 1965, Nucl. Instr. Methods **38**, 204.
Andersen, T. and Sørensen, G., 1966a, Trans. Faraday Soc. **62**, 3427.
Andersen, T. and Sørensen, G., 1966b, Proc. Cairo Conf. Solid State (Plenum Press, 1967) p. 373.
Andersen, T., Christensen, F. and Olesen, K., 1966a, Trans. Far. Soc. **62**, 248.

Andersen, T., Lundager-Madsen, H. E. and Olesen, K., 1966b, Trans. Faraday Soc. **62**, 2409.
Andersen, T., Johannsen, L. and Olesen, K., 1967, Trans. Faraday Soc. **63**, 1730.
Andersen, T., Langrad, T. and Sørensen, G., 1968a, Nature (Lond.) **218**, 1158.
Andersen, T., Langrad, T. and Sørensen, G., 1968b, Nature (Lond.) **219**, 544.
Angenberger, P. and Grass, F., 1970, Radiochim. Acta **14**, 130.
Anlauf, K., Kuntz, P., Maylotte, D., Pacey, P. and Polanyi, J., 1967, Disc. Faraday Soc. **44**, 183.
Annoni, T. and Lazzarini, E., 1970, Energia Nucleare **17**, 551.
Anselmo, V. C., 1961, Thesis, Univ. of Kansas.
Anselmo, V. C., 1965, Radiochim. Acta **4**, 203.
Anselmo, V. C., 1967, USAEC Report, COO-1618-1.
Anselmo, V. C., 1973, J. Inorg. Nucl. Chem. **35**, 1069.
Anselmo, V. C. and Sanchez, E., 1967, USAEC Report, COO-1618-2.
Anselmo, V. C. and Sanchez, E., 1969, Radiochim. Acta **12**, 17.
Apers, D. J., 1971, Sixth International Hot Atom Chemistry Symposium, Brookhaven National Lab., Upton. L. I., USA.
Apers, D. J. and Capron, P. C., 1961, Symp. Chem. Effects of Nucl. Transformations, IAEA Prague **1**, 429.
Apers, D. J. and Harbottle, G., 1963, Radiochim. Acta **1**, 188.
Apers, D. J. and Maddock, A. G., 1960, Trans. Faraday Soc. **56**, 498.
Apers, D. J., Capron, P. C. and Gilly, L. J., 1957, J. Inorg. Nucl. Chem. **5**, 23.
Apers, D. J., Dejehet, F. G., van Outryve d' Ydewalle, B. S. and Capron, P. C., 1962, J. Inorg. Nucl. Chem. **24**, 927.
Apers, D. J., Dehejet, F. G., van Outryve D'Ydewalle, B. S., Capron, P. C., Jach, J. and Moorhead, E., 1963, Radiochim. Acta **1**, 193.
Apers, D. J., Collins, K. E., Collins, C. H., Ghoos, Y. F. and Capron, P. C., 1964, Radiochim. Acta **3**, 18.
Apers, D. J., Theyskens, C. and Capron, P. C., 1967, J. Inorg. Nucl. Chem. **29**, 858.
Aras, N. K., Kahn, B. and Coryell, C. D., 1965, J. Inorg. Nucl. Chem. **27**, 527.
Arends, J., der Hartog, H. W. and Delker, A. J., 1965, Phys. Stat. Solidi **10**, 105.
Argo, J. and Maddock, A. G., 1969, Fifth International Hot Atom Chemistry Symposium, Cambridge.
Arizmendi, L. and Maddock, A. G., 1961, J. Inorg. Nucl. Chem. **17**, 191.
Armento, W. J., 1968a, Thesis, Georgia Inst. Technology, Atlanta.
Armento, W. J., 1968b, USAEC Report, TID-24870.
Arnikar, H. J. and Patil, S. F., 1971, J. Univ. Poona, Sci. Tech. **40**, 41.
Arnikar, H. J. and Rao, B. S. M., 1971, J. Indian Chem. Soc. **48**, 323.
Arnikar, H. J., Dedgaonkar, V. G. and Shrestha, K. K., 1969, Proc. Chem. Sym., Chandigarh, India **2**, 222.
Arnikar, H. J., Sharma, D. K. and Patil, S. F., 1970a, Radiochim. Acta **13**, 164.
Arnikar, H. J., Dedgaonkar, V. G. and Shrestha, K. K., 1970b, J. Univ. Poona Sci. Tech. **38**, 177.
Arnikar, H. J., Dedgaonkar, V. G. and Shrestha, K. K., 1970c, J. Univ. Poona Sci. Tech. **38**, 169.
Arnikar, J. J., Sharma, D. K. and Patil, S. F., 1971a, J. Indian Chem. Soc. **48**, 326.
Arnikar, H. J., Dedgaonkar, V. G. and Shrestha, K. K., 1971b, Proc. 2nd Chem. Symp. 1970 **2**, 87. (C.A. **76**, 14766).
Arnikar, H. J., Patil, S. F., Puntambekar, S. V. and Riebel, W., 1973a, Radiochim. Acta **19**, 173.
Arnikar, H. J., Dedgaonkar, V. G. and Barve, M., 1973b, J. Univ. Poona Sci. Tech. **44**, 31.
Asano, T., Okada, S., Sakamoto, K., Taniguchi, S. and Kobayashi, Y., 1974, J. Inorg. Nucl. Chem. **36**, 1433.

Ascoli, S. and Cacace, F., 1965, Nucl. Instr. Methods **38**, 198.
Ascoli, S., Cacace, F., Giacomello, G. and Possagno, E., 1967, J. Phys. Chem. **71**, 427.
Aten, A. H. W., 1942, Rec. Trav. Chim. Pays-Bas **61**, 467.
Aten, A. H. W., 1947, Phys. Rev. **71**, 641.
Aten, A. H. W. and Kapteyn, J. C., 1968a, Radiochim. Acta **9**, 224.
Aten, A. H. W. and Kapteyn, J. C., 1968b, Radiochim. Acta **9**, 223.
Aten, A. H. W. and van Berkum, J. B. M., 1950, J. Amer. Chem. Soc. **72**, 3273.
Aten, A. H. W., v. d. Straaten, H. and Riesebos, P. C., 1952, Science **115**, 267.
Aten, A. H. W., Koch, G. K., Wesselink, G. A. and de Roos, A. M., 1956, J. Amer. Chem. Soc. **79**, 63.
Aten, A. H. W., Beers, M. J. and de Groot, D. C., 1958, J. Inorg. Nucl. Chem. **5**, 159.
Aten, A. H. W., Heertje, I. and Polak, P., 1960, J. Inorg. Nucl. Chem. **14**, 132.
Aten, A. H. W., Lindner-Groen, M. and Lindner, L., 1965, Symp. Chem. Effects of Nucl. Transformations, IAEA, Vienna **2**, 125.
Auric, P. and Vargas, J. I., 1971, Sixth International Hot Atom Chemistry Symposium, Brookhaven National Lab., L.I., USA.
Auric, P. and Vargas, J. I., 1972, Chem. Phys. Lett. **15**, 366.
Baba, H., Yoshihara, K., Amano, H., Tanaka, K. and Shibata, N., 1961, Bull. Chem. Soc. Japan **34**, 590.
Baba, H., Tanaka, K. and Yoshihara, K., 1963, Bull. Chem. Soc. Japan **36**, 928.
Babeshkin, A. M., 1969, Vestn. Moscow Univ., Khim. **24**, 73.
Babeshkin, A. M., 1970, Vestn. Moscow Univ., Khim. **25**, 386.
Bacher, M. and Adloff, J. P., 1962, Compt. rend. **255**, 304.
Bächmann, K., Bögl, W. and Büttner, K., 1973, Z. anal. Chem. **267**, 274.
Bădiča, T., Dema, S., Gelberg, A., Ianovici, E., Ion-Mihai, R. and Zaitseva, N. G., 1970, International Conference on Hyperfine Interactions detected by Nuclear Radiation, Rehovot, Israel.
Bădiča, T., Gelberg, A., Sălăgeanu, S., Ion-Mihai, R. Ianovici, E. and Zaitseva, N. G., 1971, Radiochim. Acta **16**, 36.
Baer, M., 1969, J. Chem. Phys. **50**, 3116.
Baer, M. and Amiel, S., 1970, J. Chem. Phys. **53**, 407.
Baggio-Saitovich, E., Friedt, J. M. and Danon, J., 1972, J. Chem. Phys. **56**, 1269.
Bakker, C. J., 1937, Physica **4**, 863.
Bancroft, G. M., Dharmawardena, K. G. and Maddock, A. G., 1970, Int. J. Radiat. Phys. Chem. **2**, 45.
Baptista, J. L., 1968, Thesis, Manchester.
Baptista, J. L. and Marqués, N. S. S., 1974, J. Inorg. Nucl. Chem. **36**, 1683.
Baptista, J. L., Newton, G. W. A. and Robinson, V. J., 1968, Trans. Faraday Soc. **64**, 456.
Baptista, J. L., Newton, G. W. A. and Robinson, V. J., 1970, Trans. Faraday Soc. **66**, 213.
Barnes, G. E. and Kikuchi, C., 1970, Radiation Effects **2**, 243.
Baró, G. B., 1961, Thesis, University of Buenos Aires.
Baró, G. B. and Aten, A. H. W., 1961, Symp. Chem. Effects of Nucl. Transformations, IAEA, Prague **2**, 233.
Bartholomew, G. A. and Groshev, L. V., 1967, Nuclear Data **3A**, 367.
Bartholomew, G. A. and Groshev, L. V., 1968a, Nuclear Data **5A**, 1.
Bartholomew, G. A. and Groshev, L. V., 1968b, Nuclear Data **5A**, 243.
Bartholomew, G. A., Earle, E. D. and Gunye, M. R., 1966, Can. J. Phys. **44**, 2111.
Bartlett, N., 1962, Proc. Chem. Soc. **218.**
Batasheva, V. V., Kovalenko, Yu. A., Nesmeyanov, A. N. and Babeshkin, A. M., 1968, Vestn. Moscow Univ. Khim. **23**, 62.
Batasheva, V. V., Babeshkin, A. M. and Nesmeyanov, A. N., 1970, Vestn. Moscow Univ. Ser II

Khim. **23**, 361.

Baudler, M., 1959, Z. Naturforsch. **14b**, 464.

Baulch, D. L. and Duncan, J. F., 1957, Aust. J. Chem. **10**, 112.

Baulch, D. L. and Duncan, J. F., 1958a, Quart. Revs. **12**, 133.

Baulch, D. L. and Duncan, J. F., 1958b, Proc. Geneva Conf. Peaceful Uses of Atomic Energy U.N. **29**, 400.

Baulch, D., Duncan, J. and Thomas, F., 1961, Symp. Chem. Effects of Nucl. Transformations, IAEA, Prague **2**, 169.

Baumgärtner, F., 1961, Kerntechnik **3**, 297.

Baumgärtner, F., 1965, Symp. Chem. Effects of Nucl. Transformations, IAEA, Vienna **2**, 507.

Baumgärtner, F., 1967, Radiochim. Acta **7**, 188.

Baumgärtner, F. and Maddock, A. G., 1968, Trans. Faraday Soc.. **63**, 714.

Baumgärtner, F. and Reichold, P., 1961a, Symp. Chem. Effects of Nucl. Transformations, IAEA, Vienna **2**, 319.

Baumgärtner, F. and Reichold, P., 1961b, Z. Naturforsch. **16a**, 374.

Baumgärtner, F. and Reichold, P., 1961c, Z. Naturforsch. **16a**, 379.

Baumgärtner, F. and Reichold, P., 1961d, Z. Naturforsch. **16a**, 945.

Baumgärtner, F. and Schön, A., 1963, in Proc. Conf. Methods of Preparing and Storing Marked Molecules, Brussels p. 1331.

Baumgärtner, F. and Schön, A., 1964, Radiochim. Acta **3**, 141.

Baumgärtner, F. and Zahn, U., 1960, Z. Elektrochem. **64**, 1046.

Baumgärtner, F. and Zahn, U., 1963, Radiochim. Acta **1**, 51.

Baumgärtner, F., Fischer, E. O. and Zahn, U., 1958, Chem. Ber. **91**, 2336.

Baumgärtner, F., Zahn, U. and Seeholzer, J., 1960, Z. Naturforsch **15a**, 1086.

Baumgärtner, F., Fischer, E. O. and Zahn, U., 1961a, c, Chem. Ber. **94**, 2198.

Baumgärtner, F., Fischer, E. O. and Zahn, U., 1961b, Naturwiss. **48**, 478.

Baumgärtner, F., Fischer, E. O. and Zahn, U., 1962, Naturwiss. **49**, 156.

Baumgärtner, F., Fischer, E. O. and Laubereau, P., 1965, Naturwiss. **52**, 560.

Beeler, J. R. and Besco, D. G., 1963, J. Appl. Phys. **34**, 2873.

Belgrave, E., 1971, Radiochem. Radioanal. Lett. **7**, 325.

Belgrave, E., 1974, Thesis, Strasbourg.

Bell, R., 1964, Thesis (Jul – 181 – RC) Inst. für Radiochemie, Jülich.

Bell, R. and Herr, W., 1964, Radiochim. Acta **2**, 125.

Bell, R. and Herr, W., 1965, Symp. Chem. Effects of Nucl. Transformations, IAEA, Vienna **2**, 315.

Bell, R. and Herr, W., 1966, Radiochim. Acta **6**, 43.

Bell, R. and Stöcklin, G., 1970, Radiochim. Acta **13**, 57.

Bell, R., Rössler, K., Stöcklin, G. and Upadhyay, S. R., 1969a, Fifth International Hot Atom Chem. Symposium, Cambridge, England.

Bell, R., Rössler, K., Stöcklin, G. and Upadhyay, S. R., 1969b, Jülich Report Jul 625-RC.

Bell, R., Rössler, K., Stöcklin, G. and Upadhyay, S. R., 1972, J. Inorg. Nucl. Chem. **34**, 461.

Bellido, A. V., 1967, Radiochim. Acta **7**, 122.

Bellido, A. V. and Wiles, D. R., 1968, Radiochim. Acta **12**, 94.

Belmondi, G. and Ansaloni, A., 1956, Ricerca Sci. **26**, 3067.

Benes, P. and Garba, A., 1966, Radiochim. Acta **5**, 100.

Béraud, R., Berkes, I., Danière, J., Lévy, M., Marest, G., Rougny, R. and Vargas, J. I., 1969, Proc. Roy. Soc. **A311**, 185.

Berei, K. and Stöcklin, G., 1971, Radiochim. Acta **15**, 39.

Berger, W. G., Fink, J. and Obenshain, F. E., 1967, Phys. Lett. **25A**, 4666.

Berne, E., 1949, J. Chem. Soc., Suppl. **2**, 338.

Berne, E., 1964, Acta Chem. Scand. **6**, 1106.

Bertet, M. and Muxart, R., 1965, Symp. Chem. Effects of Nucl. Transformations, IAEA, Vienna **2**, 13.
Bertet, M., Chanut, Y. and Muxart, R., 1964, Radiochim. Acta **2**, 116.
Berthier, J., Jeandey, C., Mathieu, J. P., Oliveira, J., Sette-Camara, A. O. R. and Vargas, J. I., 1971, Sixth International Hot Atom Chem. Symposium, Brookhaven Nat. Laboratory, L. I., USA.
Bertocci, V., Jacobi, R. B. and Walton, G. N., 1961. Symp. Chem. Effects Nucl. Transformations, IAEA, Prague **2**, 337.
Betz, H. D. and Grodzins, L., 1970, Phys. Rev. Lett. **25**, 211.
Beydon, J. and Gratot, I., 1968, Saclay Report CEA-R-3493.
Blachot, J. and Carraz, L. C., 1969, Radiochim. Acta **11**, 45.
Blachot, J. and Vargas, J. I., 1967, Seminar on Isolde Chemistry Problems, CERN, Geneva.
Blais, N. C. and Truhlar, D. G., 1973, J. Chem. Phys. **58**, 1090.
Blaser, B. and Worms, K. H., 1959a, Z. anorg. allgem. Chem. **301**, 7.
Blaser, B. and Worms, K. H., 1959b, Z. anorg. allgem. Chem. **300**, 229.
Blaser, B. and Worms, K. H., 1959c, Z. anorg. allgem. Chem. **300**, 225.
Blaser, B. and Worms, K. H., 1961a, Z. anorg. allgem. Chem. **312**, 146.
Blaser, B. and Worms, K. H., 1961b, Z. anorg. allgem. Chem. **311**, 313.
Blaser, W., 1970a, Eidgenoessichen Institut für Reaktorforschung, Wuerenlingen, Switzerland, Report EIR-172.
Blaser, W., 1970b, Eidgenoessichen Institut für Reaktorforschung, Wuerenlingen, Switzerland, Report, EIR-173.
Blint, R. J. and Newton, M. D., 1975, Chem. Phys. Lett. **32**, 178.
Blomquist, J., Grappengieser, S. and Säderquist, R., 1971, Phys. Stat. Solidi. (A)**4**, 435.
Boato, G., Careri, G., Cimino, A., Molinari, E. and Volpi, C. G., 1956, J. Chem. Phys. **24**, 783.
Bochkarev, V. V., 1972, Theor. Exp. Chem. URSS **8**, 691.
Bogdanov, R. and Bondarevskii, S., 1967, Radiokhimiya **9**, 567.
Bogdanov, R. V. and Murin, A. N., 1965, Zhur. Obschei. Khim. **35**, 916.
Bogdanov, R. V. and Murin, A. N., 1967a, Phys. Tverd. Tela. **9**, 243.
Bogdanov, R. V. and Murin, A. N., 1967b, Zhur. Obschei. Khim. **37**, 2425.
Bogdanov, R. V. and Olevskii, E. B., 1969a, Radiokhimiya **11**, 604.
Bogdanov, R. V. and Olevskii, E. B., 1969b, Radiokhimiya **11**, 606.
Bogdanov, R. V. and Olevskii, E. B., 1969c, Radiokhimiya **11**, 616.
Bogdanov, R. V. and Olevskii, E. B., 1969d, Radiokhimiya **11**, 617.
Bogdanov, R. V. and Olevskii, E. B., 1973, Soviet Radiochim. **15**, 590.
Bogdanov, R. V., Olevskii, E. B. and Tamonov, A. A., 1969a, Khim. Vys. Energ. **3**, 394.
Bogdanov, R. V., Murin, A. N. and Olevskii, E. B., 1969b, Radiokhimiya **11**, 599.
Bogdanov, R. V., Murin, A. N. and Olevskii, E. B., 1969c, Radiokhimiya **11**, 612.
Bohr, N., 1940, Phys. Rev. **58**, 654.
Bohr, N., 1941, Phys. Rev. **59**, 270.
Bohr, N., 1948, Kgl. Danske Videnskab. Selskab. Mat-Fys. Medd. **18**, No. 8, 144.
Bolton, J. R. and McCallum, K. J., 1957, Can. J. Chem. **35**, 761.
Bondarevskii, S. I. and Tarasov, V. A., 1972, Soviet Radiochem. **14**, 132.
Bondarevskii, S. J., Murin, A. N. and Seregin, P. P., 1971, Russ. Chem. Rev. **40**, 51.
Bondareyskii, S. I., Tarasov, V. A. and Sherbakov, E. E., 1973, Soviet Radiochem. **15**, 913.
Borisova, N. I., Kuznetsova, M. I., Kurtchatova, L. N., Mehedov, V. N. and Tchistiakov, L. N., 1959, Zhur. Experim. Teor. Phys. **37**, 366.
Borland, J. W., Mackenzie, A. J. and Hill, W. L., 1952, Ind. Eng. Chem. **44**, 2726.
Born, J. H., 1964, Euratom. Report, Eur. 2209e.
Born, J. H., 1967, Euratom. Report, Eur. 3282d.
Borne, T. B. and Bunker, D. L., 1971, J. Chem. Phys. **55**, 4861.

Boussaha, A., Marques-Netto, A., Abbé, J. C. and Haessler, A., 1976, Radiochem. Radioanal. Lett. **43**, 24.
Bowman, J. M. and Kuppermann, A., 1971, Chem. Phys. Lett. **12**, 1.
Bowman, J. M. and Kuppermann, A., 1973a, J. Chem. Phys. **59**, 6524.
Bowman, J. M. and Kuppermann, A., 1973b, Chem. Phys. Lett. **19**, 166.
Bowman, J. M., Kuppermann, A. and Schatz, G. C., 1973, Chem. Phys. Lett. **19**, 21. See also Careless and Hyatt, 1972.
Bowman, J. M., Schatz, G. C. and Kuppermann, A., 1974, Chem. Phys. Lett. **24**, 378.
Boyd, G. E., 1964, Trans. Amer. Nucl. Soc. **7**, 346.
Boyd, G. E. and Brown, L. C., 1970, J. Phys. Chem. **74**, 1691.
Boyd, G. E. and Larson, Q. V., 1968a, J. Amer. Chem. Soc. **90**, 5092.
Boyd, G. E. and Larson, Q. V., 1968b, J. Amer. Chem. Soc. **90**, 254.
Boyd, G. E. and Larson, Q. V., 1969, J. Amer. Chem. Soc. **91**, 4639.
Boyd, G. E., Cobble, J. and Wexler, S., 1952, J. Amer. Chem. Soc. **74**, 237.
Boyd, G. E., Graham, E. W. and Larson, Q. V., 1962, J. Phys. Chem. **66**, 300.
Boyer, P., 1971, Thesis, Grenoble.
Boyer, P., Jeandey, C., Tissier, A., Vulliet, P. and Vargas, J. I., 1971, Sixth International Hot Atom Chem. Symposium, Brookhaven Nat. Lab., L. I., USA.
Brack, K. and Schwuttke, G. H., 1971, Phys. Stat. Solidi **A5**, 711.
Bračoková, V., 1968, Thesis, Prague.
Bračoková, V. and Cifka, J., 1970, J. Inorg. Nucl. Chem. **32**, 365.
Braun, W., Bass, A. M., Davis, D. D. and Simmons, J. D., 1969, Proc. Roy. Soc. (London) **A312**, 417.
Breckenbridge, W., Root, J. W. and Rowland, F. S., 1963, J. Chem. Phys. **39**, 2374.
Breslow, D. S. and Hamaker, J. W., 1942, USAEC Report, AECD 2409.
Briand, J. P. and Chevalier, P., 1970, Nucl. Instr. Methods **80**, 309.
Brinkman, J. A., 1954, J. Appl. Phys. **25**, 961.
Broda, E., 1948, J. Chim. Phys. **45**, 196.
Broda, E. and Erber, J., 1950, Monatsh. Chem. **81**, 53.
Broda, E. and Müller, H., 1950, Monatsh. Chem. **81**, 457.
Brookhaven National Laboratory Annual Report, 1970, 87.
Brooks, H., 1904, Nature (London) **170**, 270.
Brooks, L. S., 1955, J. Amer. Chem. Soc. **77**, 3211.
Brown, L. C. and Wahl, A. C., 1967, J. Inorg. Nucl. Chem. **29**, 2133.
Brown, L. C., Begun, G. M. and Boyd, G. E., 1969, J. Amer. Chem. Soc. **91**, 2250.
Brown, R. and Winkler, C. A., 1970, Angew. Chem., Int. Ed. **9**, 181.
Brown, W. L., Fletcher, R. C. and Wright, K. A., 1953, Phys. Rev. **92**, 591.
Brune, D., 1967, Acta Chem. Scand. **21**, 2087.
Bruno, M. and Belluco, U., 1956a, Ricerca Sci. **26**, 2085.
Bruno, M. and Belluco, U., 1956b, Ricerca Sci. **26**, 2384.
Bruyn, H., 1955, Proc. UN Conf. on Peaceful Uses of Atomic Energy **3**, 121.
Bulbulian, S. and Maddock, A. G., 1971, J. Chem. Soc. A. 2810.
Bulbulian, S. and Maddock, A. G., 1976, J. Chem. Soc. Dalton, 1715.
Bunker, D. L., 1970, "Molecular Beams and Reaction Kinetics" Academic Press, NY, p. 355.
Bunker, D. L., 1971, Methods Comput. Phys. **10**, 287.
Bunker, D. L., 1974, Accs. Chem. Res. **7**, 195.
Bunker, D. L. and Pattengill, M. D., 1970, J. Chem. Phys. **53**, 3041.
Bunker, D. L. and von Volkenburgh, G., 1970, J. Phys. Chem. **74**, 2193.
Burgus, W. H., 1948, BNL-C-7 Report, 71.
Burgus, W. H., Davies, J. H., Edwards, R. R., Gest, H., Stanley, C. W., Williams, R. R. and Coryell, C. D., 1948, J. Chim. Phys. **45**, 165.

Burgus, W. H. and Davies, J. H., 1951, in Radiochemical Studies: The Fission Products eds. Coryell and Sugarman, (McGraw-Hill, NY) p. 209.
Burgus, W. H. and Kennedy, J., 1950, J. Chem. Phys. **18**, 97.
Burns, W. G. and Williams, T. F., 1955, Nature (Lond.) **175**, 1043.
Bussière, P., 1969, "Les Solides Finement Divisés" ed. J. Ehretsman, La Documentation Francaise, Paris.
Butterworth, J. S., 1964, Thesis, Manchester.
Butterworth, J. S. and Campbell, I. G., 1963, Trans. Faraday Soc. **59**, 2618.
Butterworth, J. S. and Harbottle, G., 1966, Radiochim. Acta **6**, 169.
Butterworth, J. S. and Harbottle, G., 1968, Radiochim. Acta **10**, 57.
Cabral, J. M. P., 1964, J. Inorg. Nucl. Chem. **26**, 1657.
Cabral, J. M. P., 1965, Symposium Chem. Effects of Nucl. Transformations, IAEA, Vienna **2**, 325.
Cabral, J. M. P., 1966a, Thesis, Lisbon University.
Cabral, J. M. P., 1966b, J. Inorg. Nuclear Chem. **28**, 1543.
Cabral, J. M. P., 1967, Rev. Portug. Quim. **9**, 193.
Cabral, J. M. P. and Maddock, A. G., 1967, J. Inorg. Nucl. Chem. **29**, 1825.
Cacace, F., 1970, Adv. Phys. Org. Chem. (ed. V. Gold) **8**, 79.
Cacace, F. and Wolf, A. P., 1965, J. Amer. Chem. Soc. **87**, 5301.
Caillat, R. and Süe, P., 1950a, Compt. rend. **230**, 1666.
Caillat, R. and Süe, P., 1950b, Compt. rend. **230**, 1864.
Cailleret, J., 1973, Thesis, Strasbourg.
Cailleret, J. and Paulus, J. M., 1976, Radiation Effects **28**, 169.
Cailleret, J., Abbé, J. C. and Paulus, J. M., 1971, Radiochem. Radioanal. Lett. **7**, 331.
Cailleret, J., Abbé, J. C. and Paulus, J. M., 1972, Radiochem. Radioanal. Lett. **9**, 113.
Cailleret, J. Abbé, J. C. and Paulus, J. M., 1973, Radiochim. Acta **20**, 33.
Cailleret, J., Paulus, J. M. and Abbé, J. C., 1975, J. Chem. Soc. Faraday **71**, 637.
Calusaru, A., Morariu, M., Barb, D. and Rusi, A., 1973, Radiochem. Acta **19**, 203.
Campbell, D. O., 1973, Proc. 3rd Int. Conf. Mol. Sieves.
Campbell, I. G., 1959a, J. Chim. Phys. **56**, 480.
Campbell, I. G., 1959b, J. Chim. Phys. **56**, 665.
Campbell, I. G., 1960, J. Inorg. Nucl. Chem. **15**, 46.
Campbell, I. G. and Jones, C. H. W., 1968a, Radiochim. Acta **9**, 7.
Campbell, I. G. and Jones, C. H. W., 1968b, Radiochim. Acta **9**, 71.
Cann, P. K. and Hein, R. J., 1957, J. Amer. Chem. Soc. **79**, 60.
Cano, G. L., 1968, Phys. Rev. **169**, 277.
Cano, G. L. and Dressel, R. W., 1965, Phys. Rev. **139A**, 1883.
Cardin, D. J., Joblin, K. N., Johnson, A. W., Lang, G. and Lappert, M. F., 1974, Biochem. Biophys. Acta **371**, 44.
Cardito, J. M. and Diethorn, W. S., 1970, J. Inorg. Nucl. Chem. **32**, 2133.
Careless, P. M. and Hyatt, D., 1972, Chem. Phys. Lett. **14**, 358. See also Bowman et al., 1973.
Carillo, L. and Nassiff, S., 1967, Radiochim. Acta **8**, 124.
Carlier, R., 1971, Thesis, Paris.
Carlier, R. and Genet, M., 1972, Radiochim. Acta **18**, 11.
Carlson, T. A., 1960, J. Chem. Phys. **32**, 1234.
Carlson, T. A., 1963, Phys. Rev. **130**, 2361.
Carlson, T. A., 1975, in "Photoelectron and Auger Spectroscopy" (Plenum Press, NY).
Carlson, T. A. and Koski, W. S., 1955, J. Chem. Phys. **23**, 1596.
Carlson, T. A. and Krause, M. O., 1965a, Phys. Rev. **137A**, 1655.
Carlson, T. A. and Krause, M. O., 1965b, Phys. Rev. **140A**, 1057.
Carlson, T. A. and White, R. M., 1962, J. Chem. Phys. **36**, 2883.

Carlson, T. A. and White, R. M., 1963a, J. Chem. Phys. **39**, 1748.
Carlson, T. A. and White, R. M., 1963b, J. Chem. Phys. **38**, 2075.
Carlson, T. A. and White, R. M., 1963c, J. Chem. Phys. **38**, 2930.
Carlson, T. A. and White, R. M., 1966, J. Chem. Phys. **44**, 4510.
Carlson, T. A. and White, R. M., 1968, J. Chem. Phys. **48**, 5191.
Carlson, T. A., Pleasonton, F. and Johnson, C. H., 1963, Phys. Rev. **129**, 2220.
Carlson, T. A., Hunt, W. E. and Krause, M. O., 1966, Phys. Rev. **151**, 41.
Carlson, T. A., Nestor, C. W., Tucker, T. C. and Malik, F. B., 1968, Phys. Rev. **169**, 27.
Carlson, T. A., Krause, M. O. and Moddeman, W. E., 1970a, Proc. Conf. "Les Processus Electroniques Simples et Multiples du Domaine X", Paris.
Carlson, T. A., Nestor, C. W., Wasserman, N. and McDowell, J. D., 1970b, Atomic Data **2**, 63.
Carter, G. and Colligon, J. S., 1968, "Ion Bombardment of Solids" (Heinemann, London).
Carter, G., Colligon, J. S., Grant, W. A. and Whitton, J. L., 1971, Radiat. Res. Rev. **3**, 1.
Cassou, M., 1970, Thesis, Strasbourg University.
Castiglioni, M. and Volpe, P., 1975, Gazz. Chim. Ital. **105**, 247.
Cavallini, P., 1969, Thesis, Grenoble University.
Cavanagh, J. F., 1969, Phys. Stat. Solidi **36**, 657.
Cavin, P. S., Ianovici, E. and Milman, M., 1975, Radiochem. Radioanal. Lett. **22**, 71.
Cendales, M., Arroyo, A. and Nassif, S. J., 1965, Primera Conferencia Interamericana de Radioquimica, Union Pan America, Washington, USA, p. 179.
Cetini, G., Gambino, O. and Minas, B., 1963, Atti. Acad. Sci. Torino, Classe. Sci. Fis. Mat. **97**, 1137.
Cetini, G., Gambino, O., Castiglioni, M. and Volpe, P., 1965, Atti. Acad. Sci. Torino, Class. Sci. Fis. Mat. **99**, 1093.
Cetini, G., Gambino, O., Castiglioni, M. and Volpe, P., 1967, J. Chem. Phys. **46**, 89.
Cetini, G., Castiglioni, M., Volpe, P. and Gambino, O., 1969, Ricerca Sci. **39**, 3921.
Chackett, G. A. and Chackett, K. F., 1954, Nature (Lond.) **174**, 232.
Chadderton, L. T., and Torrens, I. McC., 1966, Proc. Roy. Soc. (London) **294A**, 93.
Chadderton, L. T., Morgan, D. V. and Torrens, I. McC., 1966, Phys. Lett. **20**, 329.
Chandratillake, M. R., Newton, G. W.A. and Robinson, V. J., 1976, J. Inorg. Nucl. Chem. **38**, 199.
Chanut, Y. and Muxart, R., 1962, Bull. Soc. Chim. Fr. 1783.
Chapman, S. and Suplinskas, R. J., 1974, J. Chem. Phys. **60**, 248.
Chapman, S., Valencich, T. and Bunker, D. L., 1974, J. Chem. Phys. **60**, 329.
Chappert, J., Frankel, R. B. and Misetich, A., 1969, Phys. Rev. **179**, 578.
Cherdyntsev, V. V., 1955, Izv. Akad. Nauk. SSSR, 175.
Chiotan, C., Zamfir, I. and Szabo, M., 1964, J. Inorg. Nucl. Chem. **26**, 1332.
Chiotan, C., Szabo, M., Zamfir, I. and Costea, T., 1968a, J. Inorg. Nucl. Chem. **30**, 1377.
Chiotan, C., Zamfir, I. and Costea, T., 1968b, J. Inorg. Nucl. Chem. **30**, 2857.
Chou, C. C. and Rowland, F. S., 1967, J. Chem. Phys. **46**, 812.
Christian, D., Mitchell, R. F. and Martin, D. S., 1952, Phys. Rev. **86**, 946.
Christiansen, J., Mahnke, H. E., Morfeld, U., Recknagel, E., Riegel, R. and Witthuhn, W., 1973, Z. Phys. **261**, 13.
Christman, D. R., Finn, R. D., Karlstrom, K. I. and Wolf, A. P., 1975, Int. J. App. Radiat. Isotopes **26**, 435.
Cifka, J., 1963, Radiochim. Acta **1**, 125.
Cifka, J., 1964, J. Inorg. Nucl. Chem. **26**, 683.
Cifka, J., 1965, Symposium Chem. Effects of Nucl. Transformations, IAEA, Vienna **2**, 41.
Cifka, J. and Bračoková, V., 1966, J. Inorg. Nucl. Chem. **28**, 2483.
Cifka, J. and Bračoková, V., 1970, J. Inorg. Nucl. Chem. **32**, 361.
Cifka, J. and Kliment, V., 1966, J. Inorg. Nucl. Chem. **28**, 1535.

Cifka, J. and Vesely, P., 1971, Radiochim. Acta **16**. 30.

Claasen, H. H., Selig, H. and Malm, J. G., 1962, J. Amer. Chem. Soc. **84**, 3593.

Claridge, R. F. C., 1965, Trans. Faraday Soc. **61**, 897.

Claridge, R. F. C. and Maddock, A. G., 1959, Nature (Lond.) **184**, 1932.

Claridge, R. F. C. and Maddock, A. G., 1961a, Trans. Faraday Soc. **57**, 1392.

Claridge, R.F.C. and Maddock, A.G., 1961b, Symp. Chem. Effects of Nucl. Transformation, IAEA, Vienna **1**, 475.

Claridge, R. F. C. and Maddock, A. G., 1963a, Trans. Faraday Soc. **59**, 935.

Claridge, R. F. C. and Maddock, A. G., 1963b, Radiochim. Acta **1**, 80.

Claridge, R. F., Merz, E. and Riedel, H., 1965, Nukleonik **7**, 53.

Clark, T. J. and Moser, H. C., 1963, J. Inorg. Nucl. Chem. **17**, 210.

Cleary, R. E., Hamill, W. H. and Williams, R. R., 1952, J. Amer. Chem. Soc. **74**, 4675.

Cobble, J. W. and Boyd, G. E., 1952, J. Amer. Chem. Soc. **74**, 1282.

Cogneau, M. A., 1967, Thesis, University of Louvain.

Cogneau, M. A., Apers, D. J. and Capron, P. C., 1967, Radiochim. Acta. **8**, 143.

Cogneau, M. A., Apers, D. J. and Capron, P. C., 1968, Radiochim. Acta **10**, 170.

Cogneau, M. A., Ladrielle, T. G., Apers, D. J. and Capron, P. C., 1972, Radiochim. Acta **18**, 61.

Cohen, H. and Diethorn, W. S., 1965, Phys. Stat. Solidi **9**, 251.

Collins, C. H. and Collins, K. E., 1968, Nucl. Applic. **5**, 140.

Collins, C. H. and Collins, K. E., 1971, Nature (Phys. Sec.) **232**, 109.

Collins, C. H., Collins, K. E., Ghoos, Y. F. and Apers, D. J., 1965, Radiochim. Acta **4**, 211.

Collins, C. H., Collins, K. E. and Ackerhalt, R. E., 1971, J. Radioanal. Chem. **8**, 263.

Collins, C. H., Ackerhalt, R. E. and Collins, K. E., 1972b, Radiochim. Acta **17**, 73.

Collins, K. E., 1965, Symp. Chem. Effects of Nucl. Transformations, IAEA, Vienna **2**, 275.

Collins, K. E. and Harbottle, G., 1964, Radiochim. Acta. **3**, 21.

Collins, K. E. and Willard, J. E., 1962, J. Chem. Phys. **37**, 1908.

Collins, K. E., Collins, C. H., Yang, M. H., Ke, C. N., Lo, J. M. and Yeh, S.J., 1972a, J. Radioanal. Chem. **10**, 197.

Collins, K. E., Collins, C. H., Yang, M. H., Chuang, J. T., Wei, J. C. and Yeh, S. J., 1977, J. Inorg. Nucl. Chem. In press.

Colonomos, M. and Parker, W., 1969, Radiochim. Acta **12**, 163.

Conlin, R. T., Lockhart, S. H. and Gaspar, P. P., 1975, Chem. Comm. 825.

Constantinescu, M., Constantinescu, O., Pascaru, I. and Gird, E., 1966, Rev. Roum. Phys. **11**, 249.

Constantinescu, O., Pascaru, I. and Constantinescu, M., 1968, Rev. Roum. Phys. **13**, 607.

Cook, G. B., 1960, J. Inorg. Nucl. Chem. **14**, 301.

Coryell, C. D., 1952, Ann. Rev. Nucl. Sci. **2**, 305.

Coryell, C. D. and Sugarman, N., 1951, Radiochemical Studies: "The Fission Products", National Nucl. Energy Series IV – 9 (McGraw-Hill, NY).

Costea, T., 1961a, Acad. Rep. Populare Romine, Studii Cercetari Chim. **7**, 20.

Costea, T., 1961b, Acad. Rep. Populare Romine, Studii Cercetari Chim. **9**, 27.

Costea, T., 1969, Saclay Report, CEA-R-3667.

Costea, T. and Dema, I., 1961a, Acad. Rep. Populare Romine, Studii Cercetari Chim. **9**, 109.

Costea, T. and Dema, I., 1961b, Nature (Lond.) **189**, 478.

Costea, T. and Dema, I., 1962, J. Inorg. Nucl. Chem. **24**, 1021.

Costea, T. and Dema, I., 1963, Acad. Rep. Populare Romine, Studii Cercetari Fiz. **14**, 571.

Costea, T. and Mäntescu, C., 1966, J. Inorg. Nucl. Chem. **28**, 2777.

Costea, T. and Negoescu, C., 1965, Rev. Roum. Phys. **10**, 561.

Costea, T. and Negoescu, C., 1966, J. Inorg. Nucl. Chem. **28**, 323.

Costea, T. and Podeanu, G., 1967, J. Inorg. Nucl. Chem. **29**, 2102.

Costea, T. and Podeanu, G., 1968, Radiochim, Acta **10**, 53.
Costea, T., Dema, I. and Tipluica, A., 1961, Acad. Rep. Populare Romine, Studii, Cercetari Chim. **9**, 533.
Costea, T., Dema, I. and Mäntescu, C., 1963, Rev. Roum. Phys. **8**, 59.
Costea, T., Negoescu, I., Vasudev, P. and Wiles, D. R., 1966, Can J. Chem. **44**, 885.
Costea, T., Negoescu, I. and Podeanu, G., 1970, Radiochim. Acta **14**, 87.
Costea, T., Negoescu, I. and Podeanu, G., 1971a, Radiochim. Acta **16**, 86.
Costea, T., Negoescu, I. and Podeanu, G., 1971b, Radiochim, Acta, **16**, 86.
Crasemann, B. and Stephas, P., 1969, Nucl. Phys. **A134**, 641.
Croatto, U. and Giacomello, G., 1954, Acta 45th Congress of SIPS, Naples.
Croatto, U. and Maddock, A. G., 1949, J. Chem. Soc. Suppl. **1**, 351.
Croatto, U., Giacomello, G. and Maddock, A. G., 1951, Ricerca Sci. **21**, 1788.
Croatto, U., Giacomello, G. and Maddock, A. G., 1952, Ricerca Sci. **22**, 265.
Cruset, A., 1974, Radiochem. Radioanal. Lett. **16**, 309.
Cruset, A. and Friedt, J. M., 1971a, Phys. Stat. Solidi **44B**, 633.
Cruset, A. and Friedt, J. M., 1971b, Phys. Stat. Solidi **45B**, 189.
Cruset, A. and Friedt, J. M., 1972, Radiochem. Radioanal. Lett. **10**, 353.
Cruz-Vidal, B. A. and Gomberg, H. J., 1970, J. Phys. Chem. Solids **31**, 1273.
Cruz-Vidal, B. A., Diaz-Hernandez, F. and Gomberg, H. J., 1970, J. Phys. Chem. Solids **31**, 1281.
Cumming, J. B., Schwartzschild, A. Z., Sunyar, A. W. and Porile, N. T., 1960, Phys. Rev. **120**, 2128.
Cummiskey, C., Hamill, H. H. and Williams, R. R., 1961, J. Inorg. Nucl. Chem. **21**, 205.
Cunningham, J. and Heal, H. G., 1958, Trans. Faraday Soc. **54**, 1355.
Czjzek, G. and Berger, W. G., 1970, Phys. Rev. **B1**, 957.
Czjzek, G., Ford, J. L. C., Love, J. C., Obenshain, F. E. and Wegener, H. H. F., 1968, Phys. Rev. **174**, 331.
D'Agostino, O., 1935, Gazz. Chim. Ital. **65**, 1071.
Dahl, J. B. and Birklelund, O. R., 1962, Symposium Radioisotopes in the Physical Sciences and Industry, IAEA, Vienna **2**, 471.
Dancewicz, D. and Halpern, A., 1964, Nature (London) **203**, 145.
Daniel, S. H. and Tang, Y. N., 1969, J. Phys. Chem. **73**, 4378.
Datz, S. and Taylor, A.H., 1970, "Molecular Beams and Reaction Kinetics". (Academic Press, NY.)
Datz, S., Erginsoy, C., Leibfried, G. and Lutz, H. O., 1967, Ann. Rev. Nucl. Sci. **1**, 129.
Datz, S., Moak, C. D., Lutz, H. O., Northcliffe, L. C. and Bridewell, L. B., 1971, Atomic Data **2**, 273.
Daudel, R., 1941, Compt. rend. **213**, 479.
Daudel, R., 1942, Compt. rend. **214**, 547.
Davidenko, V. A. and Kucher, A. M., 1957, Soviet Atomic Energy **2**, 405.
Davies, T., 1948, J. Phys. Coll. Chem. **52**, 595.
DeBenedetti, S., De Barros, F. and Hoy, G. R., 1966, Ann. Rev. Nucl. Sci. **16**, 31.
Debuyst, R., 1971, Thesis, Univ. of Louvain.
Debuyst, R., Apers, D. J. and Capron, P. C., 1972a, J. Inorg. Nucl. Chem. **34**, 1541.
Debuyst, R., Ladrière, J., Apers, D. J. and Capron, P. C., 1972b, J. Inorg. Nucl. Chem. **34**, 2705.
Debuyst, R., Ladrière, J. and Apers, D. J., 1972c, J. Inorg. Nucl. Chem. **34**, 3607.
de Fonseca, A. J. R., Fuller, K., Latham, A. and Shaw, P. F. D., 1969, Radiochem. Radioanal. Lett. **2**, 69.
Defrance, J. E. and Apers, D. J., 1974a, Radiochim. Acta **21**, 121.
Defrance, J. E. and Apers, D. J., 1974b, Radiochim. Acta **21**, 125.

de Halter, M. E. and Cruset, A., 1974, Radiochem. Radioanal. Lett. **17**, 163.
dehmer, J. L. and Wahl, A. C., 1969, J. Inorg. Nucl. Chem. **31**, 562.
de Jong, I. G. and Wiles, D. R., 1968, Chem. Comm., 519.
de Jong, I. G. and Wiles, D. R. 1970, Can. J. Chem. **48**, 1614.
de Jong, I. G. and Wiles, D. R., 1972, Can. J. Chem. **50**, 961.
de Jong, I. G. and Wiles, D. R., 1973, Inorg. Chem. **12**, 2519.
de Jong, I. G. and Wiles, D. R., 1974, Radiochem. Radioanal. Lett. **17**, 343.
de Jong, I. G., and Wiles, D. R., 1976, J. Fisheries Res. Board Canada **33**, 14.
de Jong, I. G., Srinivasan, S. C. and Wiles, D. R., 1969, Can. J. Chem. **47**, 1327.
de Jong, I. G., Srinivasan, S. C. and Wiles, D. R., 1971, J. Organometal. Chem. **26**, 119.
de Jong, I. G., Omori, T. and Wiles, D. R., 1974, Chem. Comm., 189.
de Kimpe, A. G., Apers, D. J. and Capron, P. C., 1969, Radiochim. Acta **12**, 113.
de Kimpe, A. G., Apers, D. J. and Capron, P. C., 1971, Sixth International Hot Atom Chem. Symp., Brookhaven Nat. Lab., L. I. USA. p. 76.
Dema, I., 1971, Radiochem. Radioanal. Lett. **7**, 245.
Dema, I. and Zaitseva, N. G., 1965, Symp. Chem. Effects Nucl. Transformations, IAEA, Vienna **2**, 385.
Dema, I. and Zaitseva, N. G., 1966a, J. Inorg, Nucl. Chem. **28**, 2491.
Dema, I. and Zaitseva, N. G., 1966b, Radiochim. Acta **5**, 113.
Dema, I. and Zaitseva, N. G., 1966c, Radiochim. Acta **5**, 240.
Dema, I. and Zaitseva, N. G., 1966d, Rev. Roum. Phys. **11**, 573.
Dema, I. and Zaitseva, N. G., 1967, Radiochim. Acta **8**, 175.
Dema, I. and Zaitseva, N. G., 1968a, Rev. Roum. Phys. **13**, 531.
Dema, I. and Zaitseva, N. G., 1968b, Rev. Roum. Phys. **13**, 857.
Dema, I. and Zaitseva, N. G., 1969a, Rev. Roum. Phys. **14**, 367.
Dema, I. and Zaitseva, N. G., 1969b, Rev. Roum. Phys. **14**, 1283.
Dema, I. and Zaitseva, N. G., 1969c, J. Inorg. Nucl. Chem. **31**, 3039.
Dema, I. and Zaitseva, N. G., 1969d, J. Inorg. Nucl. Chem. **31**, 2311.
Dema, I. and Zaitseva, N. G., 1971. Radiochem. Radioanal. Lett. **8**, 325.
Dema, I. and Zaitseva, N. G., 1972, Rev. Roum. Phys. **17**, 443.
Dema, I., Zaitseva, N. G., Salatskii, V. I. and Kobzev, A. P., 1968, Report IFA-RC-5.
de Maine, M. M., Maddock, A. G. and Tāugböl, K., 1957, Faraday Soc. Discussions **23**, 211.
Denschlag, H. O. and Gordus, A. A., 1967, Analyt. Chem. **226**, 62.
Denschlag, H. O., Henzel, N. and Hermann, G., 1963, Radiochim. Acta. **1**, 173.
de Oliveira, J., 1970, Thesis, Univ. of Grenoble.
DeVault, D. C. and Libby, W. F., 1939, Phys. Rev. **55**, 322.
DeVault, D. C. and Libby, W. F., 1941, J. Amer. Chem. Soc. **63**, 3216.
de Wieclawik, W., 1968, Comp. rend **266B**, 577.
de Wieclawik, W., 1969a, Comp. rend. **268B**, 1268.
de Wieclawik, W., 1969b, Thesis, Paris.
de Wieclawik, P. W. and Perrin, N., 1969, J. Physique **30**, 877.
Dexter, D. L., 1960, Phys. Rev. **118**, 934.
Dharmawardena, K. G. and Maddock, A. G., 1970, Chem. Comm. 549.
Diefallah, E. H. M., 1968, Thesis, Illinois.
Diefallah, E. H. M. and Kay, J. G., 1972, Indian J. Chem. **10**, 1187.
Diefallah, E. H. M. and Kay, J. G., 1973, Bull. Chem. Soc. Japan **46**, 3318.
Dienes, G. J. and Vineyard, G. H., 1957, "Radiation Effects in Solids" (Interscience Publishers, New York) p. 35.
Diestler, D. J., Truhlar, D. G. and Kuppermann, A., 1972, Chem. Phys. Lett. **13**, 1.
Diethorn, W. S., 1965, J. Appl. Rad. Isotopes **16**, 705.

Dimotakis, P. N., 1968, J. Inorg. Nucl. Chem. **30**, 29.
Dimotakis, P. N. and Kontis, S., 1963, Radiochim. Acta **2**, 85.
Dimotakis, P. N. and Maddock, A. G., 1961, Symp. Chem. Effects of Nucl. Transformations, IAEA, Prague **1**, 365.
Dimotakis, P. N. and Maddock, A. G., 1964, J. Inorg. Nucl. Chem. **26**, 1503.
Dimotakis, P. N. and Papadopoulos, B. P., 1970, J. Inorg. Nucl. Chem. **32**, 1071.
Dimotakis, P. N. and Papadopoulos, B. P., 1972, Chimica Chronika Ellas **1**, 48.
Dimotakis, P. N. and Stamouli, M. I., 1963, J. Inorg. Nucl. Chem. **25**, 473.
Dimotakis, P. N. and Stamouli, M. I., 1964, J. Inorg. Nucl. Chem. **26**, 2045.
Dimotakis, P. N. and Stamouli, M. I., 1965, Symp. Chem. Effects of Nucl. Transformations, IAEA, Vienna **2**, 237.
Dimotakis, P. N. and Stamouli, M. I., 1967, Z. Phys. Chem. (Frankfort) **55**, 197.
Dimotakis, P. N. and Stamouli, M. I., 1968, J. Inorg. Nucl. Chem. **30**, 23.
Dimotakis, P. N. and Yavas, G., 1970, J. Inorg. Nucl. Chem. **32**, 1065.
Dimotakis, P. N., Maddock, A. G. and Vassos, B., 1967, Radiochim. Acta **8**, 38.
Dobici, F. and Salvetti, F., 1964, J. Inorg. Nucl. Chem. **26**, 911.
Dodson, R. W., Goldblatt, M. and Sullivan, J. H., 1946, USAEC Report, MDDC-344.
Domeij, B., Brown, F., Davies, J. A. and McCargo, M., 1964, Can. J. Phys. **42**, 1624.
Donovan, P. F., Harvey, B. G. and Wade, W. H., 1960, Phys. Rev. **119**, 218, 275.
Dostrovsky, I., Fraenkel, F. and Hudis, J., 1961, Phys. Rev. **123**, 1452.
Dubrin, J., Mackay, C., Pandow, M. and Wolfgang, R., 1964, J. Inorg. Nucl. Chem. **26**, 2113.
Dubrin, J., Mackay, C. and Wolfgang, R., 1966, J. Chem. Phys. **44**, 2208.
Duffield, R. B. and Calvin, M., 1946, J. Amer. Chem. Soc. **68**, 1129.
Duncan, J. F. and Thomas, F. G., 1967, J. Inorg. Nucl. Chem. **29**, 869.
Dunlap, B. D., Kalvius, C. M., Ruby, S. L., Brodsky, M. B. and Cohen, D., 1968, Phys. Rev. **171**, 316.
Dupetit, G. A., 1967, Radiochim. Acta **7**, 167.
Dupetit, G. A. and Aten, A. H. W., 1967, Radiochim. Acta **7**, 165.
Duplatre, G., 1969, Thesis, Grenoble.
Duplatre, G. and Herment, J., 1976, Radiochem. Radioanal. Lett. **24**, 137.
Duplatre, G., Vargas, J. I. and Cogneau, M. A., 1972, J. Inorg. Nucl. Chem. **34**, 3021.
Durup, J. and Platzman, R. L., 1961, Disc. Faraday. Soc. **31**, 156.
Ebihara, H., 1962, Bunseki Kagaku **11**, 341.
Ebihara, H., 1965, Radiochim. Acta **4**, 167.
Ebihara, H., 1966, Radiochim. Acta **6**, 120.
Ebihara, H. and Yoshihara, K., 1960, Bull. Soc. Chem. Japan **33**, 116.
Ebihara, H. and Yoshihara, K., 1961, Bunseki Kagaku **10**, 48.
Edge, R. D., 1956, Austral., J. Phys. **9**, 429.
Edwards, R. R. and Coryell, C., 1948, USAEC Report, AECU-50.
Edwards, R. R. and Davies, T. H., 1948, Nucleonics **2**, 44.
Edwards, R. R., May, J. M. and Overman, R. F., 1953, J. Chem. Phys. **21**, 155.
Elton, L. R. B., 1968, "Nuclear Sizes", Oxford Univ. Press, Oxford.
Emery, J. F., 1965, J. Inorg. Nucl. Chem. **29**, 903.
Endo, K. and Sakanou, M., 1972, Radiochem. Radioanal. Lett. **9**, 255.
Ensling, J., 1970, Thesis, Darmstadt.
Ensling, J., Fitzsimmons, B. W., Gütlich, P. and Hasselbach, K. M., 1970, Angew. Chem. **82**, 638.
Ensling, J., Gütlich, P., Hasselbach, K. M. and Fitzsimmons, B. W., 1976, Chem. Phys. Lett. **42**, 232.
Erber, J., Rieder, W. and Broda, E., 1950, Nature (Lond.) **165**, 810.
Erdal, B. R., Wahl, A. C. and Dropesky, B. J., 1969, J. Inorg. Nucl. Chem. **31**, 3005.

Erginsoy, C., 1964, "The Interaction of Radiation with Solids", (Interscience Publishers, New York).
Erginsoy, C., Vineyard, G. H. and Englert, A., 1964, Phys. Rev. **133**, A, 595.
Erman, P., Sigfridsson, B., Carlson, T. A. and Fransson, K., 1968, Nucl. Phys. **A112**, 117.
Estrup, P. J. and Wolfgang, R., 1960, J. Amer. Chem. Soc. **82**, 2665.
Facetti, J. F., 1965, Radiochim. Acta **4**, 164.
Facetti, J. F., de Santiago, M. V. and Wheeler, O. H., 1969, Radiochim. Acta **12**, 82.
Fantola-Lazzarini, A. L. and Lazzarini, E., 1972, Energia Nucleare **19**, 407.
Fantola-Lazzarini, A. L. and Lazzarini, E., 1973a, J. Inorg. Nucl. Chem. **35**, 681.
Fantola-Lazzarini, A. L. and Lazzarini, E., 1973b, J. Inorg. Nucl. Chem. **35**, 2653.
Farkas, A. and Farkas, L., 1935, Proc. Roy. Soc. (London) **A152**, 124.
Feinberg, E. L., 1941, J. Phys. (USSR) **41**, 423.
Feinberg, E. L., 1965, J. Nucl. Phys. (USSR) **1**, 438.
Feng, D. F., Grant, E. R. and Root, J. W., 1976, J. Chem. Phys. **64**, 3450.
Fenger, J., 1964, Risø Report, 95.
Fenger, J., 1968, Radiochim. Acta **10**, 138.
Fenger, J., 1969, Radiochim. Acta **12**, 186.
Fenger, J., 1971, Radiochim. Acta **16**, 42.
Fenger, J., 1972, Radiochim. Acta **17**, 170.
Fenger, J., 1974, Risø Report, 311.
Fenger, J. and Frees, L. A., 1974, J. Chem. Soc. Dalton, 2309.
Fenger, J. and Nielsen, S. O., 1965, Symp. Chem. Effects of Nucl. Transformations, IAEA, Vienna **2**, 93.
Fenger, J. and Olsen, J., 1974, J. Chem. Soc. Dalton, 319.
Fenger, J. and Pagsberg, P. B., 1973, J. Inorg. Nucl. Chem. **35**, 31.
Fenger, J. and Pagsberg, P. B., 1975, J. Inorg. Nucl. Chem. **37**, 850.
Fenger, J., Siekierska, K. E. and Maddock, A. G., 1970a, J. Chem. Soc. (A) 1456.
Fenger, J., Maddock, A. G. and Siekierska, K. E. 1970b, J. Chem. Soc. (A) 3255.
Fenger, J., Siekierska, K. E. and Olsen, J., 1973, J. Chem. Soc. Dalton, 563.
Fermi, E., Amaldi, E. and D'Agostino, O., 1934, Proc. Roy. Soc. **A146**, 483.
Fernandez Valverde, S., Duplatre, G. and Maddock, A. G., 1978, J. Inorg. Nucl. Chem. **40**, 900.
Ferradini, C., Carlier, R., Genet, M. and Pucheault, J., 1969, Radiochim, Acta **12**, 1.
Fields, P. R., Stein, L. and Zirin, M. H., 1962, J. Amer. Chem. Soc. **84**, 4164.
Fischbeck, K. and Spingler, H. S., 1938, Z. anorg. allgem. Chem. **235**, 183.
Fischbeck, H. J., Wagner, F., Porter, F. T. and Freedman, M. S., 1971, Phys. Rev. **C3**, 265.
Fishman, L. M. and Harbottle, G., 1954, J. Chem. Phys. **22**, 1088.
Fiskell, J. G. A., 1951, Science **113**, 244.
Fiskell, J. G. A., Delong, W. A. and Oliver, W. F., 1952, Can. J. Chem. **30**, 9.
Fleisch, J. and Gütlich, P., 1976, Chem. Phys. Lett. **42**, 237.
Fleisch, J., Gütlich, P., Mohs, E. and Wolf, G. K., 1975, Proc. Int. Conf. Mössbauer Spectroscopy, Cracow **1**, 217.
Fletcher, R. C. and Brown, W. L., 1953, Phys. Rev. **92**, 585.
Flohr, K. and Appelman, E. H., 1968, J. Amer. Chem. Soc. **90**, 3584.
Flood, H. and Muan, A., 1950, Acta. Chem. Scand. **4**, 364.
Forsythe, W. E., 1954, Publ. No. 4169, Smithsonian Institution, Washington, DC.
Fox, M. and Libby, W., 1952, J. Chem. Phys. **20**, 487.
Franck, J and Rabinowitsch, 1934, Trans. Faraday Soc. **30**, 120.
Frediani, S. and Lo Moro, A., 1969, Energia Nucleare **16**, 520.
Freedman, M. S., 1974, Ann. Rev. Nucl. Sci. **24**, 209.
Freedman, M. S., 1976, in "Photoionisation and Other Probes of Many Electron Interactions", ed. Wuilleumier, F. J., Plenum Press, NY.

Freeman, J. H., Kasrai, M. and Maddock, A. G., 1967, Chem. Comm. **19**, 979.
Frey, H. M. and Walsh, R., 1969, Chem. Rev. **69**, 103.
Fried, M. and Mackenzie, A. J., 1950, Science, 492.
Friedlander, G., Kennedy, J. W. and Miller, J. M., 1964, "Nuclear and Radiochemistry", (John Wiley and Sons, Inc., NY) 2nd Edition.
Friedt, J. M., 1970a, J. Inorg. Nucl. Chem. **32**, 431.
Friedt, J. M., 1970b, J. Inorg. Nucl. Chem. **32**, 2123.
Friedt, J. M. and Adloff, J. P., 1967, Compt. rend. **264**, 1356.
Friedt, J. M. and Adloff, J. P., 1968, Comp. rend. **266**, 1733.
Friedt, J. M. and Adloff, J. P., 1969a, Compt. rend. **268**, 1342.
Friedt, J. M. and Adloff, J. P., 1969b, Inorg. Nucl. Chem. Lett. **5**, 163.
Friedt, J. M. and Asch, L., 1969, Radiochim. Acta **12**, 208.
Friedt, J. M. and Danon, J., 1970, Radiochem. Radioanal. Lett. **3**, 147.
Friedt, J. M. and Danon, J., 1972, Radiochim. Acta **17**, 173.
Friedt, J. M. and Llabador, Y., 1972, Radiochem. Radioanal. Lett. **9**, 237.
Friedt, J. M. and Vogl, W., 1974, Phys. Stat. Solidi **A24**, 265.
Friedt, J. M., Baggio-Saitovitch, E. and Danon, J., 1970, Chem. Phys. Lett. **7**, 603.
Friedt, J. M., Poinsot, R. and Sanchez, J. P., 1971, Radiochem. Radioanal. Lett. **7**, 193.
Friedt, J. M., Shenoy, G. K., Abstreiter, G. and Poinsot, R., 1973, J. Chem. Phys. **59**, 3831.
Gainer, I., 1972, Radiochem. Radioanal. Lett. **10**, 113.
Găinar, E. and Găiner, I., 1962, Acad. Rep. Populare Romine, Studii Cercetari Fiz. **13**, 309.
Găinar, I. and Ponta, A., 1968, Rev. Roum. Chim. **13**, 401.
Găinar, I. and Ponta, A., 1971a, J. Inorg. Nucl. Chem. **33**, 2291.
Găinar, I. and Ponta, A., 1971b, Radiochem. Radioanal. Lett. **7**, 79.
Gal, J., Hadari, Z., Yaniv, E., Bauminger, E. R. and Ofer, S., 1970, J. Inorg. Nucl. Chem. **32**, 2509.
Gardner, E. R., Gravenor, M. E., Harding, R. D., Raynor, J. B. and Autchakit, R., 1970a, Radiochim. Acta **13**, 100.
Gardner, E. R., Greethong, S. and Raynor, J. B., 1970b. Radiochim. Acta. **14**, 23.
Gardner, E. R., Wilson, M. E., Harding, R. D. and Raynor, J. B., 1972, Radiochim. Acta **17**, 41.
Gaspar, P. P. and Frost, J. J., 1971, Sixth Hot Atom Chemistry Symposium, Brookhaven National Laboratory, L. I., USA.
Gaspar, P. P. and Frost, J. J., 1973, J. Amer. Chem. Soc. **95**, 6567.
Gaspar, P. P. and Herold, B. J., 1971, in "Carbene Chemistry", ed. Kirmse, W., (Academic Press, NY).
Gaspar, P. P. and Markutch, P., 1970, Chem. Comm., 1331.
Gaspar, P. P., Pate, B. D. and Eckelman, W., 1966, J. Amer. Chem. Soc. **88**, 3878.
Gaspar, P. P., Bock, S. A. and Eckelman, W., 1968a, J. Amer. Chem. Soc. **90**, 6914.
Gaspar, P. P., Bock, S. A. and Levy, C. A., 1968b, Chem. Comm., 1317.
Gaspar, P. P., Levy, C. A., Frost, J. J. and Bock, S. A., 1969, J. Amer. Chem. Soc. **91**, 1573.
Gaspar, P. P., Choo, K. Y., Lam, E. Y. Y. and Wolf, A. P., 1971, Chem. Comm., 1012.
Gaspar, P. P., Markutch, P. and Holten, J. D., 1972, J. Phys. Chem. **76**, 1352.
Gaspar, P. P., Hwang, R. J. and Eckelman, W. C., 1974, Chem. Comm., 242.
Gavrilov, V. V., Isupov, V. K. and Kirin, I. S., 1975a, Soviet Radiochem. **17**, 130.
Gavrilov, V. V., Isupov, V. K. and Kirin, I. S., 1975b, Soviet Radiochem. **17**, 259.
Geissler, P. R. and Willard, J. E., 1963, J. Phys. Chem. **67**, 1675.
Genet, M., 1969, Radiochim. Acta **12**, 193.
Genet, M. and Ferradini, C., 1964, Compt. rend. **258**, 4525.
Genet, M. and Ferradini, C., 1969a, Radiochim. Acta **11**, 19.
Genet, M. and Ferradini, C., 1969b, Radiochim. Acta **11**, 25.
Gennaro, G. P. and Collins, K. E., 1970, J. Phys. Chem. **74**, 3094.

Gennaro, G. P. and Tang, Y. N., 1973, J. Inorg. Nucl. Chem. **35**, 3087.
Gennaro, G. P. and Tang, Y. N., 1974, J. Inorg. Nucl. Chem. **36**, 259.
Gennaro, G. P., Su, Y. Y., Zeck, O. F., Daniel, S. H. and Tang, Y. N., 1973, Chem. Comm., 637. See also Gaspar et al., 1974.
Gersberg, R., Krohn, K., Peak, N. and Goldman, C. R., 1976, Science **192**, 1229.
Gerth, G., Luchner, K. and Micklitz, H., 1972, Phys. Stat. Solidi **B53**. 593.
Getoff, N., 1963a, Nature (Lond.) **199**, 593.
Getoff, N., 1963b, Anz. Oesterr. Akad. Wiss. Math. Naturw. Kl., 211.
Getoff, N. and Maddock, A. G., 1963, Radiochim. Acta **2**, 90.
Gibson, J. B., Goland, A. N., Milgram, M. and Vineyard, G. H., 1960, Phys. Rev. **120**, 1229.
Gilbert, A., Roggen, F. and Rossel, J., 1944, Portugal. Phys. **1**, 43.
Gili Trujillo, P. and Canwell Morcuende, R., 1968, Ensayos Invest. **3**, 17.
Giulianelli, J. L., 1969, Thesis, Wisconsin.
Giullianelli, J. L. and Willard, J. E., 1974, J. Phys. Chem. **78**, 372.
Glass, J. C. and Klee, M., 1976, Chem. Phys. Lett. **41**, 321.
Glasstone, S., Laidler, K. J. and Eyring, H., 1941, "The Theory of Rate Processes", (McGraw-Hill Book Co., NY).
Glentworth, P. and Betts, R. H., 1961, Can. J. Chem. **39**, 1049.
Glentworth, P. and Nath, A., 1975, in "Specialist Periodical Report, Radiochemistry". ed. Newton, G.W.A., Chem. Soc. (Lond.) **2**, 74.
Glentworth, P. and Wiseall, B., 1965, Symp. Chem. Effects of Nucl. Transformations, IAEA, Vienna **2**, 483.
Glentworth, P. and Wright, C. L., 1969, J. Inorg. Nucl. Chem. **31**, 1263.
Glentworth, P., Nichols, A. L., Large, N. R. and Bullock, R. J., 1971, Chem. Comm., 206.
Glentworth, P., Nichols, A. L., Large, N. R. and Bullock, R. J., 1973, J. Chem. Soc. (Dalton), 2364.
Goldanskii, V. I. and Herber, R. H., 1968, "Chemical Applications of Mössbauer Spectroscopy", (Academic Press, New York).
Goldsmith, G. J. and Bleuler, E., 1950, J. Phys. Chem. **54**, 717.
Gordon, B. M., 1967, J. Inorg. Nucl. Chem. **29**, 287.
Gorla, P. R. and Lazzarini, E., 1967, Energia Nucleare (Milan) **14**, 537.
Gotte, H., 1946, Z. Naturforsch **1**, 377.
Gotte, H., 1948, Angew. Chem. **60A**, 19.
Govaerts, J. and Jordan, P., 1950, Experientia **6**, 329.
Gracheva, L. M., Nefedov, V. D. and Grachev, S. A., 1967, Radiokhimiya **9**, 738.
Grant, E. R. and Root, J. W., 1974, Chem. Phys. Lett. **27**, 484.
Grant, E. R. and Root, J. W., 1975, J. Chem. Phys. **63**, 2970.
Grant, E. R. and Root, J. W., 1976, J. Chem. Phys. **64**, 417.
Gratot, T. and J. Beydon, 1970, Saclay Report, CEA-R-3933.
Green, A. E. S., 1957, Phys. Rev. **107**, 1646.
Green, J. A. and Maddock, A. G., 1949, Nature (Lond.) **164**, 788.
Green, J. A. and Maddock, A. G., 1951, J. Chim. Phys. **48**, 14.
Green, J. H., Harbottle, G. and Maddock, A. G., 1953, Trans. Faraday Soc. **49**, 1413.
Grillet, S., 1966, Thesis, Paris.
Grimm, R., Gütlich, P., Link, R., Kankeleit, E., Reichenbacher, W. and Walcher, D., 1975, Proc. Int. Conf. Mössbauer Spec. Cracrow. **1**, 501.
Grodzins, L., 1969, Proc. Roy. Soc. (London) **A311**, 79.
Groening, H. R. and Harbottle, G., 1970, Radiochim. Acta **14**, 109.
Grossmann, G., 1968, Isotopenpraxis **4**, 268.
Grossmann, G., 1969a, Isotopenpraxis **5**, 262.
Grossmann, G., 1969b, Isotopenpraxis **5**, 283.

Grossmann, G., 1969c, Isotopenpraxis **5**, 203.
Grossmann, G., 1969d, Isotopenpraxis **5**, 370.
Grossmann, G., 1970, Radiochim. Acta **13**, 31.
Grossmann, G. and Krabbes, G., 1970, Isotopenpraxis **6**, 49.
Grossmann, G. and Zeuner, A., 1968, Isotopenpraxis **4**, 215.
Grossmann, G., Krabbes, G. and Tschernko, G., 1968a, Isotopenpraxis **4**, 307.
Grossmann, G., Mühl, P., Gross-Ruyken, H. and Knofel, S., 1968b, Isotopenpraxis **4**, 23.
Gunter, K. C. K., 1968, Thesis, Univ. of California, UCRL 18313.
Gunter, K., Asaro, F. and Helmholz, A. C., 1966, Phys. Rev., Lett. **16**, 362.
Gusev, Y. K., Kirin, I. S. and Isupov, V. K., 1967a, Soviet Radiochem. **9**, 697.
Gusev, Y. K., Kirin, I. S. and Isupov, V. K., 1967b, Radiokhimiya **9**, 736.
Gusev, Y. K., Isupov, V. K. and Kirin, I. S., 1967c, High Energy Chem. **1**, 531.
Gütlich, P. and Harbottle, G., 1966, Radiochim. Acta **5**, 70.
Gütlich, P. and Harbottle, G., 1967, Radiochim. Acta **8**, 30.
Gütlich, P., Odar, S. and Walcher, D., 1970, Z. Naturforsch. **25b**, 1183.
Gütlich, P. Fröhlich, K. and Odar, 1971a, J. Inorg. Nucl. Chem. **33**, 307.
Gütlich, P., Fröhlich, K. and Odar, S., 1971b, J. Inorg. Nucl. Chem. **33**, 621.
Haar, W. and Richter, F. W., 1970, Z. Physik **231**, 1.
Hackskaylo, M., Otterson, D. and Schwed, P., 1953, J. Chem. Phys. **21**, 1434.
Hager, R. S. and Seltzer, E. C., 1968, Nuclear Data **A4**, 1.
Hahn, R. L., 1963, J. Chem. Phys. **39**, 3482.
Hahn, R. L., 1964, J. Chem. Phys. **41**, 1986.
Haissinsky, M. and Cottin, M., 1948, J. Chem. Phys. **45**, 270.
Haldar, B. C., 1954, J. Amer. Chem. Soc. **76**, 4229.
Hall, D., 1958a, Proc. Australian At. En. Symposium, 580.
Hall, D., 1958b, J. Inorg. Nucl. Chem. **6**, 3.
Hall, D. and Walton, G. N., 1958, J. Inorg. Nucl. Chem. **6**, 288.
Hall, D. and Walton, G. N., 1959, J. Inorg. Nucl. Chem. **10**, 215.
Hall, D. and Walton, G. N., 1961, J. Inorg. Nucl. Chem. **19**, 16.
Hall, R. and Sutin, N., 1956, J. Inorg. Nucl. Chem. **2**, 184.
Halmann, M., 1964, Chem. Revs. **64**, 689.
Halmann, M. and Kugel, L., 1965, J. Chem. Soc., 4025.
Halpern, A., 1959, Russ. J. Inorg. Chem. **4**, 545, 1205.
Halpern, A., 1963, J. Inorg. Nucl. Chem. **25**, 619.
Halpern, A., 1971, Radiochim. Acta **15**, 83.
Halpern, A. and Dancewicz, D., 1969, Radiochim. Acta **11**, 31.
Halpern, A. and Sawlewicz., K., 1968, Nukleonika **13**, 921.
Halpern, A. and Sochacka, R., 1961, J. Inorg. Nucl. Chem. **23**, 7.
Halpern, A. and Stöcklin, G., 1974, Radiation Res. **58**, 329.
Halpern, A. and Stöcklin, G., 1977, Rad. Environm. Biophys. **14**, 167, 257.
Halpern, A., Siekierska, K. E. and Siuda, A., 1964, Radiochim. Acta **3**, 40.
Hannaford, P. and Wignall, J. W. G., 1969, Phys. Stat. Solidi **35**, 809.
Hannaford, P., Howard, C. J. and Wignall, J. W. G., 1965, Phys. Lett. **19**, 257.
Harbottle, G., 1954, J. Chem. Phys. **22**, 1083.
Harbottle, G., 1960, J. Amer. Chem. Soc. **82**, 805.
Harbottle, G., 1961, Symp. Chem. Effects of Nucl. Transformations, IAEA, Prague **1**, 301.
Harbottle, G., 1965, Ann. Rev. Nuclear Sci. **15**, 89.
Harbottle, G., 1971, Second Interamerican Conference on Radiochemistry, Mexico City, Organization of American States (Washington), 160.
Harbottle, G., 1975, Hot Atom Status Report, IAEA, Vienna, p. 19.
Harbottle G. and Hillman, M., 1971, "Radioisotope Production and Quality Control", IAEA, Vienna, Tech. Rep. 128, p. 617.

Harbottle, G. and Maddock, A. G., 1958, J. Inorg. Nucl. Chem. **5**, 249.
Harbottle, G. and Sutin, N., 1958, J. Phys. Chem. **62**, 1344.
Harbottle, G. and Sutin, N., 1959, "Advances in Inorganic Chemistry and Radiochemistry", (Academic Press, Inc., New York) **1**, 267.
Harbottle, G. and Zahn, U., 1965, Symp. Chem. Effects of Nucl. Transformations, IAEA, Vienna **2**, 133.
Harbottle, G. and Zahn, U., 1967, Radiochim. Acta **8**, 114.
Harrison, D. E., Leeds, R. W. and Gay, W. L., 1963, J. Appl. Phys. **36**, 3154.
Harvey, B. G., 1962, "Nuclear Physics and Chemistry", Prentice-Hall, NJ, USA.
Haseltine, M. W., 1967, Thesis, Kansas State University.
Haseltine, M. W. and Moser, H. C., 1967, J. Amer. Chem. Soc. **89**, 2497.
Hashimoto, T., Tamai, T., Matshuhita, R. and Kiso, Y., 1970, J. Nucl. Sci. Techn. **7**, 32.
Hassan, N. and Heitz, C., 1975, J. Inorg. Nucl. Chem. **37**, 395.
Hauser, U., Neuwirth, W., Pietsch, W. and Richter, K., 1974, Radiochem. Radioanal. Lett. **18**, 301.
Hawk, J. G. and Moir, J., 1974, Chem. Comm., 490.
Hazony, Y. and Herber, R. H., 1969, J. Inorg. Nucl. Chem. **31**, 321.
Hazony, Y. and Herber, R. H., 1971, J. Inorg. Nucl. Chem. **33**, 961.
Hegedus, L. S. and Haim, A., 1967, Inorg. Chem. **6**, 664.
Heine, K., 1961, Thesis, University of Cologne.
Heine, K. and Herr, W., 1960, Z. Elektrochem. **64**, 1037.
Heine, K. and Herr, W., 1961, Symp. Chem. Effects of Nucl. Transformations, IAEA, Prague **1**, 343.
Heitz, C., 1967a, Bull. Soc. Chim. Fr., 2439.
Heitz, C., 1967b, Bull. Soc. Chim. Fr., 2442.
Heitz, C., 1967c, Thesis, University of Strasbourg.
Heitz, C. and Adloff, J. P., 1964, Bull. Soc. Chim. Fr., 2917.
Heitz, C. and Cassou, M., 1969, Radiochim. Acta. **12**, 203.
Heitz, C. and Cassou, M., 1970, Radiochim. Acta **13**, 217.
Heitz, C. and Cassou, M., 1971, Inorg. Nucl. Chem. Lett. **7**, 47.
Heitz, C. and Ruffenach, J. C., 1971, Radiochem. Radioanal. Lett. **7**, 319.
Henglein, V. A., Drawe, H. and Perner, D., 1963, Radiochem. Acta **2**, 19.
Hennig, G., Lees, R. and Matheson, E., 1953, J. Chem. Phys. **21**, 664.
Henrich, E. and Wolf, G. K., 1969, German Report, K.F.K., 1067.
Henrich, E., Wolf, G. K., Ianovici, E. and Milman, M., 1973, German Report Heidelberg, AED-Conf-73-402-017 (C.A. **83**, 67120 W).
Henry, R. and Herczeg, C., 1958, UNESCO Conf. Paris **2**, 168.
Henry, R., Aubertin, C. and La Guéronnière, E., 1957, J. Phys. Radium **18**, 320.
Henry, R., Beydon, J. and Bardy, A., 1961, Symp. Chem. Effects of Nucl. Transformations, IAEA, Prague **2**, 291.
Herber, R. H., 1961, Symp. Chem. Effects of Nucl. Transformations, IAEA, Prague **2**, 201.
Herr, W., 1948, Z. Naturforsch. **3a**, 645.
Herr, W., 1952a, Z. Elektrochem. **56**, 911.
Herr, W., 1952b, Z. Naturforsch **7b**, 55.
Herr, W., 1952c, Z. Naturforsch **7b**, 201.
Herr, W., 1952d, Z. Naturforsch. **7a**, 819.
Herr, W., 1954a, Z. Naturforsch **9a**, 180.
Herr, W., 1954b, Angew. Chem. **66**, 340.
Herr, W. and Dreyer, R., 1957, Z. anorg. allgem. Chem. **293**, 1.
Herr, W. and Gotte, H., 1950, Z. Naturforsch **5a**, 629.
Herr, W. and Heine, K., 1960, Z. Naturforsch. **15a**, 323.

Herr, W. and Schmidt, G. B., 1962, Z. Naturforsch **17a**, 309.
Herr, W., Heine, K. and Schmidt, G. B., 1962, Z. Naturforsch **17a**, 590.
Heyder, J. and Kaul, A., 1971, Biophysik. **7**, 152.
Hieber, W. and Heusinger, H., 1957, J. Inorg. Nucl. Chem. **4**, 176.
Hillman, M. and Weiss, A. J., 1966a, Proc. of the 7th Japan Radioisotope Conference, Japan Atomic Energy Research Institute, Tokyo, 217.
Hillman, M. and Weiss, A. J., 1966b, Brookhaven Rep. BNL-10208.
Hillman, M. and Weiss, A. J., 1971, Radiochim. Acta **15**, 79.
Hillman, M., Kim, C., Shikata, E. and Weiss, A. J., 1968, Radiochim. Acta **9**, 212.
Hillman, M., Weiss, A. J. and Hahne, R., 1969, Radiochim. Acta **12**, 200.
Hillman, M., Nagy, A. and Weiss, A.J., 1973, Radiochem, Acta **19**, 9.
Hillman, M., Weiss, A. J., Ebihara, H. and Williams, K., 1975, J. Inorg. Nucl. Chem. **37**, 403.
Hinman, G. W., Hoy, G. R. and Lees, J. K., 1962, Bull. Amer. Phys. Soc. Ser II **7**, 344.
Hirschfelder, J., Eyring, H. and Topley, F., 1936, J. Chem. Phys. **4**, 170.
Hobbs, J. R. and Owens, C. W., 1966a, USAEC Report, USAMRA TR-66-05.
Hobbs, J. R. and Owens, C. W., 1966b, USAEC Report, AD-63307.
Hoff, W. F. and Rowland, F. S., 1958, J. Inorg. Nucl. Chem. **5**, 164.
Hoffman, D. C. and Martin, D. S., 1952, J. Phys. Chem. **56**, 1097.
Hoffmann, P., Bächmann, K, Bögl, W., Klenk, H. and Lieser, K. H., 1971a, Radiochem. Acta **16**, 172.
Hoffmann, P., Bächmann, K., Klenk, H. and Lieser, K. H., 1971b, Inorg. Nucl. Chem. Lett. **7**, 577.
Hoffmann, P., Bächmann, K., Klenk, H., Trautmann, W. and Lieser, K. H., 1973, Z. Anal. Chem. **267**, 277.
Hoi, B., Caussé, R., Daudel, P., Flon, M., Hertzog, C., Hoan, N. and Lacassagne, A., 1949, Compt. rend. **228**, 868.
Holmes, O. G. and McCallum, K. L., 1950, J. Amer. Chem. Soc. **72**, 5319.
Holmes, R. R., 1970, Trans. Nucl. Sci., NS-17, No. 6, 137.
Hong, K. Y., Hong, J. H. and Becker, R. S., 1974, Science **184**. 984.
Howard, J. M., Seykora, E. J. and Waltner, A. W., 1971, Phys. Rev. **A4**, 1740.
Howard, R. E. and Smoluchowski, R., 1959, Phys. Rev. **116**, 314.
Hoy, G. R. and Winterstein, P. P., 1972, Phys. Rev. Lett. **28**, 877.
Hsiung, C. and Gordus, A. A., 1962, J. Chem. Phys. **36**, 947.
Hsiung, C. and Gordus, A. A., 1965, Symp. Chem. Effects of Nucl. Transformations, IAEA, Vienna **2**, 461.
Hsiung, G., Hsiung, H. and Gordus, A. A., 1961, J. Chem. Phys. **34**, 535.
Hudis, J., 1960, Nacional Academy of Sciences Publication, No. NAS-NS 3019.
Hull, D. R. and Owens, C. W., 1975, J. Inorg. Nucl. Chem. **37**, 403.
Husain, D. and Kirsch, L., 1971, Trans. Faraday Soc. **64**, 2025.
Hwang, R. J., 1976, Thesis, Washington University.
IAEA, 1961, Chemical Effects of Nuclear Transformations, Proceedings of a symposium, Prague, 24-27 October, 1960. International Atomic Energy Agency, Vienna.
IAEA, 1965, Chemical Effects of Nuclear Transformations, Proceedings of a symposium, Vienna, 7-11 December 1964. International Atomic Energy Agency, Vienna.
IAEA, 1968, Proceedings of a symposium on Biological Effects of Transmutation and Decay of Incorporated Radioisotopes.
IAEA, 1975, Hot Atom Chemistry Status Report, Vienna.
Ianovici, E. and Taube, M., 1975, J. Inorg. Nucl. Chem. **37**, 2561.
Ianovici, E. and Zaitseva, N. G., 1969a, J. Inorg. Nucl. Chem. **31**, 2669.
Ianovici, E. and Zaitseva, N. G., 1969b, J. Inorg. Nucl. Chem. **31**, 3309.
Ianovici, E. and Zaitseva, N. G., 1970a, J. Inorg. Nucl. Chem. **32**, 3165.

Ianovici, E. and Zaitseva, N. G., 1970b, Radiokhimiya **12**, 143.
Ianovici, E. and Zaitseva, N.G., 1970c, Dubna Report, JINR-E6-4916.
Ianovici, E., Raicheva, V. and Zaitseva, N. G., 1968, Dubna Report, JINR-E6-4100.
Ikeda, N., 1963, Radiochim. Acta **1**, 129.
Ikeda, N. and Kujirai, O., 1971a, Radioisotopes (Tokyo) **20**, 36.
Ikeda, N. and Kujirai, O., 1971b, Radioisotopes (Tokyo) **20**, 56.
Ikeda, N. and Kujirai, O., 1975, Radiochem. Radioanal. Lett. **23**, 125.
Ikeda, N., Yoshihara, K. and Yamagishi, S., 1961, Bull. Chem. Soc. Japan **34**, 140.
Ikeda, N., Yoshihara, K. and Yamagishi, S., 1964, Radiochim. Acta **3**, 13.
Ikeda, N., Saito, K. and Tsuji, K., 1970, Radiochim. Acta **13**, 90.
Ilakovac, K., 1954, Proc. Phys. Soc. (London) **A67**, 601.
Ingalls, R. and Depasquali, G., 1965, Phys. Lett. **15**, 262.
Ingalls, R., Coston, C. J., Depasquali, G., Drickamer, H. G. and Pinajian, J. J., 1966, J. Chem. Phys. **45**, 1057
Irvine, J. W., 1939, Phys. Rev. **55**, 1105.
Islamova, K., Mai, N. G., Ronneau, C. and Zaitseva, N. G., 1975, J. Inorg. Nucl. Chem. **37**, 865.
Ito, A., 1974, J. Physique **35**, C6-325.
Ivanoff, N. and Haissinsky, M., 1956, J. Chim. Phys. **53**, 400.
Iyer, R. H. and Martin, G., 1961, Symp. Chem. Effects of Nucl. Transformations, IAEA, Prague **1**, 281.
Jach, J., 1968, J. Inorg. Nucl. Chem. **30**, 919.
Jach, J. and Chandra, P., 1971, J. Inorg. Nucl. Chem. **32**, 1791.
Jach, J. and Harbottle, G., 1958, Trans. Faraday Soc. **54**, 520.
Jach, J. and Sutin, N., 1958, J. Inorg. Nucl. Chem. **7**, 5.
Jach, J. and Waitz, H., 1969, Inorg. Nucl. Chem. Lett. **5**, 867.
Jach, J., Kawahara, H. and Harbottle, G., 1958, J. Chromatog. **1**, 501.
Jacob, B., 1973, 1974, Theses, Darmstadt.
Jacobs, C. G. and Hershkowitz, N., 1970, Phys. Rev. **lB**, 839.
Jaffe, R. L. and Anderson, J. B., 1971, J. Chem. Phys. **54**, 2224.
Jaffe, R. L. and Anderson, J. B., 1972, J. Chem. Phys. **56**, 682.
Jagannathan, R. and Mathur, H. B., 1968, J. Inorg, Nucl. Chem. **30**, 1663.
Jakubinek, H., Srinivasan, S. C. and Wiles, D. R., 1971, Can. J. Chem. **49**, 2175.
Jeandey, C. and Peretto, P., 1975a, Z. Physik. **261**, 13.
Jeandey, C. and Peretto, P., 1975b, Phys. Stat. Solidi **A28**, 529.
Jellinek, F., Brauer, G. and Müller, H., 1960, Nature (Lond.) **185**, 376.
Jenkins, G. M. and Wiles, D. R., 1972, Chem. Comm., 1177.
John, W., Massey, R. and Saunders, B. G., 1967, Phys. Lett. **24B** 336.
Johnson, C. H., Pleasonton, F. and Carlson, T. A., 1963, Phys. Rev. **132**, 1149.
Johnson, E. R., 1970, "The Radiation Induced Decomposition of Inorganic Molecular Ions", (Gordon and Breach, N.Y.) p. 17.
Johnson, J. A., Dema, I. and Harbottle, G., 1974, Radiochim. Acta **21**, 196.
Johnston, A. J. and Urch, D. S., 1974, J. Chem. Soc. Faraday I **70**, 369.
Johnstone, J. J., Jones, C. H. W. and Vasudev, P., 1972, Can. J. Chem. **50**, 3037.
Joliot-Curie, F., 1939, Compt. rend. **208**, 341.
Jones, C. H. W., 1967, Inorg. Nucl. Chem. Lett. **3**, 363.
Jones, C. H. W., 1970a, J. Phys. Chem. **74**, 3347.
Jones, C. H. W., 1970b, Radiochim. Acta **14**, 1.
Jones, C. H. W. and Campbell, I. G., 1968a, Radiochim. Acta **9**, 7.
Jones, C. H. W. and Campbell, I. G., 1968b, Radiochim. Acta **9**, 71.
Jones, C. H. W. and Warren, J. L., 1968, J. Inorg. Nucl. Chem. **30**, 2289.

Jones, C. H. W. and Warren, J. L., 1970a, J. Chem. Phys. **53**, 1740.
Jones, C. H. W. and Warren, J. L., 1970b, J. Inorg. Nucl. Chem. **32**, 2119.
Jones, W. M., 1949, J. Chem. Phys. **17**, 1062.
Jordan, P., 1951a, Helv. Chim. Acta **34**, 699.
Jordan, P., 1951b, Helv. Chim. Acta **34**, 715.
Jovanovic-Kovacevic, O. Z., 1969, "Chromatographic-Electrophorèses", Presses Acad. Europe, Brussels, p. 95.
Jovanovic-Kovacevic, O. Z., 1973, III Int. Symp. Chromat. Electrophor. Brussels, 1972, Presses Acad. Europe, Brussels, p. 251.
Jovanovic-Kovacevic, O. Z., 1976, J. Radioanal. Chem. **30**, 515.
Jungerman, J. and Wright, S. C., 1949, Phys. Rev. **76**, 112.
Kacena, V. and Maddock, A. G., 1965, Symp. Chem. Effects of Nucl. Transformations, IAEA, Vienna **2**, 255.
Kahn, M., 1951, J. Amer. Chem. Soc. **73**, 479.
Kamen, M. D., 1941, Phys. Rev. **60**, 537.
Kanellakopulos-Drossopulos, W. and Wiles, D. R., 1971, Can. J. Chem. **49**, 2297.
Kanellakopulos-Drossopulos, K. and Wiles, D. R., 1976, J. Inorg. Nucl. Chem. **38**, 947.
Kankeleit, E., 1975, Z. Physik. **A275**, 119.
Kanzig, W. and Cohen, M. H., 1959, Phys. Rev. Lett. **3**, 509.
Kaplan, M., 1966, J. Inorg. Nucl. Chem. **28**, 331.
Karim, H. M. A., 1973, Int. J. App. Radiat. Isotopes **24**, 599.
Karplus, M., Porter, R. N. and Sharma, R. D., 1964, J. Chem. Phys. **40**, 2033.
Karplus, M., Porter, R. N. and Sharma, R. D., 1965, J. Chem. Phys. **43**, 3259.
Karplus, M., Porter, R. N. and Sharma, R. P., 1966, J. Chem. Phys. **45**, 3871.
Kasrai, M. and Maddock, A. G., 1970, J. Chem. Soc. **A**, 1105.
Kasrai, M., Maddock, A. G. and Freeman, J. H., 1971, Trans. Faraday Soc. **67**, 2108.
Kasrai, M., Raie, M., Suh, I. S. and Maddock, A. G., 1976, J. Chem. Soc. Faraday II **72**, 257.
Kaučić, S. and Vlatkovic, M., 1963, Croatica Chemica Acta **35**, 305.
Kaufman, J. J., Harkins, J. J. and Koski, W., 1969, J. Chem. Phys. **50**, 771.
Kaufman, S., 1960, J. Amer. Chem. Soc. **82**, 2963.
Kawahara, H. and Harbottle, G., 1959, J. Inorg, Nucl. Chem. **9** 240.
Kawazu, H. and Sakanou, M., 1974, Radiochem. Radioanal. Lett. **16**, 363.
Kay, J. G., 1960, Thesis, Diss. Abs., **21**, 1078.
Kay, J. G. and Diefallah, E. H., 1966, Third International Hot Atom Chem. Symp., Purdue University.
Kay, J. G. and Rowland, F. S., 1958, J. Amer. Chem. Soc. **80**, 3165.
Kazanjian, A. R. and Libby, W. F., 1965, J. Chem. Phys. **42**, 2778.
Ke, C. N., Yeh, S. J. and Yang, M. H., 1976, Radiochem. Radioanal. Lett. **25**, 17.
Keizer, J., 1972, J. Chem. Phys. **56**, 5958.
Keizer, J., 1973, J. Chem. Phys. **58**, 4524.
Keller, K. and Lee, R. V., 1966, J. App. Phys. **37**, 1890.
Khorana, S., 1968, J. Inorg. Nucl. Chem. **30**, 2545.
Khorana, S. and Nath, A., 1967, J. Phys. Chem. Solids **28**, 1081.
Khorana, S. and Nath, A., 1969, J. Inorg. Nucl. Chem. **31**, 1283.
Khorana, S. and Wiles, D. R., 1969, J. Inorg. Nucl. Chem. **31**, 3387.
Khorana, S. and Wiles, D. R., 1971, J. Inorg. Nucl. Chem. **33**, 1589.
Khurgin, B., Ofer, S. and Rakavy, M., 1970, Phys. Lett. **33A**, 219.
Kienle, P., Weckermann, B., Baumgärtner, F. and Zahn, U., 1962a, Naturwiss **49**, 294.
Kienle, P., Weckermann, B., Baumgärtner, F. and Zahn, U., 1962b, Naturwiss **49**, 295.
Kienle, P., Baumgärtner, F., Weckermann, B. and Zahn, U., 1963a, Radiochim. Acta **1**, 84.
Kienle, P., Wien, K., Zahn, U. and Weckermann, B., 1963b, Z. Physik **176**, 226.

Kigoshi, K., 1971, Science **173**, 47.
Kikuchi, C., 1971, Bull. Inst. Chem. Res., Kyoto Univ. **49**, 14.
Kimmel, R. M. and Uhlmann, D. R., 1969, J. App. Phys. **40**, 4254.
Kinchin, G. H. and Pease, R. S., 1955a, J. Nucl. Energy **1**, 200.
Kinchin, G. H. and Pease, R. S., 1955b, Rept. Progr. Phys. **18**, I.
Kirin, I. S. and Gusev, Y. K., 1966, Dokl. Akad. Nauk. SSSR **167**, 1090.
Kirin, I. S., Gusev, Y. K., Mosewich, A. N., Kuznetsov, N. P. and Gusel'nikov, V. S. 1965, Soviet Radiochem. **7**, 137.
Kirin, I. S., Murin, A. N., Nefedov. V. D., Gusev, Y. K. and Selikov, G. G., 1966, Radiokhimiya **8**, 104.
Kirin, I. S., Zaitsev, V. M. and Gusel'nikov, V. S., 1968, Radiokhimiya **10**, 354.
Kiso, Y., Kobayshi, M. and Kitaoka, Y., 1972, in "Chromatography in Solution, Anal. Chem. of P Compounds", Chapter 3, ed. M. Halmann, (J. Wiley, NY).
Kitaoka, Y., 1973, Ann. Rep. Reactor Inst. Kyoto Univ. **6**, 74.
Kitaoka, Y., 1974, Chem. Lett. (Japan), 1247.
Kitaoka, Y., Kawamoto, K., Kobayashi, M. and Kiso, Y., 1970a, Ann. Rep. Res. Reactor Inst. Kyoto Univ. **3**, 24.
Kitaoka, Y., Kawamoto, K., Kobayashi, M. and Kiso, Y., 1970b, Ann. Rep. Res. Reactor. Inst. Kyoto Univ. **3**, 32.
Kitaoka, Y., Kobayashi, M. and Takeda, J., 1973a, Ann. Rep. Reactor Inst. Kyoto Univ. **6**, 18.
Kitaoka, Y., Kobayashi, M. and Okada, M., 1973b, Chem. Lett. (Japan), 527.
Kobayashi, K., 1956, Phys. Rev. **102**, 348.
Kobayashi, K., 1957, Phys. Rev. **107**, 41.
Kobayashi, M., Takada, J. and Kiso, Y., 1971, Sixth International Hot Atom Chem. Symp. Brookhaven Nat. Lab., L. I., USA.
Kobayashi, M., Takada, J. and Kiso, Y., 1972, Radiochem. Radioanal. Lett. **9**, 67.
Koehler, J. S. and Seitz, F., 1954, Bristol Conf. Report on "Defects in Crystalline Solids".
Koehler, J. S. and Seitz, F., 1961, Disc. Faraday Soc. **31**, 45.
Kolaczkowski, R. W. and Plane, R. A., 1964, Inorg. Chem. **3**, 322.
Kolk, B., 1975, Phys. Rev. **B12**, 4695.
König, E., Gütlich, P. and Link, R., 1972, Chem. Phys. Lett. **15**, 302.
Kornelsen, E. V., 1964, Can. J. Phys. **42**, 364.
Korshunov, I. A. and Shafiev, A. I., 1958, Zhur. Neorg. Khim. **3**, 95.
Koski, W. S., 1949a, J. Amer. Chem. Soc. **71**, 4042.
Koski, W. S., 1949b, J. Chem. Phys. **17**, 582.
Krasnoperov, V. M., Murin, A. N., Cherezov, N. K. and Yutlandov, I. A., 1969, Dokl. Akad. Nauk. USSR **186**, 296.
Krause, M. O. and Carlson, T. A., 1967, Phys. Rev. **158**, 18.
Krause, M. O., Carlson, T. A. and Dismukes, R. D., 1968, Phys. Rev. **170**, 37.
Kremer, L. N. and Spicer, L. D., 1975, J. Amer. Chem. Soc. **97**, 5021.
Krisch, R. E. and Zelle, M. R., 1969, Advs. Radiation Biol. **3**, 177.
Krishnamurthy, M. V., 1974, J. Phys. Chem. Solids **35**, 606.
Krohn, K. A., Parks, N. J. and Root, J. W., 1971, J. Chem. Phys. **55**, 5784.
Kronrád, L., 1964, Radiochim. Acta **3**, 191.
Kronrád, L. and Cífka, J., 1968, Rep. Euratom. EUR 3746, 873.
Kronrád, L. and Kačena, V., 1966, Radiochim. Acta **6**, 181.
Kronrád, L. and Kačena, V., 1967, Radiochim. Acta **8**, 93.
Kronrád, L., Ratusky, J., Malek, P., Vavrejn, B. and Kolc, J., 1970, J. Labelled Comp. **6**, 326.
Kudo, H., 1971, Thesis, Tohoku Univ. (Japan).
Kudo, H., 1972a, J. Inorg. Nucl. Chem. **34**, 453.
Kudo, H., 1972b, Bull. Chem. Soc. Japan **45**, 389.

Kudo, H., 1972c, Bull. Chem. Soc. Japan **45**, 1311.
Kudo, H. and Tanaka, K., 1975, Radiochem. Radioanal. Lett. **23**, 57.
Kudo, K. and Yoshihara, K., 1970, J. Inorg. Nucl. Chem. **32**, 2845.
Kudo, K. and Yoshihara, K., 1971, Radiochim. Acta **15**, 167.
Kujirai, O. and Ikeda, N., 1973, Radiochem. Radioanal. Lett. **15**, 67.
Kujirai, O. and Ikeda, N., 1974, Radiochem. Radioanal. Lett. **18**, 197.
Kulikov, L. A., Perfiliev, Yu. D., Afanasov, M. I. and Babeshkin, A. M., 1972a, Radiochem. Radioanal. Lett. **12**, 47.
Kulikov, L. A., Bugaenko, L. T., Perfiliev, Yu. D. and Babeshkin, A. M., 1972b, Vestn. Moscow Univ. Ser II. Khim. **13**, 347.
Kumer, L., Posch, H. and Kaltseis, J., 1972, Phys. Lett. **A40**, 59.
Kündig, W., Kobelt, M., Appel, H., Constabaris, G. and Lindquist, R. H., 1969, J. Phys. Chem. Solids **30**, 819.
Kuppermann, A., 1967, in "Fast Reactions and Primary Processes in Chemical Kinetics", ed., Claesson, S., (Interscience, Wiley, NY) p. 131.
Kuppermann, A. and Schatz, G. C., 1975, J. Chem. Phys. **62**, 2502.
Kuppermann, A., Schatz, G. C. and Baer, M., 1974, J. Chem. Phys. **61**, 4362.
Kurchatov, B., Kurchatov, I., Missovski, L. and Russinov, L., 1935, Compt. rend. **200**, 1201.
Kuzin, V. I., Nefedov, V. D., Norseev, Yu. V., Toropova, M. A., Khalkin, V. A. and Groz, P., 1970a, Soviet Radiochem. **12**, 120.
Kuzin, V. I., Nefedov, V. D., Norseev, Yu. V., Toropova, M. A., Khalkin, V. A. and Groz. P., 1970b, Radiokhimiya **12**, 137.
Kuzin, V. I., Nefedov, V. D., Norseev, Yu. V., Toropova, M. A., Filatov, E. S. and Khalkin, V. A., 1972, High Energy Chem. **6**, 161.
Ladrielle, T. G., Cogneau, M. A. and Apers, D. J., 1974, Radiochem. Acta **21**, 210.
Ladrielle, T. G., Cogneau, M. A. and Apers, D. J., 1975a, Radiochim. Acta **22**, 65.
Ladrielle, T. G., Cogneau, M. A. and Apers, D. J., 1975b, Radiochim. Acta **22**, 173.
Ladrière, J. and Apers, D. J., 1974, J. Physique **35**, C6-335.
Lam, E. Y. Y., Gaspar, P. P. and Wolf, A. P., 1971, J. Phys. Chem. **75**, 445.
Lambrecht, R. M. and Wolf, A. P., 1973, in "Radiopharmaceuticals and Labelled Compounds", IAEA, Vienna, p. 275.
Lambrecht, R. M., Măntescu, C. and Wolf, A. P., 1971, Sixth Hot Atom Chem. Symp. Brookhaven, L. I., USA.
Langhoff, H., 1971, Z. Physik. **241**, 236.
Langhoff, H., Weiss, J. and Schumacher, M., 1969, Z. Physik **226**, 49.
Langsdorf, A. and Segré, E., 1940, Phys. Rev. **57**, 105.
Lark, N. L. and Perlman, M. L., 1960, Phys. Rev. **120**, 536.
Laswick, J. A. and Plane, R. A., 1959, J. Amer. Chem. Soc. **81**, 3564.
Lawson, R. W., 1918, Sitz. Wiener Akad. Wiss. **128**, 795.
Lazzarini, E., 1967a, J. Inorg. Nucl. Chem. **29**, 7.
Lazzarini, E., 1967b, J. Inorg. Nucl. Chem. **29**, 855.
Lazzarini, E. and Fantola-Lazzarini, A. L., 1967a, Energia Nucl. **14**, 472.
Lazzarini, E. and Fantola-Lazzarini, A. L., 1967b, J. Inorg. Nucl. Chem. **29**, 895.
Lazzarini, E. and Fantola-Lazzarini, A. L., 1967c, J. Inorg. Nucl. Chem. **29**, 2161.
Lazzarini, E. and Fantola-Lazzarini, A. L., 1969, J. Inorg. Nucl. Chem. **31**, 1947.
Lazzarini, E. and Fantola-Lazzarini, A. L., 1971a, J. Inorg. Nucl. Chem. **33**, 631.
Lazzarini, E. and Fantola-Lazzarini, A. L., 1971b, Lett. Nuovo Cim. **2**, 541.
Lazzarini, E. and Fantola-Lazzarini, A. L., 1972, J. Inorg. Nucl. Chem. **34**, 817.
Lazzarini, E. and Fantola-Lazzarini, A. L., 1974a, J. Inorg. Nucl. Chem. **36**, 263.
Lazzarini, E. and Fantola-Lazzarini, A. L., 1974b, J. Inorg. Nucl. Chem. **36**, 3673.
Lazzarini, E. and Fantola-Lazzarini, A. L., 1975, J. Inorg. Nucl. Chem. **37**, 407.

Lazzarini, E., Fantola-Lazzarini, A. L. and Annoni, T., 1970, Radiochim. Acta **13**, 156.
Lazzarini, E., Annoni, T. and Fantola-Lazzarini, A. L., 1971, Radiochim. Acta **15**, 93.
Lebedev, R. A., Babeshkin, A. M. and Nesmeyanov, A. N., 1969, Vestn. Moscow Univ. Khim. **5**, 45.
Lebedev, R. A., Babeshkin, A. M., Nesmeyanov, A. N. and Popov, E. A., 1970a, Vestn. Moscow Univ. Khim. **11**, 625.
Lebedev, R. A., Babeshkin, A. M., Nesmeyanov, A. N. and Fatieva, N. L., 1970b, Radiochem. Radioanal. Lett. **5**, 83.
Lebedev, R. A., Babeshkin, A. M. and Nesmeyanov, A. N., 1971a, Vestn. Moscow Univ. Khim. **12**, 113.
Lebedev, R.A., Babeshkin, A.M., Nesmeyanov, A.N., Tsikanov, V.A. and Fatieva, H.L., 1971b, Radiochem. Radioanal. Letts. **8**, 65.
Lebedev, R. A., Babeshkin, A. M. and Nesmeyanov, A. N., 1972, Vestn. Moscow Univ. Khim. **13**, 659.
Lecington, W. C. and Owens, C. W., 1967, Radiochim. Acta **7**, 212.
Lecington, W. C. and Owens, C. W., 1971, 6th International Hot Atom Chem. Symp., Brookhaven, L.I., USA, p. 45.
Lederer, C. M., Hollander, J. M. and Perlman, I., 1969, Table of Isotopes, (Wiley, New York) 6th Edition.
Lee, E. K. C. and Rowland, F. S., 1963, J. Inorg. Nucl. Chem. **25**, 133.
Lee, J. K., Musgrave, B. and Rowland, F. S., 1960, J. Chem. Phys. **32**, 1266.
Lee, J. K., Lee, E. K. C., Musgrave, B., Tang, Y., Root, J. W. and Rowland, F. S., 1962, Anal. Chem. **34**, 741.
Lemmon, R. M., Mazetti, F., Reynolds, F. L. and Calvin, M., 1956, J. Amer. Chem. Soc. **78**, 6414.
Lemmon, R. M., Mullen, R. T. and Reynolds, F. L., 1961, Symp. Chem. Effects of Nucl. Transformations, IAEA, Prague **2**, 27.
Levey, G., Milham, R., Rice, W. and Willard, J. E., 1948, Conf. on the Chem. Effects of Nucl. Transformation, AECU-50, p. 6 USAEC Report.
Levinger, J. S., 1953, Phys. Rev. **90**, 11.
Lewis, G. N., 1907, J. Amer. Chem. Soc. **29**, 1165, 1516.
Libby, W.F., 1940, J. Amer. Chem. Soc. **62**, 1930.
Libby, W. F., 1947, J. Amer. Chem. Soc. **69**, 2523.
Lideard, A. B., 1957, Handbuch der Physik **20**, 246.
Light, J. C., 1971, Advs. Chem, Phys. **19**, 1.
Lin, C. C. and Wahl, A. C., 1973, J. Inorg. Nucl. Chem. **35**, 1.
Lin, K. C., Cotter, R. J. and Koski, W. S., 1974, J. Chem. Phys. **61**, 905.
Lin, R., San, I., Lefohn, A. S. and Garner, C. S., 1967, J. Inorg. Nucl. Chem. **29**, 1553.
Lin, T., 1969, Thesis, University of California.
Lin, T. and Matsuura, N., 1975, Bull. Chem. Soc. Japan **48**, 3450.
Lin, Y. C. and Wiles, D. R., 1970, Radiochim. Acta **13**, 43.
Lindhard, J., Scharff, M. and Schioett, H. E., 1963, Mat. Kgl. Dan sk. Vid. Fys–Medd. **33**, No. 14.
Lindner, L., 1958, Thesis, Amsterdam.
Lindner, L., 1962a, "Inorganic Isotopic Synthesis", Chap. 6, W. A. (Benjamin, NY) p. 143.
Lindner, L., 1962b, Conf. Radioisotopes, Oxford, Sept.
Lindner, L. and Harbottle, G., 1960, J. Inorg. Nucl. Chem. **15**, 386.
Lindner, L. and Harbottle, G., 1961, Symp. Chem. Effects of Nucl. Transformations, IAEA, Prague **1**, 485.
Lindner, L., Zwenk, H., Van der Ende, H., Drost-Wildschut, H. and Lasthuizen, M., 1965, Symp. Chem. Effects of Nucl. Transformations, IAEA, Vienna **2**, 109.

Lister, D. H. and Symons, M. C. R., 1970, J. Chem. Soc. (A), 782.
Livingston, R. S., 1970, Particle Accel. **1**, 51.
Llabador, Y., 1970, Radiochim. Acta **14**, 100.
Llabador, Y., 1974, J. Inorg. Nucl. Chem. **36**, 2453.
Llabador, Y. and Adloff, J. P., 1966, Radiochim. Acta **6**, 49.
Llabador, Y. and Adloff, J. P., 1967a, Radiochim. Acta **7**, 20.
Llabador, Y. and Adloff, J. P., 1967b, Radiochim. Acta **8**, 41.
Llabador, Y. and Adloff, J. P., 1968, Radiochim. Acta **9**, 171.
Llabador, Y. and Friedt, J. M., 1971, Chem. Phys. Lett. **8**, 592.
Llabador, Y. and Friedt, J. M., 1973, J. Inorg. Nucl. Chem. **35**, 2351.
Lo Moro, A. and Frediani, S., 1966, Ann. Chim. (Rome) **56**, 339.
Lo Moro, A. and Frediani, S., 1967, J. Inorg. Nucl. Chem. **29**, 1837.
Lo Moro, A. and Frediani, S., 1969, J. Inorg. Nucl. Chem. **31**, 605.
Lu, C. S. and Sugden, S., 1939, J. Chem. Soc., 1273.
Luchner, K. and Micklitz, H., 1970, J. Lumines, **1**, 368.
Lux, F. and Ammentorp-Schmidt, F., 1965, Radiochim. Acta **4**, 112.
Lux, F., Ammentorp-Schmidt, F., Dempf, D., Graw, D. and Hagenberg, W., 1970, Radiochim. Acta **14**, 57.
Maccoll, A., 1969, Chem. Rev. **69**, 33.
Machado, J. C., Machado, R. M. and Vargas, J. I., 1965, Symp. Chem. Effects of Nucl. Transformations, IAEA, Vienna **2**, 195.
Mackay, C. and Wolfgang, R., 1962, Radiochim. Acta **1**, 42.
Mackay, C. and Wolfgang, R., 1965, Science **148**, 899.
Mackay, C., Pandow, M., Polak, P. and Wolfgang, R., 1961, Symp. Chem. Effects of Nucl. Transformations, IAEA, Prague **2**, 17.
Mackay, C., Polak, P., Rosenberg, H. and Wolfgang, R., 1962, J. Amer. Chem. Soc. **84**, 308.
MacKenzie, A. J. and Borland, J. W., 1952, Anal. Chem. **24**, 176.
Maddock, A. G., 1965a, Symp. Chem. Effects of Nucl. Transformations, IAEA, Vienna **2**, 174.
Maddock, A. G., 1965b, Symp. Chem. Effects of Nucl. Transformations, IAEA, Vienna **2**, 277.
Maddock, A. G., 1965c, Symp. Chem. Effects of Nucl. Transformations, IAEA, Vienna **1**, 435.
Maddock, A. G., 1972, MTP Reviews of Inorganic Chemistry. Series 1, Chap. 6 in Vol. 8, Butterworth, London.
Maddock, A. G., 1975a, in "Physical Chemistry", Vol. 7, "Reactions in the Condensed Phase", ed. Eyring, H., (Academic Press, NY).
Maddock, A. G., 1975b, in "Hot Atom Status Report", IAEA, Vienna, p. 33.
Maddock, A. G., 1975c, MTP Reviews of Inorganic Chemistry, Series 2, Chap. 6 in Vol. 8, (Butterworth, London).
Maddock, A. G. and Collins, K. E., 1968, Can. J. Chem. **46**, 3924.
Maddock, A. G. and Del Val-Cobb, 1959, Trans. Faraday Soc. **55**, 1709.
Maddock, A. G. and De Maine, M. M., 1956a, Can. J. Chem. **34**, 441.
Maddock, A. G. and De Maine, M. M., 1956b, Can. J. Chem. **34**, 275.
Maddock, A. G. and Mahmood, A. J., 1973, Inorg. Nucl. Lett. **9**, 509.
Maddock, A. G. and Mahmood, A. J., 1976, Radiochem. Radioanal. Lett. **25**, 293.
Maddock, A. G. and Mirsky, R. M., 1965, Symp. Chem. Effects of Nucl. Transformations, IAEA, Vienna **2**, 41.
Maddock, A. G. and Mohanty, S. R., 1963, Radiochim. Acta **1**, 85.
Maddock, A. G. and Müller, H., 1960, Trans. Faraday Soc. **56**, 509.
Maddock, A. G. and Sutin, N., 1953, Research **6**, Suppl., 78.
Maddock, A. G. and Sutin, N., 1955, Trans. Faraday Soc. **51**, 184.
Maddock, A. G. and Treloar, F. E., 1962, Discussion on Nuclear Chemistry, AERE-M 1078.
Maddock, A. G. and Vargas, J. I., 1959, Nature (Lond.) **184**, 1931.

Maddock, A. G. and Vargas, J. I., 1961a, Symp. Chem. Effects of Nucl. Transformations, IAEA, Prague **I**, 375.

Maddock, A. G. and Vargas, J. I., 1961b, Trans. Faraday Soc. **57**, 992.

Maddock, A. G. and Wolfgang, R., 1968, The Chemical Effects of Nucl. Transformations, in "Nuclear Chemistry", ed. Yaffe, (Academic Press, New York) Vol. II, p. 185.

Maddock, A. G., Treloal, F. E. and Vargas, J. I., 1963, Trans. Faraday Soc. **59**, 924.

Maddock, A. G., Todesco, A.B.J.B. and Blaxell, D., 1965, Symp. Chem. Effects of Nucl. Transformations, IAEA, Vienna **2**, 337.

Maddock, A. G., Williams, A. F., Siekierska, K. E. and Fenger, J., 1976, Phys. Stat. Solidi **B74**, 183.

Magnusson, L. B., 1951, Phys. Rev. **81**, 285.

Mahieu, B. and Llabador, Y., 1974, J. Physique **35**, C-6, 329.

Mahieu, B., Apers, D. J. and Capron, P. C., 1971a, Radiochim. Acta **16**, 100.

Mahieu, B., Apers, D. J. and Capron, P. C., 1971b, J. Inorg. Nucl. Chem. **33**, 2857.

Mahieu, B., Prendez, M. and Apers, D. J., 1976, Radiochem. Radioanal. Lett. **25**, 67.

Mahmood, A. J., 1972, Thesis, Cambridge, UK.

Majer, V., 1937, Naturwiss. **25**, 252.

Majer, V., 1939, Chem. Listy **33**, 130.

Malcolme-Lawes, D.J., 1969, Thesis, Queen Mary College, London.

Malcolme-Lawes, D. J., 1972a, J. Chem. Soc. Faraday II **68**, 1613, 2051.

Malcolme-Lawes, D. J., 1972b, J. Chem. Phys. **57**, 2476, 2481.

Malcolme-Lawes, D. J., 1974a, Radiochim. Acta **21**, 105.

Malcolme-Lawes, D. J., 1974b, J. Chem. Soc. Faraday II **70**, 1942.

Malcolme-Lawes, D. J., 1975, J. Chem. Soc. Faraday II **71**, 1183.

Măntescu, C. and Costea, T., 1962, Can. J. Chem. **40**, 1232.

Măntescu, C. and Costea, T., 1963, Phys. Stat. Solidi **3**, K 290.

Măntescu, C. and Costea, T., 1966, J. Inorg. Nucl. Chem. **28**, 273.

Măntescu, C. and Genunche, A., 1963, Can. J. Chem. **41**, 3145.

Măntescu, C., Genunche, A., Cristu, D. and Costea, T., 1971, Inst. Fiz. At. (Rumania.) Report, CO26.

Măntescu, C., Genunche, A., Cristu, D. and Costea, T., 1976, Rev. Roum. Chim. **21**, 789.

Marchart, H., 1965a, Symp. Chem. Effects of Nucl. Transformations, IAEA, Vienna **2**, 276.

Marchart, H., 1965b, Symp. Chem. Effects of Nucl. Transformations, IAEA, Vienna **2**, 291.

Marchart, H., 1965c, Nature **206**, 822.

Marchart, H. and Grass, F., 1965a, Monatsh. Chem. **96**, 1117.

Marchart, H. and Grass, F., 1965b, Monatsh. Chem. **96**, 1312.

Marchant, L., Sharrock, M., Hoffman, B. M. and Munck, E., 1972, Proc. Nat. Acad. Sci. USA **69**, 2396.

Margraff, R. and Adloff, J. P., 1966, Radiochim. Acta **6**, 138.

Marqués, R. O. and Wolschrijn, R. A., 1969, Radiochim. Acta **12**, 169.

Martin, A. J. P., 1944, Symposium Biochemical Society, Cambridge, U.K. **3**, 411.

Marx, D., 1966, Z. Physik **195**, 26.

Mathur, P. K., 1969, Indian J. Chem. **7**, 820.

Matsuura, N. and Lin, T. K., 1970, J. Inorg. Chem. **32**, 353.

Matsuura, N. and Lin, T. K., 1971, J. Inorg. Chem. **33**, 2281.

Matsuura, N., Kurimura, Y. and Shinohara, N., 1967, J. of Nucl. Sci. Techn. **4**, 595.

Matsuura, N., Shinohara, N. and Lin, T., 1969, J. Inorg. Nucl. Chem. **31**, 1257.

Matsuura, T., 1965, J. Inorg. Nucl. Chem. **27**, 2669.

Matsuura, T., 1966, J. Inorg. Nucl. Chem. **28**, 313.

Matsuura, T., 1967, Int. J. Appl. Rad. Isotopes **18**, 697.

Matsuura, T., 1968, Radiochim. Acta **10**, 33.

Matsuura, T. and Hashimoto, T., 1966, Bull. Chem. Soc. Japan **39**, 2647.
Matsuura, T. and Matsuura, M., 1968, Radiochim. Acta **10**, 33.
Matsuura, T. and Sasaki, T., 1966a, Radiochim. Acta **5**, 212.
Matsuura, T. and Sasaki, T., 1966b, Inorg. Nucl. Chem. Lett. **2**, 111.
Matsuura, T. and Sasaki, T., 1967, Radiochim. Acta **8**, 33.
Matsuura, T., Sensui, Y. and Sasaki, T., 1965, Radiochim. Acta **4**, 85.
Maul, J. L., 1969, Thesis, Technische Universität Munich.
Maurer, W. and Ramm, W., 1942, Z. Physik **119**, 602.
Mayer, J. W., Erikson, L. and Davies, J. A., 1970, "Ion Implantation in Semiconductors, Si and Ge", (Academic Press, NY).
McCallum, K. J. and Holmes, O. G., 1952, Can. J. Chem. **29**, 691.
McCallum, K. J. and Maddock, A. G., 1953, Trans. Faraday Soc. **49**, 1150.
McMillan, E. M., 1939, Phys. Rev. **55**, 510.
Meier-Komor, P., 1968, Thesis, Technische Universität Munich.
Meinhold, H. and Reichold, P., 1968, Z. Naturwiss **55**, 344.
Meinhold, H. and Reichold, P., 1969, Radiochim. Acta **11**, 175.
Meinhold, H. and Reichold, P., 1970, Inorg. Nucl. Chem. Lett. **6**, 253.
Meinhold, H. and Reichold, P., 1971, Radiochim. Acta **15**, 76.
Melander, L., 1947, Acta Chem. Scand. **1**, 169.
Melander, L., 1948, Acta Chem. Scand. **2**, 290.
Melander, L. and Slatis, H., 1948, Phys. Rev. **74**, 709.
Meriadec, B. and Milman, M., 1968, J. Inorg. Nucl. Chem. **30**, 2853.
Meriadec, B. and Milman, M., 1971, J. Inorg. Nucl. Chem. **33**, 915.
Merz, E., 1964, Radiochim. Acta **2**, 172.
Merz, E., 1966, Nukleonik **8**, 248.
Merz, E. and Riedel, H. J., 1964, Radiochim. Acta **3**, 35.
Merz, E. and Riedel, H. J., 1965, Symp. Chem. Effects of Nucl. Transformations, IAEA, Vienna **2**, 179.
Metzger, F. R., 1959, Progr. Nucl. Phys. **7**, 53.
Meyer, J. P., 1970a, Radiochim. Acta **14**, 154.
Meyer, J. P., 1970b, Thesis, Strasbourg.
Meyer, J. P., 1971, Radiochim. Acta **15**, 88.
Meyer, J. P. and Adloff, J. P., 1967, Radiochim. Acta **7**, 15.
Meyer, J., Paulus, J. M. and Abbé, J. C., 1972, Radiochim. Acta **17**, 76.
Michael, J. V. and Weston, R. E., 1966, J. Chem. Phys. **45**, 3632.
Michaelov, M. and Sorantin, L., 1969, Radiochim. Radioanal. Lett. **2**, 343.
Micklitz, H., 1968, Z. Physik **215**, 302.
Micklitz, H. and Barrett, P. H., 1972, Phys. Rev. Lett. **28**, 1547.
Micklitz, H. and Luchner, K., 1969, Z. Physik **227**, 301.
Micklitz, H. and Luchner, K., 1974, Z. Physik **270**, 79.
Migdal, A., 1941, J. Phys. (USSR) **4**, 449.
Mikulaj, V., Macasek, F. and Kopunec, R., 1973, Chem. Zvesti **27**, 23.
Milenković, S. M. and Maddock, A. G., 1967, Radiochim. Acta **8**, 222.
Milenković, S. M. and Veljković, S. R., 1967, Radiochim. Acta **8**, 146.
Milenković, S. M. and Veljković, S., 1974, J. Inorg. Nucl. Chem. **36**, 683.
Milenković, S., Bingulac, S. and Veljković, S., 1971, J. Inorg. Nucl. Chem. **33**, 1187.
Milham, R. C., 1952, Thesis, Wisconsin.
Milham, R. C., Adams, A. and Willard, J. E., 1965, Symp. Chem. Effects of Nucl. Transformations, IAEA, Vienna **2**, 31.
Miller, G. and Light, J. C., 1971, J. Chem. Phys. **54**, 1635.
Miller, J. M. and Dodson, R. W., 1950, J. Chem. Phys. **18**, 865.

Miller, J. M. and Hudis, J., 1959, Ann. Rev. Nucl. Sci. **9**, 59.
Miller, J. M., Gryder, J. W. and Dodson, R. W., 1950, J. Chem. Phys. **18**, 865.
Milman, M., 1952, Thesis, Oxford.
Milstein, R., Williams, R. L. and Rowlands, F. S., 1974, J. Phys. Chem. **78**, 857.
Misroch, M. B., Schramm, C. J. and Nath, A., 1976, J. Chem. Phys. **65**, 1982.
Mitchell, R. F. and Martin, D.S., 1955, USAEC Report, ISC-567.
Mitchell, R. F. and Martin, D. S., 1956, J. Inorg. Nucl. Chem. **2**, 286.
Morinaga, H. and Zaffarano, D. J., 1954, Phys. Rev. **93**, 1422.
Mortensen, R. A. and Leighton, P. A., 1934, J. Amer. Chem. Soc. **56**, 2397.
Mosewich, A. N., Kuznetsov, N. P. and Gusev, Y. G., 1965, Soviet Radiochem. **7**, 677.
Moskvin, L. N., 1962, Radiokhimiya **4**, 514.
Muckerman, J. T., 1971, J. Chem. Phys. **54**, 1155.
Muckerman, J. T., 1972a, J. Chem. Phys. **56**, 2997.
Muckerman, J. T., 1972b, J. Chem. Phys. **57**, 8.
Mühl, P. and Grosse-Ruyken, H., 1967, Isotopenpraxis **3**, 486.
Mullen, J. G., 1965, Phys. Lett. **15**, 15.
Mullen, R. T., 1961, Thesis, University of California, UCRL. 9603.
Müller, H., 1960, Z. Elektrochemie **64**, 1045.
Müller, H., 1961, Symp. Chem. Effects Nucl. Transformations, IAEA, Prague **1**, 521.
Müller, H., 1962a, Naturwiss. **49**, 182.
Müller, H., 1962b, Z. anal. Chem. **189**, 336.
Müller, H., 1963a, Z. anorg. allgem. Chem. **321**, 124.
Müller, H., 1963b, Angew. Chem. **75**, 1132.
Müller, H., 1964a, Habilitationsschrift, Freiburg/Breisgau.
Müller, H., 1964b, International Atomic Energy Agency, Preprint No. SM-57/16, p. 21, (CONF-773023).
Müller, H., 1965a, J. Inorg. Nucl. Chem. **27**, 1745.
Müller, H., 1965b, Symp. Chem. Effects of Nucl. Transformations, IAEA, Vienna **2**, 359.
Müller, H., 1965c, Z. anorg. allgem. Chem. **336**, 24.
Müller, H., 1966a, J. Inorg. Nucl. Chem. **28**, 2081.
Müller, H., 1966b, Z. anorg. allgem. Chem. **342**, 177.
Müller, H., 1967a, Angew. Chem. **79**, 128. (Internat. Ed. **6**, 133).
Müller, H., 1967b, J. Inorg. Nucl. Chem. **29**, 2167.
Müller, H., 1967c, Z. anorg. allgem. Chem. **352**, 291.
Müller, H., 1968a, Umshau **68**, 534.
Müller, H., 1968b, Radiochim. Acta **9**, 167.
Müller, H., 1969a, Z. anal. Chem. **247**, 145.
Müller, H., 1969b, J. Inorg. Nucl. Chem. **31**, 1579.
Müller, H., 1969c, Inorg. Nucl. Chem. Lett. **5**, 761.
Müller, H., 1970, Atomkernenergie **16**, 237, 323.
Müller, H., 1976, "Nukleare Methoden in der Festkorperchemie", Jülich Jul.-Conf.-22, p. 88.
Müller, H. and Abberger, S., 1971, Angew. Chem. **83**, 921.
Müller, H. and Abberger, S., 1973, Radiochim. Acta **19**, 176.
Müller, H. and Broda, E., 1951, Monatsh. Chem. **82**, 48.
Müller, H. and Cramer, D., 1970, Radiochim. Acta **14**, 78.
Müller, H. and Martin, S., 1969, Inorg. Nucl. Chem. Lett. **5**, 761.
Müller, H. and Martin, S., 1971, Sixth International Hot Atom Chem. Symp., Brookhaven, L.I., USA.
Müller, H. and Martin, S., 1976, "Second Europhysical Topical Conf. on Lattice Defects in Ionic Crystals", Berlin, p. 141.
Müller, H. and Martin, S., 1977, 9th International Hot Atom Chem. Symp., Blacksburg, USA, p. 51.

Müller, H., Diefallah, E. H. M. and Martin, S., 1977, 9th International Hot Atom Chem. Symp., Blacksburg, p. 50.
Müller, L., 1961, Thesis, Cologne.
Munck, E. and Champion, P. M., 1974a, J. Physique **35**, C6-33.
Munck, E. and Champion, P. M., 1974b, Ann. NY Acad. **244**, 142.
Mund, W., Capron, P. C. and Jodogne, J., 1931, Bull. Soc. Chim. Belg. **40**, 35.
Mundschenk, H., 1970, Radiochim. Acta **14**, 72.
Mundschenk, H., 1971, Radiochim. Acta **15**, 193.
Münze, R., 1961, Kernenergie **4**, 808.
Murin, A.N. and Nefedov, V.D., 1955, Primenie Mechenykh Atomov v. Anal. Khim., Akad. Nauk, SSSR, Inst. Geokhim i Anal. Khim. 75.
Murin, A. N., Nefedov, V. D., Baranovsky, V. I. and Popov, D. K., 1956, Soviet Phys. Doklady **I**, 719.
Murin, A. N., Nefedov, V. D., Sinotova, E. N. and Larionov, O. V., 1958, Zhur. Neorg, Khim. **3**, 181, (Eng. Ed. **3**, 278.)
Murin, A. N., Nefedov, V. D., Zaitsev, V. M. and Grachev, S. A., 1960, Doklady Akad. Nauk SSSR **133**, 123.
Murin, A. N., Banasewitch, S. N. and Bogdanov, R. V., 1961a, Izv. Akad. Nauk SSSR Otdel. Khim. **8**, 1433.
Murin, A. N., Banasewitch, S. N. and Bogdanov, R. V., 1961b, Symp. Chem. Effects of Nucl. Transformations, IAEA, Prague **2**, 191.
Murin, A. N., Leve, B. G., Banasevich, S. N., Samosynk, G. P. and Ignatovitch, Ya. L. 1961c, Fiz. Tverd. Tela **3**, 398.
Murin, A. N., Nefedov, V. D. and Larionov, O. V., 1961d, Radiokhimiya **3**, 90.
Murin, A. N., Nefedov, V. D., Zaitsev, V. M. and Grachev, S.A., 1961e, Symp. Chem. Effects of Nucl. Transformations, IAEA, Prague **2**, 183.
Murin, A. N., Nefedov, V. D., Kirin, I. S., Leonov, V. V., Zaitsev, V. M. and Akulov, G. P., 1965a, Radiokhimiya **7**, 629.
Murin, A. N., Kirin, I. S., Nefedov, V. D., Grachev, S. A. and Gusev, Y. K., 1965b, Dokl. Akad. Nauk SSSR **161**, 611.
Murin, A. N., Nefedov, V. D., Kirin, I. S., Grachev, S. A., Gusev, Y. K. and Shapkin, G. N., 1965c, Zhur. Obshchei Khim. **35**, 2137.
Murin, A. N., Nefedov, V. D., Kirin, I. S., Leonov, V. V., Zaitsev, V. M. and Akulov, G.P., 1965d, Soviet Radiochem. **7**, 627.
Murin, A. N., Kirin, I. S., Nefedov, V. D., Grachev, S. A., Gusev, Y. K., Ivannikova, N. V. and Gusel'nikov, V. S., 1966, Radiokhimiya **8**, 449.
Murin, A. N., Bondarevski, S. I., Dzhurzka, V. V. and Tarasov, V. A., 1972, Khim. Vys. Energ. **6**, 494.
Muruyama, Y. and Idenawa, K., 1967, J. Nucl. Sci. Techn. (Tokyo) **4**, 365.
Muxart, R., Daudel, P., Daudel, R. and Haissinsky, M., 1947, Nature (Lond.) **159**, 538.
Naguib, H. M. and Kelly, R., 1970, J. Nucl. Mater. **35**, 293.
Nakamura, T., Ujimoto, K., Yoza, N. and Ohashi, S., 1970, J. Inorg. Nucl. Chem. **32**, 3191.
Naki, T. and Yajima, S., 1958, J. Chem. Soc. Japan **79**, 1267.
Narayan, S. R. and Wiles, D. R., 1969, Can. J. Chem. **47**, 1019.
Nath, A., 1961, Symp. Chem. Effects of Nucl. Transformations, IAEA, Prague **1**, 335.
Nath, A., 1964, Indian J. Chem. **2**, 332.
Nath, A., 1975, "Radiochemistry" Specialist Rep. Chem. Soc. (London) **2**, 105.
Nath, A. and Khorana, S., 1967, J. Chem. Phys. **46**, 2858.
Nath, A. and Klein, M.P., 1969, Nature (London) **224**, 794.
Nath, A. and Nesmeyanov, A. N., 1962, Radiokhimiya **4**, 122.
Nath, A. and Nesmeyanov, A.N., 1963, Indian Report, Trombay, AEET/CD/14.

Nath, A. and Shankar, J., 1961, Symp. Chem. Effects of Nucl. Transformations, IAEA, Prague **1**, 409.
Nath, A. and Vaish, S. P., 1967, J. Chem. Phys. **46**, 4660.
Nath, A., Rao, K. A. and Thomas, V. G., 1964a, Radiochim. Acta **3**, 134.
Nath, A., Rao, K. A. and Thomas, V. G., 1964b, Indian J. Chem. **2**, 331.
Nath, A., Khorana, S., Mathur, P. K. and Sarup, S., 1966, Indian J. Chem. **4**, 51.
Nath, A., Agrawal, R. D. and Mathur, P. K., 1968a, Inorg. Nucl. Chem. Lett. **4**, 161.
Nath, A., Harpold, M., Klein, M. P. and Kündig, W., 1968b, Chem. Phys. Lett. **2**, 471.
Nath, A., Klein, M. P., Kündig, W. and Lichtenstein, D., 1970, Radiation Effects **2**, 211.
Nawojska, J., 1967, Rocz. Chem. **41**, 889.
Ndiokwere, C. L. and Elias, H., 1973, Radiochim. Acta **19**, 181.
Nefedov, V. D. and Andreev, V. I., 1957, Zhur. Fiz. Khim. **31**, 563.
Nefedov, V. D. and Bel'dy, M. P., 1957, Zhur. Fiz. Khim. **31**, 986.
Nefedov, V. D. and Grachev, S. A., 1960, Radiokhimiya **2**, 464.
Nefedov, V. D. and Mikulaj, V., 1973, Radiokhimiya **15**, 846. (Eng. Trans. **15**, 289).
Nefedov, V. D. and Sinotova, E. N., 1958, Zhur. Fiz. Khim. **32**, 2392.
Nefedov, V. D. and Toropova, M. A., 1957, Zhur. Neorg. Khim. **2**, 1667.
Nefedov, V. D. and Toropova, M. A., 1958a, Zhur. Neorg. Khim. **3**, 231.
Nefedov, V. D. and Toropova, M. A., 1958b, Zhur. Neorg. Khim. **3**, 175.
Nefedov, V. D., Sinotova, E. N. and Katsapov, V. I., 1956, Zhur. Fiz. Khim. **30**, 1867.
Nefedov, V. D., Liu Yuan-Fang, Li Wang-Ch'ang and Lung Yun-Kua, 1959, Acta Chim. Sinica. **25**, 165.
Nefedov, V. D., Ryukhim, Yu. A. and Toropova, M. A., 1960a, Radiokhimiya **2**, 458 (Eng. Trans. **2**, 154).
Nefedov, V. D., Sinotova, E. N. and Trenin, V. D., 1960b, Radiokhimiya **2**, 739.
Nefedov, V. D., Bykhovtsev, V. L., Wu Chi-Lan and Grachev, S. A., 1961a, Radiokhimiya **3**, 225.
Nefedov, V. D., Ryukhin, Yu. A. and Toropova, M. A., Melnikov, V. N. and Chi Minh, L., 1961b, Symp. Chem. Effects of Nucl. Transformations, IAEA, Prague **2**, 149.
Nefedov, V.D., Kirin, I.S. and Zaitsev, V.M., 1962, Soviet Radiochem. **4**, 311.
Nefedov, V. D., Toropova, M. A., Grachev, S. A. and Grant, Z. A., 1963a, Zhur. Obschei Khim. **33**, 15.
Nefedov, V.D., Grachev, S. A. and Gluvka, S., 1963b, J. Gen. Chem. (USSR) **33**, 325.
Nefedov, V. D., Toropova, M. A., Krivokhatskaya, I. V. and Kesarov, O. V., 1964a, Radiokhimiya **6**, 112.
Nefedov, V. D., Kirin, I. S. and Zaitsev, V. M., 1964b, Radiokhimiya **6**, 78.
Nefedov, V. D., Zhuravlev, V. E., Toropova, M. A. and Levchenko, A. V., 1964c, Radiokhimiya **6**, 632.
Nefedov, V. D., Kirin, I. S. and Zaitsev, V. M., 1964d, Radiokhimiya **6**, 123.
Nefedov, V. D., Vobetskii, M. E., Sinotova, E. N. and Borak, J., 1965a, Radiokhimiya **7**, 627.
Nefedov, V. D., Vobetskii, M. E. and Borak, J., 1965b, Radiokhimiya **7**, 628.
Nefedov, V. D., Zhuravlev, V. E., Toropova, M. A. and Levchenko, A. V., 1965c, Radiokhimiya **7**, 632.
Nefedov, V. D., Gracheva, L. M., Grachev, S. A. and Petrov, L. N., 1965d, Radiokhimiya **7**, 741.
Nefedov, V. D., Vobetskii, M. E. and Borak, J., 1965e, Radiochim. Acta **4**, 104.
Nefedov, V. D., Kirin, I. S., Gracheva, L. M. and Grachev, S. A., 1966a, Radiokhimiya **8**, 98.
Nefedov, V. D., Kirin, I. S., Gracheva, L. M. and Grachev, S. A., 1966b, Radiochimiya **8**, 100, 398.
Nefedov, V. D., Kirin, I.S., Tikhonov. V. I. and Zaitsev, V. M., 1966c, Radiokhimiya **8**, 714.
Nefedov, V. D., Toropova, M. A., Levchenko, A. V. and Mosevich, A. N., 1966d, Radiokhimiya **8**, 719.

Nefedov, V. D., Grachev, S. A., Gracheva, L. M. and Petrov, L. N., 1966e, Radiokhimiya **8**, 376.

Nefedov, V. D., Toropova, M. A. and Levchenko, A. V., 1967, Radiokhimiya **9**, 138.

Nefedov, V. D., Toropova, M. A. and Sinotova, E. N., 1969, Russ. Chem. Revs. **38**, 873.

Nefedov, V. D., Toropova, M. A., Khalkin, V. A., Norseev, Y. V. and Kuzin, V. I., 1970, Radiokhimiya **12**, 194.

Nefedov, V. D., Petrov, L. N. and Avrorin, V. V., 1972, Radiokhimiya **14**, 436.

Negoesco, I. and Costea, T., 1962, Can. J. Chem. **40**, 1642.

Neiler, J. H., Walter, F. J. and Schmitt, H. W., 1966, Phys. Rev. **149**, 894.

Nelson, R. S., 1968, "The Observation of Atomic Collisions in Crystalline Solids", (North-Holland, Amsterdam).

Nelson, R. S., 1969, Proc. Roy. Soc. (London) **311A**, 53.

Nelson, D. H. and McCallum, K. J., 1958, Can. J. Chem. **36**, 979.

Nesmeyanov, A. N. and Filatov, E. S., 1961, Radiokhimiya **3**, 614.

Nesmeyanov, A. N., Korolev, B. M. and Sazanov, L. A., 1959, Radiokhimiya **1**, 694.

Nesmeyanov, A. N., Babeshkin, A. M. and Kosev, N. R., 1966, Vestn. Moscow, Univ. Ser II Khim. **4**, 51.

Neufeld, J. and Snyder, W. S., 1955, Phys. Rev. **99**, 1326.

Newton, G. W. A., 1972, 1975, "Specialist Periodical Report, Radiochemistry" Chemical Society (London), Vols. 1 and 2.

Nishi, Y. and Tominaga, T., 1974, Radioisotopes (Tokyo) **23**, 700.

Nishi, Y. and Tominaga, T., 1976, Radiochem. Radio- Anal. Lett., **24**, 249.

Norris, L. D. and Snell, A. H., 1947, Science **105**, 265.

Norris, L. D. and Snell, A. H., 1948, Phys. Rev. **73**, 254.

Norris, L. D. and Snell, A. H., 1949, Nucleonics **5**, 18.

North, A. M., 1966, Quart. Rev. **20**, 421.

Nowak, M., 1965, Int. J. Appl. Rad. Isotopes **16**, 649.

Nowak, M. and Akerman, K., 1970, Radiochem. Radioanal. Lett. **3**, 39.

Obenshain, F. E., 1968, in "Hyperfine Structure and Nuclear Radiations – Proceedings of Conference at Asilomar, Calif., 1967" (North-Holland Pub. Co., Amsterdam.) p. 655.

Oblivantsev, A. N. and Kulikov, N. F., 1973, Radiokhimiya **15**, 136.

Odru, P. and Vargas, J. I., 1971, Inorg. Nucl. Chem. Lett. **7**, 379.

Odum, R. and Wolfgang, R., 1963, J. Amer. Chem. Soc. **85**, 1050.

Ofer, S. and Schwartzchild, A., 1959a, Phys. Rev. Lett. **3**, 384.

Ofer, S. and Schwartzchild, A., 1959b, Phys. Rev. **116**, 725.

Ok, H. N. and Mullen, J. G., 1968, Phys. Rev. **168**, 550, 563.

Oleari, L., De Michelis, G. and Di Sipio, L., 1966, Mol. Phys. **10**, 111.

Omori, T. and Shiokawa, T., 1970, Radiochem. Radioanal. Lett. **3**, 39.

Omori, T., Yu-Chai Yeh and Shiokawa, T., 1969, Kakuriken Kenkyu Hokoku **2**, 139.

Omori, T., Shaw-Chii Wu, Tsurumaki, K. and Shiokawa, T., 1970a, Kakuriken Kenkyu Hokoku **3**, 123.

Omori, T., Shaw-Chii Wu and Shiokawa, T., 1970b, Radiochem. Radioanal. Lett. **3**, 40.

Ormond, D. and Rowland, F. S., 1961, J. Amer. Chem. Soc. **83**, 1006.

Ortega, J., 1966, J. Inorg. Nucl. Chem. **28**, 668.

Ottar, B., 1953, Nature (Lond.) **172**, 362.

Otterbach, J., 1972, Thesis, Cologne.

Owens, C. W. and Boyd, G. E., 1976, J. Inorg. Nucl. Chem. **38**, 935.

Owens, C. W. and Hobbs, J. R., 1969, J. Phys. Chem. **73**, 1956.

Owens, C. W. and Lecington, W. C., 1973, J. Inorg. Nucl. Chem. **35**, 685.

Owens, C. W. and Lecington, W. C., 1975, Radiochim. Acta **22**, 81.

Owens, C. W. and Rowland, F. S., 1962, J. Inorg. Nucl. Chem. **24**, 133.

Paiss, Y. and Amiel, S., 1964, J. Amer. Chem. Soc. **86**, 2332.

Palm, C., Fischer, E. O. and Baumgärtner, F., 1962a, Tetrahedron Lett. **6**, 253.

Palm, C., Fischer, E. O. and Baumgärtner, F., 1962b, Naturwiss. **49**, 279.

Palmer, D. W., Thompson, M. W. and Townsend, P. D., 1970, "Atomic Collision Phenomena in Solids", (North-Holland, Amsterdam, American Elsevier, NY).

Pappas, A. C., 1955, Proc. Int. Conf. Peaceful Uses of Atomic Energy, UN, Geneva **7**, 19.

Parker, W., 1962, Nature (Lond.) **196**, 763.

Parker, W. and Perez Alarcon, J., 1970, Radiochem. Radioanal. Lett. **3**, 223.

Parks, N. J., Peek, N. F. and Goldstine, E., 1975, Int. J. App. Radiat. Isotopes **26**, 683.

Pasternak, M., 1967, Symp. Faraday Soc., "The Mössbauer Effect", p. 119.

Pasternak, M. and Sonnino, T., 1967, Phys. Rev. **164**, 384.

Paulus, J. M., 1966, Thesis, Strasbourg.

Paulus, J. M., 1967, Radiochim. Acta **7**, 141.

Paulus, J. M. and Abbé, J. C., 1973, J. Chim. Phys. **70**, 690.

Paulus, J. M. and Adloff, J. P., 1965, Radiochim. Acta **4**, 146.

Pauly, J., 1955, Compt. rend. **240**, 2415.

Pauly, J. and Süe, P., 1955, Compt. rend. **240**, 2226.

Pauly, J. and Süe, P., 1957, J. Phys. Radium. **18**, 22.

Payne, B. R., Scargill, P. and Cook, G. B., 1958, Radioisotopes in Sci. Research, Conf., Paris **2**, 154.

Peak, D. and Corbett, J. W., 1972, Phys. Rev. **5B**, 1226.

Pearlstein, E. A., 1953, Phys. Rev. **92**, 881.

Peretrukhin, V. F., Krot, N. N., Glazunov, M. P. and Gel'man, A. D., 1967a, Dokl. Akad., Nauk., SSSR **173**, 609.

Peretrukhin, V. F., Krot, N. N., Glazunov, M. P. and Gel'man, A. D., 1967b, Radiokhimiya **9**, 665.

Perfiliev, Yu. D., Afanasov, M. I., Kulikov, L. A. and Babeshkin, A. M., 1974, Vestn. Moscow Univ. Ser. II Khim. **15**, 752.

Perlman, M. L. and Emery, G. T., 1969, Brookhaven Report, BNL-13921.

Perlow, G. J. and Perlow, M. R., 1964a, Rev. Mod. Phys. **36**, 353.

Perlow, G. J. and Perlow, M. R., 1964b, J. Chem. Phys. **41**, 1157.

Perlow, G. J. and Perlow, M. R., 1965, Symp. Chem. Effects of Nucl. Transformations, IAEA, Vienna **2**, 443.

Perlow, G. J. and Perlow, M. R., 1968, J. Chem. Phys. **48**, 955.

Perlow, G. J. and Yoshida, H., 1968, J. Chem. Phys. **49**, 1474.

Perrin, N. and De Wieclawik, W., 1966, Compt. rend. **262B**, 211.

Persky, A. and Baer, M., 1974, J. Chem. Phys. **60**, 133.

Pertessis, M. and Henry, R., 1963, Radiochim. Acta **1**, 58.

Pimental, G. C. and Spratley, R. D., 1963, J. Amer. Chem. Soc. **85**, 826.

Pimental, G. C., Spratley, R. D. and Miller, A. R., 1964, Science **143**, 674.

Pistorius, C. W. F. T., 1962, Z. Physik. Chem. **35**, 109.

Placzek, G., 1946, Phys. Rev. **69**, 423.

Pleasonton, F., 1968, Phys. Rev. **174.**, 1500.

Pleasonton, F. A. and Snell, H., 1957, Proc. Roy. Soc. (London) **A241**, 141.

Podeanu, G. and Costea, T., 1971, Radiochem, Radioanal. Lett. **6**, 57.

Pollak, H., 1962, Phys. Stat. Solidi **2**, 720.

Popov, D. K., 1959, Russ. J. Phys. Chem. **33**, 9.

Popplewell, D. S., 1963, J. Inorg, Nucl. Chem. **25**, 318.

Porter, F. T., Freedman, M. S. and Wagner, F., Jr., 1971, Phys. Rev. **C3**, 2246.

Porter, R. N., 1967, J. Chem. Phys. **45**, 2388.

Porter, R. N., 1974, Ann. Rev. Phys. Chem. **25**, 317.

Porter, R. N. and Karplus, M., 1964, J. Chem. Phys. **40**, 1105.
Porter, R. N. and Kunt, S., 1970, J. Chem. Phys. **52**, 3240.
Porter, R. N., Sims, L. B., Thompson, D. L. and Raff, L. M., 1973, J. Chem. Phys. **58**, 2855.
Porter, R. N., Thompson, D. L., Raff, L. M. and White, J. M., 1975, J. Chem. Phys. **62**, 2429.
Preetz, W. and Nadler, K., 1971, J. Inorg. Nucl. Chem. **33**, 2688.
Primak, W., 1955, Phys. Rev. **100**, 1677.
Primak, W., 1960, J. App. Phys. **31**, 1524.
Probst, C., Kienle, P., Luchner, K., Wagner, F. E. and Zahn, U., 1972, Radiochim. Acta **18**, 19.
Pronko, P. P. and Kelly, R., 1970, Radiation Eff. **3**, 161.
Prusakov, V. E., Stanko, V. I., Stukan, R. A. and Krapov, V. V., 1972, Isotopenpraxis **8**, 379.
Prusakov, V. E., Stukan, R. A., Borshagovski, B. V. and Goldanski, V. I., 1973, Teor. Eksp. Khim. URSS **9**, 707.
Prusakov, V. E., Stukan, R. A., Borshagovski, B. V. and Goldanski, V. I., 1974, Soviet Radiochem. **16**, 66.
Prussin, S. G. and Meinke, W., 1965, Radiochim. Acta **4**, 79.
Rabinovitch, B. S. and Setser, D. W., 1964, Advan. Photochem. **3**, 1.
Raff, L. M., 1974, J. Chem. Phys. **60**, 2220.
Rajman, I., 1971, Private communication.
Rajman, I., 1973, Thesis, Nucl. Res. Inst., Rez.
Ramshesh, V., 1969, J. Inorg. Nucl. Chem. **31**, 3878.
Ramshesh, V., Anthony, M. C. and Venkateswarlu, K. S., 1972b, Radiochem. Radioanal. Lett. **11**, 93.
Ramshesh, V., Venkateswarlu, K. S. and Shankar, J., 1972a, J. Inorg. Nucl. Chem. **34**, 2121.
Rankin, C. C. and Light, J. C., 1969, J. Chem. Phys. **51**, 1701.
Rao, K. A., 1969, Radiochim. Acta **12**, 11.
Rao, K. A. and Nath, A., 1966, Radiochim. Acta **5**, 162.
Rao, K. A., Nath, A. and Shankar, J., 1962, USAEC, NP-12 330, 422.
Rauscher, H. E., 1960, Thesis, Columbia University, USA.
Rauscher, H. E. and Harbottle, G., 1957, J. Inorg. Nucl. Chem. **4**, 155.
Rauscher, H., Sutin, N. and Miller, J. M., 1960, J. Inorg. Nucl. Chem. **12**, 378.
Rauscher, H., Sutin, N. and Miller, J. M., 1961, J. Inorg. Nucl. Chem. **17**, 31.
Reed, G. W. and Turkevitch, A., 1953, Phys. Rev. **92**, 1473.
Reichold, P. and Anders, H. P., 1966, Radiochim. Acta **5**, 44.
Riedel, H. J. and Merz, E., 1965, Radiochim. Acta **4**, 48.
Riedel, H. J. and Merz, E., 1966, Radiochim. Acta **6**, 144.
Rieder, W., 1951, Acta Phys. Austriaca **4**, 290.
Rieder, W., Broda, E. and Erber, J., 1950, Monatsh. Chem. **81**, 657.
Riehl, N. and Sizmann, R., 1964, Radiochim. Acta **3**, 44.
Robinson, M. T., Rössler, K. and Torrens, I. M., 1974, J. Chem. Phys. **60**, 680.
Rolfe, J., 1958, Phys. Rev. Lett. **1**, 56.
Root, J. W., 1975, US R and D Admin. Tech. Report, UCD-34P 158-74-2, Chapter X.
Root, J. W. and Rowland, F. S., 1963, J. Chem. Phys. **38**, 2030.
Root, J. W. and Rowland, F. S., 1967, J. Chem. Phys. **46**, 4299.
Root, J. W., Breckenbridge, W. and Rowland, F. S., 1965, J. Chem. Phys. **43**, 3694.
Rosenberg, H. E., 1964, Thesis, Clark University.
Rosenberg, H. E. and Sugihara, T. T., 1965, Symp. Chem. Effects of Nucl. Transformations, IAEA, Vienna **2**, 151.
Ross, R., 1969, Report ZFK-169, Dresden.
Rössler, K., 1968, Thesis, Cologne.
Rössler, K., 1970, Proc. XIII Internal. Conf. Coordination Chem. Cracow-Zakopane, Vol. K. 17.

Rössler, K. and Otterbach, J., 1971a, Radiochim. Acta **15**, 103.

Rössler, K. and Otterbach, J., 1971b, Sixth International Hot Atom Chem. Symp., Brookhaven Nat. Lab., L. I., USA. See also: Angew. Chem., Internat. Edit., **10**, 852.

Rössler, K. and Pross, L., 1974, Radiochem. Radioanal. Lett. **18**, 291.

Rössler, K. and Robinson, M. T., 1974, in "Atomic Collissions in Solids", eds. Datz, S., Appleton, B. R. and Moak, C. D., (Plenum Press, NY) p. 237.

Rössler, K., Otterbach, J. and Stöcklin, G., 6th International Hot Atom Symposium, Brookhaven Nat. Lab., L.I., USA.

Rössler, K., Otterbach, J. and Stöcklin, G., 1972, J. Phys. Chem. **76**, 2499.

Rother, P., Wagner, F. and Zahn, U., 1969, Radiochim. Acta **11**, 203.

Rowland, F. S., 1972, MTP Reviews Physical Chem., Ser. I., Vol. **9**, p. 109.

Rowland, F. S. and Coulter, P., 1964, Radiochim. Acta **2**, 163.

Rowland, F. S. and Libby, W. F., 1953, Chem. Phys. **21**, 1495.

Rowland, F. S., Cramer, J. A., Iyer, R. S., Milstein, R. and Williams, R. L., 1973, in "Radiopharmaceuticals and labelled compounds", IAEA, Vienna, p. 378.

Rowland, F. S., Hathaway, L. and Kambara, T., 1961a, Symp. Chem. Effects of Nucl. Transformations, IAEA, Prague **2**, 255.

Rowland, F. S., Lee, J. K., Musgrave, B. and White, R. M., 1961b, Symp. on Chem. Effects of Nucl. Transformations, IAEA, Prague **2**, 67.

Roy, R., 1972, Thesis, Univ. of Louvain.

Roy, R., Cogneau, M. A. and Apers, D. J., 1973, Bull. Soc. Chim. Belg. **82**, 75.

Rubinson, W., 1963, Phys. Rev. **130**, 2011.

Rudenko, N. P. and Shuvaeva, T. M. and Merz, N. I., 1964, Radiokhimiya **6**, 329.

Rudstam, G., 1956, "Spallation of Medium Weight Elements" (Upsala), p. 26.

Ruffenach, J. C., 1971, Thesis, Strasbourg.

Rutherford, E., 1951, "Radiations from Radioactive Substances," (Cambridge University Press) p. 157.

Saeki, M. and Tachikawa, E., 1973, Radiochim. Acta **20**, 27.

Saito, N., 1965, Symp. Chem. Effects of Nucl. Transformations, IAEA, Vienna **2**, 383.

Saito, N. and Ambe, F., 1970, Bull. Chem. Soc. Japan **43**, 282.

Saito, N. and Sekine, T., 1958, Bull. Chem. Soc. Japan **31**, 789.

Saito, N. and Tominaga, T., 1965, Bull. Chem. Soc. Japan **38**, 505.

Saito, N., Tomita, I. and Furokawa, M., 1959, J. Atom Energy Soc. Japan **1**, 196.

Saito, N., Sano, H. and Tominaga, T., 1960a, Bull. Chem. Soc. Japan **33**, 20.

Saito, N., Tominaga, T. and Sano, H., 1960b, Bull. Chem. Soc. Japan **33**, 120.

Saito, N., Tominaga, T. and Sano, H., 1960c, Bull. Chem. Soc. Japan **33**, 1621.

Saito, N., Sano, H. and Tominaga, T., 1961a, Chem. Ind., 1796.

Saito, N., Tominaga, T. and Sano, H., 1961b, Symp. Chem. Effects of Nucl. Transformations, IAEA, Prague **1**, 541.

Saito, N., Sano, H., Tominaga, T., Ambe, F. and Fujino, T., 1962a, Bull. Chem. Soc. Japan **35**, 744.

Saito, N., Tominaga, T. and Sano, H., 1962b, Bull. Chem. Soc. Japan **35**, 63.

Saito, N., Tominaga, T. and Sano, H., 1962c, Bull. Chem. Soc. Japan **35**, 365.

Saito, N., Tominaga, T. and Sano, H., 1962d, J. Inorg. Nucl. Chem. **24**, 1539.

Saito, N., Tominaga, T. and Sano, H., 1962e, Nature (Lond.) **194**, 466.

Saito, N., Tominaga, T. and Sano, H., 1963a, Bull. Chem. Soc. Japan **36**, 230.

Saito, N., Tominaga, T. and Sano, H., 1963b, Bull. Chem. Soc. Japan **36**, 232.

Saito, N., Sano, H. and Tominaga, 1964, Chem. Ind., 1622.

Saito, N., Ambe, F. and Sano, H., 1965a, Nature (Lond.) **205**, 688.

Saito, N., Ambe, F. and Sano, H., 1965b, Nature (Lond.) **206**, 505.

Saito, N., Itoh, S. and Tominaga, T., 1965c, Bull. Chem. Soc. Japan **38**, 504.

Saito, N., Tominaga, T. and Sano, H., 1965d, Bull. Chem. Soc. Japan **38**, 1407.
Saito, N., Ambe, F. and Sano, H., 1967a, Radiochim. Acta **7**, 131.
Saito, N., Takeda, M. and Tominaga, T., 1967b, Bull. Chem. Soc. Japan **40**, 690.
Saito, N., Ambe, F., Ambe, S. and Shimamura, A., 1970, Bull. Chem. Soc. Japan **43**, 284.
Saito, N., Takeda, M. and Tominaga, T., 1971, Radiochem. Radioanal. Lett. **6**, 169.
Sakanoue, M. and Endo, K., 1970, Radiochem. Radioanal. Lett. **4**, 99.
Sakanoue, M. and Komura, K., 1971, Nature (Lond.) **233**, 80.
Sakanoue, M., Yoneda, S., Onishi, K., Koyama, K., Komura, K. and Nakanishi, T., 1968, Geochem. J. **2**, 71.
Sanchez, J. P., Llabador, Y. and Friedt, J. M., 1973, J. Inorg. Nucl. Chem. **35**, 3557.
Sano, H. and Iwagami, H., 1971, Chem. Comm., 1637.
Sano, H. and Kanno, M., 1969, Chem. Comm., 601.
Sano, H. and Ohnuma, T., 1974a, Chem. Phys. Lett. **26**, 348.
Sano, H. and Ohnuma, T., 1974b, Chem. Lett. (Japan) **6**, 589.
Sano, H. and Ohnuma, T., 1975, Bull. Chem. Soc. Japan **48**, 266.
Sano, H., Sato, K. and Iwagami, H., 1971, Bull. Chem. Soc. Japan **44**, 2570.
Saris, F. W. and Van der Weg, W. F., 1976, eds. "Proc. Vl Int. Conf. Atomic Collisions in Solids", (North-Holland, Amsterdam).
Sarrach, D. and Vormum, G., 1961, Symp. Chem. Effects of Nucl. Transformations, IAEA, Prague **1**, 565.
Sarup, S. and Nath, A., 1967, J. Inorg. Nucl. Chem. **29**, 299.
Sasaki, T. and Shiokawa, T., 1970, Bull. Chem. Soc, Japan **43**, 2835.
Sato, S., 1955a, J. Chem. Phys. **23**, 592, 2465.
Sato, S., 1955b, Bull. Chem. Soc, Japan **28**, 450.
Sato, T. R. and Strain, H. H., 1961, Symp. Chem. Effects of Nucl. Transformations, IAEA, Prague **1**, 503.
Sato, T. R., Kisieleski, W. E., Norris, W. P. and Strain, H. H., 1953, Anal. Chem. **25**, 438.
Sato, T. R., Sellers, P. A. and Strain, H. H., 1959, J. Inorg. Nucl. Chem. **11**, 84.
Sato, J., Yokoyama, Y. and Yamazaki, T., 1966, Radiochim. Acta **5**, 115.
Saxon, R. P. and Light, J. C., 1971, J. Chem. Phys. **55**, 455.
Saxon, R. P. and Light, J. C., 1972a, J. Chem. Phys. **56**, 3874, 3885.
Saxon, R. P. and Light, J. C., 1972b, J. Chem. Phys. **57**, 2758.
Scanlon, M. D. and Collins, K. E., 1971, Radiochim. Acta **15**, 141.
Schara, M., Sĕntjure, M. Milenkovic, S. M. and Veljković, S. R., 1970, J. Inorg. Nucl. Chem. **32**, 369.
Schatz, G. C. and Kuppermann, A., 1973, J. Chem. Phys. **59**, 964.
Schatz, G. C., Bowmann, J. M. and Kuppermann, A., 1973, J. Chem. Phys. **58**, 4023.
Scheffer, F. and Ludwig, F., 1957, Naturwiss. **44**, 396.
Schleiffer, J. J. and Adloff, J. P., 1964, Radiochim. Acta **3**, 145.
Schmidt, G. B., 1961, Thesis, Cologne.
Schmidt, G. B., 1966, Radiochim. Acta **5**, 178.
Schmidt, G. B. and Herr, W., 1961, Symp. Chem. Effects of Nucl. Transformations, IAEA, Prague **1**, 527.
Schmidt, G. B. and Herr, W., 1963, Z. Naturforsch **18a**, 505.
Schmidt, G. B. and Herr, W., 1965, Symp. Chem. Effects of Nucl. Transformations, IAEA, Vienna **2**, 373.
Schmidt, G. B. and Rössler, K., 1966, Radiochim. Acta **5**, 123.
Schmidt, G. B., Heine, W. and Herr, W., 1963, Symp. Radiation Damage in Solids, IAEA, Vienna **3**, 93.
Schmidt, G. B., Herr, W. and Rössler, K., 1965, Angew. Chem. (Internat. Edit.) **4**, 990.
Schmidt-Ott, W. D., 1963, Z. Physik. **174**, 206.

Schmidt-Ott, W. D., Hoffmann, K. W., Krause, I. Y. and Flammersfeld, A., 1960a, Z. Physik. **158**, 242.

Schmidt-Ott, W. D., Hoffmann, K. W., Krause, I. Y. and Flammersfeld, A., 1960b, Z. Physik. **158**, 248.

Schmitt, H. W., Neiler, J. H. and Walter, F. J., 1966, Phys. Rev. **141**, 1146.

Schroth, F. and Adloff, J. P., 1964a, J. Chim. Physique **61**, 1373.

Schroth, F. and Adloff, J. P., 1964b, Compt rend. **258**, 5863.

Schulman, J. H. and Compton, W. D., 1962, "Colour Centres in Solids," (Pergamon Press, Oxford).

Schwartz, A., Rafaeloff, R. and Yellin, E., 1969, Int. J. Appl. Rad. Isotopes **20**, 853.

Schwarz, E., Denschlag, H. O. and Herrmann, G., 1966, Radiochim. Acta **5**, 53.

Schwartz, H. M., 1953, J. Chem. Phys. **21**, 45.

Schwarzschild, A. Z. and Warburton, E. K., 1968, Ann. Rev. Nucl. Sci. **18**, 265.

Schweinler, H. C., 1961, Symp. Chem. Effects of Nucl. Transformations, IAEA, Prague **1**, 63.

Schweitzer, G. K. and Wilhelm, D. L., 1956, J. Inorg. Nucl. Chem. **3**, 1.

Seaborg, G. T. and Kennedy, J. W., 1939, Phys. Rev. **55**, 410.

Seaborg, G. T., Livingood, J. J. and Kennedy, J. W., 1939, Phys. Rev. **55**, 794.

Seaborg, G. T., Friedlander, G. and Kennedy, J. W., 1940a, J. Amer. Chem. Soc. **62**, 1309.

Seaborg, G. T., Livingood, J. F. and Kennedy, J. W., 1940b, Phys. Rev. **57**, 363.

Seewald, D. and Wolfgang, R., 1967, J. Chem. Phys. **46**, 1207.

Seewald, D., Gersh, M. and Wolfgang, R., 1966, J. Chem. Phys. **45**, 3870.

Segré, E., Halford, R. and Seaborg, G. T., 1939, Phys. Rev. **55**, 321.

Seiler, H. and Seiler, M., 1967, Helv. Chem. Acta **50**, 2477.

Seitz, F., 1952, Physics Today **5**, 6.

Seitz, F. and Koehler, J. S., 1956, Solid State Physics, (Academic Press, New York) **2**, 305.

Sellers, P. A., Sato, T. R. and Strain, H. H., 1957, J. Inorg. Nucl. Chem. **5**, 31.

Sen Gupta, S., 1969, Indian J. Chem. **7**, 818.

Sensui, Y. and Matsuura, T., 1965, Bull. Chem. Soc. Japan **38**, 1171.

Serber, R. and Snyder, H. S., 1952, Phys. Rev. **87**, 152.

Seregin, P. P. and Savin, E. P., 1972, Soviet Phys. Solid State **13**, 2846.

Seregin, P. P. and Savin, E. P., 1973, Soviet, Phys. Solid State **15**, 541.

Setser, W. D., Moser, H. C. and Hein, R. E., 1959, J. Amer. Chem. Soc. **81**, 4162.

Seyboth, D., 1969, Proc. Roy. Soc. (London) **A311**, 119.

Shakhashiri, B. Z. and Gordon, G., 1965, J. Inorg. Nucl. Chem. **27**, 2161.

Shankar, J., 1968, Indian Atomic En. Report, BARC/348.

Shankar, J. and Gupta, S. S., 1969, Indian J. Chem. **7**, 794.

Shankar, J. and Shankar, R., 1960, J. Univ. Bombay, Phys. Sci. **29**, 63.

Shankar, J., Lal, M. and Venkateswarlu, K. S., 1960, Symposium on "Nuclear and Radiation Chemistry", Calcutta, 1960.

Shankar, J., Thomas, V. G. and Nath, A., 1961a, Symp. Chem. Effects of Nucl. Transformations, IAEA, Prague **1**, 383.

Shankar, J., Srivastava, S. B. and Shankar, R., 1961b, Symp. Chem. Effects of Nucl. Transformations, IAEA, Prague **1**, 393.

Shankar, J., Venkateswarlu, K. S. and Nath, A., 1961c, Symp. Chem. Effects of Nucl. Transformations, IAEA, Prague **1**, 309.

Shankar, J., Venkateswarlu, K. S. and Lal, M., 1961d, Symp. Chem. Effects of Nucl. Transformations, IAEA, Prague **1**, 417.

Shankar, J., Nath, A. and Rao, M. H., 1964, Radiochim. Acta **3**, 26.

Shankar, J., Venkateswarlu, K. S. and Lal, M., 1965a, Radiochim. Acta **4**, 52.

Shankar, J., Nath, A. and Vaish, S. P., 1965b, Radiochim. Acta **4**, 162.

Shankar, J., Venkateswarlu, K. S. and Lal, M., 1966, J. Inorg. Nucl. Chem. **28**, 11.

Sharman, L. J. and McCallum, K. J., 1955, J. Chem. Phys. **29**, 597.
Sharp, R. A., Schmitt, R. A., Suffredini, C. S. and Randolph, D. F., 1959, USAEC, Report, G. A.-910.
Shaviev, A. I., Vityutnev, V. M., Ivanov, V. M. and Yakovlev, G. N., 1973, Radiokhimiya **15**, 820.
Sherif, M. K., Issa, I. M., Diefallah, E. M., Mousa, M. A., Aly, H. F. and Rassoul, A. A. A., 1974, Indian. J. Chem. **12**, 712.
Shibata, N., Fujinaga, T., Yoshihara, K. and Kanchiku, Y., 1966, Radiochim. Acta **5**, 238.
Shibata, N., Amano, H., Yoshihara, K. and Ebihara, H., 1969, Atompraxis **15**, 43.
Shima, M. and Utsumi, S., 1961, J. Inorg. Nucl. Chem. **20**, 177.
Shiokawa, T. and Omori, T., 1965, Bull. Chem. Soc. Japan **38**, 1892.
Shiokawa, T. and Omori, T., 1969, Bull. Chem. Soc. Japan **42**, 696.
Shiokawa, T. and Sasaki, T., 1970, Bull. Chem. Soc. Japan **43**, 801.
Shiokawa, T., Kudo, H. and Omori, T., 1965, Bull. Chem. Soc. Japan **38**, 1340.
Shiokawa, T., Yagi, M. and Hara, M., 1967, J. Nucl. Sci. Tech. Japan **4**, 297.
Shiokawa, T., Kudo, H., and Omori, T., 1969a, Bull. Chem. Soc. Japan **42**, 436.
Shiokawa, T., Sasaki, T. and Takahashi, S., 1969b, Radiochim. Radioanal. Lett. **1**, 31.
Shiokawa, T., Yagi, M. and Sasaki, T., 1969c, Radiochim. Acta **12**, 54.
Shiokawa, T., Kudo, H. and Omori, T., 1970, Bull. Chem. Soc. Japan **43**, 2076.
Shiokawa, T., Sasaki, T. and Takahashi, S., 1971, Radiochem. Radioanal. Lett. **6**, 327.
Shukla, B. M., Singh, R. N. and Mishra, S. P., 1970, Proc. Chem. Sym., Bombay. Dep. of Atomic Energy **2**, 93.
Siegbahn, K., Nordling, C., Johannson, G., Hedman, J., Hedén, P. F., Hamrin, K., Gelius, V., Bergmark, R., Werme, L. O., Manne, R. and Baer, Y., 1969, "ESCA Applied to Free Molecules", (North-Holland, Amsterdam).
Siegel, S., 1949. Phys. Rev. **75**, 1823.
Siekierska, K. E. and Fenger, J., 1970, Radiochim. Acta **14**, 93.
Siekierska, K. E. and Halpern, A., 1966, Radiochim. Acta **5**, 51.
Siekierska, K. E. and Sokolowska, A., 1962, J. Inorg. Nucl. Chem. **24**, 13.
Siekierska, K. E., Sokolowska, A. and Campbell, I. G., 1959, J. Inorg. Nucl. Chem. **12**, 18.
Siekierska, K. E., Halpern, A. and Siuda, A., 1961, Symp. Chem. Effects of Nucl. Transformations, IAEA, Prague **1**, 171.
Siekierska, K. E., Halpern, A. and Maddock, A. G., 1968, J. Chem. Soc. A., 1645.
Siekierska, K. E., Milman, M. and Fenger, J., 1970, Radiochim. Radioanal. Lett. **4**, 251.
Siekierska, K. E., Fenger, J. and Olesen, J., 1972, J. Chem. Soc. Dalton, 2020.
Siekierska, K. E., Fenger, J. and Maddock, A. G., 1973, J. Chem. Soc. Dalton, 1086.
Silsbee, R. H., 1957, J. Appl. Phys. **28**, 1246.
Silva, R. J., 1972, MTP Reviews Series 1, Inorganic Chemistry, Vol. **8**, Butterworths, London.
Simpson, S. L., 1966, Thesis, Iowa State Univ., Ames.
Simpson, J., Taylor, D. and Anderson, D. M. W., 1958, J. Chem. Soc., 2378.
Sinotova, E. N. and Timofeev, S. A., 1968, Radiokhimiya (French Trans.) **10**, 834.
Skell, P. S., Plonka, J. H. and Engel, R. R., 1967, J. Amer. Chem. Soc. **89**, 1748.
Skorik, A. I., Boldyrev, V. V. and Komarov, V. F., 1967, Kinet. Katal. **8**, 1258.
Smoluchowski, M. V., 1916, Phys. Z. **17**, 557.
Smoluchoski, M. V., 1918, Z. Phys. Chem. **A92**, 129.
Smoluchowski, R., 1956, Proc. Intern. Conf. Peaceful Uses of Atomic Energy, UN, Geneva, **7**, 676.
Smoluchowski, R., 1964, "The Interaction of Radiation with Solids", (Interscience, NY) p. 378.
Smoluchowski, R. and Wiegand, D. A., 1961, Disc. Faraday Soc. **31**, 151.
Snediker, D. K. and Miller, W. W., 1968, Radiochim. Acta **10**, 30.
Snell, A. H., 1965, in "α, β, γ Ray Spectroscopy", ed. Siegbahn, K.. North-Holland, Amsterdam **2**, 1545.

Snell, A. H. and Pleasonton, F., 1955, Phys. Rev. **100**, 1396.
Snell, A. H. and Pleasonton, F., 1957, Phys. Rev. **107**, 790.
Snell, A. H. and Pleasonton, F., 1958, J. Phys. Chem. **62**, 1377.
Snell, A. H., Pleasonton, F. and Lenning, H. E., 1957, J. Inorg. Nucl. Chem. **5**, 112.
Snyder, W. S. and Neufeld, J., 1955, Phys. Rev. **97**, 1636.
Sokolowska, A. and Siuda, A., 1968, Radiochim. Acta **10**, 107.
Sotobayashi, T., Suzuki, T. and Noda, T., 1971, Bull. Chem. Soc. Japan **44**, 1711.
Spano, H. and Kahn, M., 1952, J. Amer. Chem. Soc. **74**, 568.
Spencer, C. D. and Schroeer, D., 1974, Phys. Rev. **B9**, 3658.
Spighel, M. and Suzor, F., 1962, Nucl. Phys. **32**, 346.
Spitsyn, V. I., Afonskii, N. S. and Tsviel'nikov, V. I., 1960, Zhur. Neorg. Khim. **5**, 1505.
Srinivasan, S. C., Wiles, D. R. and Yang, I. C., 1966, Inorg. Nucl. Chem. Lett. **2**, 399.
Srivastava, T. S. and Nath, A., 1974, Radiochem. Radioanal. Lett. **16**, 103.
Srivastava, T. S. and Nath, A., 1976, J. Phys. Chem. **80**, 529.
Srivastava, T. S., Przybylinski, J. L. and Nath, A., 1974, Inorg. Chem. **13**, 1562.
Stamouli, M. I., 1971, Thesis, Cambridge.
Stamouli, M. I., 1974, Radiochim. Acta **21**, 90.
Stamouli, M. I., 1975, Radiochim. Acta **22**, 83.
Stamouli, M. I. and Katsanos, N. A., 1968, Radiochim. Acta **9**, 13.
Stanley, C. W. and Davies, T. H., 1951, in "Radiochemical Studies: The Fission Products", eds. Coryell, C. D. and Sugarman, N., McGraw-Hill, NY **1**, 204.
Starke, K., 1940, Naturwiss. **28**, 631.
Starke, K., 1942a, Naturwiss. **30**, 107.
Starke, K., 1942b, Naturwiss. **30**, 577.
Starke, K. and Günther, E., 1964, Radiochim. Acta **2**, 159.
Statnick, R. M., Kashihara, N. and Schmidt-Bleek, F., 1969, J. Inorg. Nucl. Chem. **31**, 878.
Steigman, J., 1941, Phys. Rev. **59**, 498.
Steinwedel, H. and Jensen, J. H., 1947, Z. Naturforsch **2A**, 125.
Stenström, T., 1970, Thesis, Uppsala, Sweden, (Trans. US Nat. Tech. Inf. Service Rep. NP 18376).
Stenström, T., 1971, Sixth International Hot Atom Chem. Symp. Brookhaven Nat. Lab. L. I., U.S.A.
Stenström, T. and Jung, B., 1965, Radiochim. Acta **4**, 3.
Stephas, P., 1969, Phys. Rev. **186**, 1013.
Stephas, P. and Crasemann, B., 1967, Phys. Rev. **164**, 1509.
Stevović, J. and Muxart, R., 1968, Radiochim. Acta **9**, 76.
Stevović, J., Jacimović, L. and Veljković, S. R., 1965, J. Inorg. Nucl. Chem. **27**, 29.
Stewart, G., 1971, Thesis, University of Idaho.
Stewart, G. W. and Hower, C. O., 1972, J. Inorg. Nucl. Chem. **34**, 39.
Stewart, G. W., Henis, J. M. S. and Gaspar, P. P., 1973, J. Chem. Phys. **58**, 890.
Stewart, G. W., Dymerski, P. P. and Hower, C. O., 1974, J. Chem. Phys. **61**, 483.
Stöcklin, G., 1969, "Chemie heisser Atome", Verlag Chemie, Weinheim.
Stöcklin, G., 1975, "Hot Atom Chem. Status Report", IAEA, Vienna, p. 161.
Stöcklin, G. and Tornau, W., 1966, Radiochim. Acta **6**, 86.
Stöcklin, G. and Vogelbruch, K., 1968, Radiochim. Acta **10**, 177.
Stone, J. A. and Pillinger, W. L., 1964, Phys. Rev. Lett. **13**, 200.
Stone, J. A. and Pillinger, W. L., 1966, Bull. Amer. Phys. Soc. **11**, 809.
Straatmann, M. and Welch, M., 1973, Radiochim. Acta **20**, 124.
Stranks, D. R. and Baker, F. B., 1963, Inorganic Syntheses **7**, 201.
Stratton, J. A., 1941, "Electromagnetic Theory", (McGraw-Hill, New York) p. 15.
Strickert, R. G., Amiel, S. and Wahl, A. C., 1974, Inorg. Nucl. Chem. Lett. **10**, 129.

Stucky, G. L. and Kiser, R. W., 1969, Radiochim Acta **11**, 5.
Su, H. Y. and White, J. M., 1975, J. Chem. Phys. **63**, 499.
Su, H. Y., White, J. M., Raff, L. M. and Thompson, D. L., 1975, J. Chem. Phys. **62**, 1435.
Süe, P., 1948, J. Chim. Phys. **45**, 177.
Süe, P. and Kayas, G., 1948, J. Chim. Phys. **45**, 188.
Süe, P. and Melander, L., 1947, Compt. rend. **225**, 413.
Suess, H., 1940, Z. Physik. Chem. **B45**, 312.
Su[illegible] N. and Dodson, R. W., 1958, J. Inorg. Nucl. Chem. **6**, 91.
S[illegible]., 1960, J. Physique **21**, 233, and additional references given there.
S[illegible] H. B., Tumosa, C. S. and Ache, H. J., 1971, Radiochim. Acta **16**, 112.
Sy[illegible]kov, N. G., 1965, Atomnaya Energia **19**, 169.
Szilard, L. and Chalmers, T.A. 1934a, Nature (Lond.) **134**, 462.
Szilard, L. and Chalmers, T. A., 1934b, Nature (Lond.) **134**, 494.
Szucs, S. and Delfosse, J. M., 1965, Phys. Rev. Lett. **15**, 163.
Tamaki, Y., Omori, T. and Shiokawa, T., 1975, Radiochem. Radioanal. Lett. **20**, 255.
Tanaka, K., 1964a, Bull. Chem. Soc. Japan **37**, 1032.
Tanaka, K., 1964b, Bull. Chem. Soc. Japan **37**, 1346.
Tanaka, K., 1964c, Bull. Chem. Soc. Japan **37**, 1730.
Tang, K. T. and Karplus, M., 1968, J. Chem. Phys. **44**, 1676.
Tang, K. T. and Karplus, M., 1971, Phys. Rev. **A4**, 1244.
Tang, Y. N., Gennaro, G. P. and Su, Y. Y., 1972, J. Amer. Chem. Soc. **94**, 4355.
Tenorio, D., Bulbulian, S. and Adloff, J. P., 1976, Radiochem. Radioanal. Lett. **24**, 61.
Teofilovski, C., 1966, Bull. Boris Kidric Inst. Nucl Sci. **17**, 17.
Terell, J., Scott, W. E., Gilmore, J. S. and Minkkinen, O. C., 1953, Phys. Rev. **92**, 1091.
Thomas, V. G., 1967, Thesis, Banares Hindu University.
Thomas, V. G., 1969, Indian J. Chem. **7**, 247.
Thomas, V. G., 1972, Radiochem. Radioanal. Lett. **12**, 97.
Thomas, W. D. E., and Nicholas, D. J. D., 1949, Nature (Lond.) **163**, 719.
Thompson, J. L. and Miller, W. W., 1963, J. Chem. Phys. **38**, 2477.
Thompson, J. L., Ching, J. and Fung, E. Y., 1972, Radiochim. Acta **18**, 57.
Thompson, M. W. and Nelson, R. S., 1962, Phil. Mag. (8) **7**, 2015.
Thomsen, P. V., 1968, J. Chem. Phys. **49**, 756.
Tissier, A., 1970, Thesis, Univ. of Grenoble.
Tominaga, T., 1973a, Radioisotopes (Tokyo) **22**, 8.
Tominaga, T., 1973b, Radioisotopes (Tokyo) **22**, 87.
Tominaga, T., 1973c, Radioisotopes (Tokyo) **22**, 411.
Tominaga, T. and Fujiwara, K., 1970, Bull. Chem. Soc. Japan **43**, 2279.
Tominaga, T. and Nishi, Y., 1971, Radiochem. Radioanal. Lett. **8** 151.
Tominaga, T. and Nishi, Y., 1972a, Radiochem. Radioanal. Lett. **11**, 289.
Tominaga, T. and Nishi, Y., 1972b, Bull. Chem. Soc. Japan **45**, 3213.
Tominaga, T. and Nishi, Y., 1972c, Radiochem. Radioanal. Lett. **11**, 289.
Tominaga, T. and Rowland, F. S., 1968, J. Phys. Chem. **72**, 1399.
Tominaga, T. and Sakai, T., 1972, Bull. Chem. Soc. Japan **45**, 1237.
Tominaga, T., Hosaka, A. and Rowland, F. S., 1969, J. Phys. Chem. **73**, 465.
Tominaga, T., Ishii, M. and Saito, N., 1971a, Radioisotopes (Tokyo) **20**, 579.
Tominaga, T., Sakai, T. and Fujiwara, K., 1971b, Bull. Chem. Soc. Japan **44**, 3036.
Tominaga, T., Nishi, Y. and Motohashi, E., 1974, Radiochem. Radioanal. Lett. **18**, 15.
Toropova, M. A., 1957, J. Inorg. Chem. (USSR) **2**, 1201.
Toropova, M. A., Nefedov, V. D., Levchenko, A. V. and Saikov, Yu. P., 1968a, Soviet Radiochem. **10**, 601.
Toropova, M. A., Nefedov, V. D., Levchenko, A. V. and Matveev, O. G., 1968b, Radio-

khimiya **10**, 613.
Torrens, I. McC. and Chadderton, L. T., 1967, Phys. Rev. **159**, 671.
Triftshäuser, W. and Craig, P. P., 1966, Phys. Rev. Lett. **16**, 1161.
Triftshäuser, W. and Craig, P. P., 1967, Phys. Rev. **162**, 274.
Triftshäuser, W. and Schroerr, D., 1969, Phys. Rev. **187**, 491.
Truhlar, D. G. and Kuppermann, A., 1970, J. Chem. Phys. **52**, 3841.
Truhlar, D. G. and Kuppermann, A., 1971, Chem. Phys. Lett. **9**, 269.
Truhlar, D. G. and Kuppermann, A., 1972, J. Chem. Phys. **56**, 2232.
Truhlar, D. G., Kuppermann, A. and Adams, J. T., 1973, J. Chem. Phys. **59**, 395.
Tucker, C. W. and Senio, P., 1956, J. Appl. Phys. **27**, 207.
Tumosa, C. S. and Ache, H. J., 1970, Radiochim. Acta **14**, 83.
Turco, A. and Scatena, M., 1955, Ricerca Sci. **25**, 2651.
Tyson, J. J., Saxon, R. P. and Light, J. C., 1973, J. Chem. Phys. **59**, 363.
Udupa, M. R., Hariharan, P. V. and Aravamudan, G., 1970, Curr. Sci. (India) **39**, 230.
Ueta, M., 1967, J. Phys. Soc. Japan **23**, 1265.
Ujimoto, K., Nakamura, T., Asada, M., Yoza, N., Takashima, Y. and Ohashi, S., 1970, J. Inorg. Nucl. Chem. **32**, 3177.
Urch, D. S., 1972, in MTP Reviews Inorganic Chem., Ser. I, Vol. **8**, Radiochemistry, p. 149.
Urch, D. S., 1975, in Specialist Rep. Chem. Soc. (London), Radiochemistry **2**, 1.
Urch, D. and Welch, M., 1966, Radiochem. Acta **5**, 2021.
Urch, D. and Wolfgang, R., 1961a, J. Amer. Chem. Soc. **83**, 2982.
Urch, D. and Wolfgang, R., 1961b, Symp. Chem. Effects of Nuclear Transformations, IAEA, Prague **2**, 67.
Vaish, S. P., 1970, J. Inorg. Nucl. Chem. **32**, 2082.
Valencich, T. and Bunker, D. L., 1973a, J. Chem. Phys. **61**, 21.
Valencich, T. and Bunker, D. L., 1973b, Chem. Phys. Lett. **20**, 50.
van de Leest, 1965, Thesis Univ. of Louvain.
van Dulmen, A. A. and Aten, A. H. W., 1971, Radiochim. Acta **15**, 26.
vand, V., 1943, Prog. Phys. Soc. **55**, 222.
van Herk, G. and Aten, A. H. W., 1972, Radiochim. Acta **17**, 214.
van Ooij, W. J., 1971, Thesis Delft.
van Ooij, W. J., 1973, J. Chromat. **81**, 190.
van Ooij, W. J. and Houtman, J. P. W., 1967, Radiochim. Acta **7**, 115.
van Ooij, W. J. and Houtman, J. P. W., 1973a, Radiochim. Acta **20**, 21.
van Ooij, W. J. and Houtman, J. P. W., 1973b, Radiochim. Acta **20**, 47.
van Ooij, W. J. and Houtman, J. P. W., 1974a, Radiochim. Acta **21**, 136.
van Ooij, W. J. and Houtman, J. P. W., 1974b, Radiochim. Acta **21**, 142.
van Urk, P., 1970, Thesis, University of Amsterdam.
van Vliet, D., 1970, Contemp. Phys. **11**, 173.
van Wazer, J.R., 1958, "Phosphorus and its Compounds", (Interscience, NY).
Vargas, J. I., 1972, MTP Reviews, Ser. I, Inorganic Chemistry, **8**.
Vargas, J.I., Berthier, J., Hocquenghem, J.C., Ribot, J.J. and Boyer, P., 1969, Proc. Roy. Soc. **A311**, 191.
Varley, J. H. O., 1962, J. Phys. Chem. Solids **23**, 985.
Vasudev, P., 1965, Thesis, Carleton University, Canada.
Vasudev, P. and Jones, C. H. W., 1972, Radiochim Acta **17**, 121.
Veljković, S. R., 1970, Phil. Mag. **21**, 627.
Veljković, S. R. and Harbottle, G., 1961a, J. Inorg. Nucl. Chem. **23**, 159.
Veljković, S. R. and Harbottle, G., 1961b, Nature (London) **191**, 1287.
Veljković, S. R. and Harbottle, G., 1962, J. Inorg. Nucl. Chem. **24**, 1517.
Veljković, S. R., Milenković, S. M. and Ratković, M. R., 1965a, J. Inorg. Nucl. Chem. **27**, 266.

Veljković, S. R., Milenković, S. M. and Ratković, M. R., 1965b, Symp. Chem. Effects of Nucl. Transformations, IAEA, Vienna **2**, 267.
Venkateswarlu, K. S., 1969a, Indian Atomic En. Report, BARC/446.
Venkateswarlu, K. S., 1969b, Proc. Chem. Symp., Bombay.
Venkateswarlu, K. S. and Kishore, K., 1968, J. Inorg. Nucl. Chem. **30**, 320.
Venkateswarlu, K. S. and Kishore, K., 1971, Radiochim. Acta **15**, 70.
Venkateswarlu, K. S., Anthony, M. C. and Ramshesh, V., 1971, Radiochem. Radioanal. Lett. **7**, 259.
Venkateswarlu, K. S., Lal, M. and Shankar, J., 1962, Proc. Nucl. Phys. Symposium, Madras, NP 12330.
Vineyard, J. H., 1961, Disc. Faraday Society **31**, 7.
Vlatković, M. and Aten, A. H. W., 1961, Symp. Chem. Effects of Nucl. Transformations, IAEA, Prague **1**, 551.
Vlatković, M. and Aten, A. H. W., 1962, J. Inorg. Nucl. Chem. **24**, 139.
von Egidy, T., 1969, Symposium – "Neutron Capture Gamma-Ray Spectroscopy", IAEA, Vienna, 127.
Vosko, S. and Smoluchowski, R., 1961, Phys. Rev. **122**, 1406.
Vulliet, P., 1970, Thesis, Grenoble.
Wahl, A. C., 1958, J. Inorg. Nucl. Chem. **6**, 263.
Wahl, A. C. and Bonner, N. A., 1957, "Radioactivity Applied to Chemistry", (Wiley, NY).
Wahl, A. C., Fergusson, R. L., Nethaway, D. R., Troutner, D. E. and Wolfsberg, K., 1962, Phys. Rev. **126**, 1112.
Waite, T. R., 1957a, Phys. Rev. **107, 463.**
Waite, T. R., 1957b, Phys. Rev. **107,** 471.
Waite, T. R., 1960, J. Chem. Phys. **32**, 21.
Wall, F. T. and Porter, R. N., 1963, J. Chem. Phys. **39**, 3112.
Wall, F. T., Hiller, I. A. and Mazur, J., 1958, J. Chem. Phys. **29**, 255.
Wall, F. T., Hiller, I. A. and Mazur, J., 1961, J. Chem. Phys. **33**, 1284.
Walton, G. N., 1957, Progr. Nucl. Phys. **6**, 192.
Walton, G. N., 1964, Radiochim. Acta **2**, 108.
Walton, G. N. and Croall, I. F., 1955, J. Inorg. Nucl. Chem. **1**, 149.
Warburton, E. K., Olness, J. W. and Poletti, A. R., 1967, Phys. Rev. **160**, 938.
Warren, J. L. and Jones, C. H. W., 1971, Radiochem. Radioanal. Lett. **7**, 97.
Warren, J. L., Jones, C. H. W. and Vasudev, P., 1971, J. Phys. Chem. **75**, 2867.
Webber, I. and Wiles, D. R., 1976, J. Inorg. Nucl. Chem. **38**, 1103.
Weber, M., Stelter, A. and Herrmann, G., 1972, Radiochim. Acta **18**, 80.
Wei, M. H., Takeda, M. and Saito, N., 1974, Radiochem. Radioanal. Lett. **18**, 311.
Weigel, F., 1959, Angew. Chem. **71**, 289.
Weiner, R. M., 1966, Phys. Rev. **144**, 127.
Welch, M., 1968, Chem. Comm., 1354.
Welch, M. J., 1969, "The Year Book of Nuclear Medicine", ed. Quinn, J. L., (Year Book Publishers, NY).
Welch, M. J. and Lifton, J., 1971, J. Amer. Chem. Soc. **93**, 3385.
Welch, M. J. and Straatmann, M. G., 1973, Radiochim. Acta **20**, 124.
Welch, M. J. and Ter-Pogossian, M. M., 1968, Radiation Res. **36**, 580.
Welch, M., Lifton, J. and Gaspar, P., 1971, J. Nucl. Med. **12**, 405.
Welch, M., Withnell, R. and Wolf, A. P., 1967, Anal. Chem. **34**, 275.
Wendell, K., Jones, C. A., Kaufman, J. J. and Koski, W. S., 1975, J. Chem. Phys. **63**, 750.
Wertenstein, L., 1914, Ann. Physik **1**, 347.
Wertheim, G. K., 1961, Phys. Rev. **124**, 764.
Wertheim, G. K., 1971, Accounts Chem. Res. **4**, 373.

Wertheim, G. K. and Buchanan, D. N. E., 1969, Chem. Phys. Lett. **3**, 87.
Wertheim, G. K., Guggenheim, H. J. and Buchanan, D. N. E., 1969, J. Chem. Phys. **51**, 1931.
Weston, R. E., 1959, J. Chem. Phys. **31**, 892.
Wexler, S., 1959, J. Inorg. Nucl. Chem. **10**, 8.
Wexler, S., 1962, J. Chem. Phys. **36**, 1992.
Wexler, S., 1965a, Act. Chim. Biol. Radiat. **8**, 107.
Wexler, S., 1965b, in "Actions chimique et biologique des Radiations", ed. Haissinsky, M., Masson, Paris **8**, 105.
Wexler, S., 1967, Science **156**, 901.
Wexler, S. and Anderson, G. R., 1960, J. Chem. Phys. **33**, 850.
Wexler, S. and Davies, T. H., 1952, J. Chem. Phys. **20**, 1688.
Wexler, S. and Hess, D. C., 1958, J. Phys. Chem. **62**, 1382.
Wexler, S., Anderson, G. R. and Singer, L. A., 1960, J. Chem. Phys. **32**, 417.
Wheeler, O. H. and McClin, M. L., 1967a, Int. J. App. Rad. Isotopes **18**, 788.
Wheeler, O. H. and McClin, M. L., 1967b, Radiochim. Acta **7**, 181.
Wheeler, O. H. and McClin, M. L., 1967c, Radiochim. Acta **8**, 179.
Wheeler, O. H. and Trabal, J. E., 1970, J. Appl. Rad. Isotopes **21**, 241.
Wheeler, O. H., Trabal, J. E. and McClin, M. L., 1968, Radiochim. Acta **9**, 49.
Wheeler, O. H., Gonzalez, C. L. and Lopez-Alonso, H., 1970a, J. Appl. Rad. Isotopes **21**, 244.
Wheeler, O. H., Trabal, J. E. and Wiles, D. R., 1970b, Can. J. Chem. **48**, 3609.
Wheeler, O. H., Lopez-Alonso, H., Trabal, J. E. and Wiles, D. R., 1970c, Can. J. Chem. **48**, 3609.
Wickman, H. W. and Wertheim, G. K., 1968, in "Chemical Applications of Mössbauer Spectroscopy", ed. Goldanski, V. I. and Herber, R. H., (Academic Press, NY) p. 604.
Wiechmann, W. and Biersack, J., 1967, Nucleonik **9**, 399.
Wiles, D. R., 1973, Adv. Organomet. Chem. **11**, 207.
Wiles, D. R., 1974, J. Inorg. Nucl. Chem. **36**, 2405.
Wiles, D. R. and Baumgärtner, F., 1972, Fortschr. Chem. Forsch. **32**, 63.
Wiles, D. R. and Coryell, C. D., 1954, Phys. Rev. **96**, 696.
Wiles, D. R. and Wong, W. H., 1967, Can. J. Chem. **45**, 1813.
Wilkin, J. and Wolfgang, R., 1968, J. Phys. Chem. **72**, 2631.
Wilkins, R. L., 1972, J. Chem. Phys. **57**, 912.
Willard, J. E., 1965, Symp. Chem. Effects of Nucl. Transformations, IAEA, Vienna **I**, 221.
Williams, L. L., Sutin, N. and Miller, J. M., 1961, J. Inorg. Nucl. Chem. **19**, 175.
Williams, R. R., 1948a, J. Chem. Phys. **16**, 513.
Williams, R. R., 1948b, J. Phys. Chem. **52**, 603.
Williams, R. R., Jenks, G. H., Leslie, W. B., Richter, J. W. and Larson, Q. V., 1951, "Radiochemical Studies: The Fission Products", eds. Coryell, C.D. and Sugarman, N., (McGraw-Hill, NY) **1**, 184.
Winsberg, L. and Alexander, J. M., 1961, Phys. Rev. **121**, 518.
Winther, A., 1952, Kgl. Danske Videnskab. Selskab. Mat-Fys. Medd. **27**, No. 2.
Wolf, A. P., 1960, Ann. Rev. Nucl. Sci. **10**, 259.
Wolf, A. P., 1961, Symp. Chem. Effects of Nucl. Transformations, IAEA, Prague **2**, 3.
Wolf, A. P., 1964, "Advances in Physical Organic Chemistry", (Academic Press, NY) **2**, 201.
Wolf, G. K., 1966, Radiochim. Acta **6**, 39.
Wolf, G. K. and Fritsch, T., 1969, Radiochim. Acta **11**, 194.
Wolfgang, R., 1956, J. Inorg. Nucl. Chem. **2**, 180.
Wolfgang, R., 1963, J. Chem. Phys. **39**, 2983.
Wolfgang, R., 1965a, Progress in Reaction Kinetics **3**, 97.
Wolfgang, R., 1965b, Ann. Rev. Phys. Chem. **16**, 15.
Wolfgang, R., 1969, Accounts Chem. Research **2**, 248.

Wolfgang, R. and Rowland, F. S., 1958, Anal. Chem. **30**, 403.
Wolfgang, R., Pratt, T. and Rowland, F. S., 1956, J. Amer. Chem. Soc. **78**, 5132.
Wolfsberg, M., 1956, J. Chem. Phys. **24**, 24.
Wolken, G. and Karplus, M., 1974, J. Chem. Phys. **60**, 351.
Wood, A. B., 1914, Phil. Mag. **28**, 808.
Wong, W. H. and Wiles, D. R., 1968, Can. J. Chem. **46**, 3201.
Wu, C. S. and Moszkowski, A. A., 1966, in "Beta Decay", (John Wiley and Sons, New York).
Yaffe, L., 1949, Radiochemistry at Chalk River; Brookhaven Conf. Report, BNL-C-9.
Yajima, S., Shiba, K. and Handa, M., 1963a, Bull. Chim. Japan **36**, 253.
Yajima, S., Shiba, K. and Handa, M., 1963b, Bull. Chim. Japan **36**, 258.
Yajima, S., Shiba, K. and Handa, M., 1965, Bull. Chim. Soc. Japan **38**, 278.
Yang, I. C. and Wiles, D. R., 1967, Can. J. Chem. **45**, 1357.
Yang, J. W. and Wolf, A. P., 1960, J. Amer. Chem. Soc. **82**, 4488.
Yang, M. H., 1969, Radiochim. Acta **12**, 167.
Yang, M. H., Yoshihara, K. and Shibata, N., 1970a, Radiochim. Acta **14**, 16.
Yang, M. H., Kudo, H. and Yoshihara, K., 1970b, Radiochim. Acta **14**, 52.
Yang, M. H., Yoshihara, K. and Shibata, N., 1971, Radiochim. Acta **15**, 17.
Yang, M. H., Wei, J. C., Chuang, J. T., Yeh, S. J., Collins, C. H. and Collins, K. E., 1973, Radiochem. Radioanal. Lett. **13**, 173.
Yankwich, P. E., 1956, Can. J. Chem. **34**, 301.
Yankwich, P. E., Rollefson, G. K. and Norris, T. H., 1946, J. Chem. Phys. **14**, 131.
Yasukawa, T., 1966, J. Inorg. Nucl. Chem. **28**, 17.
Yasukawa, T., 1967, J. Inorg. Nucl. Chem. **29**, 605.
Yasukawa, T. and Saito, N., 1965, J. Inorg. Nucl. Chem. **27**, 1433.
Yeh, S. J., Shibata, N., Amano, H., Yoshihara, K., Yang, M. H., Chen, P. Y., Ke, C. N. and Kudo, H., 1969, J. Nucl. Sci. Techn. (Tokyo) **6**, 75.
Yeh, S. J., Shibata, W., Amano, H., Yoshihara, K., Yang, M. H., Chen, T. F., Chen, C. T. and Kudo, H., 1970, J. Nucl. Sci. Techn. (Tokyo) **7**, 300.
Yoshida, H. and Herber, R. H., 1969, Radiochim. Acta **12**, 14.
Yoshihara, K., 1964, Nature (Lond.) **204**, 1296.
Yoshihara, K., 1970, Nippon Genshiryoku Gakkaishi **12**, 206.
Yoshihara, K. and Ebihara, H., 1964, Radiochim. Acta **2**, 219.
Yoshihara, K. and Ebihara, H., 1966, J. Chem. Phys. **45**, 896.
Yoshihara, K. and Harbottle, G., 1963, Radiochim. Acta **1**, 68.
Yoshihara, K. and Kudo, H., 1969, Nature (Lond.) **222**, 1060.
Yoshihara, K. and Kudo, H., 1970, J. Chem. Phys. **52**, 2950.
Yoshihara, K. and Mizusawa, T., 1972, Radiochem. Radioanal. Lett. **9**, 263.
Yoshihara, K. and Yang, M. H., 1968, Radiochim. Acta **9**, 168.
Yoshihara, K. and Yang, M. H., 1969a, Radiochem. Radioanal. Lett. **1**, 37.
Yoshihara, K. and Yang, M. H., 1969b, Inorg. Nucl. Chem. Lett. **5**, 389.
Yoshihara, K. and Yokoshima, T., 1961, Bull. Chem. Soc. Japan **34**, 123.
Yoshihara, K., Huang, T. C., Ebihara, H. and Shibata, N., 1964, Radiochim. Acta **3**, 185.
Yoshihara, K., Yang, M. H. and Shiokawa, T., 1970, Radiochem. Radioanal. Lett. **4**, 143.
Yoshihara, K. Kishimoto, K., Takahashi, M. and Suzuki, S., 1974, Radiochim. Acta **21**, 148.
Yoshihara, K., Fujita, A. and Shiokawa, T., 1976, Res. Report Lab. Nucl. Sci. Tohoku Univ. **9**, 193.
Yoshihara, Y., Baba, H. and Tanaka, K., 1963, Bull. Chem. Soc. Japan **36**, 928.
Yosim, S. and Davies, T. H., 1952, J. Phys. Chem. **56**, 599.
Zaborenko, K. B., 1973, Bull. Soc. Chim. Fr., 2581.
Zaborenko, K. B. and Babeshkin, A. M., 1961, Symp. Chem. Effects of Nucl. Transformations, IAEA, Prague **2**, 157.

Zahn, U., 1967a, Radiochim. Acta **7**, 170.
Zahn, U., 1967b, Radiochim. Acta **8**, 177.
Zahn, U. and Harbottle, G., 1966, J. Inorg. Nucl. Chem. **28**, 925.
Zahn, U., Collins, C. H. and Collins, K. E., 1969, Radiochim. Acta **11**, 33.
Zahn, U., Potzel, W. and Wagner, F. E., 1973, in "Perspectives in Mössbauer Spectroscopy", Eds. Cohen, S. G. and Pasternak, M., (Plenum Press, NY).
Zaitsev, V. M., Kirin, I. S. and Tikhonov, V. I., 1968, Khim. Vys. Energ. **2**, 322.
Zaitseva, N. G. and Mo-Ling Chou, 1963, Radiokhimiya **5**, 614.
Zaitseva, N. G. and Ianovici, E., 1970, Radiokhimiya **12**, 143.
Zaitseva, N. G. and Ianovici, E., 1971, Radiokhimiya **13**, 107.
Zaitseva, N. G. and Ianovici, E., 1972, Radiokhimiya **14**, 319.
Zaitseva, N. G., Ianovici, E. and Raitseva, V., 1969, Radiokhimiya **11**, 366.
Zaitseva, N. G., Islamova, K. M., Mai, N. H. and Ronneau, K., 1973, Dubna Report, JINR,-P6-7203.
Zaitseva, N. G., Islamova, K. M. and Mai, N. H., 1974a, Radiochem. Radioanal. Lett. **16**, 155.
Zaitseva, N. G., Islamova, K. M. and Mai, N. H., 1974b, Radiochem. Radioanal. Lett. **19**, 361.
Zeck, O. F., Gennaro, G. P. and Tang, Y. N., 1974a, Chem. Comm., 52.
Zeck, O. F., Gennaro, G. P. and Tang, Y. N., 1974b, J. Amer. Chem. Soc. **96**, 5967.
Zeck, O.F., Su, Y.Y. and Tang, Y.N., 1975a, Chem. Comm., 156.
Zeck, O. F., Gennaro, G. P. and Tang, Y. N., 1975b, J. Amer. Chem. Soc. **97**, 4498.
Zicha, G., 1966, 1969, Theses, Technische Universität, Munich.
Zicha, G., Zahn, U., Kienle, P. and Wien, K., 1967, in "Diskussionstagung uber Neutronenphysik on Forschungsreaktoren" Jülich, p. 24.
Zuber, A. V., 1954a, Thesis, Columbia University, USA.
Zuber, A., 1954b, USAEC. Document NYO-6142 BNL.
Zuber, A., Rauscher, H. B., Miller, J. M. and Sutin, N., 1961, Symp. Chem. Effects of Nucl. Transformations, IAEA, Prague **1**, 359.
Zvara, I., Tarasov, L. K., Kozhivanek, M., Su Hung-Knei and Zvarova, T. S., 1963, Soviet Phys. Doklady **8**, 63.
Zvara, I., Oganesyan, Yu. T., Lobanov, Yu. V., Kuznetsov, V. I., Druin, V. A., Perelygin, V. P., Gavrilov, K. A., Tretiakova, S. P. and Plotko, V. M., 1964, Atomn. Energiya **17**, 310.
Zvara, I., Zvarova, T. S., Caletka, R., Chuburkov, Yu. T. and Shalaevskii, M. R., 1967, Radiokhimiya **9**, 231.
Zvara, I., Chuburkov, Yu. T., Caletka, R. and Shalaevsky, M. R., 1969, Radiokhimiya **11**, 163.
Zvara, I., Chuburkov, Yu. T., Belov, V. Z., Buklanov, G. V., Zakhvataev, B. B., Zvarova, T. S., Maslov, O. D., Caletka, R. and Shalaevsky, M. R., 1970, J. Inorg. Nucl. Chem. **32**, 1885.

AUTHOR INDEX

SUBJECT INDEX*

* Page numbers in italics indicate that the subject continues over a number of pages.